W9-BJV-480

A HISTORY

OF

MECHANICS

BY

RENÉ DUGAS

MAÎTRE DE CONFÉRENCES OF THE ECOLE POLYTECHNIQUE, PARIS

FOREWORD BY LOUIS DE BROGLIE

TRANSLATED INTO ENGLISH

BY

J. R. MADDOX

DOVER PUBLICATIONS, INC.
New York

This Dover edition, first published in 1988, is an unabridged republication of the work first published by Éditions du Griffon, Neuchâtel, Switzerland, in 1955. The Table of Contents, which appeared at the end of the first edition, is published here preceeding the Foreword.

Manufactured in the United States of America
Dover Publications, Inc., 31 East 2nd Street, Mineola, N.Y. 11501

Library of Congress Cataloging-in-Publication Data

Dugas, René.
  A history of mechanics.

  Translation of: Histoire de la mécanique.
  Includes index.
  1. Mechanics—History. 2. Wave mechanics. I. Title.
QA802.D8613 1988   531'.09   88-11836
ISBN 0-486-65632-2 (pbk.)

TABLE OF CONTENTS

Pages

PREFACE   11

PART ONE

THE ORIGINS

CHAPTER ONE. — HELLENIC SCIENCE   19
  1. Aristotelian mechanics   19
  2. The Statics of Archimedes   24

CHAPTER II. — ALEXANDRIAN SOURCES AND ARABIC MANUSCRIPTS . .   32
  1. The " mechanics " of Hero of Alexandria   32
  2. Pappus' theories of the inclined plane and of the centre of
    gravity   33
  3. The fragments attributed to Euclid in arabic writings   36
  4. The book of Charistion   37

CHAPTER III. — THE XIIIth CENTURY. THE SCHOOL OF JORDANUS .   38
  1. Jordanus of Nemore and " gravitas secundum situm "   38
  2. The anonymous author of " Liber Jordani de ratione ponderis. "
    The angular lever. The inclined plane   41

CHAPTER IV. — THE XIVth CENTURY. THE SCHOOLS OF BURIDAN AND
ALBERT OF SAXONY. NICOLE ORESME AND THE OXFORD SCHOOL   47
  1. The doctrine of " impetus " (John Buridan)   47
  2. The sphericity of the earth and the oceans—Albert of Saxony
    and the aristotelian tradition   51
  3. Albert of Saxony's theory of centre of gravity   53
  4. Albert of Saxony's kinematics. The acceleration of falling
    bodies   56
  5. The discussion of action at a distance   57
  6. Nicole Oresme—a disciple of Buridan   58
  7. Oresme's rule in kinematics. (Uniformly accelerated motion.) .   59
  8. Oresme as a predecessor of Copernicus   62
  9. The Oxford School   66
  10. The tradition of Albert of Saxony and of Buridan   68

Pages

CHAPTER V. — XVth AND XVIth CENTURIES. THE ITALIAN SCHOOL. BLASIUS OF PARMA. THE OXFORD TRADITION. NICHOLAS OF CUES AND LEONARDO DA VINCI. NICHOLAS COPERNICUS. THE ITALIAN AND PARISIAN SCHOOLMEN OF THE XVIth CENTURY. DOMINIC SOTO AND THE FALL OF BODIES 69

1. Blasius of Parma and his treatise on weights 69
2. The Italian tradition of Nicole Oresme and the Oxford School . 70
3. Nicholas of Cues (1404-1464) and the doctrine of "impetus impressus" 71
4. Leonardo da Vinci's contribution to mechanics 72
5. Nicholas Copernicus (1472-1543). His system of the world and his ideas on attraction 82
6. John Fernel (1497-1558) and the figure of the earth 86
7. Italian scholasticism in the XVIth century 86
8. Parisian scholasticism in the XVIth century 87
9. The attack of the humanists 90
10. Dominic de Soto (1494-1560) and the laws of falling bodies . . 91

CHAPTER VI. — XVIth CENTURY (continued) THE ITALIAN SCHOOL OF NICHOLAS TARTAGLIA AND BERNARDINO BALDI 95

1. Nicholas Tartaglia 95
2. Jerome Cardan (1501-1576) 96
3. Julius-Caesar Scaliger and Buridan's doctrine 99
4. Bento Pereira (1535-1610). The classical reaction 99
5. The "Mechanicorum Liber" of Guido Ubaldo (1545-1607) . . 100
6. J.-B. Villalpand (1552-1608) and the polygon of sustentation . 101
7. J.-B. Benedetti (1530-1590). Statics. Figure of the earth. Doctrine of "impetus" 103
8. Giordano Bruno (1548-1600) and the composition of motion . . 105
9. Bernardino Baldi (1553-1617). Statics and gravity "ex violentia" 106

CHAPTER VII. — XVIth CENTURY (continued). XVIIth CENTURY. TYCHO-BRAHE AND KEPLER 108

1. The system due to Tycho-Brahe (1546-1601) 108
2. Kepler (1571-1631). The general character of his contribution 110
3. The origin of the law of areas 111
4. Origin of the law of the ellipticity of planetary trajectories . . . 114
5. Kepler's third law 116
6. Kepler and the concept of inertia 117
7. Kepler and the doctrine of attraction 118

PART II

THE FORMATION OF CLASSICAL MECHANICS

CHAPTER I. — STEVIN'S STATICS. SOLOMON OF CAUX 123

1. The statics of Stevin (1548-1620) 123
2. Stevin and the principle of virtual work 127
3. Stevin's hydrostatics 128
4. Solomon of Caux (1576-1630) and the concept of work 128

Pages

CHAPTER II. — GALILEO AND TORRICELLI 129

1. Galileo's statics 129
2. Galileo and the fall of bodies 132
3. Galileo and the motion of projectiles 141
4. Galileo and hydrostatics 143
5. Galileo and the Copernican system 144
6. Torricelli's principle 145
7. Torricelli and the motion of projectiles 146
8. Torricelli's experiment 147
9. Torricelli's law flow through an orifice 147

CHAPTER III. — MERSENNE (1588-1648) AS AN INTERNATIONAL
GO-BETWEEN IN MECHANICS. ROBERVAL (1602-1675) 149

1. The arrival of foreign theories in France. The part played by
Mersenne 149
2. Roberval and compound motion 150
3. Roberval's treatise on mechanics 150
4. Roberval and the law of composition of forces 151

CHAPTER IV. — DESCARTES' MECHANICS. PASCAL'S HYDROSTATICS . 154

1. Descartes' statics 154
2. Descartes and the fall of heavy bodies 158
3. Descartes and the conservation of quantities of motion . . . 160
4. Descartes and the impact of bodies 162
5. The discussion between Descartes and Roberval on the centre
of agitation 163
6. The quarrel about geostatics 166
7. Pascal's hydrostatics 169

CHAPTER V. — THE LAWS OF IMPACT (WALLIS, WREN, HUYGHENS,
MARIOTTE). THE MECHANICS OF HUYGHENS (1629-1697) . . . 172

1. The mechanics of Wallis (1616-1703) 172
2. Wren (1632-1723) and the laws of elastic impact 175
3. Huyghens (1629-1697) and the laws of impact 176
4. The plan of Huyghens' fundamental treatise 181
5. Huyghens and the fall of bodies 182
6. The isochronism of the cycloidal pendulum 185
7. The theory of the centre of oscillation 186
8. The theory of centrifugal force 194
9. Huyghens and the principle of relativity 198
10. Mariotte and the laws of impact 199

CHAPTER VI. — NEWTON (1642-1727) 200

1. The newtonian method 200
2. The newtonian concepts 201
3. The newtonian laws of motion 205
4. Newton and the dynamical law of composition of forces . . . 207
5. The motion of a point under the action of a central force . . . 209
6. Newton's explanation of the motion of the planets 210
7. The universal attraction 213

Pages

CHAPTER VII. — LEIBNIZ AND LIVING FORCE 219
1. The " vis motrix " in the sense of Leibniz 219
2. Leibniz and the laws of impact 220
3. Living and dead forces 221

CHAPTER VIII. — THE FRENCH-ITALIAN SCHOOL OF ZACCHI AND VARIGNON 222
1. Zacchi and Saccheri. Lamy and the composition of forces . . 222
2. The statics of Varignon (1654-1722) 224
3. Varignon and Torricelli's law of flow 226

PART III

THE ORGANISATION AND DEVELOPMENT
OF THE PRINCIPLES OF CLASSICAL MECHANICS
IN THE XVIIIth CENTURY

CHAPTER I. — JEAN BERNOULLI AND THE PRINCIPLE OF VIRTUAL WORK (1717). DANIEL BERNOULLI AND THE COMPOSITION OF FORCES (1726) 231
1. Jean Bernoulli and the principle of virtual work 231
2. Daniel Bernoulli and the composition of forces 233

CHAPTER II. — THE CONTROVERSY ABOUT LIVING FORCES 235

CHAPTER III. — EULER AND THE MECHANICS OF A PARTICLE (1736) . 239

CHAPTER IV. — JACQUES BERNOULLI AND THE CENTRE OF OSCILLATION (1703). D'ALEMBERT'S TREATISE ON DYNAMICS (1743) 243
1. Jacques Bernoulli and the centre of oscillation 243
2. The introductory argument of d'Alembert's Treatise on dynamics 244
3. D'Alembert and the concept of accelerating force 248
4. D'Alembert's principle 248
5. D'Alembert's solution of the problem of the centre of oscillation 250
6. The priority of Herman and Euler in the matter of d'Alembert's principle 251
7. D'Alembert and the laws of impact 252
8. D'Alembert and the principle of living forces 253

CHAPTER V. — THE PRINCIPLE OF LEAST ACTION 254
1. Return to Fermat 254
2. Cartesian objections to Fermat's principle 258
3. Leibniz and the path of " least resistance " for light 260
4. Maupertuis' law of rest 260
5. The principle of least action in Maupertuis' sense (1744) . . . 260
6. The application of the principle of least action to the direct impact of two bodies 264
7. The principle of least action in Maupertuis' work 267
8. D'Alembert's condemnation of final causes 269
9. The polemic on the principle of least action 270
10. Euler's judgement on the controversy on least action 272
11. Euler and the law of the extremum of $\int mvds$ 273
12. Final remark 274

Pages

CHAPTER VI. — EULER AND THE MECHANICS OF SOLID BODIES (1760) . 276

CHAPTER VII.—CLAIRAUT AND THE FUNDAMENTAL LAW OF HYDROSTATICS 279
1. Clairaut's principle of the duct 279
2. The condition to be satisfied by the law of gravity to assure the conservation of the shape of a rotating fluid mass 282
3. Condition for the equilibrium of a fluid mass 283

CHAPTER VIII. — DANIEL BERNOULLI'S HYDRODYNAMICS. D'ALEMBERT AND THE RESISTANCE OF FLUIDS. EULER'S HYDRODYNAMICAL EQUATIONS. BORDA AND THE LOSSES OF KINETIC ENERGY IN FLUIDS 286
1. Return to the hydrodynamics of the xviiith century 286
2. Daniel Bernoulli's hydrodynamics 287
3. D'Alembert and the motion of fluids 290
4. D'Alembert and the resistance of fluids — His paradox . . . 295
5. Euler and the equilibrium of fluids 299
6. Euler and the general equations of hydrodynamics 301
7. Borda and the losses of kinetic energy in fluids 305

CHAPTER IX. — EXPERIMENTS ON THE RESISTANCE OF FLUIDS (BORDA, BOSSUT, DU BUAT). COULOMB AND THE LAWS OF FRICTION . . . 309
1. Borda's experiments and newtonian theories 309
2. The abbé Bossut's experiments 313
3. Du Buat (1734-1809) : Hydraulics and the resistance of fluids . 316
4. Coulomb's work on friction 319

CHAPTER X. — LAZARE CARNOT'S MECHANICS 323
1. Carnot and the experimental character of mechanics 323
2. The concepts and postulates of Carnot's mechanics 324
3. Carnot's theorem 327

CHAPTER XI. — THE " MÉCANIQUE ANALYTIQUE " OF LAGRANGE . . 332
1. The content and purpose of Lagrange's " Mécanique analytique " 332
2. Lagrange's statics 333
3. Lagrange and the history of dynamics 339
4. Lagrange's equations 342
5. The conservation of living forces as a corollary of Lagrange's equations 344
6. The principle of least action as a corollary of Lagrange's equations 345
7. On some problems treated in the " Mécanique analytique " . . 347
8. Lagrange's hydrodynamics 347

PART IV

SOME CHARACTERISTIC FEATURES
OF THE EVOLUTION
OF CLASSICAL MECHANICS AFTER LAGRANGE

FOREWORD 353

CHAPTER I. — LAPLACE'S MECHANICS (1799) 354
1. Laplace and the principles of dynamics 354
2. The general mechanics compatible with an arbitrary relation between the " force " and the velocity 357
3. Laplace and the significance of the law of universal gravitation 359

Pages

CHAPTER II. — FOURIER AND THE PRINCIPLE OF VIRTUAL WORKS (1798)   361

CHAPTER III. — THE PRINCIPLE OF LEAST CONSTRAINT (1829)   367

CHAPTER IV. — RELATIVE MOTION : RETURN TO A PRINCIPLE OF CLAI-
RAUT. CORIOLIS' THEOREMS. FOUCAULT'S EXPERIMENTS . . .   370

1. Return to a principle of Clairaut (1742)   370
2. Coriolis' first theorem   374
3. Coriolis' second theorem   377
4. The experiments of Foucault (1819-1868)   380

CHAPTER V. — POISSON'S THEOREM (1809)   384

1. Poisson's theorem and brackets   384
2. The Lagrange-Poisson square brackets   387

CHAPTER VI. — ANALYTICAL DYNAMICS IN THE SENSE OF HAMILTON
AND JACOBI   390

1. Hamilton's optics. Its double interpretation in terms of emission
and wave propagation   390
2. The dynamical law of varying action in Hamilton's sense . . .   394
3. The significance of the hamiltonian dynamics   399
4. Jacobi's criticism   401
5. Jacobi's fundamental theorem   403
6. The canonical equations and Jacobi's multiplier   406
7. Geometrisation of the principle of least action   407

CHAPTER VII. — NAVIER'S EQUATIONS   409

1. The molecular hypothesis   409
2. Equilibrium of fluids   410
3. The molecular forces in the motion of a fluid   411
4. Remark on the origin of the general equations of elasticity . .   414

CHAPTER VIII. — CAUCHY AND THE FINITE DEFORMATION OF CONTI-
NUOUS MEDIA   418

CHAPTER IX. — HUGONIOT AND THE PROPAGATION OF MOTION IN CON-
TINUOUS MEDIA   423

1. Nature of the problem   423
2. Compatibility of two solutions. Velocity of propagation of one
solution in another. Hugoniot's theorem   425
3. Discontinuities in the propagation of motion   429

CHAPTER X. — HELMHOLTZ AND THE ENERGETIC THESIS DISCUSSION
OF THE NEWTONIAN PRINCIPLES (SAINT-VENANT, REECH, KIRCH-
HOFF, MACH, HERTZ, POINCARÉ, PAINLEVÉ, DUHEM)   434

1. Helmholtz and the energetic thesis   434
2. Barré de Saint-Venant   436
3. Reech and the " School of the thread "   438
4. Kirchhoff and the logistic structure of mechanics   441
5. Mach   443
6. Hertz   444
7. Poincaré— Criticism of the principles and discussion of the
notion of absolute motion   447

Pages

8. Poincaré and the energetic thesis. 451
9. Painlevé and the principle of causality in mechanics 453
10. Duhem and the evolution of mechanics 456
11. Conclusion of this chapter 457

PART V

THE PRINCIPLES
OF THE MODERN PHYSICAL THEORIES
OF MECHANICS

FOREWORD 461
CHAPTER I. — SPECIAL RELATIVITY 463

A. PRESENTATION

1. Immediate antecedents of the special theory of relativity . . . 463
2. Michelson's experiment and Lorentz's hypothesis of contraction 464
3. The Lorentz transformation 466
4. Introduction to Einstein's electrodynamics 472
5. Definition of simultaneity 473
6. Relativity of lengths and times 474
7. Transformation of the coordinates of space and time 475
8. Contraction of lengths and correlative dilation of times 478
9. Composition of velocities 478
10. Transformation of Maxwell's equations in the vacuum. Electrodynamic relativity 479
11. Transformation of Maxwell's equations including convection currents 480
12. Dynamics of the slowly accelerated electron 481
13. Space-time in the sense of Minkowski 482

B. ANALYSIS AND INTERPRETATION

2a. On Michelson's experiment 487
3a. Dynamics of the electron in Poincaré's sense 488
3b. From Lorentz to Einstein 489
4a. The ether made superfluous 490
5a. Difficulties of Einstein's notion of simultaneity 490
6a. Field of validity of the principle of special relativity— " Galilean " systems of reference 491
7a. On different mathematical ways of obtaining the Lorentz transformation 494
8a. Pseudo-paradoxes in special relativity 495
9a. Composition of velocities and Fizeau's experiment 497
12a. Return to the dynamics of variable mass in Painlevé's sense . 497
13a. On the meaning of space-time 500
CHAPTER II. — GENERALISED RELATIVITY 502

A. PRESENTATION

1. Statement of the principle of generalised relativity 502
2. Remark on the mathematical tools of generalised relativity . . 505
3. The equations of motion of a free particle in a gravitational field 506

                                                                Pages
4. Equations of the gravitational field in the absence of matter . 507
5. General form of the equations of gravitation 509
6. Reversion to Newton's theory in the first approximation . . . 510
7. The conduct of measurements of space and time in a static gravitational field. Deviation of light rays. Displacement of the
   perihelion of the planets 512
8. The spatially closed universe 514
9. Gravitation and electricity 517

B. ANALYSIS AND INTERPRETATION

1a to 5a. General observations 521
6a. On the geometrisation of classical mechanics 522
7a. Expression and interpretation of Schwartzschild's " ds^2 " . . 524
7b. Generalised relativity in comparison with experiment 526
7c. Painlevé's criticism— The semi-einsteinien theory of gravitation 528
8a. Remark on the universes of Einstein and de Sitter 532
    Generalised relativity and Riemann spaces 533

CHAPTER III. — THE DYNAMICS OF QUANTA IN BOHR'S SENSE 535

A. PRESENTATION

1. Bohr's first dynamical model 535
2. Generalisation of the first model— The " quantum conditions "
   for a system of several degrees of freedom 538
3. Example of application— Theory of fine structure 544
4. Bohr's correspondence principle 547

B. ANALYSIS AND INTERPRETATION

1a. Return to Planck's frequency relation 549
1b. Criticism of Bohr's models 551
2a. Remarks on Bohr's theorem 552
2b. Antinomy between Planck's frequency relation and classical
    dynamics (Poincaré) 552

CHAPTER IV. — WAVE MECHANICS IN THE SENSE OF LOUIS DE BROGLIE
AND SCHRÖDINGER 554

A. PRESENTATION

1. The phase wave 554
2. Recapitulation of the principles of Hamilton and Maupertuis . 555
3. Application of the principle of least action to the dynamics of the
   electron 556
4. The vector " world wave " 557
5. Extension of the quantum relation 558
6. Return to the Bohr-Sommerfeld quantum conditions 559
7. Diffraction of electrons by matter 560
8. Schrödinger's wave equation 561

B. ANALYSIS AND INTERPRETATION

1a. On the guiding ideas and the origin of wave mechanics . . . 567
8a. On Schrödinger's equation 570

Pages

CHAPTER V. — QUANTUM MECHANICS IN THE SENSE OF HEISENBERG
AND DIRAC 571

A. *PRESENTATION*

1. Quantum analogue of classical mechanics (Heisenberg) . . . 571
2. Introduction of " matrices " (Born, Jordan) 575
3. Formulation of matrix mechanics by Heisenberg, Born and Jordan 577
4. Dirac's formulation 578

B. *ANALYSIS AND INTERPRETATION*

1*a*. On the origin of quantum mechanics in Heisenberg's sense . . 581
2*a*. Cooperation of physicists and mathematicians 582

CHAPTER VI. — DEVELOPMENT OF THE PRINCIPLES OF QUANTUM
MECHANICS 583

A. *PRESENTATION*

1. Mathematical identity of wave mechanics and quantum
   mechanics 583
2. Probability interpretation of Schrödinger's wave function . . . 586
3. Heisenberg's " uncertainty relationships " 587
4. Ehrenfest's theorem 591
5. Dirac and the general theory of observation 592
6. Wave formulation of the general theory of quantisation (Louis
   de Broglie) 600
7. The relativistic and quantised electron in Dirac's sense 603
8. The Dirac electron and experiment 606

B. *ANALYSIS AND INTERPRETATION*

1*a*. On the identity of wave mechanics and quantum mechanics . 608
3*a*. Reciprocal limitations of the validity of the corpuscular and
   wave-like representation (Heisenberg) 609
7*a*. On the Dirac electron— relativity and quanta 611

CHAPTER VII. — DISCUSSION OF THE PRINCIPLES OF QUANTUM
MECHANICS 614

1. Complementarity in the sense of Bohr 614
2. Classical "legality" and the "semi-legality" of quantum mechanics 616
3. The " probabilities of presence " in classical mechanics, in the
   old quantum theory, in wave mechanics and in Dirac's mechanics 622
4. On the " reality " of quantum mechanics 627
5. On the difference between abstract reasoning and intuitive
   reasoning in quantum mechanics 634
6. Conclusion 636

SOME REMARKS BY WAY OF A GENERAL CONCLUSION 637
NOTES I.   On the mechanics of the Middle Ages 641
NOTES II. Henri Poincaré and the principles of mechanics 643
INDEX 651

FOREWORD

The history of mechanics is one of the most important branches of the history of science. From earliest times man has sought to develop tools that would enable him to add to his power of action or to defend himself against the dangers threatening him. Thus he was unconsciously led to consider the problems of mechanics. So we see the first scholars of ancient times thinking about these problems and arriving more or less successfully at a solution. The motion of the stars which, from the Chaldean shepherds to the great Greek and Hellenistic astronomers, was one of the first preoccupations of human thought, led to the discovery of the true laws of dynamics. As is well known, although the principles of statics had been correctly presented by the old scholars those of dynamics, obscured by the false conceptions of the aristotelian school, did not begin to see light until the end of the Middle Ages and the beginning of the modern era. Then came the rapid development of mechanics due to the memorable work of Kepler, Galileo, Descartes, Huyghens and Newton; the codification of its laws by such men as Euler, Lagrange and Laplace; and the tremendous development of its various branches and the endlessly increasing number of applications in the Nineteenth and Twentieth Centuries. The principles of mechanics were brought to such a high degree of perfection that fifty years ago it was believed that their development was practically complete. It was then that there appeared, in succession, two very unexpected developments of classical mechanics—on the one hand, relativistic mechanics and on the other, quantum and wave mechanics. These originated in the necessity of interpreting the very delicate phenomena of electromagnetism or of explaining the observable processes on the atomic scale. Whereas reativistic mechanics, while upsetting our usual notions of time and space only, in a sense, completed and crowned the work of classical mechanics, the quantum and wave mechanics brought us more radically new ideas and forced us to give up the continuiuty and absolute determinism of elementary phenomena.

Relativistic and quantum mechanics today form the two highest peaks of the progress of our knowledge in the whole field of mechanical phenomena.

To appraise the evolution of mechanics from its origin up to the present time would be obviously a difficult task demanding a considerable amount of work and thought. Few men would be tempted to write such a history of mechanics; for its compilation would require not only a wide and thorough knowledge of all the branches of mechanics ancient and modern, but also a great patience, a well-informed scholarship and an acute and critical mind. These varied qualities M. René Dugas—who has already become known for his fine studies on certain particular themes in the history of dynamics and for his critical essays on different matters in classical, relativistic and quantum mechanics—unites these to a high degree. More than this, he has tackled this overwhelming task and, after several years, has brought it to a successful conclusion. The important work that he now publishes on the history of mechanics constitutes a comprehensive view of the greatest interest which will be highly appreciated by all those who study the history of scientific thought.

Mr. Dugas' book is in certain ways comparable with Ernest Mach's famous book " Mechanics, A historical and critical presentation of its development. " Certainly the reading of Mach's book, so full of original ideas and profound comments, is still extremely instructive and absorbing. But Mr. Dugas' history of mechanics has the advantage of being less systematic and more complete. Mach's thought was in fact dominated by the general ideas which secured his adherence in Physics to the energetic school and in Philosophy to the positivistic thesis. He frequently sought to find an illustration, in the history of mechanics, of his own ideas. Often this gives his book a character which is a little too systematic, that of a thesis in which the arguments in favour of preconceived ideas are rehearsed. Mr. Dugas' attitude is quite different. A scrupulous historian, he has patiently followed all the vagaries of thought of the great students of the subject, collating their texts carefully and always preserving the strictest objectivity.

More impartial than Mach, Mr. Dugas has been helped by the development of historical criticism on the one hand, by the progress of science on the other, and has been able to be more complete. He has given us a much more detailed picture of the efforts that were made and the results obtained in Antiquity and, especially, in the Middle Ages. It is particularly to the authoritative work of Pierre Duhem that Mr. Dugas owes his ability to show us the important contributions made to the development of the principles of mechanics by masters like Jordanus of Nemore, Jean Buridan, Albert of Saxony, Nicole Oresme and a great artist of universal interest like Leonardo da Vinci. Of Duhem's magnificent researches —which are often a little difficult to study in that eminent and erudite physicist's original text, usually lengthy and somewhat vague—Mr. Dugas has been able to make, in a few pages, a short presentation that the reader will read easily and with the greatest profit.

Well informed of the most recent progress of the science, the author, accustomed to reflect of the new contemporary forms of mechanics, has devoted the last part of his book to relativistic mechanics and wave and quantum mechanics. This very accurate presentation made by following closely, as is the author's practice, the ideas of the innovators and the text of their writings, naturally makes Mr. Dugas' history of mechanics much more complete that those of his predecessors.

The central part of the book, devoted to the developments of mechanics in the Seventeenth, Eighteenth and Nineteenth Centuries, has demanded a great amount of work, for the material is immense. Being unable to follow all the details of the development of mechanics in the Eighteenth Century, and especially in the Nineteenth Century, Mr. Dugas has selected for a thorough study certain questions of special importance, either in themselves or because of the extensions which they have had into the contemporary period. It seems to me that this selection has been made very skillfully and has enabled the author, without losing himself in details, to outline the principal paths followed by scientific thought in this domain.

Perhaps, in reading Mr. Dugas' so clear text, the reader will not appreciate the work that the writing of such a book represents. Not only has Mr. Dugas had to sift various questions to select those which would most clearly illustrate the decisive turning-points in the progress of mechanics, but he has always referred to the original texts themselves, never wanting to accomplish the task at second hand. When, for example, he summarises for us the work of Kepler in a few pages, it is after having re-examined and, in some way, rethought these arguments —often complicated and a little quaint and, moreover, written in a bad Latin whose meaning is often difficult to appreciate—which enabled the great astronomer to discover the correct laws of the motion of the planets. It is this necessary conjunction of the procedures of a patient erudition and a wide knowledge of the past and present results of the science which makes the history of science particularly difficult and restricts the number of those who can, with profit, devote themselves to it.   Therefore Mr. René Dugas should be warmly thanked for having placed at the service of the history of science, qualities of mind and methods of work rarely united in one man, and for having given us a remarkable work which will remain a document of the first rank for the historian of mechanics.

Louis de Broglie

of the Académie française
permanent secretary of the Académie des sciences.

PREFACE

Mechanics is one of the branches of physics in which the number of principles is at once very few and very rich in useful consequences. On the other hand, there are few sciences which have required so much thought— the conquest of a few axioms has taken more than 2000 years.

As Mr. Joseph Pérès has remarked, to speak of the miracle of Greece or of the night of the Middle Ages in the evolution of mechanics is not possible. Correctly speaking, Archimedes was able to conquer statics and knew how to construct a rational science in which the precise deductions of mathematical analysis played a part. But hellenic dynamics is now seen to be quite erroneous. It was however, in touch with every-day observation. But, being unable to recognise the function of passive resistances and lacking a precise kinematics of accelerated motion, it could not serve as a foundation for classical mechanics.

The prejudices of the Schoolmen, whose authority in other fields was undisputed, restricted the progress of mechanics for a long period. Annotating Aristotle was the essential purpose of teaching throughout the Middle Ages. Not that the mediaeval scholars lacked originality. Indeed, they displayed an acute subtlety which has never been surpassed, but most often they neglected to take account of observation, preferring to exercise their minds in a pure field. Only the astronomers were an exception and accumulated the facts on which, much later, mechanics was to be based.

The Thirteenth Century had, however, an original school of statics which advocated, in the treatment of heavy bodies, a new principle —under the name of *gravitas secundum situm*—that was to develop into the principle of virtual work; moreover, this principle solved, long before Stevin and Galileo, the problem of the equilibrium of a heavy body on an inclined plane, which Pappus had not succeeded in doing correctly. In the fourteenth century, Buridan formulated the first theory of energy under the name of *impetus*. This theory explicitly departs from the Peripatetic ideas, which demanded the constant intervention of a mover to maintain violent motion in the Aristotelian sense. Incorporated into a continued tradition in which it was deformed in order to conform to an animist doctrine, which in the hands of the German metaphysicians of the fifteenth century was to subsist with Kepler, the theory of *impetus* became, in the hands of Benedetti, an early form of the principle of inertia, while one of its other aspects was to become, after a long polemic, the doctrine of *vis viva*. And in the Fourteenth Century, the Oxford School, which in other respects indulged in such artificial quibbling, was to clarify the laws of the kinematics of uniformly accelerated motion.

The mechanics of the Middle Ages received something of a check during the Renaissance, which caused a return to classical traditions. The Schools were attacked by the humanists. Yet, before Galileo, Dominico Soto successfully formulated the exact laws of heavy bodies even if he did not verify them experimentally.

Under what may seem an ambitious title, *A History of Mechanics*, we shall deal with the evolution of the principles of general mechanics, while we shall omit the practical applications and, *a fortiori*, technology.

As far as possible we shall follow a chronological order, in the manner of elementary text books which begin with early history and end with the latest war. After considerable reflexion, this order has seemed to us preferable to the one'adopted by important critics, which consists in choosing a given principle or problem and analysing its evolution throughout the centuries. This latter method offers the advantage of isolating a theory and treating it closely ; it lends itself to short cuts which are striking but which sometimes seem a little artificial : the different key problems of mechanical science evolved in fact along parallel lines, profiting by the progress made in mathematical language ; what is more, these problems were interconnected. We have preferred here to follow the elementary order in time. Each century will thus appear in full light, with its own mentality and atmosphere. So we jump necessarily from one theme to another, but the works find once more their unity and their natural background, without the distortion caused by juxtaposing them in time.

The present book is divided into five parts. The first treats of the originators, the precursors, from the beginning up to and including the Sixteenth Century ; the second, of the formation of classical mechanics. In this domain the Seventeenth Century deserves to be considered as the great creative century, with three great peaks formed by Galileo, Huyghens and Newton. The third part is devoted to the Eighteenth Century, which emerges as the century of the organisation of mechanics and finds its climax in the work of Lagrange, immediately preceded by Euler and d'Alembert. The development of mathematical analysis enabled mechanics to take a form which, for a considerable time, appeared to be finally established, and which is still a part of the classical teaching.

We have found ourselves somewhat embarrassed in the writing of the fourth part, which is concerned with classical mechanics after Lagrange. Indeed, nothing would be gained by duplicating the textbooks. Therefore we have confined ourselves to a selection from the work of the Nineteenth Century and the beginning of the Twentieth Century. This selection may appear somewhat arbitrary to the well-informed reader. Moreover, it is rather interesting to observe that the uneasiness about the structure of classical mechanics which is evident in the writings of the critics did not prepare the way for the relativistic and quantum revolutions, dated 1905 and 1923 respectively. These came from outside, imposed by the necessities of optics, electromagnetism and the theory of radiation. However, these reflections of the critics were not valueless, for they showed that the lagrangian science was not intangible and that the axioms of mechanics, to use Mach's words, " not only assumed but also demanded the continual control of experiment."

Thus the determinism—or, if it is preferred, legality—of which the students of mechanics were so proud and which made their science the model of all physical theories, now appears, after the development of wave mechanics, as justified in the macroscopic domain because of statistical compensation between the individuals of a large assembly, without there being an underlying determinism.

The fact of collecting in one book the stammerings of the early students, the creation and organisation of the classical science and the rudiments of the new mechanics—the object of the fifth part of this book—may appear a wager. It is only so in appearance. Indeed, the original works never had that codified aspect which is, of necessity, lent them by the textbooks. Just as the French currency remained stable for more than a century, Lagrange's mechanics was able to appear as complete for a period of roughly the same length. Despite its mathematical perfection, it had no other foundation that that of common experiment. The double revolution of 1905 and 1923—the second much more radical than the first—profoundly disturbed the classical structure. For these new doctrines intended to rule over the whole of physics, only admitting the validity of classical mechanics in the limits at which the velocity of light can be considered infinite and Planck's universal constant negligible. As regards the principles, certain thinkers have made the mistake of incorporating into a system what was only a stability of fact, a stage in evolution, as long as this stability appeared to be verified and however important were, and still are, its conquests.

For the historian, who is only a spectator and who, by his profession, is aware of the fragility of our anthropomorphic science, these revolutions are not an object of scandal or even a surprise. This does not mean that innovations must be accepted without analysis, or new dogmas professed without reserve.

It is difficult for us to indicate in detail what this work owes to the great historians and critics of mechanics. From Duhem we have taken what material he could extract from the manuscripts of the Middle Ages, at the same time bringing to this semi-darkness the light of his particularly profound and alert mind. We must confess however that we have had to disagree with several of his opinions, which appeared too categorical to us. Duhem had undergone the polarisation of the investigator which leads him to attach too great an importance sometimes to the original he has just discovered. Besides, Duhem's attachment to energetics made him somewhat biassed. For example, the principle of virtual work triumphed over that of virtual velocity which, up to the the early writings of Galileo, remained inspired by peripatetic doctrines, but it already explained correctly the laws of the equilibrium of simple machines, and long preceded the former. Curiously enough, Duhem makes Leonardo da Vinci the centre of his studies of the Precursors, to the extent of considering the Unknown Man of the Thirteenth Century, that same one who discovered the law of the equilibrium of a heavy body on an inclined plane, as a precursor of Leonardo, whereas this latter was always in error on this problem, and of qualifying as plagiaries or successors of Leonardo a large number of authors of the Sixteenth Century who may very well not have come under the influence of the great painter who in mechanics was scarcely more than an amateur of genius.

These few reserves, for which we beg to be excused, do not prevent Duhem's historical work from being of the greatest importance. An indefatigable reader, he succeeded not only in bringing to life works of the Scholastics of the Middle Ages, that were until then little known, but in establishing between them and the classical period filiations of indisputable interest. It is certain that, on more than one point, this Scholastic sheds light on and prefigures Descartes.

From Emile Jouguet, who honoured us with his teaching, we have borrowed several of his *Lectures*—given conscientiously and with great regard for the original authors—and a number of opinions that were, out of modesty, consigned to notes, lest they should hide his perfect knowledge of the Ancients.

In many places we have cited the very personal, and sometimes very judicious, observations of Mach. His *Mechanics* was one of the first systematic works of its kind and represented both a very wide reading and a critical mind of remarkable independence.[1]

Strictly speaking, Painlevé did not treat the history of mechanics. On his own account, with the analytical mind that he applied to everything, he rethought the evolution of mechanics. The lectures which he gave us at the *École polytechnique* were revised and developed in his *Axiomes de la Mécanique*. This contains not only an original discussion of the classical principles, but an often constructive and always valuable criticism of relativistic doctrines.

In spite of the contributions of the great critics we have just mentioned, in spite of the several researches of the original authors themselves which this book contains, we do not conceal its imperfections and its omissions. Certain of these omissions, especially from the classical field after Lagrange, have been accepted deliberately ; others may be unknown to us and, for this reason, more serious. We have not sought to restrict ourselves too narrowly to our subject, and have made some incursions into the domain of astronomy and that of hydrodynamics when it seemed that these served our purpose. But a presentation of a system of the world, or a complete history of the mechanics of fluids, should not be looked for here ; these subjects themselves would require whole volumes.

This book will only be read with profit by those who already have some knowledge of the didactic aspect of things. It also presupposes, as does all mechanics, a somewhat extensive mathematical background. Our purpose will have been achieved if the reader finds in it, with less effort than it has cost us to unite and explain the original texts, a reflection of the joy of knowledge that we have found.

I must thank the " Éditions du Griffon " for having applied all their recourses to the production of this book.

---

[1] Throughout this book, DUHEM's *Origines de la Statique* are indicated by the initials *O. S.*, MACH's *Mechanics* by the initial *M.*, and JOUGUET's *Lectures de Mécanique* by the initials *L. M.*

PART ONE

THE ORIGINS

CHAPTER ONE

HELLENIC SCIENCE

1. ARISTOTELIAN MECHANICS.

For lack of more ancient records, history of mechanics starts with Aristotle (384-322 B. C.) or, more accurately, with the author of the probably apocryphal treatise called *Problems of Mechanics (Μηχανικὰ προβλήματα)*. This is, in fact, a text-book of practical mechanics devoted to the study of simple machines.

In this treatise the *power* of the agency that sets a body in motion is defined as the product of the weight or the mass of the body—the Ancients always confused these concepts—and the velocity of the motion which the body acquires. This law makes it possible to formulate the condition of equilibrium of a straight lever with two unequal arms which carry unequal weights at their ends. Indeed, when the lever rotates the velocities of the weights will be proportional to the lengths of their supporting arms, for in these circumstances the powers of the two opposing powers cancel each other out.

The author regards the efficacy of the lever as a consequence of a magical property of the circle. " Someone who would not be able to move a load without a lever can displace it easily when he applies a lever to the weight. Now the root cause of all such phenomena is the circle. And this is natural, for it is in no way strange that something remarkable should result from something which is more remarkable, and the most remarkable fact is the combination of opposites with each other. A circle is made up of such opposites, for to begin with it is made up of something which moves and something which remains stationary. . . . "

In this way *Problems of Mechanics* reduces the study of all simple machines to one and the same principle. " The properties of the balance are related to those of the circle and the properties of the lever to those of the balance. Ultimately most of the motions in mechanics are related to the properties of a lever. "

To Aristotle himself, just as much in his *Treatise on the Heavens*, *(Περὶ οὐρανοῦ)* as in his *Physics*, concepts belonging to mechanics were not differentiated from concepts having a more general significance. Thus the notion of movement included both changes of position and changes of kind, of physical or chemical state. Aristotle's law of powers, which he called δύναμις or ἰσχύς, is formulated in Chapter V of Book VII of his Physics in the following way.

" Let the motive agency be α, the moving body β, the distance travelled γ and the time taken by the displacement be δ. Then an equal power, namely the power α, will move half of β along a path twice γ in the same time, or it will move it through the distance γ in half the time δ. For in this way the proportions will be maintained. "

Aristotle imposed a simple restriction on the application of this rule—a small power should not be able to move too heavy a body, " for then one man alone would be sufficient to set a ship in motion. "

This same law of powers reappears in Book III of the *Treatise on the Heavens*. Its application to statics may be regarded as the origin of the *principle of virtual velocities* which will be encountered much later.

In another place Aristotle made a distinction between *natural motions* and *violent motions*.

The fall of heavy bodies, for example, is a natural motion, while the motion of a projectile is a violent one.

To each thing corresponds a natural place. In this place its substantial form achieves perfection—it is disposed in such a way that it is subject as completely as possible to influences which are favourable, and so that it avoids those which are inimical. If something is moved from its natural place it tends to return there, for everything tends to perfection. If it already occupies its natural place it remains there at rest and can only be torn away by violence.

In a precise way, for Aristotle, the position of a body is the internal surface of the bodies which surround it. To his most faithful commentators, the natural place of the earth is the concave surface which defines the bottom of the sea, joined in part to the lower surface of the atmosphere, the natural place of the air.[1]

Concerning the natural motion of falling bodies, Aristotle maintained in Book I of the *Treatise on the Heavens* that the " relation which weights have to each other is reproduced inversely in their durations of fall. If a weight falls from a certain height in so much time, a weight which is twice as great will fall from the same height in half the time. "

In his *Physics* (Part V), Aristotle remarked on the acceleration of

---

[1] *Cf.* DUHEM, *Origines de la Statique*, Vol. II, p. 21. Throughout the present book this work of Duhem will be indicated by the letters *O. S.*

falling heavy bodies. A body is attracted towards its natural place by means of its heaviness. The closer the body comes to the ground, the more that property increases.

If the natural place of heavy bodies is the centre of the World, the natural place of light bodies is the region contiguous with the Sphere of the Moon. Heavenly bodies are not subject to the laws applicable to terrestrial ones—every star is a body as it were divine, moved by its own divinity.

We return to terrestrial mechanics. All *violent* motion is essentially impermanent. This is one of the axioms which the Schoolmen were to repeat—*Nullum violentum potest esse perpetuum.* Once a projectile is thrown, the motive agency which assures the continuity of the motion resides in the air which has been set in motion. Aristotle then assumes that, in contrast to solid bodies, air spontaneously preserves the impulsion which it receives when the projectile is thrown, and that it can in consequence act as the motive agency during the projectile's flight.

This opinion may seem all the more paradoxical in view of the fact that Aristotle remarked, elsewhere, on the resistance of the medium. This resistance increases in direct proportion to the density of the medium. " If air is twice as tenuous as water, the same moving body will spend twice as much time in travelling a certain path in water as in travelling the same path in air. "

Aristotle also concerned himself with the *composition of motions.* " Let a moving body be simultaneously actuated by two motions that are such that the distances travelled in the same time are in a constant proportion. Then it will move along the diagonal of a parallelogram which has as sides two lines whose lengths are in this constant relation to each other. " On the other hand, if the ratio between the two component distances travelled by the moving body in the same time varies from one instant to another, the body cannot have a rectilinear motion. " In such a way a curved path is generated when the moving body is animated by two motions whose proportion does not remain constant from one instant to another. "

These propositions relate to what we now call kinematics. But Aristotle immediately inferred from them dynamical results concerning the composition of forces. The connection between the two disciplines is not given, but as Duhem has indicated, it is easily supplied by making use of the law of powers—a fundamental principle of aristotelian dynamics. In particular, let us consider a heavy moving body describing some curve in a vertical plane. It is clear that the body is actuated by two motions simultaneously. Of these, one produces a vertical descent while the other, according to the position of the body on its trajectory, results in an increase or a decrease of the distance from the centre. In Aristotle's sense, the body will have a natural falling motion due to gravity, and will be carried horizontally in a violent motion. Consider different moving bodies unequally distant from the centre of a circle and on the same radius. Let this radius, in falling, rotate about the centre. Then it may be inferred that for each body the relation of the natural to the violent motion remains the same. " The contemplation of this equality held Aristotle's attention for a long time. He appears to have seen in it a somewhat mysterious correlation with the law of the equilibrium of levers. " [1]

Aristotle believed in the *impossibility of a vacuum* (*Physics*, Book IV, Chapter XI) on the grounds that, in a vacuum, no natural motion, that is to say no tendency towards a natural place, would be possible. Incidentally this idea led him to formulate a principle analogous to that of inertia, and to justify this in the same way as that used by the great physicists of the XVIIIth Century.

" It is impossible to say why a body that has been set in motion in a vacuum should ever come to rest ; why, indeed, it should come to rest at one place rather than at another. As a consequence, it will either necessarily stay at rest or, if in motion, will move indefinitely unless some obstacle comes into collision with it. "

Aristotle's ideas on gravitation and the figure of the Earth merit our attention, if only because of the influence which they have had on the development of the principles of mechanics. First we shall quote from the *Treatise on the Heavens* (Book II, Chapter XIV). " Since the centres of the Universe and of the Earth coincide, one should ask oneself towards which of these heavy bodies and even the parts of the Earth are attracted. Are they attracted towards this point because it is the centre of the Universe or because it is the centre of the Earth ? It is the centre of the Universe towards which they must be attracted.... Consequently heavy bodies are attracted towards the centre of the Earth, but only fortuitously, because this centre is at the centre of the Universe. "

If the Earth is spherical and at the centre of the World, what happens if a large weight is added to one of the hemispheres ? The answer to this question is the following. " The Earth will necessarily move until it surrounds the centre of the World in a uniform way, the tendencies to motion of the different parts counterbalancing one another." Duhem points out that the centre, $\tau\grave{o}$ $\mu\acute{e}\sigma o\nu$, that in every body is attracted to the centre of the Universe, was not defined in a precise

---

[1] DUHEM, *O. S.*, Vol. I, p. 110. Note here that, *for the same fall*, the longer the lever the less the natural motion will be disturbed. It is therefore natural to assume that a weight has more power at the end of a long lever than at the end of a short one.

way by Aristotle. In particular, Aristotle did not identify it with the centre of gravity, of which he was ignorant.[1]

In this same treatise Aristotle repeatedly enumerates the arguments for the spherical figure of the Earth. He distinguishes *a posteriori* arguments, such as the shape of the Earth's shadow in eclipses of the Moon, the appearance and disappearance of constellations to a traveller going from north to south, from *a priori* arguments, of which he says—

" Suppose that the Earth is no longer a single mass, but that, potentially, its different parts are separated from each other and are placed in all directions and attracted similarily towards the centre. Then let the parts of the Earth which have been separated from each other and taken to the ends of the World be allowed to reunite at the centre ; let the Earth be formed by a different procedure—the result will be exactly the same. If the parts are taken to the ends of the World and are taken there similarily in all directions, they will necessarily form a mass which is symmetrical. Because there will result an addition of parts which are equal in all directions, and the surface which defines the mass produced will be everywhere equidistant from the centre. Such a surface will therefore be a sphere. But the explanation of the shape of the Earth would not be changed in any way if the parts which form it were not taken in equal quantities in all directions. In fact, a larger part will necessarily push away a smaller one which it finds in front of it, for both have a tendency towards the centre and more powerful weights are able to displace lesser ones. "

To Aristotle, heaviness does not prove rigorously that the Earth will be spherical, but only that it will tend to be so. On the other hand, for the surface of water, this is obvious if it is admitted that " it is a property of water to run towards the lowest places, " that is, towards places which are nearest the centre.

Let $\beta\varepsilon\gamma$ be an arc of a circle with centre α ; the line $\alpha\delta$ is the shortest distance from α to $\beta\gamma$. " Water will run towards δ from all sides until its surface becomes equidistant from the centre. It therefore follows that the water takes up the same length on all the lines radiating from the centre. It then remains in equilibrium. But

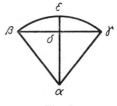

Fig. 1

the locus of equal lines radiating from a centre is a circumference of a circle. The surface of the water, $\beta\varepsilon\gamma$, will therefore be spherical. "

Adrastus (360-317 B. C.), commenting on Aristotle, made the pre-

[1] Duhem, *O. S.*, Vol. II, p. 11.

ceding proof precise in the following terms. " Water will run towards
the point δ until this point, surrounded by new water, is as far from α
as β and γ. Similarily, all points on the surface of the water will be at
an equal distance from α. Therefore the water exhibits a spherical
form and the whole mass of water and Earth is spherical. "

Adrastus supplemented this proof with the following evidence,
which was destined to become classical.[1]

" Often, during a voyage, one cannot see the Earth or an approaching ship from the deck, while sailors who climb to the top of a mast can
see these things because they are much higher and thus overcome the
convexity of the sea which is an obstacle. "

We shall say no more about aristotelian mechanics. However inadequate they may seem now, these intuitive theories have their origin in
the most everyday observations, precisely because they take the passive
resistances to motion into account. To an unsophisticated observer,
a horse pulling a cart seems to behave according to the law of powers,
in the sense that it develops an effort which increases regularly with
the speed. In order to break away from Aristotle's ideas and to construct the now classical mechanics, it is necessary to disregard the various ways in which motion may be damped, and to introduce these
explicitly at a later stage as frictional forces and as resistances of the
medium. However it may be, aristotelian doctrines provided the
fabric of thought in mechanics for nearly two thousand years, so that
even Galileo, who was to become the creator of modern dynamics,
made his first steps in science in commenting Aristotle, and proved in
his early writings to be a faithful Peripatetic ; which, it may be said in
passing, in no way diminishes his glory as a reformer, on the contrary,
it only adds to it.

2. THE STATICS OF ARCHIMEDES.

Unlike Aristotle, whose mechanics is integrated into a theory of
physics which goes so far as to incorporate a system of the world, Archimedes (287-212 B.C.) made of statics an autonomous theoretical science,
based on postulates of experimental origin and afterwards supported
by mathematically rigourous demonstrations, at least in appearance.

Here we shall follow the treatise *On the Equilibrium of Planes or on
the Centres of Gravity of Planes* [2] in which Archimedes discussed the
principle of the lever.

---

[1] This thesis of ADRASTUS is known to us by means of *A Collection of Mathematical
Knowledge useful for the Reading of Plato*, by THEON OF SMYRNA.
[2] Translation by PEYRARD, Paris, 1807. The reader should also refer to that of
P. VER EECKE, Paris and Anvers, Desclée de Brouwer, 1938.

Archimedes made the following postulates as axioms—

1) Equal weights suspended at equal distances (from a fulcrum) are in equilibrium.

2) Equal weights suspended at unequal distances cannot be in equilibrium. The lever will be inclined towards the greater weight.

3) If weights suspended at certain distances are in equilibrium, and something is added to one of them, they will no longer be in equilibrium. The lever will be inclined towards the weight which has been increased.

4) Similarily, if something is taken away from one of the weights, they will no longer be in equilibrium, but will be inclined towards the weight which has not been decreased.

5) If equal and similar plane figures coincide, their centres of gravity will also coincide.

(The concept of Centre of Gravity appears to have been defined by Archimedes in an earlier manuscript, of which no trace remains.)

6) The centres of gravity of unequal but similar figures are similarily placed.

7) If magnitudes suspended at certain distances are in equilibrium, equivalent magnitudes suspended at the same distances will also be in equilibrium.

8) The centre of gravity of a figure which is nowhere concave is necessarily inside the figure.

With this foundation, Archimedes proved the following propositions.

Proposition I. — When weights suspended at equal distances are in equilibrium, these weights are equal to each other.
(Proof by *reductio ad absurdum* based on Postulate 4).)

Proposition II. — Unequal weights suspended at equal distances will not be in equilibrium, but the greater weight will fall.
(Proof based on Postulates 1) and 3).)

Proposition III. — Unequal weights suspended at unequal distances may be in equilibrium, in which case the greater weight will be suspended at the shorter distance.
(Proof based on Postulates 4), 1) and 2). This proof only confirms the second part of the proposition, and does not demonstrate the *possibility* of the equilibrium of two unequal weights. This must be regarded as an additional postulate of experimental origin.)

Proposition IV. — If two equal magnitudes do not have the same centre of gravity, the centre of gravity of the magnitude made up of these two magnitudes is the point situated at the middle of the line which joins their centres of gravity.

(The proof, based on Postulate 2), is a demonstration by *reductio ad absurdum* which, moreover, assumes that the centre of gravity of the combined magnitude lies on the line joining the centres of gravity of the component magnitudes.)

Proposition V. — If the centres of gravity of three magnitudes lie on the same straight line, and if the magnitudes are equally heavy and the distances between their centres of gravity are equal, the centre of gravity of the combined magnitude will be the point which is the centre of gravity of the central magnitude.

(This is a corollary of Proposition IV, which Archimedes later extended to the case of *n* magnitudes. The enunciation is suitably modified if *n* is even.)

Proposition VI. — Commensurable magnitudes are in equilibrium when they are reciprocally proportional to the distances at which they are suspended.

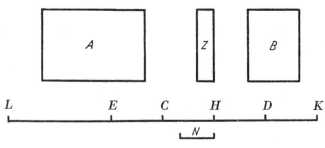

Fig. 2

" Let the commensurable magnitudes be *A* and *B*, and let their centres of gravity be the points *A* and *B*. Let *ED* be a certain length and suppose that the magnitude *A* is to the magnitude *B* as the length *DC* is to the length *CE*. It is necessary to prove that the centre of gravity of the magnitude formed of the two magnitudes *A* and *B* is the point *C*.

" Since *A* is to *B* as *DC* is to *CE* and since the areas *A* and *B* are commensurable, the lengths *DC* and *CE* will also be commensurable. . . . Therefore the lengths *EC* and *CD* have a common measure, say *N*.

Suppose that each of the lengths DH and DK is equal to the length EC, and that the length EL is equal to the length DC. Since the length DH is equal to the length CE, the length DC will be equal to the length EH, and the length LE will be equal to the length EH. Hence the length LH is twice the length DC, and the length HK twice the length CE. Therefore the length N is a common measure of the lengths DH and HK since it is a common measure of their halves. But A is to B as DC is to CE, so that A is to B as LH is to HK. Let A be as many times greater than Z as LH is greater than N. The length LH will be to the length N as A is to Z. But KH is to LH as B is to A. Therefore, by equality, the length KH is to the length N as B is to Z. Then B is as many times greater than Z as KH is a multiple of N. But it has been arranged that A is also a multiple of Z. Therefore Z is a common measure of A and B. Consequently, if LH is divided into segments each equal to N, and A into segments each equal to Z, A will contain as many segments equal to Z as LH contains segments equal to N. Therefore, if a magnitude equal to Z is applied to each segment of LH in such a way that its centre of gravity is at the centre of the segment, all the magnitudes [1] will be equal to A. Further, the centre of gravity of the magnitude made up of all these magnitudes will be the point E, remembering that they are an even number and that LE is equal to HE (Proposition V). Similarly it could be shown that if a magnitude equal to Z was applied to each of the segments of KH, with its centre of gravity at the centre of each segment, all those magnitudes [1] would be equal to B and that the combined centre of gravity would be D. But the magnitude A is applied at the point E and the magnitude B at the point D. Therefore certain equal magnitudes are placed on the same line, their centres of gravity are separated from each other by the same interval and they are an even number. It is therefore clear that the centre of gravity of the magnitude composed of all these magnitudes is the point at the middle of the line on which the centres of gravity of the central magnitudes lie (Proposition V). But the length LE is equal to the length CD and the length EC to the length CK. Thus the centre of gravity of the magnitude composed of all these areas is the point C. Therefore, if the magnitude A is applied to the point E and the magnitude B to the point D, the two areas will be in equilibrium about the point C. "

Archimedes then extended this proposition to the case of magnitudes A and B which were incommensurable. This demonstration depended on the method of exhaustion. We have reproduced this proof of

---

[1] Read, " the combination of all these magnitudes. "

Proposition VI in its entirety in order to illustrate the nature of Archimedes' logical apparatus. This should not be allowed, however, to create too great an illusion of power.

Indeed, Archimedes assumes in this proof that the load on the fulcrum of a lever is equal to the sum of the two weights which it supports.[1] Further, he made use of the principle of superposition of equilibrium states, without emphasising that this was an experimental postulate. Finally, and this is the most telling objection to the preceeding analysis, Archimedes, together with those of his successors who tried to improve his proof, tacitly made the hypothesis that the product PL measures the effect of a weight P placed at a distance L from a horizontal axis. In fact, in the case of complete symmetry which is envisaged in Archimedes' first postulate, equilibrium obtains *whatever* law of the form $Pf(L)$ is taken as a measure of the effect of the weight P. It is therefore impossible, with the help of Archimedes' postulates alone, to substantiate Proposition VI in a logical way.[2]

For the rest, the treatise *On the Equilibrium of Planes* is concerned with the determination of the centres of gravity of particular geometrical figures. After having obtained the centres of gravity of a triangle, a parallelogram and a trapezium, Archimedes determined the centre of gravity of a segment of a parabola by means of an analysis which is a milestone in the history of mathematics (Book II, Proposition VIII).

We shall now concern ourselves with Archimedes' treatise on *Floating Bodies*. The author starts from the following hypothesis—

" The nature of a fluid is such that if its parts are equivalently placed and continuous with each other, that which is the least compressed is driven along by that which is the more compressed. Each part of the fluid is compressed by the fluid which is above it in a vertical direction, whether the fluid is falling somewhere or whether it is driven from one place to another. "

From this starting point, the following propositions derive in a logical sequence.

Proposition I. — If a surface is intersected by a plane which always passes through the same point and if the section is a circumference (of a circle) having this fixed point as its centre, the surface is that of a sphere.

---

[1] This is a point which can be established rigorously by considerations of symmetry alone, as FOURIER was to show, much later, in his perfection of a similar attempt due to D'ALEMBERT.

[2] *Cf.* MACH, *Mechanics*, p. 21. Throughout this work, Mach's treatise will be indicated by the letter M.

Proposition II. — The surface of any fluid at rest is spherical and the centre of this surface is the same as the centre of the Earth.

This result had already, as we have seen, been enunciated by Aristotle.

Proposition III. — If a body whose weight is equal to that of the same volume of a fluid (α) is immersed in that fluid, it will sink until no part of it remains above the surface, but will not descend further.

We shall reproduce the proof of this proposition, which is the origin of Archimedes' Principle.

" Let a body have the same heaviness as a liquid. If this is possible, suppose that the body is placed in the fluid with part of it above the surface. Let the fluid be at rest. Suppose that a plane which passes through the centre of the Earth intersects the fluid and the

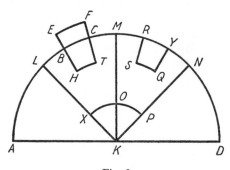

Fig. 3

body immersed in it in such a way that the section of the fluid is *ABCD* and the section of the body is *EHTF*. Let *K* be the centre of the Earth, *BHTC* be the part of the body which is immersed in the fluid and *BEFC* the part which projects out of it. Construct a pyramid whose base is a parallelogram in the surface of the fluid (α) and whose apex is the centre of the Earth. Let the intersection of the faces of the pyramid by the plane containing the arc *ABCD* be *KL* and *KM*. In the fluid, and below *EFTH* draw another spherical surface *XOP* having the point *K* as its centre, in such a way that *XOP* is the section of the surface by the plane containing the arc *ABCD*. Take another pyramid equal to the first, with which it is contiguous and continuous, and such that the sections of its faces are *KM* and *KN*. Suppose that there is, in the fluid, another solid *RSQY* which is made of the fluid and is equal and similar to *BHTC*, that part of the body

EHTF which is immersed in the fluid. The portions of the fluid which are contained by the surface *XO* in the first pyramid and the surface *OP* in the second pyramid are equally placed and continuous with each other. But they are not equally compressed. For the portions of the fluid contained in *XO* are compressed by the body *EHTF* and also by the fluid contained by the surfaces *LM, XO* and those of the pyramid. The parts contained in *PO* are compressed by the solid *RSQY* and by the fluid contained by the surfaces *OP, MN* and those of the pyramid. But the weight of the fluid contained between *MN* and *OP* is less than the combined heaviness of the fluid between *LM* and *XO* and the solid. For the solid *RSQY* is smaller than the solid *EHTF—RSQY* is equal to *BHTC*—and it has been assumed that the body immersed has, volume for volume, the same heaviness as the fluid. If therefore one takes away the parts which are equal to each other, the remainder will be unequal. Consequently it is clear that the part of the fluid contained in the surface *OP* will be driven along by the part of the fluid contained in the surface *XO*, and that the fluid will not remain at rest. Therefore, no part of the body which has been immersed will remain above the surface. However, the body will not fall further. For the body has the same heaviness as the fluid and the equivalently placed parts of the fluid compress it similarily. "

Proposition IV. — If a body which is lighter than a fluid is placed in this fluid, a part of the body will remain above the surface.
(Proof analogous to that of Proposition III.)

Proposition V. — If a body which is lighter than a fluid is placed in the fluid, it will be immersed to such an extent that a volume of fluid which is equal to the volume of the part of the body immersed has the same weight as the whole body.
The diagram is the same as the preceding one (Proposition III).
" Let the liquid be at rest and the body *EHTF* be lighter than the fluid. If the fluid is at rest, parts which are equivalently placed will be similarly compressed. Then the fluid contained by each of the surfaces *XO* and *OP* is compressed by an equal weight. But, if the body *BHTC* is excluded, the weight of fluid in the first pyramid is equal, with the exclusion of the fluid *RSQY*, to the weight of fluid in the second pyramid. Therefore it is clear that the weight of the body *EHTF* is equal to the weight of the fluid *RSQY*. From which it follows that a volume of fluid equal to that of the body which is immersed has the same weight as the whole body. "

Proposition VI. — If a body which is lighter than a fluid is totally and forcibly immersed in it, the body will be thrust upwards with a force equal to the difference between its weight and that of an equal volume of fluid.

Proposition VII. — If a body is placed in a fluid which is lighter than itself, it will fall to the bottom. In the fluid the body will be lighter by an amount which is the weight of the fluid which has the same volume as the body itself.

The first Book of the treatise *On Floating Bodies* concludes with the following hypothesis— " We suppose that bodies which are thrust upwards all follow the vertical which passes through their centre of gravity. "

In Book II, Archimedes modified the principle which is the subject of Proposition V, Book I, to the following form—

" If any solid magnitude which is lighter than a fluid is immersed in it, the proportion of the weight of the solid to the weight of an equal volume of fluid will be the same as the proportion of the volume of that part of the solid which is submerged to the volume of the whole solid. "

We shall pass over the proof of this proposition, in which Archimedes once more deploys that powerful logical apparatus with which we are now familiar. The rest of Book II is devoted to a detailed study of the equilibrium of a floating body which is shaped like a right segment of a " parabolic conoid. " In Archimedes' language (in the treatise *On Conoids and Spheroids*), this term refers to the figure which we would now call a parabolic cylinder. It may be surmised that Archimedes was interested in this problem for a most practical reason, for this surface is similar to that of the hull of a ship.

It is of interest that, throughout this study, Archimedes approximated the free surface of a fluid by a plane, and that he treated verticals as if they were parallel. This is necessary if the concept of centre of gravity is to be utilised. Thus Archimedes must have understood the necessity and the practical importance of this approximation, even though his principle was based on the convergence of the verticals at the centre of the Earth, the spherical symmetry of fluid surfaces and a rather vague hypothesis about the pressures obtaining in the interior of a fluid.

CHAPTER TWO

ALEXANDRIAN SOURCES AND ARABIC MANUSCRIPTS

1. The " mechanics " of Hero of Alexandria.

It seems that Hero of Alexandria lived at some time during the IInd Century A. D. His treatise *Mechanics* discusses certain simple machines—the lever, pulley-block and the screw—alone or in various combinations, and is only available to us in the form of an arabic version which has been translated and published by Carra de Vaux.[1]

As far as it concerns the history of mechanics, the essential importance of this work lies in the fact that its author used the now classical idea of moment in his discussion of a lever which was not straight. Whether or not this conception was an original one remains doubtful. Indeed alexandrian learning had access to a treatise of Archimedes that was devoted to levers *(Περὶ ζυγῶν)* and in which the problem of the angular lever had been treated. No trace of this writing is extant.

Fig. 4

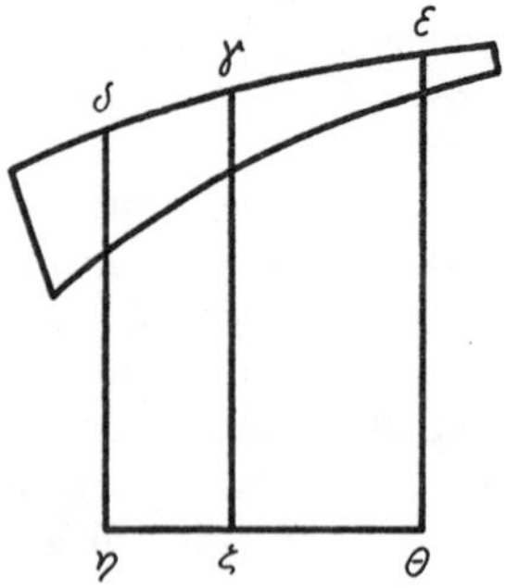

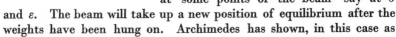

However this may be, we shall quote from Hero's *Mechanics*.

" Consider an arm of a balance which does not have the same thickness or heaviness throughout its length. It may be made of any material. It is in equilibrium when suspended from the point γ —by equilibrium we understand the arrest of the beam in a stable position, even though it may be inclined in one direction or the other. Now let weights be suspended at some points of the beam—say at δ and ε. The beam will take up a new position of equilibrium after the weights have been hung on. Archimedes has shown, in this case as

[1] *Journal asiatique*, 1894.

well, that the relation of the weights to each other is the same as the inverse relation of the respective distances.[1]   Concerning these distances in the case of irregular and inclined beams, it should be imagined that a string is allowed to fall from γ towards the point ζ.   Construct a line which passes through the point ζ—the line $\eta\zeta\theta$—and which should be arranged to intersect the string at right angles.   When the beam is at rest the relation of $\zeta\eta$ to $\zeta\theta$ is the same as the relation of the weight hung at the point ε to the weight hung at the point δ. "

Hero employed a similar argument in his discussion of the wheel and axle.   In fact, in reducing the study of these machines to the *principle of circles* he was implicitly using the notion of moment.   Thus it is clear, though not explicitly stated, that in his discussion of the axle Hero understands that a weight ζ can be replaced by an equal force applied tangentially at A, because AF has the same moment as ζ.[2]

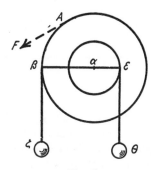

Fig. 5

2. PAPPUS'S THEORIES OF THE INCLINED PLANE AND OF THE CENTRE OF GRAVITY.

Pappus (IVth Century A. D.) appears to be the only geometer of Antiquity who took up the problem of the motion and equilibrium of a heavy body on an inclined plane.' The proof that we shall analyse now is taken from Book VIII of his *Collections (From among the varied and delightful problems of mechanics)*.

Pappus assumes that a certain effort γ is necessary to move a weight α on the horizontal plane $\mu\nu$, and that a power θ is necessary to draw it

---

[1] CARRA DE VAUX's surmise that Hero is referring to the treatise $\Pi\varepsilon\varrho\grave{\iota}$ $\zeta\upsilon\gamma\tilde{\omega}\nu$ is probably correct.

[2] *Cf.* JOUGUET, *Lectures de Mécanique*, Vol. I, p. 215.   Throughout the present book this treatise will be indicated by the letters *L. M.*

up the inclined plane $\mu\varkappa$.   He sets out to determine the relation between γ and θ.

The weight α on the plane $\mu\varkappa$ has the form of a sphere with centre ε. Pappus reduces the investigation of the equilibrium of this sphere on

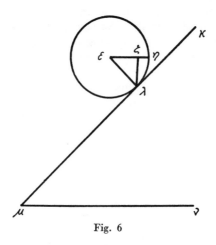

Fig. 6

the inclined plane to the following problem.   A balance supported at λ carries the weight α at ε and the weight β which is necessary to keep it in equilibrium at η—the end of the horizontal radius $\varepsilon\eta$.   The law of the angular lever, which Pappus borrows from Archimedes' $\Pi\varepsilon\varrho\grave{\imath}\ \zeta\upsilon\gamma\tilde{\omega}\nu$ or from Hero's *Mechanics*, provides the relation

$$\beta = \alpha\,\frac{\varepsilon\zeta}{\eta\zeta}.$$

On the horizontal plane where the power necessary to move α is γ, the power necessary to move along β will be

$$\delta = \gamma\,\frac{\varepsilon\zeta}{\eta\zeta}.$$

Pappus then assumes that the power θ that is able to move the weight α on the inclined plane $\mu\varkappa$ will be the sum of the powers δ and γ, that is

$$\theta = \gamma\left(1 + \frac{\varepsilon\zeta}{\eta\zeta}\right) = \gamma\,\frac{\varepsilon\eta}{\eta\zeta}.$$

Evidently the necessity of a power γ for pulling the weight α on the horizontal plane derives from Aristotle's dynamics, in which all unnatural motion requires a motive agency.   The argument by which Pappus introduces the lever $\varepsilon\lambda\eta$ supporting the two weights α and β is rather a natural one, even though it does not lead to a correct evaluation of β. The last hypothesis, concerning the addition of δ and γ, the powers that are necessary to move β and α respectively on the horizontal plane is, on the other hand, most strange.

However incorrect it may have been, this proof was destined to inspire the geometers of the Renaissance.   Guido Ubaldo was to adopt it and Galileo was to be occupied in demonstrating its falsehood.

Archimedes certainly formulated a precise definition of centre of gravity, but there is no trace of anything of this kind in those of his writings that are available to us.   Therefore it is of some value to record the definition which is due to Pappus.

Imagine that a heavy body is suspended by an axis $\alpha\beta$ and let it take up its equilibrium position.   The vertical plane passing through $\alpha\beta$ " will cut the body into two parts that are in equilibrium with each other and which will be hung in such a way, on one side of the plane and on the other, as to be equal with respect to their weight. " Taking another axis $\alpha'\beta'$ and repeating the same operation, the second vertical plane passing through $\alpha'\beta'$ will certainly cut the

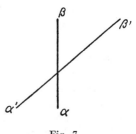

Fig. 7

first—if it were parallel to it " each of these two planes would divide the body into two equal parts which would be at the same time of equal weight and of unequal weight, which is absurd. "

Now suspend the body from a point γ and draw the vertical $\gamma\delta$ through the point of suspension when equilibrium is established.   Take a second point of suspension γ' and, in the same way, draw the vertical $\gamma'\delta'$.   The two lines $\gamma\delta$, $\gamma'\delta'$ necessarily intersect.   For if not, through each of them a plane could be drawn so as to divide the body into two parts in equilibrium with each other, and in such a way that these two planes were parallel to each other.   This is impossible.

All lines like $\gamma\delta$ will therefore intersect at one unique point of the body that is called the *centre of gravity*.

Pappus did not distinguish clearly, as Guido Ubaldo was to do in his *Commentary on Archimedes' two books on the equilibrium of weights* (1588) between "equiponderant" parts, that is parts that are in equilibrium in the positions which they occupy, and parts which have the same weight.

3. THE FRAGMENTS ATTRIBUTED TO EUCLID IN ARABIC WRITINGS.

Greek antiquity does not attribute any work on mechanics to Euclid. However his name occurs frequently in this connection in the writings of arabic authors.

Euclid's book on the balance, an arabic manuscript of 970 A. D. which has been brought to light by Dr. Woepke,[1] seems to have remained unknown to the western Middle Ages. This relic of greek science may be contemporaneous with Euclid and may thus antedate Archimedes. It contains a geometrical proof of the law of levers which is independent of Aristotle's dynamics and which makes explicit appeal to the hypothesis that the effect of a weight P placed at the end of an arm of a lever is expressed by the product PL. We have had occasion to emphasise the necessity of this hypothesis in our analysis of Archimedes' proof.

The treatise *Liber Euclidis de gravi et levi*, often simply called *De ponderoso et levi*, has been known for a long time. It includes a very precise exposition of aristotelian dynamics which is arranged, after Euclid's style, in the form of definitions and propositions. The latin text renders the terms δύναμις and ἰσχύς, by which Aristotle meant " power ", as *virtus* and *fortitudo*. Bodies that travel equal distances in the same medium—air or water—in times which are equal to each other, are said to be equal in *virtus*. Bodies that travel equal distances in unequal times are of different *virtus*, and that which takes the shorter time is said to have the greater *virtus*. Bodies of the same *kind* are those that, volume for volume, are equal in *virtus*. That which has the greater *virtus* is said to be *solidius* (more dense).

The *virtus* of bodies of the same kind is proportional to their dimensions ; that is, the bodies fall with velocities which are proportional to their volume. If two heavy bodies are joined together, the velocity with which the combination will fall will be the sum of the velocities of the separate bodies.

Duhem has found, in a XIVth Century manuscript,[2] four propositions on questions in statics which complete *De ponderoso et levi*. This manuscript contains a theory of the roman balance, and shows that the fact that the balance is a heavy homogenous cylinder does not alter the relation of the weights to each other.

Finally, in a XIIIth Century manuscript, Duhem has unearthed a text called *Liber Euclidis de ponderibus secundum terminorum circonferentiam* [3] which connects the law of levers with aristotelian dynamics and also contains a theory of the roman balance.

---

[1] *Journal asiatique*, Vol. 18, 1851, p. 217.
[2] *Bibliothèque Nationale*, Paris, latin collection, Ms. 10,260.
[3] *Ibid.*, Ms. 16,649.

4. THE BOOK OF CHARISTION.

Liber Charastonis is the latin version of an arabic text due to the geometer Thâbit ibn Kurrah (836-901). The original greek version remains unknown, and the question of whether *karaston* (in Arabic— *karstûn*) refers merely to the roman balance or to the name of the greek geometer Charistion (a contemporary of Philon of Byzantium in the IInd Century B. C.) has been the subject of much scholarly debate.

We shall follow Duhem[1] in summarising the theory of the roman balance which is found in *Liber Charastonis*.

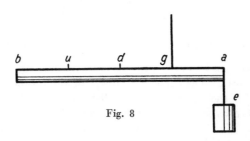

Fig. 8

A heavy homogeneous cylindrical beam *ab* whose arms *ag* and *bg* are unequal may be maintained in a horizontal position by means of a weight *e* hung from the end of the shorter arm *ag*. If *bd* is the amount by which the longer arm exceeds the shorter arm and *u* is the centre of *bd*, the weight *e* will be to *bd* as *gu* is to *ga*. If *p* is the total weight of the beam

$$e = p \, \frac{db}{2ga}.$$

If this weight were known it could be represented exactly by a scale-pan hung from the shorter arm, and the *karaston* arranged in this way could be treated as a weightless beam.

We must also mention, as one of the sources of statics, the treatise *De Canonio*,[2] a latin translation of a greek text which adds nothing essential to *Liber Charastonis*.

[1] *O. S.*, Vol. I, p. 90.
[2] *Bibliothèque Nationale*, Paris, latin collection, Ms. 7378 A.

CHAPTER THREE

THE XIIIth CENTURY
THE SCHOOL OF JORDANUS

1. JORDANUS OF NEMORE AND " GRAVITAS SECUNDUM SITUM. " '

The Middle Ages had access to the *Problems of Mechanics* and to the works of Aristotle. They had also inherited the fragments attributed to Euclid—with the exception of the *Book on the Balance*—as well as the *Liber Charastonis* from arabic learning. They had no knowledge of Archimedes, Hero of Alexandria and Pappus.

In spite of the researches of the scholars, the personality of Jordanus remains mysterious. At least three XIIIth Century manuscripts on statics have been attributed to him, although these are clearly in the style of different authors. Neither Jordanus's nationality nor the period in which he lived is known with any certainty. Daunou believes him to have lived in Germany about 1050, Chasles associates him with the XIIIth Century while Curtze places him about 1220 under the name of Jordanus Saxo. Michaud has identified him with Raimond Jordan, provost of the church of Uzès in 1381 which is clearly too late. With Montucla, we shall here adopt the intermediate view that associates Jordanus of Nemore with the XIIIth Century.

Like Duhem, we shall follow the *Elementa Jordani super demonstrationem ponderis*.[1] This work comprised seven axioms or definitions followed by nine propositions. The essential originality of Jordanus lay in the systematic use, in his study of the motion of heavy bodies, of the *effective path in a vertical direction* as a measure of the effect of a weight, which was usually placed at the end of a lever and described a circle in consequence. Thus his statics stems, implicitly, from the principle of *virtual work*. The word *work*, taken in the modern sense, is to be con-

---

[1] *Bibliothèque Nationale*, Paris, Ms. 10,252, dated 1464. There also exists an incomplete manuscript of the same work, dating from the XIIIth Century, in the *Bibliothèque Mazarine*, Ms. 3642.

trasted with the word *velocity* and with the concept of *virtual velocities* which may be traced in the arguments of *Problems of Mechanics*. Of course Jordanus never used the word " work " itself. He considered the heaviness of a particle relative to its situation *(gravitas secundum situm)* without making clear the relation that exists between this quantity and the heaviness in the strict sense.

Jordanus formulated his principle in a picturesque Latin which merits quotation.

" *Omnis ponderosi motum esse ad medium, virtutemque ipsius potentiam ad inferiora tendendi et motui contrario resistendi.*

" *Gravius esse in descendendo quando ejusdem motus ad medium rectior.*

" *Secundum situm gravius, quando in eodem situ minus obliquus est descensus.*

" *Obliquiorem autem descensum in eadem quantitate minus capere de directo.* "

Or—

" The motion of all heavy things is towards the centre,[1] its strength being the power which it has of tending downwards and of resisting a contrary motion.

" A moving body is the heavier in its descent as its motion towards the centre is the more direct.

" A body is the heavier because of its situation as, in that situation, its descent is the less oblique.

" A more oblique descent is one that, for the same path, takes less of the direct. "

Thus a certain weight placed at b, at the end of the lever cb, has a smaller gravity *secundum situm* than the same weight has when it is at a, at the end of the horizontal radius ca. Indeed, on the circumference of the circle with centre c and radius $ca = cb$, if the body falls from b to h along the arc $\widehat{bh}$ the effective path in a vertical direction is $b'\,h'$. On the other hand if the body starts from a and falls along an arc $\widehat{az}$, which is equal to the arc $\widehat{bh}$, the effective vertical path

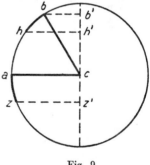

Fig. 9

is cz' and is greater than $b'\,h'$. Thus the descent $\widehat{bh}$, equal to the descent $\widehat{az}$, is more oblique than that and *takes less of the direct.*

---

[1] Understood as the common centre of all heavy things in Aristotle's sense.

This idea led Jordanus to a proof of the rule of the equilibrium of the straight lever whose originality cannot be contested.

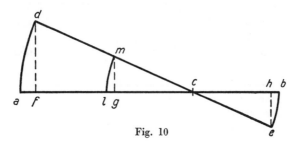

Fig. 10

" Let *acb* be the beam, *a* and *b* the weights that it carries, and suppose that the relation of *b* to *a* is the same as that of *ca* to *cb*. I maintain that this rule will not change its place. Indeed, if the arm supporting *b* falls and the beam takes up the position *dce*, the weight *b* will descend by *he* and *a* will rise by *fd*. If a weight equal to the weight *b* is placed at *l*, at a distance such that *cl* = *cb*, this will rise in the motion by *gm* = *he*. But it is clear that *df* is to *mg* as the weight *l* is to the weight *a*. *Consequently, what is sufficient to bring* a *to* d *will be sufficient to bring* l *to* m. But we have shown that *b* and *l* counterbalance each other exactly, so that the supposed motion is impossible. This will also be true of the inverse motion. "

Duhem writes in this connection [1]—

" Underlying this demonstration of Jordanus the following principle is clearly evident—that which can lift a weight to a certain height can

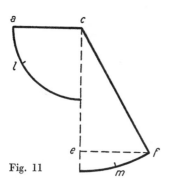

Fig. 11

also lift a weight which is *k* times as great to a height which is *k* times less. This principle is then the same as that which Descartes took as a basis for his complete theory of statics and which, thanks to John Bernoulli, became the principle of virtual work. "

Jordanus was less fortunate when he turned his attention to the angular lever. He considered a lever *acf* carrying equal weights at *a* and *f* which were placed in such positions that *ac* = *ef*. Jordanus was of the opinion that, under these conditions, *a* would dominate *f*. He arrived at this conclusion by considering two equal arcs $\widehat{al}$ and $\widehat{mf}$. It is clear that the " direct " taken by the weight *a* is greater than the " direct " taken by the weight *f*. This incorrect conclusion is obtained because, since the linkages are rigid, the two displacements $\widehat{al}$ and $\widehat{fm}$ are not simultaneously possible. Jordanus thus misunderstood the idea of moment.

As early as the XIIIth Century the *Elementa Jordani* were generally united by the copyists with *De Canonio* to form the *Liber Euclidis de ponderibus*.[1] This artificial associations and this imaginative titles are the despair of the scholars and it has needed all the learning of Duhem to elucidate them.

Every truly novel idea evokes a reaction. The *Elementa Jordani* did not provide an exception to this rule. In the XIIIth Century a critic wrote a commentary of Jordanus which Duhem calls the *Peripatetic Commentary*.[2] This author at every turn invokes the authority of Aristotle and has scruples about applying the *gravitas secundum situm* to a motionless point—in modern language, about making appeal to a virtual displacement. It does not appease his conscience to consider that rest is the end of motion. " The scientific value of the *Commentary* is nothing, " declares Duhem.[3] " But its influence did not disappear for a very long time, and even the great geometers Tartaglia, Guido Ubaldo and Mersenne had not entirely freed themselves from it. "

2. THE ANONYMOUS AUTHOR OF " LIBER JORDANI DE RATIONE PONDERIS. " THE ANGULAR LEVER. THE INCLINED PLANE.

We now come to an especially noteworthy work which figures in the same manuscript as the *Peripatetic Commentary* under the title *Liber Jordani de ratione ponderis*, and which did not remain unknown in the Renaissance. Tartaglia sent it to Curtius Trojanus who published it in 1565. This work supercedes and rectifies the *Elementa Jordani* on many important points. All the same, it is based on the same principle of *gravitas secundum situm*.

Duhem, who brought this manuscript to light, terms the anonymous author a " Precursor of Leonardo da Vinci. " Indeed, in many respects this precursor surpassed Leonardo, who, for example, spent himself in fruitless efforts to evaluate the apparent weight of a body on an inclined plane. It seems more natural to simply speak of an anonymous

---

[1] *Bibliothèque Nationale*, Paris, latin collection, Mss. 7310 and 10,260.
[2] *Ibid.*, Ms. 7378 A.
[3] *O. S.*, Vol. I, p. 134.

author of the XIIIth Century, a disciple of Jordanus who had out-stripped his master.

In connection with the bent lever this author corrected Jordanus's error. As before, let a lever *acf* carry equal weights at *a* and *f* and be placed in such a position that $aa' = ff'$.

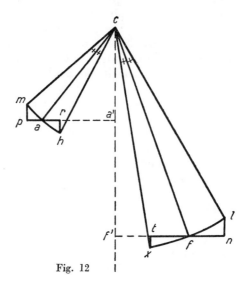

Fig. 12

It is impossible that the weight *a* should dominate the weight *f*. For if two arcs $\widehat{ah}$, $\widehat{fl}$, are considered on the two circles drawn through *a* and *f* and corresponding to equal angles $\widehat{ach}$ and $\widehat{fcl}$, the descent of *a* along *rh* necessitates that the equal weight at *f* should rise through a distance *ln* which is greater than *rh*. This is impossible.

In the same way it can be seen that *f* will not dominate *a*. For if the arcs $\widehat{fx}$ and $\widehat{am}$ correspond to equal angles $\widehat{fcx}$ and $\widehat{acm}$, the descent of *f* along *tx* makes it necessary that the equal weight placed at *a* should rise by *pm*, which is greater than *tx*. This is impossible. Therefore there is equilibrium in the position considered, in which $aa'=ff'$.

The anonymous author generalised this result to an angular balance carrying unequal weights at *a* and *b*,

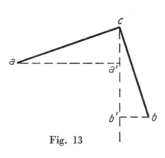

Fig. 13

and obtained the result that in equilibrium it is necessary that the distances aa' and bb' from a and b to the vertical drawn through the point of support, c, are in inverse ratio to the weights a and b.

We see that this author knew and used the notion of moment. Elsewhere he wrote on this subject, " If a load is lifted and the length of its support is known, it can be determined how much this load weighs in all positions. The weight of the load carried at e by the support be will be to the weight carried at f by fb as el is to fr or as pb is to xb. A weight placed at e, at the end of the lever be, will weigh as if it were at u on the lever bf. "

Thus the idea of *gravitas secundum situm*, which Jordanus had used qualitatively, became precise.

Our anonymous author also concerned himself with the stability of the balance, and rectified certain errors which were contained in the relevant parts of *Problems of Mechanics*.

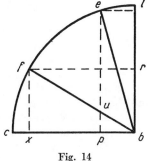

Fig. 14

More than this, he resolved the problem of the equilibrium of a heavy body on an inclined plane, a problem which had eluded the wisdom of the greek and alexandrian geometers.

In order that this may be done, it is first observed that the *gravitas secundum situm* of a weight on an inclined plane is independent of its position on the plane. The author then attempts a comparison of the value that that gravity takes on differently inclined planes. We shall quote from Duhem's translation of this same XIIIth Century manuscript.

" If two weights descend by differently inclined paths, and if they are directly proportional to the declinations, they will be of the same strength in their descent.

" Let ab be a horizontal and bd, a vertical. Suppose that two oblique lines da and dc fall on one side and on the other of bd, and that dc has the greater relative obliquity. By the relation of the obliquities I understand the relation of the declinations and not the relation of the angles ; this means the relation of the lengths of the named lines counted as far as their intersection with the horizontal, in such a way that they take similarily of the direct.

" In the second place, let e and h be the weights placed on dc and da respectively, and suppose that the weight e is to the weight h as

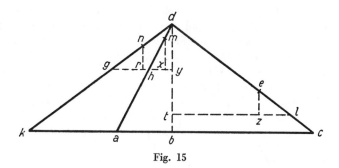

Fig. 15

dc is to *da*. I maintain that in such a situation the two weights will have the same strength.

" Indeed, let *dk* be a line having the same obliquity as *dc* and, on that line, let there be a weight *g* which is equal to *e*.

" Suppose that the weight *e* should descend to *l*, if that is possible, and that it should draw the weight *h* to *m*. (It is clear that the author imagines the two weights to be connected by a thread which passes over a pulley at *d*.) Take *gn* equal to *hm*, and consequently equal to *el*. Draw a perpendicular to *db* which passes through *g* and *h*, say *ghy*. Drop a perpendicular *lt* from the point *l* onto *db*. Then drop [the perpendiculars] *nr*, *mx*, and *ez*. The relation of *nr* to *ng* is that of *dy* to *dg* and also that of *db* to *dk*. Therefore *mx* is to *nr* as *dk* is to *da*; that is to say, as the weight *g* is to the weight *h*. But as *e* is not able to pull g up to *n* (*nr* = *ez*), it is no better able to pull *h* up to *m*. The weights therefore remain in equilibrium. "

This demonstration, which leads to the correct law of the apparent heaviness of a body on an inclined plane, was directly inspired by that of Jordanus concerning the equilibrium of a straight lever. Like that, it implicitly proceeds according to the principle of virtual work.

We shall now give some indication of the ideas on dynamics which were used by the author of *Liber Jordani de ratione ponderis*.

The environment's resistance to the motion of a body depends on the shape of the body, which penetrates the environment the better as its shape is the more acute and its figure the more smooth. It depends, in the second place, on the density of the fluid traversed. All media are compressible; the lower strata, compressed by the upper ones, are the denser and those which hinder motions more. At the front of the moving body will be a part of the medium compressed on, and sticking to it. But the other parts of the medium, which are displaced by the moving body, curl round behind to occupy the space which the body has left empty. This motion of lateral parts of the medium may be compared to the bending of an arc.

The more heavy the medium is at traversal, the slower is the descent of a heavy body.

The descent is slower in a fluid which is more dense.

Greater width diminishes the gravity.

A heavy thing will move more freely as the duration of its fall increases. " This is more true in air than in water, because air is suited to all kinds of motion. Thus a falling body drags with it, from the outset of its motion, the fluid that lies behind it and sets in motion the fluid in its immediate contact. The parts of the medium set in motion in this way, in their turn move those that adjoin them, in such a way that the latter, which are already in motion, present a lesser obstacle to the falling body. For this reason the body becomes heavier and imparts a greater impulsion to the parts of the medium which it displaces until these are no longer simply pushed by the body, but drag it along with them. Thus it happens that the gravity of a moving body is increased by their traction and that, reciprocally, their motion is multiplied by this gravity so that it continually increases the velocity of the body. "

The shape of a heavy body affects the strength of its weight.

The strength of a motive agency seems to be equally baulked by a body's very large or very small weight.

Rotation of a propellant increases its strength, and does so more effectively as the radius is greater.

A body whose parts are coherent is thrown directly backwards if it is stopped by a collision during its motion. " The parts of a moving body A that lie in front are the first to meet the obstacle C. They are therefore compressed by the mass and the impetuosity of the parts which lie behind them, and are forced to condense. The impetuosity of the parts behind is annulled in this way. The parts in front now assume their original volume again and recoil backwards, thus communicating an impulsion to the others. If the parts which are compressed in this way are able to detach themselves from each other they will be thrown off in one direction and another. "

If the heaviness of a body is not uniform, the denser part will place itself in front, whatever the part to which the impulsion is given.[1]

· These ideas on dynamics held by Jordanus's School are much less interesting and moreover, less original than its statics. We have cited them here as curiosities.

---

[1] *Cf.*, DUHEM, *Études sur Léonard de Vinci*, Series I (Hermann), 1906.

From the historical point of view it must be remarked that Duhem, in writing his first studies on the origin of statics, first believed the work of this unknown disciple of Jordanus to be entirely original. But the later discovery of a XIIIth Century manuscript[1] led him to a treatise *De Ponderibus* which was more complete than the *Liber Jordani de ratione ponderis*.

Now this treatise, divided into four books, seems to be a complex in which various works have been artificially united. There is first a book, of indisputable Medieval origin, that repeats the demonstrations Jordanus used and supplements them by the condition for the equilibrium of the angular lever and the determination of the apparent weight of a body on an inclined plane. The second book appears to have been inspired by *De Canonio* while the third treats the concept of moment and the conditions for the stability of the balance. Finally there is a fourth book devoted to dynamics.

The last two books are closely related to *Problems of Mechanics* although they alter, correct and complete this work in many places. Certain indications led Duhem to surmise that the two books are a relic of greek science and were probably handed on by the Arabs— this because no latinised greek terms are found in them.[2] Accordingly it is possible that our unknown author did not discover the idea of moment himself. This limits the originality of his work, but it remains that *gravitas secundum situm* properly belongs to the XIIIth Century School, and that it was used by this School to obtain a correct solution of the problem of the inclined plane long before Stevin and Galileo did so.

---

[1] *Bibliothèque Nationale*, latin collection, Ms. 8680 A.

[2] *Cf.* DUHEM, *O. S.*, Vol. II, note F, p. 318. LEONARDO DA VINCI himself seems to have been unaware of the three last books of *De Ponderibus*—another argument for not regarding the unknown author as his precursor.

CHAPTER FOUR

THE XIVth CENTURY
THE SCHOOLS OF BURIDAN AND ALBERT OF SAXONY
NICOLE ORESME AND THE OXFORD SCHOOL

1. THE DOCTRINE OF " IMPETUS " (JOHN BURIDAN).

The idea of attributing a certain energy to a moving body solely on account of its motion is entirely foreign to aristotelian dynamics.

In Antiquity John of Alexandria—surnamed Philopon—who lived in the Vth Century A. C., was alone in disputing Aristotle's belief in this matter. Thus he held that the air which was set in motion could not become the motive agency of a projectile, whose motion was, on the contrary, easier in a vacuum than in air. " Whoever throws a projectile embodies in it a certain action, a certain power of self-movement which is incorporated. . . . Nothing prevents a man from throwing a stone or an arrow even when there is no other medium than the vacuum. A medium hinders the motion of projectiles, which cannot advance without dividing it—nevertheless they can move through these media. Nothing therefore prevents an arrow, a stone or any other body from being thrown in the vacuum. Indeed, the motive agency, the moving body and the space that will receive the projectile are all present." [1]

Philopon's thesis was handed on to the Middle Ages by the Arabs —in particular, by the astronomer Al Bitrogi. But while assuming the existence of a " property which remains attached to a stone or an arrow after the projectile has been thrown, " he held that this property decreased at such a rate and to such an extent as the projectile was separated from its motive agency.

Albertus Magnus and Saint Thomas Aquinas knew of this tradition but did not give the least credit to John Philopon's argument. For example, Saint Thomas Aquinas believed that if the existence of an

---

[1] *Erudissima commentaria in primos quatuor Aristotelis de naturali auscultatione libros*, Venice (1532), Trans. DUHEM.

apparent property impressed on a moving body were assumed, " violent motion would arise from an intrinsic property of a moving body, which is contrary to the very notion of violent motion. Moreover, it would follow from this that a stone would be altered in its substantial form by the very fact that it moved from place to place, which is contrary to common sense. " [1]

Roger Bacon, Walter Burley and John of Jandun all adopted Aristotle's doctrine on this matter. The first Schoolman to oppose this opinion was William of Ockham (1300-1350). He asked himself where the motive agency might be. It cannot reside in the apparatus or organism that has thrown the projectile, for this apparatus can be destroyed immediately after the launching without interupting the progress of the projectile.

Nor can the motive agency be the air which is set in motion. For the arrows of two archers which are shot towards each other can be arranged to collide with each other, which requires that the same air produces two different motions at the same time.

There cannot be distinguished elsewhere a cause that could provide the motive power. Such a cause cannot reside in the launching apparatus nor in the motion of the projectile itself. If something which is its own motive agency is thrown, that which moves the body cannot be distinguished from the moving body itself. Moreover, motion from place to place is not something which is renewed at each instant, requiring the constant presence of a motive cause. It is true that the projectile passes through a different region at each instant, but this does not in itself constitute anything novel. It is only novel with respect to the moving body.[2]

Thus William of Ockham decided to reject Aristotle's axiom which requires the continuous existence of a motive agency in contact with, yet not part of, the projectile. He did not, however, replace it by any new principle.

We now arrive at the doctrine of *impetus* that was conceived by Buridan. John I. Buridan, of Béthune, was rector of the University of Paris in 1327 and canon of Arras in 1342. He died in Paris after 1358.[3]

In a memoir called *Quaestiones octavi libri physicorum*,[4] Buridan

---

[1] *Opera omnia*, Vol. III—*Commentaria in libros Aristotelis de Caelo et Mundo*, Book III, lect. VII.

[2] *Cf.* DUHEM, *Études sur Léonard de Vinci*, Series II, p. 192.

[3] DUHEM, who has studied BURIDAN's works in detail, including those concerning free will, says that he has found no trace of the parable of the ass, which apart from his status in the history of mechanics, has made Buridan's name classical.

[4] *Bibliothèque Nationale*, Paris, latin collection, Ms. 14,723, fol. 106-107. In the text we are following DUHEM's translation.

discussed the scholastic thesis of the motion of projectiles. Aristotle, he says, mentions two opinions on this matter.

The first invokes ἀντιπερίστασις. As a projectile moves rapidly away from its position, Nature, who does not allow the existence of a vacuum, makes the air behind the projectile rush in towards this position with the same velocity. This air pushes the projectile and the same effect is reproduced, at least for a certain distance. This opinion is rejected by Aristotle—if no other principle than ἀντιπερίστασις is invoked, it is necessary that all bodies which are behind the particle, including the sky itself, participate in the projectile's motion. Indeed, the air will also leave its position. It is then necessary that another body must replace it, and so on in an indefinite sequence, unless it is assumed that a certain rarefaction of bodies behind the projectile is produced.

According to the second opinion, which Aristotle seems to have supported, the launching of a projectile disturbs the ambient air at the same time. This air, violently set in motion, has, in its turn, the power to move the projectile. The first mass of air moves the projectile until this comes to a second mass of air. This second one moves it until a third is reached, and so on. Further, Aristotle is heard to say, there is not merely a single moving body, but successive moving bodies, a series of consecutive or contiguous motions.

Buridan set the following observations against these theories. A top or a grindstone will turn for a very long time without leaving its position, in such a way that the air does not have to follow it to fill an abandoned place. Further, the wheel will continue to turn if it is covered and thus separated from the surrounding air. A javelin whose following end is armed with a point as sharp as its tip will move as rapidly as if this were not tapered at the back. Now since the air is easily divided by the javelin's sharpness, it cannot push strongly on this backward pointed part. A ship will continue to move for a long time after towing has been stopped, and a boatman will not feel the air pushing it—on the contrary, he feels the air slowing down the ship's motion.

The air set in motion should be able to move a feather more easily than a stone. Now we are not able to throw a feather as far as a stone.

Buridan himself put forward the following thesis.

" Whenever some agency sets a body in motion, it imparts to it a certain *impetus*, a certain power which is able to move the body along in the direction imposed upon it at the outset, whether this be upwards, downwards, to the side or in a circle. The greater the velocity that the body is given by the motive agency, the more powerful will be the *impetus* which is given to it. It is this *impetus* which moves a stone after it has been thrown until the motion is at an end. But because of the resistance of the air and also because of the heaviness, which inclines the motion of the stone in a direction different from that in which the *impetus* is effective, this *impetus* continually decreases. Consequently the motion of the stone slows down without interruption. Finally the *impetus* is overcome and destroyed at the point where gravity dominates it, and henceforth the latter moves the stone towards its natural place. . . .

" All natural forms and dispositions are received by matter in proportion to itself. Consequently the more matter a body contains, the more *impetus* can be imparted to it, and the greater is the intensity with which it can receive the *impetus*. . . . A feather receives such a weak *impetus* that this is immediately destroyed by the resistance of the air. In the same way, if someone throws projectiles and sets in motion with equal velocities a piece of wood and a piece of iron, which have the same volume and the same shape, the piece of iron will travel further because the *impetus* which is imparted to it is stronger. It is for the same reason that it is more difficult to stop a large blacksmith's wheel, moving rapidly, than a smaller one. . . . "

In short the *impetus*, in Buridan's sense, increases with the velocity. In addition, it is proportional to the density and to the volume of the body concerned. Further, in Buraidan's view, the existence of *impetus* explained the acceleration of falling bodies.

" The existence of *impetus* seems to be the cause by which the natural fall of bodies accelerates indefinitely. At the beginning of the fall, indeed, the body is moved by gravity alone. Therefore it falls more slowly. But before long this gravity imparts a certain *impetus* to the heavy body—an *impetus* which is effective in moving the body at the same time as gravity does. Therefore the motion becomes more rapid. But the more rapid it becomes, the more intense the *impetus* becomes. Therefore it can be seen that the motion will be accelerated continuously. "

Further, Buridan applied the notion of *impetus* to stars as well as to terrestrial bodies.

" In the Bible there is no evidence of the existence of intelligences charged with communicating their appropriate motion to the heavenly bodies. It is therefore permissible to show that there is no necessity to suppose the existence of such intelligences. Indeed it can be said that when He created the World, God set each of the heavenly bodies in motion in the way that he had chosen—that He imparted to each of them an *impetus* which has kept it moving ever since. Thus God no longer has to move these bodies, except for a general influence similar to that by which He gives his consent to all things that occur. It is for this reason that, on the seventh day, He was able to rest from the tasks which He had accomplished and to confine himself to the creation of things concerning mutual actions and feelings. The *impetus* that God imparted to the heavenly bodies is neither weakened nor destroyed by the passage of time. For in these heavenly bodies there are no tendencies towards other motions and because, moreover, there is no longer any resistance which could corrupt and repress this *impetus*. I do not say all this with complete assurance. I would only ask the theologians to show me how all these things happen. "

As a true Scholastic Buridan believed himself obliged to defend the doctrine of *impetus* from the metaphysical objections that could be advanced against it. The motion of a projectile is a violent one in Aristotle's sense. Now, according to the *Ethics* (Book III), violent phenomena stem from an extrinsic, not an intrinsic, cause. To this Buridan replied that the *impetus* of a moving body is effectively violent, not natural. The nature of heavy things favours a different motion and the destruction of the *impetus*. On the question of whether the *impetus* is distinct from the motion and whether it is of a permanent kind, Buridan replied that *impetus* could not itself be motion because all motion requires a motive agency ; that *impetus* was a permanent reality, distinct from the local motion of the projectile ; and that it was probable that the *impetus* was a quality whose nature was to actuate the body to which it was imparted.

These subtleties add nothing to Buridan's positive doctrine. It is more important to remark that Buridan maintained that the *impetus lasted indefinitely* if it was not diminished by a resistance of the medium or modified by some agency affecting the moving body. This is the germ of the modern principle of inertia.

2. THE SPHERICITY OF THE EARTH AND THE OCEANS—ALBERT OF SAXONY AND THE ARISTOTLETIAN TRADITION.

In the first chapter of this book we referred to the *a priori*, or physical proofs, and the *a posteriori* proofs which Aristotle gave of the sphericity of the Earth and the oceans. For better or worse, tradition preserved and enriched these proofs.

Pliny the Elder, in his *Natural History*, supplemented Aristotle's evidence with facts that strictly derive from capillarity—the sphericity of drops of water, the meniscuses of liquids, etc. . . . Ptolemy only retained the *a posteriori* proofs which Aristotle had given. Simplicius, in his commentary on *De Caelo*, corrected the dimensions attributed to the Earth after Erastosthenes' evaluation. Averroës confined himself to an elaboration of Aristotle's evidence.

John Sacro Bosco—the author of a treatise called *De Sphaera* which became the most widely known cosmography in the XIIIth Century—reproduced Ptolemy's account. Albertus Magnus firmly excluded the evidence depending on the sphericity of water drops. Saint Thomas Aquinas limited himself to Aristotle's proofs alone, while Roger Bacon supplemented them with the following corollary which was acclaimed by the Schoolmen—any given vessel will contain a smaller quantity of liquid as it is taken further from the centre of the Earth.

We now come to Albert of Rickmersdorf, called Albert of Saxony. Though his biography is somewhat mysterious, it is certain that he was enlisted at the Sorbonne from 1350 to 1361 and that he was rector of the University of Paris from 1353. His *Acutissimae Quaestiones* on Aristotle's *Physics* had considerable repercussions, and his influence was felt by most students of mechanics, including Galileo himself.

Albert of Saxony suggested going back to the measurement of a degree of meridian at different latitudes in order to determine the true figure of the Earth. (This idea was applied by John Fernel at the beginning of the XVIth Century and, of necessity, repeated in the XVIIth Century.) " If these two paths are found to be equal this is a certain indication that the Earth is circular from north to south. If on the contrary, it were found that they lacked equality this would be an indication that the Earth was not round from north to south. "

Like Albertus Magnus, Albert of Saxony excluded the evidence provided by small drops, which is common to all liquids, like mercury, and is especially noticeable in small quantities.

Albert of Saxony stated the following corollaries, which were to become popular among the Schoolmen.

" 1. From the fact that the Earth is round it follows that lines normal to the surface of the Earth will approach each other continuously, and meet at the centre.

" 2. It follows that if two vertical towers are built, the higher they become, the further away from each other they will be ; and that the deeper they are, the nearer together they will be.

" 3. If a well is dug with a plumb-line, it will be larger near the opening than at the bottom.

" 4. Any line such that all its points are at an equal distance from the centre is a curved line. If a straight line touches the Earth's surface at its middle point, this point will be nearer to the centre than the ends of the line. It follows that if a man goes along this straight line, he will descend for a time and then will rise. He will descend, indeed, until he has come to the point which is nearest to the centre of the Earth and will rise from the moment that he leaves that point behind him.

" From this it can be concluded that a body which describes a trajectory between two fixed ends, a trajectory which either rises or falls without interruption, must necessarily travel a shorter distance than if the path went from one point to the other without rising or falling. This is seen clearly if it is supposed that the first trajectory is a diameter of the Earth and the second is a half-circumference having this diameter as chord.

" 5. When a man walks on the surface of the Earth his head moves more quickly than his feet.... One can conceive of a man so tall that his head moves in the air twice as quickly as his feet move over the ground. "

These paradoxes are typical of the scholastic attitude of mind and it is for this reason that we have quoted them. Undoubtedly they were intended to stimulate the minds of students, and perhaps, too, to confuse those who were not scholars. The dialectic of the Schoolmen was not in the least concerned with orders of magnitude. It was amusing to proliferate the consequences of the convergence of verticals and their practical parallelism was of no concern—that was a notion suitable for craftsmen. And these, in their turn, were not much worried by the comments of the Schoolmen when they were building their towers and digging their wells according to the simple rules of their practice.

3. Albert of Saxony's theory of centre of gravity.

When commenting on that thesis of Aristotle according to which there exists, in each heavy body, a centre of gravity *(τὸ μέσον)* which tends to be carried towards the centre of the Universe, Albert of Saxony specified that " each of the parts of a heavy body is not moved in such a way that its own centre would come to the centre of the World, for this would be impossible. Rather it is the whole body which falls in such a way that its centre would become the centre of the World. It is false, and contrary to observation, to say that a large body falls more slowly than a lighter body, or that ten stones which are united together hinder each other's fall. "

The Earth, limited partly by the concave surface of the water and partly by the concave surface of the air, is in its natural position when its *centre of gravity* is at the centre of the World. If this is not so, it will start falling and will move until the centre of the aggregate which it forms with all the other heavy bodies becomes the centre of the World.

It should be remarked, as Jouguet has done in this connection,[1] that Albert of Saxony's concept of *centre of gravity*, the point of a body at which all the weight appears to be concentrated, was a purely experimental one, to him and his School. It was not the same as the modern conception of centre of gravity, which depends on the approximation that verticals are parallel. On the contrary, it was developed together with a systematic consideration of the convergence of verticals which was carried to the point of paradox, as we have seen. This co-existence was at the root of several fallacies which were to perplex people, even such eminent ones as Fermat, until the XVIIth Century.[2]

We return to Albert of Saxony. Suppose that the Earth is displaced from its natural place—for example, to the concavity of the orbit of the Moon—and held there by force. Suppose that, there, a heavy body is allowed to fall. Then this body will be attracted towards the centre of the World, not towards the centre of the Earth. " If heavy bodies move towards the ground, this is in no way caused by the Earth, but happens because they approach the centre of the World by going towards the Earth. "

The Earth does not have a uniform gravity— " the part which is not covered by sea, being exposed to the rays of the Sun, is more dilated than the part the waters cover. Besides, if its *geometrical centre* were to coincide with its *centre of gravity*, and consequently with the centre of the World, it would be entirely covered by the waters. "

Here, in Albert of Saxony's writings, is the trace of an argument that had preoccupied some of his immediate predecessors. If all elements, declared Walter Burley (1275-1357), had the form of spheres with centres at the centre of the Universe, each would be in its natural place— but then the Earth would be completely covered with water. John Duns Scot (1275-1308) resolved this difficulty, in his *Doctor Subtilis*, with a finalist explanation—to witt, a part of the Earth is uncovered with a view to the safety of living beings.

Albert of Saxony believed therefore that it was the Earth's centre, of gravity, not its geometrical centre, that was placed at the centre of the World. Furthermore, the Earth was not fixed in position. A host of reasons, such as heating by rays of the Sun, could produce a variation of the distribution of gravity in the terrestrial mass, and could

---

[1] JOUGUET, *L. M.*, Vol. I, p. 60.

[2] This question was at the root of the controversy on *Geostatics*, to which we shall return.

displace its centre of gravity. As a more substantial mechanism, Albert of Saxony mentioned erosion.

The question arose as to how the mass of the waters could be introduced. On this point Albert of Saxony's opinion was somewhat variable.

In commenting on the *Physics* he wrote, " What I have written about the Earth alone may be understood equally for the whole aggregate formed by the Earth and the waters. These two elements undoubtedly form a total and unique gravity whose centre of gravity is at the centre of the World. " At this same centre of the World was also to be found the *centre of lightness* of light bodies.

It is this that explains the following picture which he boldly painted.

" Since the cold is especially intense at the poles, the layer of igneous element there must be thinner than at the equator if fire, which is continuously created at the equator is not to run towards the poles. In the same way that water constantly runs towards lower places in order that the centre of all gravity shall be at the centre of the World, so we must assume that fire travels, without interruption, from the equator towards the poles in order that its centre of lightness shall be at the centre of the World.

" It should be imagined that, at the poles, fire is constantly being transformed into air, and at the equator air is constantly being transformed into fire ; and that fire continually runs from the equator towards the poles in order that the centre of all lightness, like the centre of gravity, shall be found at the centre of the World. "

In short, as Duhem has observed,[1] " the common centre of heavy bodies—both the closed earth and the water—and the common centre of light bodies—both air and fire—are placed at the centre of the World. "

However, in commenting on *De Caelo*, Albert of Saxony took a different view.

" We reply by denying that the centre of the World coincides with the centre of the aggregate formed by the earth and the water. Indeed, if it is imagined that all the water were lifted off, the centre of gravity of the Earth would still be at the centre of the World. . . . For, essentially, the earth is heavier than water. Therefore, whatever may be the quantity of water which is found on one side of the Earth and not on the other, this part of the Earth will in no way receive more help than previously in counterpoising and pushing away the other part. . . . "

It is explained " that one part of the Earth rises out of the waters. The Earth, indeed, is not uniformly heavy, so that its centre of gravity

---

[1] DUHEM, *O. S.*, Vol. II, p. 28.

is placed at a great distance from its geometrical centre. The centre of gravity is much nearer one of the convex hemispheres that define the Earth than the other. Therefore the water, which is unifirmly dense and tends towards the centre of the World, runs towards that part of the terrestrial sphere that is nearest the centre of gravity of the Earth, so that the other hemisphere, that which is further distant from the centre of gravity, remains uncovered. "

The weakness of this argument is clear. But undoubtedly the theory was in harmony with the belief, common at that time, in the existence of a terrestrial hemisphere completely covered by a vast ocean. It is paradoxical to see Albert of Saxony thus holding that the waters of the sea do not exert any heaviness, but this is in accord with a more general thesis that is indicated below.

Albert of Saxony distinguished between heaviness in the *potential state*, that of a heavy body occupying its natural place, and the *actual heaviness* that sets a body in motion when it has been displaced from its natural place (or shows itself as a resistance to obstacles which oppose the body's motion).

We shall quote from Duhem's commentary.

" The parts of a heavy body, be they solid or liquid, do not push the adjacent parts when they are in their natural place, since their heaviness remains in its potential state. Thus the bottom of the sea does not support any load or any pressure that is due to the water above it. In all circumstances the strength of the heaviness, whether it is *habitual* or *actual*, has the same magnitude in the same heavy body. A part of the earth inclines towards its natural place just as much if it is placed higher up than if it is lower down. "

It is clear that this thesis contradicts the fundamental axiom of Jordanus—*Gravius esse in descendendo quando ejusdem motus ad medium rectior*. Moreover, it is not surprising that Albert of Saxony should have rejected the idea of *gravitas secundum situm*, and have substituted for it the concept of a greater or smaller resistance of the supporting medium to the fall of a moving body.

4. ALBERT OF SAXONY'S KINEMATICS. THE ACCELERATION OF FALLING BODIES.

Whether explicitly or not, the physicists and astronomers of Antiquity treated only the simple *uniform* motions of translation and rotation, and confined themselves to a simple qualitative description of accelerated motion.

In a XIIIth Century manuscript[1] there is the statement that it is correct to ascribe the velocity of its mid-point to a radius turning about its centre. This text is mentioned by Thomas Bradwardine, proctor of the University of Oxford, in his *Tractatus Proportionum* (1328). Bradwardine denies this statement and attributes the velocity of its most rapidly moving point to a body in uniform rotation.

Albert of Saxony stated these two opinions and supported that of Bradwardine. To set against this, he gave a correct definition of the angular velocity of rotation *(velocitas circuitionis)*. Further, he distinguished between *deformed* motions, in which the velocity of a moving body varies from one point to another, and *irregular* motions, in which the velocity varies from one instant to another.

In Book II, paragraph XIII of his *Quaestiones*, Albert of Saxony examined two possible laws which might govern the fall of bodies—an increase of velocity which is proportional to the distance travelled and an increase proportional to the time taken.

In another place, he rejects these two laws, which lead to velocities which become infinite with the distance travelled or the time taken, and contemplates a law which would necessitate that the velocity approach a finite limit when the time increases indefinitely. On this occasion, Albert of Saxony declares himself a supporter of the doctrine of *impetus* in order to explain the acceleration of falling bodies. He observes, however, that the resistance, increasing more quickly than the *impetus* is acquired, will limit the velocity of the moving body. It is important to notice this connection between Albert of Saxony and John Buridan's doctrine, and to recognise the considerable authority which the latter's work had over this long period.

5. THE DISCUSSION OF ACTION AT A DISTANCE.

In commenting on Aristotle, Averroës and Albertus Magnus had held that the weight of a heavy body did not vary with its distance from the centre of the World. On the other hand Saint Thomas Aquinas, arguing from the acceleration of falling bodies, assumed that a heavy body increased in weight as it approached this centre.

In the XIVth Century John of Jandun, in his commentary on *De Caelo* (Book IV, para. XIX), declared himself for the first opinion. The natural place cannot be the " motor " of a heavy body, because the motor must always accompany the moving body. *It is not possible to have action at a distance.* The attraction of iron by a magnet presumes

[1] *Bibliothèque Nationale*, Paris, latin collection, Ms. 8680 A.

the alteration of the medium, the propagation of a *species magnetica*.

On the other hand, William of Ockham denied that the motive agency should always accompany the moving body. He declared that iron is attracted at a distance by a magnet without the intermediary of any quality, either in the medium or the iron. He assumed that the magnet was, in itself, the *total cause* of the effect.

In his *Quaestiones* (Book III, para. VII), Albert of Saxony held that the effect of a body's natural place on a body was different from the action of a magnet on iron. It is true that a heavy body accelerates in falling, " *but its initial[1] velocity is not greater when it is close to its natural place than when it is separated from it.* " Thus the gravity does not depend on the distance from the centre of the World—the attraction of a magnet, on the other hand, vanishes at some distance.

The Schoolmen of the XIVth Century therefore rejected the hypothesis that ascribed weight to an attraction at a distance exerted by the centre of the Earth. But, as Duhem has observed, " in order to prove this hypothesis wrong, it was necessary to work out its consequences. They [the Schoolmen] discovered that, on the basis of this supposition, the weight of a body would vary with its distance from the centre of attraction. From this, it was argued that the body would have, in falling, an initial velocity[1] which was less if its starting-point were further away from the centre. "[2]

These discussions on the attraction and even the more metaphysical argument about the plurality of worlds made the copernican revolution, to a certain extent, possible.

6. NICOLE ORESME—A DISCIPLE OF BURIDAN.

From 1348, Nicole Oresme, of the diocese of Bayeux, was a student of theology. In 1362 he was grand master of the College of Navarre. He became Bishop of Lisieux on August 3rd, 1377 and died there on July 11th, 1382.

Charles V entrusted Oresme with the task of translating (into French) and annotating certain of Aristotle's works which had previously only been accessible to the scholars. The four books of *On the Heavens and the World* were included in the commission, though this particular part

---

[1] The question here is one of a free fall—the initial velocity in the modern sense is therefore zero. The velocity which ALBERT OF SAXONY intended, however, is the velocity acquired after a very short time if the body starts from rest. This velocity is proportional to the weight and can serve as a measure of the gravity.

[2] DUHEM, *Études sur Léonard de Vinci*, Series II, p. 90.

of the translation was never printed. At the beginning of the XVIth Century the remainder of the translation *(Ethics, Politics* and *Economics)* was published, together with Oresme's *Treatise on the Sphere.*

In dynamics, Oresme was a disciple of Buridan and adopted from him the doctrine of *impetus.* Thus he maintains, in his *Treatise on the Heavens and the World,*[1] written about 1377, that the acceleration of falling bodies is not, strictly speaking, accompanied by an increase of the heaviness of the body. Rather there is an increase of an " accidental property which is caused by the reinforcement of the *isnelté* (velocity) and this property can be called impetuosity *(impetus).* "

This property is not the same as the heaviness " because if a hole were to be dug to the centre of the earth and then out the other side, and a heavy thing were to fall in this hole, when it came to the centre it would pass beyond it and rise, through the agency of this acquired and accidental property. Then it would fall back and go and come in the way that we see in a heavy thing hanging by a long string. Therefore this is not strictly heaviness, since it is able to make [the body] ascend. "

7. ORESME'S RULE IN KINEMATICS. (UNIFORMLY ACCELERATED MOTION.)

Oresme was above all a mathematician, and in this capacity he emerges as Descartes' forerunner in the matter of the invention of co-ordinates. As will be seen, and as Moritz Cantor has pointed out,[2] we shall not stray from the subject in hand if we emphasise this aspect of his work.

We shall follow the *Tractatus de figuratione potentiarum et mensurarum difformitatum.*[3] Oresme starts from the principle that every measurable thing can be thought of as a continuous quantity. Each intensity can be represented by means of a straight line erected vertical from each point of the " subject " which affects the intensity.

Extension *(longitudo)* is represented diagrammatically by a horizontal line drawn in the direction of the subject. At each point of this line a vertical is erected whose height *(altitudo* or *latitudo)* is proportional to the intensity *(intensio)* of the property at the point corresponding to the subject.

Thus the triangle of figure 16 represents a uniformly deformed quality *(uniformiter difformis)* terminated at a value zero. The

---

[1] *Bibliothèque Nationale,* Paris, french collection, Ms. 1083.
[2] *Vorlesungen über die Geschichte der Mathematik,* 2nd Ed., Vol. II, p. 129. (Leipzig, 1900.)
[3] *Bibliothèque Nationale,* Paris, latin collection, Ms. 7371, Trans. DUHEM.

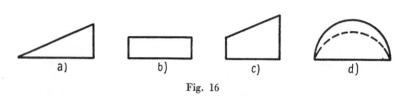

Fig. 16

rectangle *b* represents a uniform quality and the trapezium *c* a uniform-ely deformed quality terminated by certain values at one end and at the other. Any other quality is said to be deformably deformed—that is, non-uniformly deformed or non-uniformly varying *(difformiter difformis)*. Such a one can be represented in the same way by erecting a vertical proportional to the intensity from each point of the extension. Oresme pointed out explicitly that the scale of such a representation could be chosen at will. Therefore the same quality can be represented by diagrams whose verticals are in a given relation to each other. Thus Oresme understands that the same quality can be represented by a diagram which is, for example, either circular or elliptical. He then proceeds to a classification of deformities according to the direction of their concavity and according to whether they are rational (circular) or not. In this way he was able to enumerate 62 different kinds of deformity.

Oresme came very near to modern analytical geometry when he wrote, " A uniformly deformed quality is such that when any three points of the subject are given, the relation of the interval between the first and the second to the interval between the second and the third is the same as the relation of the excess of intensity of the first over the second to the excess intensity of the second over the third. " This statement expresses the relation between the co-ordinates of three points on a straight line explicitly.

Oresme went further than this in envisaging *superficial qualities* —qualities which had two dimensions with respect to the subject and whose intensity must be represented by a normal to a plane surface which defines the extension. Similarly, he put the question of how a *corporeal quality*—one having three dimensions with respect to the sub-ject—can be represented. This passage merits quotation.

" A superficial quality is represented by a solid figure. Now a fourth dimension does not exist and it is impossible to conceive of one. Nevertheless a corporeal quality may be thought of as having a *double corporeity*. One in a real extension, through the effect of the extension of the subject, has a locus in all dimensions. But there is also another which is only imagined and which arises from the intensity of the qua-

lity. This quality is repeated an infinite number of times by the multi-
tude of surfaces which may be traced with respect to the subject. "

In kinematics, Oresme accepted Albert of Saxony's ideas but expressed them with the help of his graphical representation. Velocity
is susceptible of a double extension, either in time or with respect to
the subject. It can be uniform or deformed with respect to each of
these two extensions.

Further, Oresme defined the *total quality* or the *measure* of a quality
which was linear (or superficial) with respect to the subject as the area
(or the volume) of the diagram which represents it.

It is clear that if time is taken as the variable of extension, the
measure or total quality of the velocity of a uniform motion is equal
to the distance travelled. Oresme
did not confine himself to this
instance. He contemplated a succession of uniform motions in the
following way.

He divides the time t into
proportional parts which form a
geometrical progression with ratio
$\frac{1}{2}$ and the first term $\frac{t}{2}$. The velocity has intensity ni in the nth

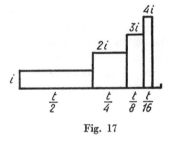

Fig. 17

interval. Under these conditions Oresme states that the total distance
travelled is equal to four times the first rectangle, that is $4\left(\frac{it}{2}\right)$.[1]

Oresme stated the following general rule for uniformly deformed qualities— " *Omnis qualitas, si fuerit uniformiter difformis, secundum gradum
puncti medii ipsa est tanta quanta qualitas ejusdem subjecti.* " That is,
any uniformly deformed quality has the same total quality (measure) as if it were related to the
subject with the value which it
takes at the middle point.

Oresme verifies this rule for a
linear uniformly deformed quality
that starts with an intensity AC
at A and finishes with zero value
at B. If D is the centre of the
line AB which represents the

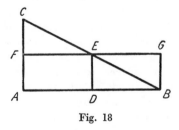

Fig. 18

[1] It should be remarked in passing that this fact shows that Oresme knew how to
calculate the sum of the series whose general term is $\frac{n}{2^n}$.

subject *(subjectiva linea)*, the corresponding intensity is *DE*. The uniform quality that has the rectangle *AFGB* as its measure has the same measure as the uniformly deformed quality represented by the triangle *ACB*, because the area *ACB* is equal to the area *AFGB*.

Oresme declares, " Any uniformly deformed quality or velocity is thus found to be equivalent to a uniform velocity, " but he does not explain, *at this point*, the identity of the measure and the distance travelled. So that he does not apply the rule which he has just formulated to a uniformly deformed motion, although this rule *includes the law of distance travelled* in such a motion. Indeed, in modern language we may write simply

$$e = \frac{1}{2} gt^2 = \left(\frac{1}{2} gt\right) t = v_{mean} \cdot t.$$

We have seen that some of the Schoolmen discussed the fall of bodies while others were concerned with kinematics, and that each of these sections developed a representation which contained a key to the law of the distance travelled by a moving body. But the union of these two problems was not effected. Undoubtedly the reason for this lies in the fact that the Schoolmen were satisfied when they had constructed abstract systems, whose niceties distracted their attention from the rudimentary experimental basis which they possessed.

8. ORESME AS A PREDECESSOR OF COPERNICUS.

Except to the extent that it impinges on the history of the principles of mechanics, we are not here concerned with the history of world-systems. However, by considering this aspect of Oresme's most original work, we shall see him to have been a prophet of Copernicus.[1]

The following quotations are taken from Nicole Oresme's *Treatise on the Heavens and the World*.

Aristotle had established, in the second book of *De Caelo*, that the Earth remained motionless at the centre of the World. Oresme declared, " no observation could prove that the Heavens moved with a diurnal [2] motion, and that the Earth did not. "

In this connection he made use, in an especially complete way, of the relativity of all motion.

" If a man were placed in the Heavens, suppose that he were moved with a diurnal motion. Then if this man who is carried above the Earth sees the Earth clearly, and picks out the mountains, the valleys,

---

[1] *Cf.* DUHEM, *Revue générale des Sciences*, Nov. 15, 1909.
[2] *Lit.*, " journal. "

rivers, towns and castles, it will seem to him that that the Earth is moved diurnally just as, to us on the Earth, the Heavens seem to move. And similarily, if the Earth is moved with a diurnal motion and the Heavens not, it will seem to us that the Earth is still and that the Heavens move. "

Oresme discussed Aristotle's argument according to which a stone thrown vertically upwards should fall to the west if the Earth is not at rest. In this connection, he declares that a stone thrown vertically in this way would be carried very rapidly towards the east, " together with the air through which it passes and with all the mass of the lower part of the World " which participates in the diurnal motion. In short, Oresme believed that the stone links its motion with that oft the Earth, from which it originally obtained its *impetus*. This thesis is correct, at least in its essentials.[1]

In a similar way, Oresme gave the lie to Ptolemy's argument that an arrow shot vertically from the deck of a ship moving very rapidly towards the east will fall far to the west of the ship. He then discussed the following reasons which had been given in support of the hypothesis that the Heavens moved and the Earth was stationary.

1) Any simple body can only have a simple motion. The Earth can only have a natural falling motion.

2) Apart from this natural falling motion could not have a circular motion—this motion, " which is violent, could not be perpetual. "

3) Averroës holds that any motion from place to place can be related to a body at rest and, for this reason, he assumes that the Earth is necessarily fixed at the centre of the Heavens.

4) All motion supposes a " motive virtue. " Now the Earth cannot be moved circularly by means of its heaviness. " And if it is moved in this way by some outside agency, such a motion will be violent and not perpetual. "

5) " If the Heavens were not moved with a diurnal motion, all Astrology [2] would be false. "

---

[1] We now know that, in a free downward fall, a heavy moving body suffers a small deflection towards the East as a result of the rotation of the Earth. The complete calculation requires that account be taken of the compound centrifugal force, but a very simple intuitive argument can give the direction of this deflection, by means of the hypothesis that the motion of the body, starting from rest, proceeds according to the law of areas

$$r^2\dot{\theta} = \text{Constant}$$

Initially, $r = r_o$ and $\dot{\theta} = \omega$, the velocity of rotation of the Earth. During the motion r will decrease. The inequality $r < r_o$ requires, by the law of areas, that $\dot{\theta} > \omega$, which shows that the body is diverted towards the East.

[2] Read " Astronomy. "

6) Motion of the Earth would contradict the Holy Scriptures— " *Oritur sol et occidit, et ad locum suum revertitur . . . Deus firmavit orbem Terrae qui non commovebitur.* "

7) The Scriptures also say that the Sun stopped in Joshua's time, and that is started its journey again in the time of King Hezekiah. And if it was the Earth which moved, not the Heavens, " such a cessation would have been inversed. "

Oresme's replies to these arguments were the following.

1) It is more reasonable to believe that every simple body and part of the World, except for the Heavens, is activated by a rotational motion in its natural place. And that, if a part of such a body is displaced from its place in the whole it will, if this is allowed, return there as directly as possible.

2) The rotational motion of the Earth is certainly a natural one, but parts of the Earth that are displaced from their accustomed position have a natural, ascending or descending, motion.

3) " Supposing that circular motion requires the presence of some other body at rest, " it does not follow that the body at rest should be inside the body which moves, for there is nothing at rest inside a grindstone " except a single mathematical point, which is not a body. "

4) The virtue which produces the rotational motion of the lower part of the World is the nature of this part. It is the same thing that also produces the motion of the Earth towards its natural place when it has been displaced from it, in the same way that iron is drawn towards a magnet.

5) All appearances, all conjunctions, oppositions, constellations and influences in the Heavens remain unchanged when it is supposed that the motion of the Heavens is apparent and the motion of the Earth is real.

6) The Holy Scriptures are consistent here, " in the manner of ordinary human speech, " to the same extent that they agree in many other places. Things are not " as the letter sounds. " Thus it is written that God covered the Heavens with clouds— " *Qui operit Caelum nubibus.* " Now, on the contrary, all the evidence shows that it is the Heavens which cover the clouds. Here again the words indicate the appearance and not the truth. It is the same for the motion of the Earth and the Heavens.

7) The stopping of the Heavens in the time of Joshua was an illusion—in fact, it was the Earth which stopped, and which started its motion again, or accelerated, in Hezekiah's time.

Oresme then gives us several " good reasons " intended to show that the Earth has a diurnal motion and the Heavens do not.

Every thing which benefits from another thing must set itself to receive, by its own motion, the benefit which it obtains from the other. Thus each element is moved towards its natural place, where it is kept. On the other hand, the natural place does not move towards the element.

From which it follows that the Earth and the elements on the Earth, which benefit from the heat and the influence of the Heavens, must arrange themselves by their own motion to duly receive this benefit ; " also, to speak familiarly, as something which is roasted at the fire receives the heat of the fire about itself because it is turned, and not because the fire is turned towards it. "

It is natural that the motions of the simple bodies of the World should have the same direction. Now, according to the Astronomers, it is impossible that all these motions should take place from east to west. On the contrary, if it is assumed that the Earth moves from west to east then this will agree with the other motions, " the Moon in one month, the Sun in a year, Mars in about two years and similarly for the others. "

In this way, the part of the Earth which is habitable will be at the top and on the left of the World. " And it is reasonable that human habitation should be found in the most noble place that there is in the Earth. "

According to Aristotle, the most noble thing which is and can be has its perfection at rest. Terrestrial bodies are set in motion towards their natural place in order to rest there. We pray that God should give the dead rest—*Requiem aeternam.* . . . The Earth, the most common thing, is displaced more rapidly than air, the Moon or the stars.

In the hypothesis of the stationary Earth, the velocities which must be assigned to the stars, because of their distance from the centre, are inadmissible.

The constellation of the North Wind—*Major Ursa*—does not turn round, with the chariot in front of the cattle, as it would if it partook of a diurnal motion.

All the appearances can be saved by means of a minor change—the diurnal motion of the Earth, whose size is so small in comparison with the Heavens—and without demanding so many different and incredible processes that God and Nature would have created for no purpose. By this means, the introduction of a IXth sphere is also made unnecessary.

When God accomplishes a miracle, he does so " without changing the common course of nature, except to the extent that this must be done. " Thus it is natural that the arrest of the Sun in Joshua's time, and the start again in Hezekiah's time, should be the result of terrestrial motion alone.

Oresme concluded that considerations such as these " are valuable for the defence of our Faith. " More astute than Galileo, and safe from the thunderbolts that were to be hurled at this thesis later, he was nominated Bishop of Lisieux in reward for his work.

Since, as we have said, Oresme's *Treatise on the Heavens and the World* was never printed, it is very unlikely that his ideas on the diurnal motion of the Earth could have become available to Copernicus.

Perhaps the reader will decide that we have devoted too much attention to this early philosopher. But we have seen the best and the worst of Oresme's arguments about the system of the World. We have felt the mood of the time, at once naive and acute, fantastic and serious, familiar and dogmatic. Of the originality of Oresme as a mathematician, and of the vigour of his penetrating thought, there is no doubt. The prejudices of the Schools and the accepted ideas of the time did not imprison him. In the field of mechanics he was one of the first to address himself to the great French public, or, as he said himself, more accurately, " to all men of free condition and noble intellect. "

9. THE OXFORD SCHOOL.

In his study of the representation of qualities Oresme invoked the authority of certain *veteres* whom he did not name. It is reasonable to believe that the ancients who preceeded Oresme in the general study of forms were the logicians of the Oxford School.

One of the most eminent masters of this school was William Heytesbury, or Hentisberus. It is said that Heytesbury was a fellow of Merton College in 1330, that he belonged to Queen's College about 1340 and that he was Chancellor of the University of Oxford in 1371.

Primarily Heytesbury was a logician of the most acute kind. But he was also concerned with kinematics, and it is in this connection that he claims our attention. In his *Regulae solvendi sophismata* the following rule is given without proof— when the velocity of a moving body increases with the time in such a way that it is uniformly deformed, in a given time the body travels the same path as if it had moved uniformly with the velocity acquired half way through this time. This is Oresme's rule applied specifically to distance.

To set against this, Heytesbury supported Thomas Bradwardine's opinion—referred to earlier—that the effective velocity of a rotating body was that of its most rapidly-moving point.

The most remarkable feature of Heytesbury's work is the appearance, albeit shrouded in obscurity, of the *concept of acceleration*. This was unknown to the Paris School.

In fact, in his treatise *De Tribus praedicamentis*, Heytesbury distinguished between the *latitudo motus* (velocity) and the *velocitas intensionis vel remissionis motus* whose value was the increase or the decrease of the former. This quantity corresponds to acceleration.

" For a moving body which starts from rest there can be imagined a range of velocity which increases indefinitely. In the same way can be imagined a range of acceleration or of slowing down *(latitudo intensionis vel remissionis)* according to which a body can accelerate or slow down its motion with an infinitely variable quickness or slowness. This second range is related to the range of motion (velocity) as the motion (velocity) is related to the magnitude (distance) that may be travelled in a continous manner. " [1]

Through this obscure language we catch the first glimpse of quantities that have become familiar tools of our trade, the vector representing the distance travelled, S; the velocity (vector derivative), $\dfrac{dS}{dt}$; and the acceleration (vector derivative of the velocity), $\dfrac{d^2S}{dt^2}$.

We must also refer to the *Liber Calculationum*. In order to avoid the labyrinths of the Oxford School, we shall confine ourselves to a mention of Swineshead (variously called Suincet, Suisset, Suisseth, . . . by the continental copyists). The tradition of the XVth and XVIth Centuries, on the publication of *Liber Calculationum* in 1488, 1498 and 1520, added the epithet " Calculator " to this name. The document is the most typical of the Oxford dialectic that is available to us, and in spite of the relentless attacks of the Humanists, it was very highly regarded until the XVIIth Century. Thus Leibniz, writing to Wallis, could express his wish to see it republished.

Unfortunately, this work has only been attributed to Swineshead in error. Duhem, that tireless investigator, found a manuscript [2] of this work which goes back to the XIVth Century, and in which the copyist attributes the work to Ricardus of Ghlymi Eshedi. (This must refer to William Collingham, a Master of Arts of Oxford.)

This treatise is concerned with the general theory of forms, and sophist discussions make up the essential part of it. We shall quote a single extract from Chapter XV, which is called *De medio uniformiter difformi*.

---

[1] Venice Edition, 1494.
[2] *Bibliothèque Nationale*, Paris, latin collection, Ms. 6558.

" If the motion of a body is uniformly accelerated and starts with a value zero, the body will travel three times further in the second half of the time than in the first half. "

This is a direct corollary of the law of distances in uniformly varying motion.

In this way then, in XIVth Century Oxford, the kinematics of uniformly varying motion was known and commonly taught. The English School has the merit of having stated the law of distances more precisely than the School at Paris—on the other hand, it seems to have neglected Oresme's remarkable representation of uniformly deformed qualities.

Just as much at Oxford as at Paris, these developments in kinematics, and the related general study of properties in intensity and extension, had no influence on the study of the fall of bodies. The description of these phenomena remained completely qualitative.

10. The tradition of Albert of Saxony and of Buridan.

The tradition of Albert of Saxony and of Buridan was preserved in France and Germany by Themon, a son of Jew who taught at Paris in 1350, and Marsile of Inghen, who became rector of Heidelberg in 1386 after having been at the University of Paris in 1379. It is noteworthy that Marsile modified Buridan's doctrine in a somewhat unfortunate way. Thus he held that the *impetus* was at first weak in those parts of a body that were not in contact with the motive agency, and that it was strengthened there as the whole *impetus* became uniformly distributed throughout the moving body.

We must also refer to Pierre d'Ailly (1330-1420) who was high master of the College of Navarre in 1384 and who added the following original items to Albert of Saxony's paradoxes.

" Someone who owns a field adjoining another piece of land, and who excavates his earth in such a way that the area of the cavity remains constant, is defrauding his neighbour.

" If the Earth is cut by a plane surface whose centre is at the centre of the World, when water is poured on this plane it will tend to assume the form of a hemisphere.

" In the second place, if the bottom of a pool is flat, this pool will certainly be deeper in the middle than at the sides. . . . "

Pierre d'Ailly also gave Roger Bacon's paradox which has been quoted above (p. 52).

With such intellectual games did the Schoolmen of the XIVth Century delight themselves. So alive was this tradition that it maintained itself for over two hundred years.

CHAPTER FIVE

XVth AND XVIth CENTURIES
THE ITALIAN SCHOOL
BLASIUS OF PARMA   THE OXFORD TRADITION
NICHOLAS OF CUES AND LEONARDO DA VINCI
NICHOLAS COPERNICUS
THE ITALIAN AND PARISIAN SCHOOLMEN
OF THE XVIth CENTURY
DOMINIC SOTO AND THE FALL OF BODIES

1. BLASIUS OF PARMA AND HIS TREATISE ON WEIGHTS.

Blasius of Parma (Biagio Pelacani), who became a doctor at Padua in 1347, taught at Padua and Bologna.   He went to Paris about 1405 and died at Parma in 1416.

His *Treatise on Weights* is known to us through a copy made by Arnold of Brussels and dated 1476.

This treatise derives from Jordanus' School and links up the idea of *gravitas secundum situm*—a first principle of XIIIth Century statics— with the tendency of a heavy body to fall along a chord rather than along an arc of a circle, and thus to take the shortest path, in Aristotle's sense, to its natural place.

Blasius of Parma observed that when a balance with equal arms supporting equal weights is moved away from the centre of the World, these weights will appear to become heavier.   Indeed, the line along which each of the weights tends to fall makes an angle with the vertical through the point of support which is the more acute as the balance is the further away from the centre of the World.   This embellishment adds nothing useful to the positive statics of the authors of *De Ponderibus*.

In a general way, Blasius of Parma was a critic and a sceptic who was content to multiply the objections to his predecessors' theories.   For example, he observed that it is necessary to take account of passive resistances, though the correctness of the propositions of statics depends on the process of neglecting all resistances occasioned by the medium. In a very naive way Blasius of Parma attempted to take these resistances into consideration.

Apart from his *Treatise on Weights*, Blasius of Parma also wrote *Quaestiones super tractatu de latitudinibus formarum*, which was printed in 1486. In it he appears as an unsophisticated commentator of Oresme's doctrine.

Although a critic of no great originality, Blasius of Parma was one of the means by which the statics of the XIIIth Century and the kinematics of the XIVth Century were handed on to the Italian School, which was destined to dominate mechanics during the period that we are going to study.

2. The Italian tradition of Nicole Oresme and the Oxford School.

Together with Blasius of Parma, we must refer to Gaëtan of Tiène. Like Blasius, this author taught at Padua and died there in 1465, and he is responsible for having preserved the tradition of William Heytesbury and the " Calculator " in Italy. One by one, he annotated the sophisms of the Oxford School, and his work was printed in Venice in 1494, together with the works of Heytesbury.

In particular, Gaëtan of Tiène emphasised the distinction between *latitudo motus* (velocity) and *latitudo intensionis motus* (acceleration). In this way the Italian School explained, more clearly than Heytesbury had done, the fact that a uniformly deformed motion corresponds to a constant *latitudo intensionis motus*—that is, to a constant acceleration ; and that a deformably deformed motion corresponds to a uniformly deformed *latitudo intensionis motus*.

Bernard Torni, a Florentian physician who died about 1500, carried on the work of Gaëtan of Tiène, and published *Annotata* to Heytesbury's treatise which made frequent mention of the " Calculator. " He was equally enthusiastic about Oresme's analysis, though he was only concerned with the arithmetical procedures contained in this work.

John of Forli, who taught medicine at Padua about 1409 and died there in 1414, wrote a treatise *De intensione et remissione formarum* which was printed at Venice in 1496. In it he refuted W. Burley, rejected Oresme's rule for the evaluation of a uniformly deformed quality, and attempted to introduce into medicine a terminology which was inspired by the Oxford School. The Humanists, especially Vivès, made him their target.

It may be inferred, as Duhem has remarked,[1] " that thanks to

---

[1] DUHEM, *Études sur Léonard de Vinci*, Series III, p. 509.

Nicole Oresme, William Heytesbury and the ' Calculator ', at the middle of the Quattrocento the Italian masters were well-acquainted with all the laws of uniformly accelerated or uniformly retarded motion. But it seems that none of them was inspired to assume that the fall of bodies was uniformly accelerated or, for this reason, to apply these laws to that phenomenon. "

3. Nicholas of Cues (1404-1464) and the doctrine of " impetus impressus. "

Nicholas of Cues studied at Heidelberg from 1416, and later at Padua in 1424. On returning to Germany he devoted himself to theology and science. He became Bishop of Brixen (Tyrol) in 1450 and died at Todi (Umbria) on August 11th, 1464.

His works were published in three parts between 1500 and 1514, and later reprinted at Basle in 1575.

Nicholas of Cues was primarily a metaphysician. In *De docta ignorantia* he maintained that it was impossible to accept the idea of absolute truth, and argued the identity of the absolute maximum and the absolute minimum, as well as the existence of an Universe at once finite and unlimited.

In mechanics, Nicholas of Cues has left us the dialogues *De ludo globi.* He is concerned with a game where a hemisphere is thrown in such a way that it meets some pins which are arranged in a spiral. The problem is to explain the trajectory of the body. Further, in the dialogue *De Possest,*[1] he concerned himself with the gyroscopic motion of a toy top.

" A child takes up this dead toy, devoid of motion, and wishes to make it live. For this purpose, by a procedure which he has invented and which is the instrument of his intelligence, he impresses on the toy the permanence of the idea which he has conceived. By a motion of his hands which is at once straight and oblique, consisting simultaneously of a pressure and a traction, he impresses on it a motion which is, for a top, supernatural. Naturally the plaything would have no other motion than the downward motion common to all heavy bodies—the child gives it the opportunity to move circularly, like the Heavens. This *motive spirit*, imparted by the child, is invisibly present in the material of the toy—the length of time for which it remains there depends on the impressive force which communicates this property. When this spirit ceases to animate the toy, it resumes its motion towards the centre, as at the beginning. Do we not have here an image of what happened when the Creator wanted to give the spirit of life to an inanimate body? "

[1] *Dialogus trilocutorius de Possest*, translated into French by Duhem.

The World-system accepted at that time assumed that the motion of the different celestial spheres was maintained by that of the outermost sphere, itself activated by a Prime Mover. Nicholas of Cues held that is was sufficient that the Creator should have imparted an *impetus* to the spheres *at the beginning*, and that the *impetus* would then be conserved indefinitely. Thus we come across the " chiquenaude, " the fillip, of which Pascal talked in connection with Descartes. This was also the doctrine of the Parisian Schoolmen of the XIVth Century, of Buridan and Albert of Saxony—in these celestial bodies there is no influence which can corrupt the initial *impetus*. In a manner which, for Nicholas of Cues, is very precise, the rotational motion of any perfect sphere is a natural motion. The impression of *impetus* on a moving body is comparable with the creation of a soul in the body.

Nicholas of Cues became one of the inspirations of Copernicus and of Kepler, as well as of Leonardo da Vinci.

4. LEONARDO DA VINCI'S CONTRIBUTION TO MECHANICS.

In mechanics, Leonardo da Vinci cuts the figure of a gifted amateur. Though he had read and meditated upon the Schoolmen that preceded him, his bold imagination was not inhibited as theirs was. He tackled all kinds of problem, often with more faith than success. Frequently he returned to the same problem by very different paths, and did not scruple to contradict himself.

Leonardo made no concessions to systematics. But it seems that the original ideas which he threw off throughout his manuscripts were taken over by more than one of his successors.

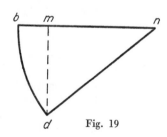

Fig. 19

His work in mechanics is quite unique, and the few pages which we are able to devote, in this book, to an attempt at analysing its objective content can only provide a feeble echo of the torrent of ideas which flowed from this " autodidacte[1] par excellence ."

a) *Leonardo da Vinci's concept of moment*. — Leonardo da Vinci grasped the idea of moment and applied it in a most complete way to a heavy body turning about a horizontal axis, a body which he described as being "convolutable. " Thus, for a lever *nb* turning about a point *n*, Leonardo stated the following rule.

---

[1] Autodidacte = one who is self-taught.

" The ratio of the distance (length) *mn* to the distance *nb* is such that it is also the ratio of the falling weight at *d* to the (same) weight at the position *b*. " [1]

That is, the effect of a heavy body suspended at *d* is the same as if the body were suspended from the arm of the *horizontal* lever, *nm*, that is obtained by projecting *nd* onto the horizontal *nb*. Leonardo called this arm of a horizontal lever, equivalent to the inclined arm *nd*, the " *arm of the potential lever.* " It would seem here that he had read the XIIIth Century statists. [2]

b) *The motion of a heavy body on an inclined plane.* — This is one of the problems that captured Leonardo da Vinci's interest, and which evoked some rather strange arguments from him. Thus we find the following passage among his writings.

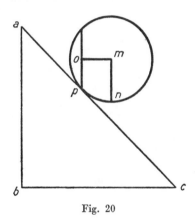

Fig. 20

" A heavy spherical body will assume a motion which is all the more rapid as its contact with its resting place is further separated from the perpendicular through its central line. *The more ab is shorter than ac, the more slowly the ball will fall along the line ac* [than along the vertical *ab*] ... because, if *p* is the pole of the ball, the part *m* which is outside *p* would fall more rapidly if there were not that small resistance provided by the counterpoising of the part *o*. And if there were not this counterpoise, the ball would fall along the line *ac* more quickly if *o* divided into *m* more often. That is, if the part *o* divides into *m* one hundred times, and the part *o* is missing throughout the

---

[1] *Les Manuscrits de Léonard de Vinci*, published by Ch. RAVAISSON-MOLLIEN, Paris, 1890, Ms. E, fol. 72.
[2] See above p. 42.

rotation of the ball, this will fall more quickly on *n* by one hundredth of the ordinary time. . . . *If* p *is the pole at which the ball touches the plane, the greater the distance between* n *and* p, *the more rapid the ball's journey will be.* " [1]

Elsewhere Leonardo wrote on the same subject in the following terms.

" *On motion and weight.* All heavy bodies seek to fall to the centre, and the most oblique opposition provides the smallest resistance.

" If the weight is at *A*, its true and direct resistance will be *AB*. But the pole is at the place where the circumference touches the earth, and the portion which is furthest outside the pole falls. If *SX* is the pole, it is clear that *ST* will weigh more than *SR*, from which it follows that the part *ST* falls, that it dominates over *SR* and lifts it up, and then moves along the slope with fury. If the pole were at *N*, the more often *AN* divided into *AC*, the more quickly the wheel would run along the slope than if it were at *X*. " [2]

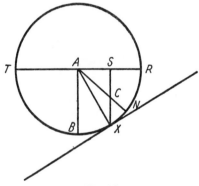

Fig. 21

I am reluctant to comment on these texts and to attribute to Leonardo things that he did not intend. Certainly he, like the aristotelians, did not differentiate between dynamics and statics. It is also true to say that he reproduced and repeated the law of powers which Aristotle had formulated (see above, page 20). Further, Duhem, arguing from the relation of velocities that Leonardo gave, believes that he immediately applied this same relation to powers—that is, to the apparent weights of a given body on differently inclined planes—and that he arrived in this way at the accurate law which we now accept.

[1] Ms. A, fol. 52.
[2] *Ibid.*, fol. 21.

I believe that it is more accurate to take a view of this kind—that Leonardo only sketched the solution of a problem which his rich imagination had formulated, but that he never gave it a final form. It certainly seems that Leonardo was unaware of the solution of the same problem that the unknown author of *Liber Jordani de ratione ponderis* had given, and which was based on the single concept of *gravitas secundum situm*, considered as a first principle of the statics of heavy bodies.[1]

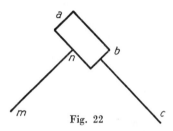

Fig. 22

Leonardo has, moreover, the merit of having attempted to solve this same problem of the inclined plane by another method, one which was unknown to his predecessors. Thus he observes[2] that a uniform heavy body which falls obliquely divides its weight into two different aspects, along the line *bc* and along the line *nm*. But here again we are left in suspense—he does not carry out the resolution of the weight into its two components along *bc* and *nm* (normal to *bc*).

c) *Leonardo da Vinci and the resolution of forces.* — Leonardo asked himself how the weight of a heavy body, supported by two strings, was apportioned between these two. He was of the opinion that the weight of the body suspended at *b* was divided between the strings *bd* and *ba* as the ratio of the lengths *ea* and *de*. This guess contradicts the now classical rule of the parallelogram.

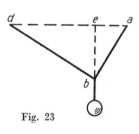

Fig. 23

However Leonardo used, at least implicitly, the following rule.

" With respect to a point taken on one of the components of a force, the moment of the other component is equal to the moment of the total force with respect to the same point. "

---

[1] See above, p. 43.
[2] Ms. G, fol. 75.

Thus, through the intermediary of the concept of moment, Leonardo arrived at the resolution of forces. Indeed, on different occasions he drew the figure opposite, in which the weight N is hung from two strings CB, CA, which are equally inclined to the vertical through N. He wrote—

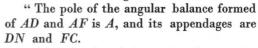

Fig. 24

" The pole of the angular balance formed of AD and AF is A, and its appendages are DN and FC.

" As the angle of the string that carries the weight N at its centre increases, the length of its potential lever decreases and the length of the potential counter-lever which carries the weight increases. "

This remains somewhat mysterious, but, like Duhem,[1] one may believe that the tension of the string CB, and the weight N, would maintain the rigid body formed of the two potential arms AB and AF in equilibrium, if the body were able to turn about the point A.

" A confusion of ideas poured from Leonardo's mind but, to a high degree, he lacked the power of discriminating between the true and the false. Also, as an inevitable consequence, a truth which might emerge from the surface of incomplete or false beliefs and become clear to him at one instant, was thrown back again, to await the future which would finally return it to the shore. "

d) *Leonardo and the Energy of moving bodies.* — Leonardo was aware of Buridan's doctrine through the intermediary of Albert of Saxony, who had adopted it. Moreover, he had read Nicholas of Cues.

Leonardo reconciled these doctrines in the following way.

At the outset he defined a quantity called *impeto*, analogous to *impetus* in Buridan's sense— " *Impeto* is a virtue created by motion and transmitted from the motor to the moving body, which has as much motion as the *impeto* has life. "[2]

Elsewhere he says, " *Impeto* is the impression of motion which is transmitted from the motor to the moving body. . . . All impression desires permanance, as is shown us by the similarity of the motion impressed on the body. "[3]

Leonardo regarded the motion of a projectile as being separated

[1] *O. S.*, Vol. I, p. 181.
[2] Ms. E, fol. 22.
[3] Ms. G, fol. 73.

into three phases. In the first the motion is *purely violent* and is effected as if the projectile had no mass and was subject only to the initial *impeto*.

In the third period, the *impeto* has completely disappeared. The moving body has a *purely natural* motion under the sole influence of gravity.

Between these two extreme phases Leonardo assumed the existence of an intermediate period in which the motion was mixed, part violent, part natural. This is the period of *compound impeto*.

The following quotation will illustrate this idea.

" A stone or other heavy thing, thrown with fury, changes the direction of its travel half way along its path. And if you are able to shoot a cross-bow for 200 yards, place yourself at a distance of 100 yards from a tower, aim at a point above the tower and shoot the arrow. You will see that 100 yards from the tower the arrow will be driven in perpendicularly. And if you find it thus, it is a sign that the arrow has finished its violent motion and has started the natural motion, that is, that being heavy, it falls freely towards the centre. "[1]

Or better still—

" *On convolutory motion.* A top which loses the power which the inequality of its weight has about the centre of its convolution because of the speed of this convolution, because of the effect of the *impeto* which dominates the body, is one which will never have that tendency to fall lower, which the inequality of the weight seeks to do, as long as the power of the body's motive *impeto* does not become less than the power of the inequality.

" But when the power of the inequality surpasses the power of the *impeto*, then it becomes the centre of the motion of convolution and the body, brought to a recumbent position, expends the remainder of the aforesaid *impeto* about this centre.

" And when the power of this inequality becomes equal to the power of the *impeto*, then the top is inclined obliquely, and the two powers struggle with each other in a *compound motion*, both moving in a wide circuit, until the centre of the second kind of convolution is established. In this the *impeto* expends its power. "[2]

Following the example of Nicholas of Cues, Leonardo concerned himself with the " game of the sphere ." He wrote—

" *On compound impeto.* A compound motion is one in which the *impeto* of the motor and the *impeto* of the moving body participate

[1] Ms. A, fol. 4.
[2] Ms. E, fol. 50.

together, as in the motion *FBC* which is intermediate between two simple motions. One of these is close to the beginning of the motion and the other close to the end. But the first is determined solely by the motor, and the second only by the shape of the body.

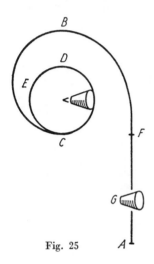

Fig. 25

" *On decomposed impeto.* Decomposed impeto is associated with a moving body which has three kinds of *impeto*. Two of these arise from the motor and the third arises from the moving body. But the two that arise from the motor are the rectilinear motion due to the motor and the curved motion of the moving body, and are mixed together. The third is the simple motion of the moving body, which only tends to turn round with its centre of convexity in contact with the plane on which it turns and lies. "[1]

Here da Vinci's imagination is given free rein. Our author becomes even more lyrical when he defines the *forza*.

" *As for the forza—* I say that the *forza* is a spiritual quality, an invisible power which, by means of an external and accidental violence, is caused by the motion and introduced, fused, into the body ; so that this is enticed and forced away from its natural behaviour. The *forza* gives the body an active life of magical power, it constrains all created things to change shape and position, hurtles to its desired death and changes itself according to circumstances. Slowness makes it powerful and speed, weak—it is born of violence and dies in freedom. The stronger it is, the more quickly it consumes itself. It furiously drives away anything that opposes it until it is itself destroyed—it seeks to defeat and kill anything that opposes it and, once victorious, dies. It becomes more powerful when it meets great obstacles. Every thing willingly avoids its death. All things which are constrained constrain themselves. Nothing moves without it. A body in which it is born does not increase in weight or size. No motion that it creates is lasting. It grows in exertion and vanishes in rest. A body on which it is impressed is no longer free. "[2]

Or again, " I say that *forza* is a spiritual, incorporeal, invisible power which is created in bodies which, because of an accidental violence, are

[1] Ms. E, fol. 35.
[2] Ms. A, fol. 35.

in some other state that their natural being and rest. I have said spiritual because there is in this *forza* an active incorporeal life, and I have said invisible because bodies in which it is born change neither in weight nor in shape ; of short life, because it always seeks to overcome its cause, and having done so, to die. " [1]

We shall attempt, without too much quotation, to indicate Leonardo's ideas on *forza*, ideas which were inspired by the metaphysics of Nicholas of Cues.

Forza can be born of the " expansion undergone by a tenuous body in one that is dense, like the multiplication of fire during the firing of cannons. " It can also be born of a deformation as in a cross-bow. Finally, one *forza* can engender another—this is the case of impact.

Leonardo returned to a pythagorean doctrine according to which a heavy body that is detached from a star to which it belongs tends to return there, in order to reconstitute the completeness of the star. He contrasted weight with *forza*, saying that these oppose each other.

" Weight is natural and seeks stability, then rest—*forza* seeks killing and death for itself. " Weight is indestructible. When a heavy body arrives on the ground it exerts a pressure on it, " and penetrates, from one support to another, to the centre of the World. " A weight embodies power, *forza*, motion and impact at the same time. But the fall of a body is itself preceeded by an accidental ascent. To be precise, at the origins of all actions in mechanics there must be a prime mover. And Leonardo, seduced by metaphysics, concludes—all motion arises from the mind.

Further comment on this adventurous thesis of Leonardo seems, to us, unnecessary—its qualities are more of poetry than of precision, of eloquence than solidity, more metaphysical than positive.

e) *Leonardo da Vinci and perpetual motion.* — Leonardo denied the possibility of perpetual motion on the grounds that *forza* continually expends itself. On the other hand, gravity seeks to produce equilibrium, all motions which are set in train by gravity have rest as their ultimate end.

f) *Leonardo and the Figure of the Earth.* — Having read and meditated Albert of Saxony, Leonardo wrote in connection with the figure of the Earth—

" Every heavy body tends downwards, and things which are at a height will not remain there, but will all, in time, fall down. Thus, in time, the World will become spherical and in consequence, will be com-

[1] Ms. B, fol. 63.

pletely covered with water. "[1]   And, without hesitation, he adds,—the Earth will be uninhabitable.

In Leonardo's belief, the seas exerted no pressure on the part of the globe which they covered. Quite the contrary. " A heavy body weighs more in a lighter medium. Therefore the Earth, that is covered by air is heavier than that covered by water. " [2]

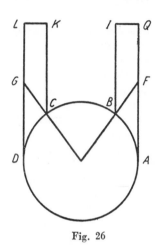

Fig. 26

g) *Leonardo da Vinci and the theory of centre of gravity. The flight of birds.* — Leonardo considered two towers *ABIQ*, *CDLK*, in " continual uprightness, " erected parallel to each other from the bases *AB*, *CD*, on the Earth. He predicted that " the two towers will tumble down towards each other if their construction is continued above a certain height in each case. "

Here is his argument. " Let the two verticals through *B* and *C* be produced in ' continual straightness. ' If they cut one of the towers in *GC* and the other in *BF*, it follows that these lines do not pass through the centre of gravity of the lengths of the towers. Therefore *KLCG*, a part of one tower, weighs more than the remainder, *GCD*, and of these unequal things, one will be dominant over the other, in such a way that, of necessity, the greatest weight of the tower will carry away all the opposite tower. And the other tower will do the same, in a way which is inverse to the first. " [3]

To recapitulate, Leonardo asserted that the vertical from the centre of gravity should not pass outside the base. This is, implicitly, the now classical theorem of the polygon of sustentation, but it contains the error, common to all the Schoolmen, that the convergence of the verticals has not been neglected.

In this connection, Leonardo almost goes as far as to suggest that a measurement of the distance apart of two verticals at the top and the bottom of a tower should be used to deduce the length of the Earth's radius.[4]

[1] Ms. F, fol. 70.
[2] *Ibid.*, fol. 69.
[3] *Ibid.*, fol. 83.
[4] Duhem, *O. S.*, Vol. II, p. 84.

Going over from statics to dynamics, Leonardo, guided by his bold and ubiquitous imagination, affirmed that " any heavy body moves towards the side on which it weighs more. . . . The heaviest parts of bodies which move in air become guides for their motion. " [1]

He also wrote, " Every thing which moves on a perfectly plane ground in such a way that its pole is never found between parts of equal weight, never comes to rest. An example is provided by those who slide on ice, and who never stop if their parts do not become equidistant from their centres. " [2]

In his Treatise on Painting, Leonardo applied the preceding ideas to the flight of birds.

" Any body that moves by itself will do so with greater velocity if its centre of heaviness is further removed from its centre of support.

" This is mentioned principally in connection with the motion of birds. These, without any clapping of wings or assistance from the wind, move themselves. And this occurs when the centres of their heaviness are displaced from the centres of their support, that is, away from the middle of the extension of their wings. Because, if the middle of the two wings is in front of or behind the middle, or the centre, of the heaviness of the whole bird, then the bird will carry its motion upwards or downwards [and this all the more so] as the centre of heaviness is more distant from the middle of the wings. . . . "

h) *Leonardo and the fall of bodies.* — It was inevitable that Leonardo should have become interested in the fall of heavy bodies. After having hesitated for some time between the two laws of velocity that were mentioned by Albert of Saxony (see above, page 57), Leonardo declared himself entirely in favour of the correct law $v = kt$. To set against this, the content of the studies on the latitude of forms (Oresme, Heytesbury) completely escaped him. Throughout he believed that motion *(moto)* was proportional to velocity *(velocitas)* and, in consequence, was mistaken about the law of distances.

In this connection we shall confine ourselves to a single quotation.

" *On motion.* A heavy body which falls freely acquires one unit of motion in each unit of time ; and one unit of velocity for each unit of motion.

" Let us say that in the first unit of time it acquires one unit of velocity. In the second unit of time it will acquire two units of motion and two units of velocity, and so on in the way described above. " [3]

[1] Ms. E, fol. 57.
[2] Ms. A, fol. 21.
[3] Ms. M, fol. 45.

i) *Leonardo's hydrostatics.* — Like the ancients, Leonardo set out to explain how water could appear in springs at the tops of mountains. He wrote, " It must be that the cause which keeps blood at the top of a man's head is the same as that which keeps water at the tops of mountains. " Leonardo sought this mechanism in the nature of heat. " There are veins which thread throughout the body of the Earth. The heat of the Earth, distributed throughout this continuous body, keeps the water raised in these veins even at the highest summits. "[1] To be accurate, Albert of Saxony, in his commentary of the relevant parts of Aristotle's treatise *Meteores*, had already invoked the intervention of heat in this matter.

Leonardo was more fortunate when he gave a complete formulation of the law of the flow of currents.

" All motion of water of uniform breadth and surface is stronger at one place than at another according as the water is shallower there than at the other. " Leonardo also outlined a theory of hydraulic pumps in the writing *Del moto e misura dell'acqua* in which a hint of Pascal's principle can be discovered.[2]

j) *Leonardo da Vinci and the geocentric hypothesis.* — On looking for it in Leonardo's writings, there can always be found evidence of the kind that Duhem indefatigably sought. Thus there is the following passage, which is aimed at the geocentric hypothesis. " . . . Why the Earth is not at the centre of the circle of the Sun nor at the centre of the World, but rather at the centre of its elements, which accompany it and with which it is united. "[3]

5. NICHOLAS COPERNICUS (1472-1543). HIS SYSTEM OF THE WORLD AND HIS IDEAS ON ATTRACTION.

In this book we can only discuss the different World-systems to the extent that they have had an influence on the development of mechanics. The copernican system that was, in the hands of Kepler and Newton, to play a fundamental part in the creation of dynamics, had no immediate influence on the scientists of the Renaissance. On the whole, these remained faithful to aristotelian ideas. We remark, for example, that the Sorbonne in the XVIth Century remained closed to copernican ideas and continued to teach Ptolemy's system.

Our attention, therefore, should only be held for a short time by

[1] Ms. A, fol. 56.
[2] *Cf.* DUHEM, *Études sur Léonard de Vinci*, Series I, p. 198.
[3] Ms. F, fol. 41.

Copernicus' ideas on dynamics and the circumstances which facilitated the copernican revolution.

From Antiquity there had been writers whose opinions were similar to those of Copernicus. Philolaus of Crete (a disciple of Pythagoras), Nicete of Syracuse and Aristarchus of Samos had attributed to the Earth both a daily and an annual motion, circular and oblique, about the Sun. (There was also supposed to an invisible earth which was symmetrical with ours with respect to the Sun.)

In the Middle Ages William of Ockham, Buridan and Albert of Saxony assumed that the Earth could have a rotational motion which was not necessarily identical with the apparent motion of the stars. Albert of Saxony was not alone in attributing the precession of the equinoxes to a slow displacement of the Earth. We have seen in detail how Nicole Oresme, who was certainly unknown to Copernicus, had defended the theories of a fixed Heaven and an Earth which had a diurnal motion. The appeal to the doctrine of *impetus* in Oresme's thesis, which was used to destroy that of Aristotle, is especially important.

In the religious fiels the Church in the XIIIth Century, tolerant because of its power, had the wisdom to brush aside the *a priori* questions which could be opposed to every doctrine that deviated from the geocentric hypothesis. As early as 1277 Etienne Tempier, Bishop of Paris, made the assumption that the question of whether the Heavens had a translation motion, or not, could be discussed. Thus the Church in the XIIIth Century assumed that the study of world-systems could be pursued as a piece of contingent research. In fact, a century later, Nicole Oresme did not compromise his ecclesiastical career in any way by believing in the motion of the Earth.

In the field of metaphysics, even the debate on the *plurality of worlds* helped the copernican revolution. In reaction against the pythagorean doctrine, Aristotle explicitly understood the term " Heavens " *(Οὐρανός)* in the sense of " All " or of " Universe. " The absolute fixity of the Earth and the perpetual rotation of the Heavens constituted a dogma of science. The Universe is unique and each body has a unique *natural place*, to which it returns of its own accord if it is violently displaced from it. Any other world which can exist must necessarily be made of the same elements as ours. To Aristotle, this meant that the co-existence of several worlds implied a contradiction. Beyond this Eight Sphere there can be neither space nor time.

This thesis was to be attacked in the XIIIth Century on the very grounds of the omnipotence of God.

Michael Scot (1230) was of the opinion that God could have created several worlds, but that Nature would not have been able to accommodate them. Saint Thomas Aquinas attempted to reconcile Aristotle's doctrine with the principle of divine omnipotence—the creation of similar worlds would be superfluous, the creation of dissimilar worlds would detract from the perfection of each of them, for only the ensemble can be perfect.

In 1277 the theologians of Paris, at the request of Etienne Tempier, condemned the anti-pluralist thesis.

William of Ockham intervened in the same direction—he argued that identical elements could simultaneously be directed towards different places. Thus a fire at Oxford would not move towards the same place as if it had been lit at Paris. The direction of a *natural* motion could therefore depend on the initial position of the element. Albert of Saxony decided against the plurality of worlds except " in a supernatural way, to the liking of God. " On the other hand, towards the end of the XVth Century Joannes Majoris asserted, in his *De infinito*, not only the plurality of worlds but the existence of an infinite number of worlds.

These discussions in no way lessen the originality of Copernicus' work, but to a certain extent they explain why he ventured to present his thesis. Being by profession Canon of Thorn, he protected himself with certain cautious declarations. Thus, in dedicating his works to Pope Paul III, he wrote, " I have believed that I would be readily permitted to examine whether, in supposing the motion of the Earth, something more conclusive *(firmiores demonstrationes)* might not be found in the motion of the celestial bodies. "

Strictly speaking, the doctrinal opposition only came much later with, for example, Melanchton and Father Riccioli. The latter was able to enumerate 77 arguments against the motion of the Earth and to refute 49 of the copernican arguments. As far as the Congregation of Cardinal Inquisitors was concerned, it only officially condemned Copernicus' writings on March 5th, 1616. In order to fix certain essential dates, we recall that Copernicus was born at Thorn on January 19th, 1472. He received his doctorate at Krakov, and made his way to Bologna and then to Rome, where he devoted himself to astronomy.

Copernicus gave himself up to a thorough study of the different world-systems which had been proposed by the Ancients, and used the motions of Mercury and Venus in order to place the Sun at the centre of the planets. In referring to the Pythagoreans, he proposed that the Sun should be placed at the centre of the World. Not wishing to advance anything without evidence, he started observation of planetary motions. The account of this task, completed in 1530, was only printed on his death in 1543.

If, in not making the centre of the Earth coincide with that of the Universe, he dispensed with the aristotelian doctrine on an essential point, he kept, for the rest, most of the ideas of the Schoolmen.   However, he did dispose of the distinction which Albert of Saxony had made between the centre of gravity and the geometrical centre of the Earth. We shall quote from Copernicus' *De revolutionibus orbium caelestium*.

" The Earth is spherical because, on all sides, it strives towards the centre.   The element of the Earth is the heaviest of all, and all heavy bodies are carried towards it and seek its intimate centre.

" To my mind, gravity is nothing else than a certain natural quality given to the parts of the Earth by the divine providence of He who made the Universe, in order that they should converge towards their unity and integrity, by uniting in the form of a globe.   It is probable that this property also belongs to the Sun, the Moon and to the wandering lights so that these too, by its virtue, keep that round shape in which we see them. "

And here Copernicus attacks Albert of Saxony.

" Because of their gravity, water and earth both tend towards the same centre....   One should not heed the Aristotelians when they claim that the centre of gravity is separate from the geometrical centre.... It is clear that both earth and water strive towards a unique centre of gravity at the same time, and that this centre is in no way different from the centre of the Earth. "

Copernicus' doctrine on the figure of the Earth agreed perfectly with all the geographical observations.   More simple than that of Albert of Saxony—an abstraction founded on prejudices opposed to the motion of the Earth—it was destined to triumph.   But at this point the copernican ideas came up against a scholastic tradition whose root must be found in Aristotle's *Meteores*—the four elements earth, water, air and fire have equal masses and therefore occupy volumes which are inversely proportional to their densities.   Moreover, the Aristotelians held that when a given mass of an element became " corrupted " in order to produce the next element in the succession, its volume increased. Aristotle himself mentioned this relationship for the single instance of the transformation of water into air, but his commentators applied it without hesitation to the transformation of earth into water and, carrying the argument to the limit, said that the total volume of water was greater than the total volume of earth.   Gregory Reisch, prior of Fribourg, put forward opinions of this kind in his *Margarita philosophica* (1496), a small encyclopedia that was widely circulated in the XVIth Century.   Twelve years after the end of Magellan's navigation of the globe—which should have clarified the scholastic opinion of the face of the Earth—Mauro of Florence (1493-1566) took up Reisch's thesis again, and held that the volume of the closed earth was ten times less than that of the waters. Copernicus felt himself obliged to refute this author. We have only stressed this geophysical issue in order to show the kind of objection which the great reformer of the system of the world met during his lifetime.

6. JOHN FERNEL (1497-1558) AND THE FIGURE OF THE EARTH.

John Fernel, chief physician to Henry II, deserves to be mentioned in a history of mechanics for having been the first among the moderns who had the initiative to measure a degree of terrestrial meridian. This he did by counting the number of revolutions of the wheels of his carriage between Paris and Amiens. In his *Cosmotheoria*, published at Paris in 1528, Jean Fernel disputed Albert of Saxony's doctrine, and decided in favour of the existence of a unique spherical surface for the combined mass of earth and water. He imagined the Earth to be absolutely immobile and to present the shape of a globe which had been hollowed out in places and whose cavities had been filled with water.

If one is to believe the chronicles,[1] John Fernel, who was a distinguished astronomer and mathematician, would certainly have written other things " if his wife had not compelled him, so to speak, to leave the sterile study of mathematics. "

7. ITALIAN SCHOLASTICISM IN THE XVIth CENTURY.

At the beginning of the XVIth Century the Italian Schoolmen were divided into three camps ; there were the *Averroïsts*, the *Alexandrists* —those who made appeal to Alexander of Aphrodisias—and the *Humanists*.

As an example of the first school, we shall cite Agostino Nifo, to whom we owe a commentary of *De Caelo et Mundo* dated 1514. In this manuscript there is nothing but scorn for the parisian school of the XIVth Century, whose representatives are called *Juniores*, terminalists (nominalists), Sorticoles (disciples of Sortes, that is, of Socrates) and *Captiunculatores* (a corruption of *Calculatores*). Albert of Saxony is ridiculed with the title of *Albertutius* or *Albertus Parvus*.

The Averroïsts rejected the doctrine of *impetus* and returned to Aristotle's explanation of the motion of projectiles. Concerning the

---

[1] LALANDE, *Astronomie*, Vol. 1, p. 189.

fall of heavy bodies, Nifo, like Saint Thomas Aquinas,[1] held that proximity to its natural place contributed towards a body's acceleration. He added to this an " instrumental cause, " and a quality belonging to the moving body.

Among the Alexandrists Peter Pomponazzi of Mantua, who is said to have taught at Bologna, devoted a treatise *De intensione et remissione formarum* (1514) to an attack on the Oxford School. In his *De reactione* (1515) he called William Heytesbury " the greatest of the Sophists " and contrasted him with " the clear and great voice of Aristotle. "

The thesis of Alexander of Aphrodisias, to whom this faction gave allegiance, has been preserved for us by Simplicius. It consisted of the assumption that a heavy body which was placed at a height became lighter. This lightness obtained at the beginning of the fall and then, continuously, became less apparent.

The Italian Humanists reproached the Schoolmen for their " parisian manner, " which was described as barbarous, sordid, gross and uncultured, but approved of their religious orthodoxy. In addition the Humanists, in the person of Giorgio Valla who taught at Padua (1470) and at Venice (1481), made an especial attack on the Averroïsts. These were taken to task for their language—studded with arabic terms—and for their exclusive cult of Aristotle and consequent neglect of Plato. Valla went as far as to consider Averroës, in Latin, of course, as a " primitive creature emerging from the mud " and as " pigheaded. " In dynamics Valla echoed the thesis of intermediate rest *(quies intermedia)* between the ascent and the descent of a body, which compromised the continuity of the motion. He assumed the existence, in every moving body, of a *vis insita*. This quantity, however, has no connection with Buridan's *impetus*, but rather is accounted for by the proximity of a motive agency or the natural place, according as a violent or a natural motion is in question.

These violent polemics added nothing new to mechanics, and we have only described them in order to illustrate the atmosphere of the time, with which original thinkers had to contend.

8. PARISIAN SCHOLASTICISM IN THE XVIth CENTURY.

The teachings of Buridan and Albert of Saxony were preserved at the College of Montaigu under the Scotsmen Joannes Majoris and George Lockhart. Jean Dullaert de Gand and the Spaniard Luiz Coronel taught at the same college. Another Spaniard, Jean de Celaya, taught at Sainte-Barbe.

[1] See above, p. 57.

This tradition was eclectic. It declared that it followed " the triple voice of Saint Thomas Aquinas, the Realists and the Nominalists. " Nevertheless, this School quibbled and argued much more than the masters of the XIVth Century, and lacked their originality.

Joannes Majoris, who was primarily a great teacher, taught Oresme's work on the latitude of forms and, in 1504, had Buridan's *Summulae* printed. In his *Disputationes Theologiae* Majoris argued, more explicitly than Buridan had dared to do, the identity of the dynamics of celestial and terrestrial bodies. Thus, like Nicholas of Cues, he prepared the way for Kepler. But it was left to him to fight the Reformation and to defend the dialectic which the students had begun to neglect. It was a time when, " covered with threadbare garments and with empty purses, the unhappy logicians of the University of Paris mused sadly on chairs which were no longer surrounded by pupils. They listened to the raillery that was poured on their learning, which they had only acquired with great effort, and to which they had consecrated their working lives. " [1] Already attacked by the Humanists, Scholasticism no longer paid.

In 1509 Jean Dullaert de Gand (1471-1513) continued the printing of Buridan's works. In 1506 he himself published some *Quaestions* on Aristotle's *De Caelo* and *Physics*. At Montaigu he taught the doctrine of *impetus*. He assumed that the *impetus* was modified by the shape of the projectile and supported the notion of intermediate rest, which he took to be at the moment when the *impetus* of the ascending motion was overcome by the gravity. He came to no conclusion as to the nature of *impetus*, whether it was a distinct property of a moving body or not. Concerning the fall of heavy bodies, he assumed that the *impetus* increased continually, though he did not know whether it should be taken as proportional to the size of the body. Similarily, Dullaert taught Oresme's rule on uniformly deformed motion through the reading of Bernard Torni, though he confined the treatment to the algebraic form of the rule. He lost himself in discussions on the nature of motion, a " successive entity truly distinct from all permanent things. "

Luiz Nuñez Coronel, of Segovia, published *Physicae Perscrutationes* in 1511. In dynamics, he believed in the gradual weakening of the *impetus* of all violent motion, which was a serious regression from Buridan's thesis. As for intermediate rest, he " imagined instances in which a stone thrown in the air remains there at rest for as much as an hour, two hours or even three, " without being dismayed by the objection that such rest was never seen. " This objection is not conclusive. The

---

[1] DUHEM, *Études sur Léonard de Vinci*, Series III, p. 179.

Celaya was rather reticent on the subject of the plurality of worlds. The Catholic faith provides no argument from which the existence of several worlds can be deduced, and the Philosopher (Aristotle) saw objections to such a happening. All the same, " from the supernatural point of view, there can exist several worlds, either simultaneously or successively, either concentrically or excentrically. " For " God can do all things that do not imply a contradiction, " and here there is none. The opinion of the Philosopher according to which the World contains all possible matter " is heretical, and the Philosopher would not be able to prove it. " Finally, we remark that though Celaya taught the work of Nicole Oresme and the Oxford School, like Jean Dullaert, he only knew them through the Italian tradition.

9. THE ATTACK OF THE HUMANISTS.

The rapid sketch which we have given above is sufficient to show that Parisian Scholasticism in the XVIth Century was in regression from the original work of the XIVth Century. The Humanists who were to proclaim its decadence were pupils of the College of Montaigu— Didier Erasme and Jean Luiz Vives.

Moreover, in mechanics, these Humanists preserved the tradition which they had received from Majoris and Dullaert. Thus, in his immensely successful *Colloquia* (1522), Erasme discussed the oscillation of a heavy body that travelled through to the centre of the Earth—this is, as we know, a problem that had already been raised by Oresme—in terms that his masters would not have disowned. Erasme's *Eulogy of madness*, which antedates the *Colloquia* and was published in 1508, includes a determined attack on the theologians, " those quibblers who are so puffed up with the wind and smoke of their empty and quite verbal learning that they will not give way on any point. "

Jean Luiz Vives (1492-1540) was born at Valence and was a pupil of Jean Dullaert before becoming a professor himself at Louvain. In *De prima philosophia* (1531) he discussed " intermediate rest " at great length, in terms which were in complete conformity with the pure scholastic doctrine. He had therefore retained traces of the teaching of Montaigu. His violent diatribes were directed at the Parisian masters and at the Oxford School with its XVth Century tradition, which sought to extend the Calculator's dialectic to medicine.

In *De philosophiae naturae corruptione* (1531) Vives wrote, " How can there be learning in subjects so divorced, so completely separated, from God on the one hand and from sensibility and spirit on the other ?

great distances may prevent one from seeing the rest, or it may e happen that the stone remains motionless for a time which is imp ceptible. "

To Coronel, *impetus* was an aptitude of the moving body, a cert; " actual entity, " produced in it by means of a repeated series of lo motions. *Impetus* was thus identified with a cognition acquired by t repetition of the same perception, like that of handwriting to the finge of the hand. This physiological model, however arbitrary it may b was taken up again by Kepler. In the theory of gravitation, Coron showed himself to be singularily naive. If weight is a property emana ing from the natural place, in order to prevent this property from passin through the surface of the earth, it will be sufficient if this is covere with a garment. . . . Elsewhere Coronel attributes the generation, i a free fall, of an *impetus* of greater or lesser intensity, exclusively t gravitation or to the substantial form of the heavy body.

In the motion of projectiles, Coronel assumed a mixture of decreasing *impetus* and progressive agitation of the air, which resulted in a certain compensation and assured a maximum violence at the middle of the trajectory.

In 1517 Jean de Celaya published *Expositio in libris Physicorum*, a literal commentary on Aristotle. The relevant discussion only appeared later under the title *Sequitur glosa*, and is distinguished by having explained, rather clearly, a law of inertia in the following terms.

" It would follow from the theory that a body which is projected will move forever. However, this result is false and the reason is clear. The theory does not include anything which will destroy the *impetus*, and it will therefore move the projectile forever.

" To this we reply by refusing to recognise the validity of the argument, and this because we deny the antecedent. Indeed, this *impetus* is sometimes destroyed by the resisting medium, sometimes by the shape or the property of the projectile that exerts a resisting action, sometimes, finally, by an obstacle. "

Celaya assumes that in the absence of these three mechanisms of destruction, *impetus* lasts indefinitely. " It is not necessary to suppose as many intelligences as there are heavenly bodies. It is sufficient to say that there is in each star an *impetus*, that this *impetus* was put there by the Prime Cause, and that it is this which moves the star. This *impetus* is not modified for the very reason that the heavenly body has no inclination towards a different motion. "

This is entirely in agreement with Buridan's thesis.

In the general sense, *impetus* was a second quality to Celaya. He compared it to " knowledge and dispositions of the soul. "

In a domain in which, founded on nothing, there is seen a vast structure of contradictory assertions concerning the increase and decrease of intensity, the dense and the tenuous, uniform motion, non-uniform motion, uniformly varying motion and non-uniformly varying motion ? It is not possible to count those who, without any limit, discuss instances which never occur, which could never turn up in nature ; who talk of infinitely tenuous and infinitely dense bodies; who divide an hour into proportional parts for this reason or that, and consider, in each of these parts, a motion, or an acceleration, or a rarefaction, varying in a given way. . . . ''

Further, in *De medicina* we find, '' the young people and adolescents who have been educated by means of these specious and tricky discussions know nothing of plants, of animals, nor of nature in the round. They have been brought up with no experience of natural things, without knowledge of reality. They have no prudence. Their judgement and their counsel are excessively weak, and yet they are expected to be able to win honour for themselves ! ''

And he concludes *(In pseudodialecticos)*, '' For myself, I have a great gratitude to God, and I thank him that I have at last left Paris, that I have emerged from the Cimmerian darkness, have come out into the light, that I have discovered the truly dignified studies of mankind— those which have earned the name Humanities. ''

10. DOMINIC DE SOTO (1494-1560) AND THE LAWS OF FALLING BODIES.

At the very moment that Scholasticism appeared to be discredited by the attacks of the Humanists, there intervened an original work which succeeded in formulating the laws of falling bodies correctly. We shall now analyse this work in some detail.

Dominic de Soto was born in 1494, the son of a gardener at Segovia. He attended the University of Alcala of Henares, and then took himself to the University of Paris where the Spaniards were already rather numerous. He returned to Alcala in 1520 and gave up the chair which he had obtained in order to take the habit of a preaching friar. From 1532 to 1548 he taught theology at Salamanca. As confessor to Charles V, he followed his king to Germany. Later he returned to Salamanca and taught theology there from 1550 until his death in 1560.

Soto had been a witness of the furious attacks of the Humanists upon the Paris School but remained, for his part, a Schoolman. However, he eschewed nominalism and attacked it in his *Quaestiones* (1545) on Aristotle's *Physics*.

We shall not discuss the metaphysical content of Soto's work, in which he rejected the concept of an actual infinity in favour of a virtual one, and shall only be concerned with his contribution to mechanics. In the first place we note, in passing, that Soto adopted Albert of Saxony's opinion on the equilibrium of the earth and the seas. In connection with the motion of projectiles, he taught the doctrine of *impetus*, and presented it in the following way.

" *First Conclusion.* — It cannot be denied that a man or a mechanism sets the air in motion when throwing a projectile, just as we see the circular agitation of water around a stone which has been thrown into it. The truth of this conclusion is especially evident for cannons, from which the air is driven in the form of a very violent explosion at the same time as the shot.

" *Second Conclusion.* — Air is not the only cause of the motion of projectiles. Whatever has thrown the moving body is also a cause, through the intermediary of the *impetus* which it has impressed on the body. " [1]

Thus Soto sought to reconcile Aristotle's doctrine with that of *impetus* by assuming that the agitation of the air played some part in the motion of projectiles. However, he summarily dismissed Marsile Inghen's opinion (see above, page 68).

" Observation proves that air too is a cause of the motion of projectiles. Indeed, we know that an arrow does not hit an object which is near with as much violence as it hits one that is a little more distant. This is why Aristotle says, in the second book of the Heavens, that natural motion is more intense towards the end, while the greatest intensity of the motion of a projectile is attained neither at the beginning nor at the end, but near the centre.

" Some suppose that the reason for this happening is the following one— the *impetus* is not all imparted to the arrow at the first instant. Later it becomes more intense, or else distributed through the extension of the arrow, so that it moves it in a more urgent way. But this is not very easy to understand. Indeed, one cannot see what could increase the intensity of the *impetus* after the arrow has been separated from the ballista, for an accident does not, of itself, become more intense. On the other hand, as the arrow is a continuous body, the *impetus* is simultaneously imparted to the whole body. Therefore it cannot distribute itself further later. " [2]

[1] *Quaestiones in libros Physicorum*, Vol. II, fol. 100.
[2] *Ibid.*

Soto regarded *impetus* as a " property distinct from the subject in which it is encountered, " like gravity or lightness. Conversely, he saw gravity as a " natural *impetus*. "

In his desire to reconcile Aristotle and Buridan, Soto went as far as to argue that Aristotle did not doubt the doctrine of *impetus*, but that he must have taken it as obvious, from the analogy with heavy and light bodies, and passed over it in silence.

But the essential part of Soto's work is that which concerns the fall of bodies. Some of the Schoolmen who had preceeded him had discussed the fall of bodies, albeit in a purely qualitative manner; others had discussed uniformly varying motion in the field of pure kinematics; but these studies had remained separate. It has now been established that the synthesis of these discussions was accomplished in Soto's time. He himself does not describe this achievement as a personal success. Is this modesty on his part or, on the other hand, the reflection of a movement which had already been completed by the Schoolmen ? The answer to this question is of little importance—what does matter is the law which was clearly expressed by this Spanish master.

We shall quote Soto's own text, as translated by Duhem.[1]

" Motion which is uniformly deformed with respect to time is that in which the deformity is so— if it is divided according to time, that is according to intervals which succeed each other in time, in each part the motion at the central point exceeds the weaker terminal motion in this part by an amount equal to that by which it itself exceeded by the more intense terminal motion.

" This kind of motion is one which is appropriate to bodies which have a natural motion and to projectiles *(Haec motus species proprie accidit naturaliter motis et projectis)*.

" Indeed, each time that a mass falls from the same height in a homogeneous medium, it moves more quickly at the end than at the beginning. On the contrary, the motion of bodies which are projected [upwards] is weaker at the end than at the beginning. And similarily the first motion is uniformly accelerated and the second, uniformly retarded. "

Soto was concerned with the law of distances for uniformly varying motion, and in his writings the ideas of Nicole Oresme and of the Oxford School may be clearly identified. After some hesitation he declared himself for the correct law.

" Uniformly deformed motion with respect to time follows almost the same law as uniform motion does. If two bodies travel equal

---

[1] *Theologi ordinis praedicatorum super octo libri Physicorum Aristotelis Quaestiones*, Salamanca, 1572, fol. 92 d.

distances in a given time, even though one moves uniformly and the other in any deformed manner—for example, in such a way that it covers one foot in the first half-hour and two feet during the second— from the moment that the latter covers as many feet as the former, which moves uniformly, in the whole hour, the two moving bodies will move equally.

" But here an uncertainty arises. Should the velocity of a body in uniformly varying motion be *denominated* by its most intense degree ? If for example, the velocity of a falling body increases in one hour from degree zero to degree eight, should it be said that this body has a motion of degree eight ? It seems that the affirmative reply is the correct one, for this is the law which seems to be followed by uniformly varying motion *with respect to a subject moving body*. Nonetheless we reply that the velocity of uniformly varying motion is evaluated by the mean degree and should be given the denomination of that degree. One should not argue in this respect as in the case of uniformly varying motion with respect to the subject. Indeed, in the latter case the reason for the rule adopted is the following— each part of the moving body describes the same line as the most rapidly moving point, in such a way that the whole moves as quickly as this point. Whereas a body which moves with a motion that is uniformly deformed with respect to time does not describe a path as great as if it were moving uniformly with the velocity which it attains at its supreme degree. This goes without saying. Therefore we believe that uniformly deformed motion should be denominated by its mean degree. Example— *If the moving body* A *moves for one hour and constantly accelerates its motion from degree zero to degree eight, it will travel just as great a path as the moving body* B *which moves uniformly with degree four for the same time.*

" It follows from this that when bodies move with a deformed motion, these motions should be reduced to uniform ones. " [1]

[1] *Ibid.*, fol. 93 and 94.

CHAPTER SIX

XVIth CENTURY
(Continued)
THE ITALIAN SCHOOL
OF NICHOLAS TARTAGLIA AND BERNARDINO BALDI

1. Nicholas Tartaglia.

Nicholas Fontana, called Tartaglia, derived his surname—which indicates stammering—from an injury obtained when he was wounded, while still an infant, in the sack of Brescia. He was born at Brescia at the beginning of the XVIth Century and died at Venice in 1557.

Tartaglia was one of the means by which the original statics of the XIIIth Century, which had been forgotten, was preserved for the Italian School of the XVIth Century. Indeed, Tartaglia entrusted Curtius Trojanus with the publishing of the work of the unknown author of the XIIIth Century. This appeared in 1565 under the title of *Jordani opusculum de ponderosite Nicolai Tartaleae studio correctum.*[1] Instead of giving his predecessors credit for their work, Tartaglia, who was not very scrupulous in matters of scientific propriety, claimed their demonstrations as his own.

Dynamics is treated in two of Tartaglia's works, *Nova Scientia* (1537) and *Quesiti et inventioni diversi* (1546).

In the first of these works Tartaglia attributes the acceleration of falling bodies to their approach to their natural place. " A heavy body hastens towards its proper nest, which is the centre of the World, and if it comes from a place which is more distant from this centre it will travel more quickly in approaching it. "

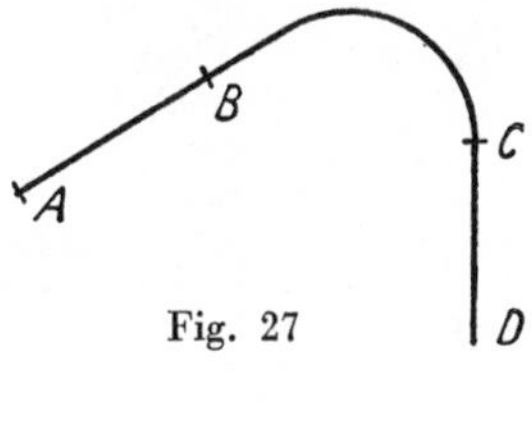

Elsewhere he distinguished three phases in the trajectory of a projectile— *AB* (rectilinear), *BC* (a curved join) and *CD* (vertical). He held that the velocity was

Fig. 27

[1] See above, pp. 41 to 46.

least at *C*, at the point at which the violent motion finished and the natural motion began.

This dynamics is improved a little in the *Quesiti*, in which he asserts that, except for the case in which the particle is thrown vertically, the trajectory of a shot has no rectilinear portion. It is the natural gravity which makes the trajectory curve downwards. The more rapidly a heavy body is thrown in the air, the less heavy it is and the straighter it travels through the air, which supports a lighter body more effectively. The more the velocity decreases the more the gravity increases, and this gravity continually acts upon the body and draws it towards the earth.[1]

Tartaglia adds that the motion of a projectile starts with an acceleration. He says that, for the same cannon with the same charge of powder and the same elevation, a second shot will go further than the first because it will find the air already divided and more easily penetrable.

2. JEROME CARDAN (1501-1576).

Jerome Cardan was born at Padua in 1501, died at Rome in 1576, and was at once physician, astrologer, algebraist and a student of mechanics.

Cardan's two works which are relevant to mechanics are the *De Subtilitate* (1551)—which was translated into French by Richard le Blanc in 1556—and the *Opus novum* (1570).

In statics Cardan believed that that he had surpassed Archimedes, whom he had read and admired, by treating the weight of the two arms of a balance. Indeed he wrote, " The heavinesses [moments] of the two arms of a beam [horizontal, cylindrical and homogeneous] have the same proportion to each other as that of the squares of the lengths of the two arms ... *Hoc est quod Archimedes reliquit intactum.* "[2]

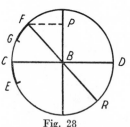

Fig. 28

Cardan uses the concept of moment fully.

" It is clear that, in balances and in things which lift loads, the further the burden is from the fulcrum the heavier it is. Now the weight at *C* is separated from the fulcrum by the length of the line *CB* and that at *F*, by the length of the line *FP*. "[3]

---

[1] *Cf.* DUHEM, *Études sur Léonard de Vinci*, Series III, p. 188. It may be that Tartaglia used Leonardo da Vinci's notes without acknowledgement.

[2] *Opus novum*, Proposition XCII.

[3] *De subtilitate*, translated by RICHARD LE BLANC, p. 16.

Like his predecessors of the XIIIth Century, Cardan then consi-
dered equal arcs $\widehat{FG}$ and $\widehat{CE}$ starting from the points F and C, but
he directed his attention to the *velocities* and not to the paths, by
observing that the fall from F to G was more " tardy " than the fall
from C to E.

He concludes, " then this argument is general—that the further
the weights are from the end, or the line fall of along the straight line
or the oblique, that is to say along the angle, the heavier they are. . . .
Thus the intention of the weight is to be carried directly towards the
centre. But because it is prevented from doing this by the *linkages*,
it moves as best it can. " [1]

Duhem interprets this rather obscure passage in the following way.

" When a heavy body falls vertically the power of the body is
measured, as Aristotle intended, by the velocity with which it falls.
But through the agency of the mechanism that carries it, because
of the linkages or constraints—to use the modern term—it may happen
that the body does not move vertically. Therefore in order to reckon
its motive power it is necessary to take account, not of the body's
total velocity, but only of the vertical component of this velocity,
or in other words, of the velocity of fall.

" If then a given weight is suspended from some point of a solid
which can move about a horizontal axis, the power of this weight
will be greater as the point of suspension falls more rapidly when a
given rotation is applied to the support. Therefore it will be greater
as the point of suspension is further from the vertical plane containing
the axis. " [2]

Like Leonardo, Cardan investigated the pulley-block, together with the screw and the jack.

Cardan was of the opinion that on an
inclined plane the heaviness of a given body
is proportional to the velocity with which it
moves down the plane. Therefore this heaviness is zero on a horizontal plane and
increases with the angle of inclination. Cardan assumes that the apparent weight is
proportional to this angle.

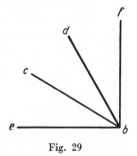

Fig. 29

" Let a sphere a, of weight g, be placed at
the point b and suppose that it is desired to
draw it along the plane bc. The vertical plane is bf. On the horizontal
plane be the force needed to move a may be taken as small as desired. . . .

---

[1] *Ibid.*
[2] Duhem, *O. S.*, Vol. I, p. 46.

Consequently, according to the consensus of opinion, the force which will move a along be will be zero. On the other hand, a will be moved towards f by a constant force equal to g ; in the direction bc by a constant force equal to k ; in the direction bd by a constant force equal to h. Since the motion along be is produced by a zero force, the relation of g to k will be as the relation of the force which moves a along bf to the force which moves a along bc, and as the relation of the right angle $\widehat{ebf}$ to the angle $\widehat{ebc}$. In the same way the force which moves a along bf is to the force which moves a along bd as the angle $\widehat{ebf}$ is to the angle $\widehat{ebd}$. " [1]

There is no clearer distinction between statics and dynamics in Cardan's work than can be discovered in that of Leonardo da Vinci. Like Leonardo, Cardan asserted the impossibility of perpetual motion unless natural motions were in question. We shall quote *De Subtilitate* on this subject.

" Either the continuity of motion will arise from the fact that the motion is in conformity with nature, " (hereby Cardan excepts the motion of the Heavens), " or else this continuity will not be maintained equal to itself. Now that which continually diminishes and is not augmented by some external action, cannot be perpetual. . . .

" The motions that bodies can have are of three kinds ; they may essentially tend to the centre of the World ; they may not be directed towards the centre in a simple way, like the running of water ; or they may stem from a particular characteristic, like the motion of iron towards a magnet. Patently, perpetual motion should be sought in motions of the first two kinds. Now when a weight is pulled more strongly, or held back more energetically, than is consistent with its nature its motion is natural, it is true, but not free of violence. Examples of these two conditions are seen in the weights of clocks. As for motion in a circle, this only belongs naturally to the sky and the air, and the latter is not actuated by an ever-present mechanism. For other bodies, it [motion in a circle] always has its root in vertical motion. Thus in rivers, at the rate and to the extent that the waters are generated by the source, they continually descend along the slope of the bed. Now in order that a motion should be perpetual, it would be necessary that the bodies which were displaced and came to the end of their path should be carried back to their initial position. But they can only be carried there by means of a certain excess of motive power. . . . "

[1] *Opus novum*, Proposition LXXII.

3. Julius-Caesar Scaliger and Buridan's doctrine.

Julius-Caesar Scaliger was a supporter of the parisian Scholasticism and one of Cardan's opponents. The latter, in Book XVI of *De Subtilitate*, had had the naive audacity to make a classification of genius, in order of decreasing merit, in the following way— Archimedes, Aristotle, Euclid, John Duns Scot, Swineshead the Calculator, Apollonius of Pergum, Archytas of Tarento, etc. . . . Scaliger replied on this matter. " You have given a simple artisan the place above Aristotle, who was not less erudite than he in these same mechanical skills ; above John Duns Scot, who was like the file of truth ; above Swineshead the Calculator, who almost surpassed the limits imposed on the human intelligence ! You have passed over Ockham in silence, that genius who outwitted all previous geniuses. . . . You have placed Euclid after Archimedes, the torch after the lantern. . . . " [1]

Scaliger explicitly refused to consider the agitated air as the seat of the motive agency of projectiles, and accepted Buridan's doctrine in all but form. In this he differed from Cardan, who remained an Aristotelian in this matter and who added nothing to the work of Tartaglia and da Vinci.

" The *motio* (here synonymous with *impetus*) is an entity which implanted in the moving body and which can remain there even when the prime mover is taken away. By prime mover I mean that which causes this entity to penetrate into the body. For it is not necessary that the efficient cause should continue to exist with its effect. " [2]

Scaliger continued—

" Heavy bodies, stones for example, have nothing which favours their being set in motion. They are, on the contrary, quite opposed to it. . . . Why then does a stone move more easily after the motion has started ? Because the stone has already received the impression of motion. To a first part of the motion a second succeeds, and each time the first remains. So that, rather than a single motor exerting its action, the motions which it imparts in this continuous succession are multiplied. For the first *impetus* is kept by the second, and the second by the third. . . . " [3]

4. Bento Pereira (1535-1610). The classical reaction.

In 1562 Bento Pereira published at Rome a treatise called *De communibus omnium rerum naturalium principiis* which became very po-

---

[1] *Exotericarum exercitationum libri*, Paris 1557, Exerc. 324. Translated by Duhem.
[2] *Ibid.*, Exerc. 76.
[3] *Ibid.*, Exerc. 77.

pular and which was studied by Galileo himself.   Bento Pereira knew
of Scaliger's *Exercitationes* but adhered, himself, to Aristotle's doctrine
on the motion of projectiles.   Cesalpin and Borro may also be cited as
representatives of this classical reaction.

5. The " Mechanicorum Liber " of Guido Ubaldo (1545-1607).

We now come to a student of mechanics who was a great authority
until the beginning of the XVIIIth Century, and who was one of Galileo's
masters.   Descartes, who gave few references, recalled having read him
and even Lagrange quoted him often in the historical part of his *Mécanique Analytique*.   To the classical impedimenta of the Medieval authors,
Guido Ubaldo added a reading of Archimedes and of Pappus, and
through the latter achieved a partial knowledge of Hero of Alexandria.
His *Mechanicorum Liber* is dated 1577.

Guido Ubaldo, who was Marquis del Monte, lived in seclusion in his
Castle del Monte Barrochio, and devoted all his leisure to study.

In his writings on statics he reproached the Schoolmen of the XIIIth
Century, with good reason, for having made a first principle of *gravitas
secundum situm* without having justified this action in any way.   He
wished to see substituted for this concept, the effect of the reaction of
the support.

" The mind cannot be at peace while the variation of gravity has not
been attributed to some other cause than this.   Indeed, it seems that
[the variation *secundum situm*] is a symbol rather than a true reason.

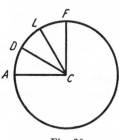

Fig. 30

" The line *CD* resists a weight placed at
D less than the line *CL* resists a weight at *L*.
Thus, then, the same weight can be heavier
or lighter in virtue of the effect of the position it occupies ; not that by the very fact
of this situation it really acquires a new
gravity or that it loses its original gravity—
rather it always keeps the same gravity in
whatever place it may be ; *but because it
always weighs more or less on the circumference.*"

Guido Ubaldo confined himself to this
qualitative statement, for he did not have at his disposal the law of
the composition of forces.

Nevertheless, he used the concept of moment to substantiate the
condition for the equilibrium of a lever, by means of an argument whose
form is directly inspired by Archimedes.   He corrected certain errors in

the XIIIth Century discussion of the stability of the balance, but made the mistake of using the same treatment when the verticals were assumed parallel as he used when they were supposed to converge.

Guido Ubaldo favoured Pappus' solution of the problem of the inclined plane—we have already seen the weakness and superficial character of this solution. Thus he was led to attribute a gravity to a moving body situated on a horizontal plane, contrary to the content of XIIIth Century statics. However, he in general preferred to consider *virtual displacements* than *virtual velocities*. He said that it is necessary to deploy a greater power in order to move a body than is necessary to maintain it in equilibrium, which shows that he did not understand the part played by the passive resistances.

Guido Ubaldo took over Pappus' definition of the centre of gravity and supplemented it with the following commentary, which was to have a great influence on the authors of the XVIIth Century.

" The rectilinear fall of bodies shows clearly that heavy bodies fall according to their centres of gravity. . . . Strictly speaking, a heavy body weighs through its centre of gravity. The very name centre of gravity seems to declare this truth. Clearly, all the force, all the gravity of the weight is massed and united at the centre of gravity ; it seems to run from all sides towards this point. Because of its gravity, indeed, the weight has a natural desire to pass through the centre of the Universe. But it is the centre of gravity that properly tends to the centre of the World. "

Thus to Guido Ubaldo, just as much as to the writers of the XIVth Century, the concept of centre of gravity was a purely experimental one. It was not linked in any way with the parallelism of verticals.

Guido Ubaldo's works, " sometimes erroneous, always mediocre, were often a regression from the ideas that had inspired the writings of Tartaglia and Cardan. " [1] However, this work is a milestone in the history of mechanics in that it had a direct stimulating influence on the great founders of mechanics, to whom it brought the content of the researches of Antiquity and the Middle Ages. Its value was at least that of a link with the past.

6. J.-B. VILLALPAND (1552-1608) AND THE POLYGON OF SUSTENTATION.

J.-B. Villalpand was born at Cordoba in 1552 and belonged to the Society of Jesuits. He became concerned with mechanics because of the diversion of an archeological mission to Jerusalem. He took it

[1] DUHEM, *O. S.*, Vol. I, p. 226.

upon himself to refute certain of Ezechiel's commentators, who had claimed that, because of its physical geography, Judea offered better possibilities for agriculture and construction than a plain of the same area would have done.

This explains the title of Villalpand's book, *Apparatus Urbis ac Templi Hierosolymitani*, which was printed at Rome in 1603.

In it Villalpand states, among others, the following proposition—

" A heavy body that rests on the ground and covers a certain area remains in equilibrium when the vertical drawn through the centre of this area passes through the centre of gravity ; or, otherwise, when a vertical drawn through the edge of this area passes through the centre of gravity or leaves it on the same side as the area. But if it leaves the centre of gravity on the other side of the area, the heavy body will necessarily fall. "

Here is his proof—

" If the line *FC*, when produced, leaves the centre of gravity *L* of the body on the opposite side to the area *BC* upon which the heavy body rests, the body will necessarily fall. Indeed, the weight *CLG* is

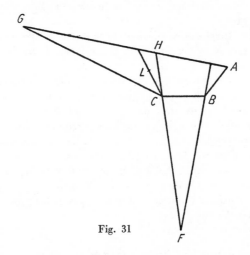

Fig. 31

equal to the weight *CLA*. The weight *CGH* will be greater than the weight *CHA*. The heavier volume will drag the less heavy one along..., and the body will fall on the side of *G*. "

It is quite probable that Villalpand, either directly or otherwise, borrowed this result, together with his later considerations on the walking of living beings and the flight of birds, from da Vinci. However it may be, we are indebted to P. Mersenne for having made the preceding theorem on the polygon of sustentation classical. That tireless scholar was able to extract it from the religious exposition in which it had been lost, and to reproduce it in his collection *Synopsis mathematica* (Mechanicorum libri), published at Paris in 1626.

7. J.-B. BENEDETTI (1530-1590). STATICS. FIGURE OF THE EARTH. DOCTRINE OF " IMPETUS. "

From the start of his scientific career in 1553 Benedetti denied the truth of the following proposition of Aristotle, a proposition which had been adopted by Jordanus's School— Let two bodies, A and B, be made of the same substance and let A have twice the volume of B. The velocity of fall for A is twice that for B.

More generally, Benedetti rejected Aristotle's statics. " The laws of the lever, " he wrote, " do not depend in any way on the rapidity or on the extent of the motion. " This does not mean that he adopted Jordanus's doctrine, or in other words, that he substituted the concept of virtual work for that of virtual velocities. In fact, he reduced the whole of statics to the single rule of the lever and the concept of moment.

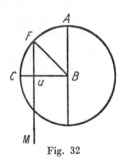

Fig. 32

" The ratio of the gravity of the weight placed at C to the gravity of the weight placed at F is equal to the ratio of BC to Bu. . . . This will appear evident to us if we imagine a vertical thread Fu, and if we imagine that the weight at F hangs from the end of the thread at M. It is clear that the weight hung in this way would produce the same effect if it were placed at F. " It seems that Benedetti had an inkling of the general utilisation of moments for measuring the effects of weights or of any motive powers whatever.

To a certain extent then this criticism of Benedetti's was useful and constructive. On the other hand his rejection of the solution of the problem of the inclined plane, due to the unknown author of the XIIIth Century, and his repetition of Leonardo da Vinci's errors concerning the division of a weight between two convergent supports, was less fortunate.

In the matter of the figure of the Earth and the separation between the continents and the oceans, Benedetti found his inspiration in Copernicus. In 1579 he denied the truth of Albert of Saxony's opinions in the following terms.

" We are certain that the spherical surface of the water is everywhere equidistant from the centre of the Universe, the point sought by all heavy bodies. Moreover, because of the numerous islands, because of the different countries which navigation has discovered in all regions, we can be sure and certain that the water and the earth comprise one globe, and that the geometrical centre of the Earth, together with the centre of its gravitation, is at the centre of the Universe. "

We must add that Benedetti considered the copernican system to be a plausible one, though he did not accept it himself.

It is said that Benedetti's works, united under the title *Diversarum speculationum mathematicarum et physicarum* and published in 1585, covered all the branches of mechanics. It remains to us to speak of Benedetti's important contribution to the doctrine of *impetus*.

At the outset Benedetti maintained that a constant motive agency produced an accelerated motion. " In natural and rectilinear motion the *impressio*, the *impetuositas recepta*, increases continually, for the moving body contains in itself the motive cause, that is to say the propensity to take itself to the place to which it is assigned. Aristotle should not have said that a body moves more rapidly as it approaches its goal, but rather that a body moves more rapidly as it becomes further separated from its point of departure. For the *impressio* increases proportionally as the natural motion is prolonged, the body continually receiving a new *impetus*. Indeed it contains in itself the cause of motion, which is the tendency to regain the natural place from which it has been torn by violence. " This quotation shows that even if Benedetti remained impregnated with Aristotle's ideas, he was not imprisoned by them. As we shall see, he was also able to amend Buridan's thesis.

Benedetti believed that the entity which was conserved in motion was the *impetus in a straight line*. In his opinion a horizontal wheel, as exactly symmetrical as possible and resting on a single point, cannot have a perpetual motion of rotation. He gives four different reasons for this.

The first is " that such a motion is not natural for the wheel. "

The second is because of the friction at the support.

The third, because of the resistance of the air.

The fourth reason, which is the only truly important one, we shall quote from Benedetti's text.[1]

" We consider each of the corporeal parts which moves on its own by means of the *impetus* which has been imparted to it. This part has a natural tendency to rectilinear motion, not to a curvilinear one. If a

---

[1] Translated into French by DUHEM.

particle chosen on the circumference of the aforesaid wheel was cut off
from this body, there is no doubt that, at a certain time, this detached
part would move in a straight line through the air. We can see this in
the example of the slings which are used to throw stones. In these
slings the *impetus* of motion which has been imparted to the projectile
describes, by a kind of natural propensity, a rectilinear path. The stone
which is thrown sets out on a rectilinear path along the line which is
tangent to the circle which it describes at the outset, and which touches
this circle at the point at which the stone was released, as it is reasonable
to assume. "

In short, Benedetti was the first to have clarified the idea that
the *impetus* was conserved in a straight line. From this correct idea,
however, he formed an incorrect conclusion. Thus he maintained
that the motion of a wheel must slow down spontaneously, because its
particles do not follow the rectilinear paths which they have an innate
tendency to take.

In fact, when the Schoolmen of the XIVth Century applied their
doctrine of *impetus* indiscriminately to rectilinear and curvilinear
motions, they confused two notions which a classical science should
have distinguished ; the *principle of inertia*, or the conservation, in
certain privileged connections, of the rectilinear uniform motion of
an isolated material point ; and the *principle of energy*, which entails
the conservation of the living force when these forces do no work.
If he had the essential merit of having caught a glimpse of the principle
of inertia, Benedetti, on the other hand, misunderstood part of the
truth of Buridan's thesis.

8. Giordano Bruno (1548-1600) and the composition of motion.

Giordano Bruno is best known as a metaphysician. A remote
disciple of Nicholas of Cues, he believed at the same time in the unity
and the infinity of worlds. He illustrated this by means of a system
of *Monads* which were at once material and spiritual, which were not
born and did not perish, but combined with and separated from each
other. He was burnt alive at Rome on February 17th, 1600 for his
lampooning of the Papacy rather than, it seems, for his metaphysical
ideas.

Bruno, who taught at the College of France and accepted Copernicus' system, was a determined adversary of aristotelian ideas. Thus
he rebutted, in his *Cena de le Ceneri* (1584), Aristotle's objection to
the motion of the Earth which had depended on the fact that a stone
thrown vertically upwards fell again at its starting-point. This he

accomplished by an argument which was analogous to, but more precise than, that of Oresme.

For this purpose, he visualises two men, one on the deck of a ship and the other on the bank, and each holding a stone in his hand. It is arranged that, at some instant, the hands are in sensibly the same position and that, then, the stones are allowed to fall simultaneously on the deck of the ship. The second man's stone will fall behind that of the first. For the stone belonging to the man on the ship " moves with the same motion as the ship. It has therefore a certain *virtus impressa* which the other does not possess . . . even though the stones have the same gravity ; though they traverse the same air ; though they start from points which are, as nearly as can be arranged, the same ; though they are subject to the same initial impact. "

9. BERNARDINO BALDI (1553-1617). STATICS AND GRAVITY " EX VIOLENTIA. "

Bernardino Baldi was at once a theologian, archeologist, linguist and geographer. He was a familiar of Guido Ubaldo and, in mechanics, seems to have been influenced by Leonardo da Vinci and others. In 1582 he wrote *Exercitationes in mechanica Aristotelis problemata* which was not printed until 1621. Bernardino Baldi rejected the point of view of virtual velocities—that of Aristotle—in statics. " We cannot be sure that the admirable effect of a lever has as its cause the velocity which follows from the lengths of the arms. Indeed, what is the velocity of something that does not move ? Now the lever and the balance do not move when they are in equilibrium and nevertheless a small power can then support a large weight. It will be retorted that if a very great velocity is not apparent in very long arms, it will at least be potentially present. Now the force which maintains [the lever] maintains the action. "

In a more positive way Baldi concerned himself with the equilibrium of a tripod and, in this connection, gave the rule of the polygon of sustentation. He took the product of the weight of a body and the height of the centre of gravity as a measure of the effort necessary to overturn the body. He also discovered the correct law for the stability of the balance and made a study of the sensitivity of the balance. He accepted Leonardo's solution of the problem of the inclined plane without rectifying it.

In dynamics Baldi distinguished between gravity *by nature* and gravity *by violence*, in which the influence of an external motive agency was concerned. In a projectile animated by a simple motion of translation, the centre of the natural gravity, *B*, coincides with the centre of the gravity *ex violentia*, under the influence of an impulsion with direction *BD*. These two centres are " only distinct by reasoning and not in reality. " And Baldi adds—

" Projectiles cease to move because the impression whose nature and impetuosity governs them is in no way natural, but purely accidental and violent. Now nothing which is violent is perpetual. . . . As long as violence predominates, violent motion is entirely similar to natural motion—it is lower at the start; later, by the very fact of the motion, it becomes more rapid ; then, as the impressed violence weakens bit by bit, it slows down ; finally the motion disappears at the same time as the *impetus* and the moving body comes to rest. "

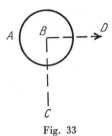

Fig. 33

As Duhem has remarked,[1] " this opinion is strange and not very logical. If one can assume that the natural gravity, which is a permanent motive agency, creates at each instant a new *impetus*, one cannot conclude from this that the artificial gravity, that is the *impetus* imparted by the motive agency, engenders in its turn an *impetus* of a second kind. " However strange it may be this thesis, handed on by Mersenne, was to be taken over by Roberval. Duhem has even followed its trail as far as Descartes. " [2]

[1] *Études sur Léonard de Vinci*, Vol. I, p. 139.
[2] *Cf.* a letter from DESCARTES to MERSENNE on April 26th, 1643, which discusses the question of whether a sword thrust is more effective if it is made with the point, the central part or that near the hilt of the sword.

CHAPTER SEVEN

XVIth CENTURY
(Continued)
XVIIth CENTURY

TYCHO-BRAHE AND KEPLER

1. THE SYSTEM DUE TO TYCHO-BRAHE (1546-1601).

While the students of mechanics of the Renaissance remained faithful to the Schoolmens' tradition and rehearsed their arguments without taking account of the observations that were available to them the astronomers were patiently accumulating a host of data that were to be seized by classical science for the formulation of the laws of dynamics. Tycho-Brahe occupies a prominent place among these observers because of the volume and the precision of his observations, which were the foundation upon which Kepler's laws were based. We must say a word here of his system of the World and of his ideas on dynamics.

Tycho-Brahe rejected Ptolemy's system because of the complexity of its epicycles. He rejected Copernicus' system on the grounds that the comets observed in opposition to the Sun were not affected by the annual motion of the Earth.

In his *Astronomiae instauratae progymnasmata* (1582) he wrote, " That heavy mass of the earth, so ill-disposed towards motion, cannot be displaced and agitated in this way without conflicting with the principles of physics. The authority of the Holy Scriptures opposes it. . . . I have set out to examine seriously whether there is any hypothesis which is completely in accord with the phenomena and the mathematical principles without being repugnant to physics and without incurring the censures of theology. It has turned out as I had hoped. . . .

" I believe, firmly and without reservation, that the motionless earth must be placed at the centre of the World, in accord with the feelings of ancient astronomers or physicists and the testimony of the Scriptures. I in no way assume, like Ptolemy and the Ancients, that the earth is the centre of the orbits of the secondary moving bodies. Rather I believe that the celestial motions are arranged in such a way that only the Moon and the Sun and the Eighth Sphere—the most distant of all—have the centres of their motions at the earth. The five other planets turn round the Sun as round their Chief and King, and the Sun is always at the centres of their spheres and is accompanied by them in its annual motion. Thus the Sun will be the law and the end of all these revolutions and, like Apollo among the Muses, it alone will determine all the celestial harmony of the motions which surround it. "

Tycho-Brahe's initial faith in his system is embodied in the following formula. " *Nova mundani systematis hypotyposis ab authore nuper adinventa qua tum vetus illa Ptolemaica redundantia et inconcinnitas, tum etiam recens Coperniana in motu terrae physica absurditas excluduntur, omniaque apparentiis caelestibus aptissime correspondent.* "

However, this assurance is less obvious in a letter written to Roth-mann and dated February 21st, 1589. " If you prefer to make the earth and the seas, together with the moon, revolve ; if you wish that the earth, however ill-suited to motion and far below the stars it may be, behave like a star in the ethereal regions, you are certainly the master. . . . But are not earthly things being confused with celestial things ? is not the whole order of nature being turned upside-down? "

Fundamentally it was religious prejudice that dictated the form of Tycho-Brahe's thesis, for he was too wideawake not to admit the super-iority of the copernican system over that of Ptolemy. " I acknowledge that the revolution of the five planets, which the Ancients attributed to epicycles, are easily and at little cost explained by the simple motion of the Earth ; that the mathematicians have adopted many absurdities and contradictions which Copernicus set aside ; and that his system even agrees a little more accurately with celestial phenomena. "

In order that the planets might turn about the Sun, Tycho-Brahe was obliged to assume that the rotation of the Sun round the Earth was due to an attraction that was different from that between the planets and the Sun.

In dynamics he opposed the motion of the Earth with the objection that a stone dropped from the top of a tower fell at the bottom. Thus he did not appreciate the fallacies in this argument, which Oresme and Giordano Bruno had already indicated, though he was almost certainly unaware of their writings.

2. KEPLER (1571-1631). THE GENERAL CHARACTER OF HIS CONTRIBUTION.

It may seem strange that this review of the origin of mechanics should finish with Kepler's work. But if he is numbered among the classics for his three fundamental laws on the motion of the planets, his metaphysical tendencies and his ideas on dynamics place him in the scholastic tradition. Though a forerunner of Newton, his own inspiration were the writings of Nicholas of Cues.

Kepler's character is most complex. A tireless calculator, he returned to the interpretation of observations without ever being discouraged, and rejected every law that allowed the slightest imprecision. With great wisdom he remarked that in the domain of Astronomy innovations were apt to lead to absurdities. By this he meant that the observations of the Ancients, however rough, should not be neglected. Though a disciple of Tycho-Brahe, he had no less respect for Ptolemy and was alive to the necessity of not adhering to the copernican system. His preconceived ideas, his errors, inconsistencies and illusions are not hidden from the reader. Occasionally his writings have the air of the confessional— thus he declares that his desire to succeed makes him blind, " *cum essem caecus pro cupiditate.* " [1] He compares scientific truth to a nymph who steals away after allowing herself to be seen, and quotes Virgil [2] in this connection. We see him sacrificing himself to metaphysics, seeking the reflection of preordained harmonies on every occasion, and even lending himself to astrology. Should we regard this as a fashion of the time, or as evidence of difficulties of quite another kind ? Indeed, the following declaration is attributed to Kepler— Astronomy would die of hunger if her daughter, Astrology, did not earn enough bread for two. . . .

Kepler's first scientific work, *Mysterium Cosmographicum,* was published at Tubingen in 1596. The pythagorean influence which was to become apparent in all Kepler's thought emerges clearly from this youthful work. Thus he sought to incorporate the dimensions of the different planetary orbits into the copernican system by comparing them with the radii of spheres inscribed or circumscribed to five regular polyhedra. He assumed that the planets moved under the influence of an *anima motrix* localised in the Sun, whose action on the planets

---

[1] *Astronomia nova*, p. 215.
[2] " *Malo me Galataea petit, lasciva puella*
    Et fugit ad salices, et se cupit ante videri. "
The question here is the discovery of the elliptic trajectories of the planets, *Astronomia nova*, p. 283.

was greater as they came nearer to the Sun. This property, confined to the plane of the ecliptic, is therefore inversely proportional to the distance. The same is true of the velocity produced, in accordance with the aristotelian dynamics to which Kepler remained faithful.

At least *Mysterium Cosmographicum* had the merit of attracting the interest of Tycho-Brahe, who thereupon used Kepler in the analysis of planetary observations and, if the tradition is to be believed, charged Kepler with the task of preparing a new table of the planets. Kepler completed this task in 1627, with the publication of *Tabulae Rudolphinae*.

3. THE ORIGIN OF THE LAW OF AREAS.

We shall now follow Kepler's fundamental work in theoretical astronomy— *Astronomia nova* αιτιολογητος,[1] *seu Physica Caelestis tradita commentariis de motibus stellae Martis ex observationibus G. V. Tychonis Brahe*, Prague, 1609.

In this work Kepler seeks a theory of Mars which will take account of the observations in a precise way and which will be, at the same time, compatible with the systems of Ptolemy, Copernicus and Tycho-Brahe.

The following is a very much abbreviated version of Kepler's demonstration of the law of areas in the case of eccentrics.[2]

In the figure α is the *centre of the World*, that is, the Sun in the copernican system and the Earth in the other astronomical systems.

The *centre of the eccentric* which the planet describes is at β. (This term refers to the Earth in the copernican system and the Sun in the others.)

The point γ is the *equant (punctum aequantis)*, or the point about which, according to Ptolemy's hypothesis, the " planet " appears to describe a circle with uniform velocity. Kepler draws this circle as a dotted line with the point γ as centre and radius equal to that of the eccentric of centre β.

Further, like Ptolemy, Kepler assumes the *bisection of the eccentricity*, or that αβ is equal to βγ.

Starting from the aphelion (or apogee) δ and the perihelion (or perigee) ε, two *very small* arcs $\widehat{δψ}$ and $\widehat{εω}$ are drawn in such a way that the points ψ, α and ω are colinear. Then the lines γψ and γω are drawn to cut the dotted circle in χ and τ respectively.

" According to Ptolemy, since the entire circle νφ (equal to the eccen-

---

[1] That is, " concerning the search for causes, " meaning at once dynamical and metaphysical causes.

[2] *Astronomia nova*, Chapter XXXII, p. 165.

tric but with centre γ (is a measure of the planet's period, then the arc $\widehat{\nu\chi}$ will be a measure of the time the planet *(mora)* spends on the arc $\widehat{\delta\psi}$ of the eccentric. " Kepler calls the arc $\widehat{\delta\psi}$ *arcus itineris* and the arc $\widehat{\nu\chi}$, *arcus temporis*. The same is true for the arcs $\widehat{\varepsilon\omega}$ and $\widehat{\varphi\tau}$.

Having supposed the angle $\widehat{\delta\alpha\psi}$ to be very small, Kepler writes

(1) $$\frac{\gamma\nu}{\gamma\delta} = \frac{\text{the arc }\widehat{\nu\chi}}{\text{the arc }\widehat{\psi\delta}} \quad \text{and} \quad \frac{\gamma\varepsilon}{\gamma\varphi} = \frac{\text{the arc }\widehat{\varepsilon\omega}}{\text{the arc }\widehat{\tau\varphi}}.$$

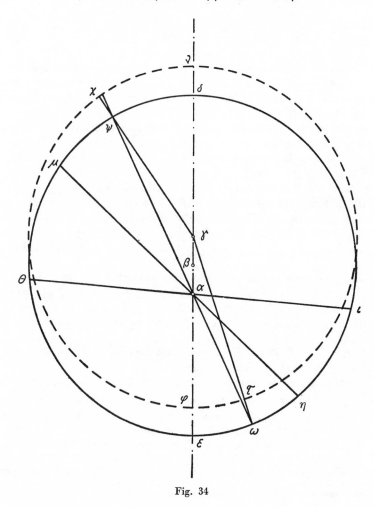

Fig. 34

Because of the bisection of the eccentricity, the length $\beta\delta$ is the arithmetic mean of $\gamma\delta$ and $\alpha\delta$. But the arithmetic mean of two quantities which are nearly equal to each other is *just greater* [1] than their geometric mean—this Kepler verifies by means of a numerical example. Then

$$\frac{\beta\delta \ (\text{or } \gamma\nu)}{\gamma\delta} \approx \text{ and } > \frac{\alpha\delta}{\beta\delta}.$$

From which follows through eq (1)

(2) $$\frac{\text{the arc } \widehat{\nu\chi}}{\text{the arc } \widehat{\psi\delta}} \approx \text{ and } > \frac{\alpha\delta}{\beta\delta}.$$

In the same way, if $\beta\varepsilon$ is the arithmetic mean between and $\gamma\varepsilon$ and $\alpha\varepsilon$, it is found that

$$\frac{\gamma\varepsilon}{\beta\varepsilon \ (\text{or } \gamma\varphi)} \approx \text{ and } < \frac{\beta\varepsilon}{\alpha\varepsilon}.$$

Hence, by (1),

$$\frac{\text{the arc } \widehat{\varepsilon\omega}}{\text{the arc } \widehat{\varphi\tau}} \approx \text{ and } < \frac{\beta\varepsilon}{\alpha\varepsilon}.$$

If then one considers, on the eccentric, two very small arcs $\widehat{\delta\psi}$ and $\widehat{\varepsilon\omega}$, *assumed to be equal to each other*, each of them will be the mean proportional between the arc $\widehat{\nu\chi}$—the time spent at the aphelion—and the arc $\widehat{\varphi\tau}$—the time spent at the perihelion. Further, the ratio of the arc $\widehat{\nu\chi}$ to the arc $\widehat{\varphi\tau}$ will be very nearly equal to $\left(\dfrac{\beta\varepsilon}{\alpha\varepsilon}\right)^2$.

Or again, more clearly, if two very small and equal arcs $\widehat{\delta\psi}$ and $\widehat{\varepsilon\omega}$ are taken on the eccentric, the ratio of the times spent on the arcs will be the ratio of the arcs $\widehat{\nu\chi}$ and $\widehat{\varphi\tau}$, and will be equal to $\dfrac{\alpha\delta}{\alpha\varepsilon}$, since, to the square in the eccentrics, $\dfrac{\alpha\delta}{\alpha\varepsilon} = \left(\dfrac{\varepsilon\beta}{\varepsilon\alpha}\right)^2$.

Now Kepler is in a position to state the law of areas for eccentrics.

" *Quanto longior est $\alpha\delta$ quam $\alpha\varepsilon$, tanto diutius moratur Planeta in certo aliquo arcui excentrici apud δ, quam in aequali arcu excentrici apud ε.*"

That is, the greater $\alpha\delta$ is than $\alpha\varepsilon$, the longer the planet will remain on a certain arc in the immediate neighbourhood of δ than on an equal arc of the eccentric in the neighbourhood of ε.

---

[1] In the modern sense, " approximately equal to and greater than " or " $\approx$ and $>$."

In the neighbourhoods of other points on the eccentric which are opposite to each other with respect to the centre of the world α, the behaviour of the planet is analogous, " *quanto evidentior in demonstratione, tanto minor in effectu.* "

In fact, Kepler confined himself to the remark that the proportion of αμ to αν is smaller, and that of αθ to αι is much smaller, than the proportion of αδ to αε. (See fig. 34.)

Kepler translated this purely geometrical and kinematic analysis into dynamical terms in the very title of the chapter which we have analysed— *Virtutem quam Planetam movet in circulum attenuari cum discessu a fonte.* (The strength—understood as the force—by means of which the Planet moves circularly falls off with the distance from the source [of motion].) This is evidence of the fact that Kepler remained faithful to aristotelian dynamics. Indeed, the force is measured in Kepler's mind by the inverse of a duration of sojourn on an arc, that is, by the *velocity to which it corresponds*. We see here the continuity of Kepler's views from his *Mysterium Cosmographicum* to his *Astronomia nova*.

4. ORIGIN OF THE LAW OF THE ELLIPTICITY OF PLANETARY TRAJECTORIES.

Tycho-Brahe and Longomontanus had prepared a table of the oppositions of Mars since 1580. Tycho-Brahe, who had started Kepler on his study of the theory of Mars, himself represented the orbit of that planet by an eccentric whose geometrical centre *did not bisect the eccentricity*.

Now Kepler, either by tradition or because of his metaphysics, was attached to the hypothesis of the bisection of the eccentricity, which Ptolemy had put forward in connection with the major planets alone. He even went as far as to extend it to the Earth's orbit (in the context of Copernicus' system) and to that of the Sun (in the other systems).

Kepler immediately started a methodical refinement of the values assigned to the radii of the Earth's orbit, which determined the scale of all the other interplanetary distances.

He then turned his attention to Mars. Being unable to follow him through all the various detours that he made, we shall only record that he succeeded in accounting for all of twelve oppositions of Mars to within 2′ of arc. This was accomplished by a painful method of trial and error in which four longitudes of Mars in opposition were used simultaneously. This necessitated, on Kepler's own confession,[1] no

---

[1] *Astronomia nova*, Chapter XVI, p. 95.

less than seventy repetitions of the calculation. But the longitudes of
Mars in position other than opposition invalidated the eccentric calculated in this way. Moreover, the eccentric did not satisfy the hypothesis
of the bisection of the eccentricity—it transpired that the distances
from the geometrical centre to the Sun and to the equant were not
equal, but were in the ratio $\dfrac{7\cdot232}{11\cdot332}$.

Returning to the hypothesis of the bisection of the eccentricity for
the orbit of Mars, and relying on the observation of the opposition of
Mars in 1613, Kepler found an error of about 8' in the annual parallax
of the planet. Fortunately for theoretical astronomy and for the development of newtonian mechanics, Kepler
refused to neglect such a disparity
between calculation and observation.
He proceeded to evaluate the distances
from Mars to the Sun in terms of the
distances from the Earth to the Sun.
The accompanying diagram shows how
a knowledge of the longitudes of Mars
and of the Earth, together with a
knowledge of the two radii (SB and SC)
of the Earth's orbit, allow the distance
from the Sun to Mars (SM) to be determined. In the diagram the circle
with centre O represents the Earth's
orbit, S the Sun and M, Mars, while B

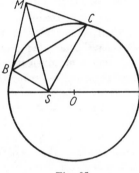

Fig. 35

and C are two positions of the Earth for the same position of Mars.
In particular, Kepler proceeded in this way for the distances from Mars
to the Sun in the neighbourhood of the aphelion and of the perihelion,
and thus obtained the eccentricity of the planet.

Kepler compared these observations with a circular eccentric satisfying the principle of the *bisection of the eccentricity*. He established a
systematic failure of distances with respect to the circumference of the
circle, " *Itaque plane hoc est ; orbita planetae non est circulus, sed ingre-
diens at latera utraque paulatim, iterumque ad circuli amplitudinem in
perigeo exiens, cujusmodi figuram itineris ovalem appellitant.* " [1]

Only observation could make Kepler give up the hypothesis of the
circle, which was based on the authority of the ancients and, for the
rest, agreed with his own metaphysics.

At first Kepler was reluctant to make an ellipse of this oval orbit,

---

[1] *Astronomia nova*, Chapter XLIV, p. 213.

though he did investigate whether a particular ellipse that he had
chosen could reconcile the data, only to discover that this was not so. [1]
Finally, however, after many unsuccessful attempts, he wrote, " *Inter
circulum vero et ellipsin, nihil mediat nisi ellipsis alia* " (between a circle
and an ellipse there can be nothing but a second ellipse). And he concludes that " *Ergo ellipsis est Planetae iter.* " [2]

5. KEPLER'S THIRD LAW.

The extremely important positive success of the theory of Mars did
not turn Kepler's interest away from astrology and metaphysics. Thus
more than ten years elapsed before *Harmonices Mundi* was published
at Linz in 1619. Of the first five books in this work only the last makes
mention of astronomy, and even this is confused with strange metaphysical conceptions. For example, we see him develop an analogy
between the angular velocities of the planets about the Sun and the
frequencies of musical notes, and expressing the oscillation of these
angular velocities during the course of a revolution by means of a musical
notation. The question is really one of pythagorean harmony, with
the reservation that it remains abstract and that Kepler did not pretend
that it was perceptible by our senses.

In following the *Astronomia nova* we have seen that Kepler came
across the law of areas before that of ellipticity. It has however become
customary to reverse the order in which these two laws are presented,
and to forget that Kepler, without justification, extended the law of
areas to elliptical trajectories allthough he had only established it for
eccentrics.

Kepler's third law is stated in Chapter III of Book V of *Harmonices
Mundi*.[3] He recalls the fruitless efforts that he had made, since the
beginning of his scientific career, to establish a connection between the
periods of the planets and the dimensions of their orbits. Not until
March 8th, 1618, did he come across the characteristic ratio in this law
—a gross error of calculation made him reject it at first. Finally he
persuaded himself of its correctness— " *Res est certissima exactissimaque,
quod proportio quae est inter binorum quorumcunque Planetarum tempora
periodica, sit praecise sesquialtera proportionibus mediarum distantiarum,
id est Orbium ipsorum.* " (One thing is absolutely certain and correct,
that the ratio between the periods of any two planets is, to the power $\frac{3}{2}$,

---

[1] *Astronomia nova*, Chapter XLV.
[2] *Ibid.*, Chapter LV, p. 285.
[3] P. 189.

exactly that of their mean distances, that is, of their orbits.) This quite empirical result may be written

$$\frac{T'}{T} = \left(\frac{a'}{a}\right)^{\frac{3}{2}}.$$

6. KEPLER AND THE CONCEPT OF INERTIA.

Kepler had the merit of having emphasised the concept of inertia more completely than his predecessors had done—indeed it is sometimes maintained that he actually formulated the principle of inertia. This is not true in the sense that Kepler's concept of inertia remained linked with Aristotle's mechanics and with Buridan's doctrine as modified by the German School of the XVth Century.

" The proper characteristic of material which forms the greatest part of the Earth is the inertia. Motion is repugnant to it, and more so as a great quantity of material is confined in a smaller volume. " [1]

Kepler adds—

" This material inertia of a terrestrial body, this density of the same body, constitute exactly the subject on which the *impetus* of rotational motion is impressed. It is impressed there exactly as in a top which turns because of violence. The heavier the material of the top is, the better it assimilates the motion impressed by the external force and the more lasting this motion is. " [2]

Kepler's dynamics follows directly from the ideas of Nicholas of Cues, whom he called " *divinus mihi Cusanus.* " He took up the example of the toy top which Nicholas of Cues had given, and applied the doctrine of *impetus impressus* to celestial bodies.

" Could not God have produced [such an *impetus impressus*] in the Earth, as from the exterior, at the beginning of time ? It is this impression which has produced all the past rotations of the Earth and which maintains them even now, though their number already exceeds two millions. Indeed, this impression keeps all its vigour because the rotation of the Earth is not hindered by impact or by any external roughness ; or by the ethereal fluid, which is devoid of density. No more is it hindered by any weight, or by any internal gravity. As for the inertia of the material, that is the very subject which receives the *impetus* and conserves it as long as the motion continues. " [3]

---

[1] *Opera omnia*, Vol. VI, p. 174.
[2] *Ibid.*, p. 175.
[3] *Ibid.*, p. 176.

Kepler believed that the material of the Earth was separated into circular fibres whose centres were aligned with the axis of rotation. " This arrangement of the Earth into circular fibres predisposes it to the motion that it receives. All the same, it appears that these fibres are the instruments of the motive cause rather than the motive cause itself. " [1]

The *impetus* communicated to the Earth by the Creator becomes a soul. " It is a soul of a strange kind. It confers on the Earth neither growth nor discursive reason *(sic)*— it merely moves it. But, better than a simple corporeal faculty, this motive soul assures the perfect regularity of diurnal motion. This motion, indeed, is no longer a violent motion, in any sense, for the Earth. What is there, indeed, more natural to a material than its form, to a body than its faculty or soul ? " [2]

7. KEPLER AND THE DOCTRINE OF ATTRACTION.

Following the example of Copernicus,[3] Kepler showed himself to be a Pythagorean in the matter of gravitation. He therefore denied the thesis that Albert of Saxony had made classical since the XIVth Century.

" The doctrine of gravitation is erroneous. A single mathematical point, whether it be the centre of the World or any other point, cannot effectively move heavy bodies, nor be the object towards which they tend. Therefore let them, the Physicists, prove that such a force can belong to a point, which is not a body and which is only conceived in an entirely relative way.

" It is impossible that the [substantial] force of a stone, which sets the body in motion of itself, should seek a mathematical point, the centre of the World without regard to the body in which that point may be situated. Therefore let them, the Physicists, establish that natural things have sympathy for that which does not exist. " [4]

And Kepler expounds " the true doctrine of gravity. "

" Gravity is a mutual affection between parent bodies *(Gravitas est affectio corporea, mutua inter cognata corpora)* which tends to unite them and join them together. The magnetic faculty is a property of the same kind. It is the Earth which attracts the stone, even though it might not tend towards the Earth. In the same way, if we place the centre of the Earth at the centre of the World, it is not towards the centre of the World that bodies are carried, but rather towards the

[1] *Opera omnia*, Vol. VI, p. 178.
[2] *Ibid.*, p. 179.
[3] See above, p. 85.
[4] *Astronomia nova, Introductio*, para. VIII.

centre of the body around which they belong, that is to say, the Earth. Also, the heavy bodies will be carried towards whatever place the Earth is carried to, because of the faculty which animates it.

" If the Earth was not round, heavy bodies would not move directly towards the centre from all directions. But according to whether they come from one place or another, they will be carried to different points.

" If, in a certain position in the World, two stones are placed near each other and outside the sphere of attraction of all other bodies which could attract them, these stones, like two magnets, will tend to unite in an intermediate position and the distances they will travel in order to unite will be in inverse ratio to their masses. " [1]

[1] *Ibid.*

PART TWO

THE FORMATION OF CLASSICAL MECHANICS

CHAPTER ONE

STEVIN'S STATICS
SOLOMON OF CAUX

1. THE STATICS OF STEVIN (1548-1620).

Stevin's first work on statics was published in Flemish at Leyden in 1586, under the title *De Beghinselen der Weegconst*. A more complete version appeared in 1605. Finally, in 1608, Stevin united these works under the title of *Hypomnemata Mathematica*. This work was translated into French as early as 1634.

Stevin's statics is developed geometrically in a manner similar to that used by Archimedes.

In it, the author systematically neglects " the motions of machines, formed of wood or iron, in which certain parts are lubricated with oil or lard, others are swollen by the humidity of the air or corroded with rust, in which these varied circumstances and also many others sometimes facilitate the motion, sometimes hinder it. "

Moreover, Stevin refuses to consider the excess of motive power which motion demands, " for the obstacles to motion have no certain and unique relation with the object moved. "

Still more rigorously, Stevin rejected the consideration of arcs of a circle described by the ends of the arms in the problem of the equilibrium of a lever. And he justified this by means of a syllogism. " Something which does not move does not describe a circle. Two weights in equilibrium do not move. Therefore two weights in equilibrium do not describe circles. "

We see that Stevin eschewed the point of view of virtual velocities in order to romp in the field of pure statics. At least he imposed this restriction on the form of his writing. He was not, however, to maintain it exclusively, as we shall show.

On the subject of the lever, Stevin added some further refinements to Archimedes' demonstrations which we shall pass over.

He solved the problem of the equilibrium of a heavy body on an inclined plane by a method that was completely original and which was based on the impossibility of perpetual motion.

This is his demonstration, taken from the French edition of 1634.

" *Given.* — Let *ABC* be a triangle whose plane is perpendicular to the horizon and whose base *AC* is parallel to the horizon. Let a weight *D* be placed on the side *AB*, which is to be twice *BC*, and a weight *E*, equal to *D*, be placed on the side *BC*.

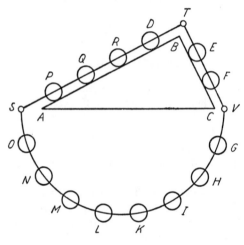

Fig. 36

" *The Requirement.* — It is necessary to show that the power (or capacity of exerting power) of the weight *E* is to that of the weight *D* as *AB* is to *BC*, that is, as 2 is to 1.

" *Construction.* — Round the triangle let there be arranged a system of fourteen spheres equal in weight, size, and equidistant from each other at the points *D, E, F . . . . P, Q, R*, and threaded on a cord passing through their centres in such a way that there are two spheres on *BC* and four on *BA. . . .* Let *S, T, V* be three fixed points on which the cord can run freely without being caught.

" *Demonstration.* — If the power of the weights *D, R, Q, P* were not equal to the power of the weights *E, F*, one of the sides would be heavier than the other. Suppose then that the four *D, R, Q, P* are heavier than the two *E, L*. Now the four *O, N, M, F* are equal to the four *G, H, I, K*. Now the side with eight spheres *D, R, Q, P, O, N, M, L* will be heavier than that of the six spheres E, F, G, H, I, K and, since the heavier part will dominate the lighter, the eight spheres will fall and the six will rise. Thus D will come to where O is at present and the others will do the same. That is, that E, F, G, H come to the positions where P, Q, R, D are now and I, K to where E, F are. However the effect of the spheres will have the same disposition as previously and for the same reason the eight spheres will weigh more and, when they fall, will make eight others come in their place. *Thus this motion will have no end, which is absurd.* The demonstration will be the same in the opposite case. Therefore the part D, R .... L of the ring will be in equilibrium with the part E, F, ... K. If there be taken away from both sides the heavinesses which are equal and similarily arranged, the four spheres O, N, M, L on the one hand and the four G, H, I, K on the other, the four D, R, Q, P which will be left will be in equilibrium with the two E, F. Hence E will have a power twice that of D. Therefore the power of E is to that of D as the side BA, let it be 2, is to the side BC, let it be 1. "

Stevin had read Cardan and referred to his *Opus novum*. He could not have been ignorant of *De Subtilitate* in which Cardan, after Leonardo da Vinci, held that perpetual motion was impossible.

Stevin was legitimately proud of his demonstration. He reproduced the associated diagram in the frontispiece of *Hypomnemata Mathematica* with the legend " *Wonder en is gheen wonder* " (The magic is not

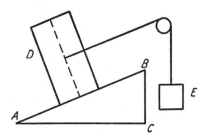

Fig. 37

magical), no doubt intending to indicate that he had logically explained a fact instead of invoking magic as the Greeks had done, before Archimedes, in connection with levers.

From this theorem on the inclined plane Stevin deduced the value of the weight E which could support the column D on an inclined plane by means of a thread parallel to the plane and starting from the centre of gravity of the column.   The result was given by

$$\frac{D}{E} = \frac{AB}{BC}.$$

He then studied a series of more complicated examples, like the following one in which the *direct elevation*, M, in equilibrium with the column D is compared with an *oblique elevation*, such as E, which is able to hold the column on the inclined plane.

In these circumstances M is equal to D, and the preceding result is applicable : $\dfrac{D}{E} = \dfrac{AB}{CB}.$

Now $\dfrac{BC}{AB} = \dfrac{DI}{DL}.$ Therefore $\dfrac{M}{E} = \dfrac{DL}{DI}.$

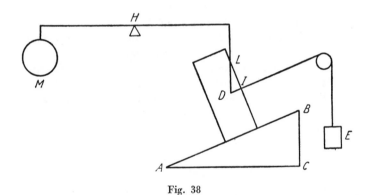

Fig. 38

Thus Stevin, after many false starts, arrived at an enunciation and even a verification of the rule of the parallelogram of forces for the particular instance in which the two forces are at right angles.

Let a column of centre of gravity C be hung from the points D and E by means of two strings CD, CE.   Complete the parallelogram CHIK whose diagonal CI is vertical.

" The *direct elevation* is to the *oblique elevation* as CI is to CH.   But the direct elevation CI is equal to the weight of the column.   Therefore the weight of the whole column to the weight which occurs at D is as CI is to CH.   In the same way, the weight which occurs at E will be found by producing the line IK from I parallel to DC to meet CE.

In words, the weight of the column which occurs at *E* will be as the straight elevation *CI* is to the oblique elevation *CK*. "

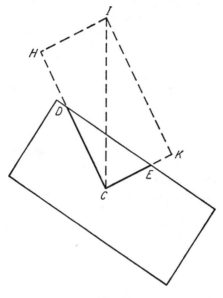

Fig. 39

From these considerations Stevin deduced the tension of the threads of a funicular polygon and thus became the originator of graphical statics.

2. STEVIN AND THE PRINCIPLE OF VIRTUAL WORK.

Returning to the point of view which had turned statics into a purely deductive science, and starting only from the assumption of the impossibility of perpetual motion, Stevin quite clearly stated the principle of virtual work. This occurs in volume IV of his *Hypomnemata* in connection with Stevin's work on the equilibrium of systems of pulleys.

" The distance travelled by the force acting is to the distance travelled by the resistance as the power of the resistance is to that of the force acting. (*Ut spatium agentis ad spatium patientis, sit potentia patientis ad potentiam agentis*). "

3. STEVIN'S HYDROSTATICS.

Stevin's contribution to hydrostatics is quite remarkable. He clearly stated *the principle of solidification* according to which a solid body of any shape and of the same density as a given fluid can remain in it at equilibrium whatever its position may be, and without the pressures in the rest of the fluid being modified. He used this principle to determine the pressure on each element of the base by solidifying all the liquid except that in a narrow channel abutting on this element, and verified that this pressure was independent of the shape of the receptacle and depended only on the weight of the column of liquid which filled the channel. This led him to state the *hydrostatic paradox*— that a fluid, by means of its pressure, can exert a total effort on the bottom of a vessel which can be considerably greater than the total weight of the fluid. He also determined the resultant of the pressures on an inclined plane boundary wall by dividing this surface into horizontal slices and passing to the limit by increasing the number of slices indefinitely.

Finally, he related Archimedes' principle to the impossibility of perpetual motion. Thus he was guided by the same idea as in the problem of the inclined plane.

4. SOLOMON OF CAUX (1576-1630) AND THE CONCEPT OF WORK.

Solomon of Caux was a practical Norman who was concerned with the construction of hydraulic screws. In 1615, after having read Cardan, he published at Frankfurt a work entitled *The reasons of moving forces together with various machines, as much useful as pleasant, to which are added some designs of grottoes and fountains.*

It is to this author that we owe the term *work* in the sense that it is used in mechanics now.

CHAPTER TWO

GALILEO AND TORRICELLI

1. GALILEO'S STATICS.

Galileo (1564-1642) started his scientific career in the way that was customary in his time, by annotating Aristotle's *De Caelo*. His manuscript remained unpublished until 1888 and is a typical scholastic document, even though it refers to certain moderns like Cardan and Scaliger.

Nevertheless, holding a chair of mathematics at the University of Pisa at the early age of twenty-five, Galileo was not long in causing a scandal by publicly experimenting on the fall of heavy bodies, by attacking his elders, and by offending a natural protector like John de Medici by wounding his pride in his inventions. However, his fierce intellectual independence, which, in its turn, was to earn him many rebuffs, developed very rapidly. He did not long remain a slave to the scholastic discipline.

We shall not concern ourselves here with Galileo's biography, which belongs to the general history of science, but shall make an analysis of his contribution to statics and dynamics.

First we shall follow the *Mechanics of Galileo* in the French version which Mersenne published at Paris in 1634. Chronologically this work lies between the manuscript of Galileo's lectures at Padua in 1594 and the treatise *Della Scienza meccanica* which was printed at Ravenna in 1649, seven years after its author's death.

At the beginning of the *Mechanics* Galileo emphasises " that machines are useful for manoeuvring great loads without dividing them, because often there is much time and little force . . . but he who would shorten the time and use only a little force will deceive himself. " That is, Galileo considers the product of the force and the velocity in conformity with Aristotle's thesis.

To Galileo the heaviness of a body was a " natural inclination of the body to take itself to the centre of the Earth. "

The *moment* was the inclination of the same body considered in the situation which it occupied on the arm of a lever or a balance. It is " made up of the absolute heaviness of the body and its separation from the centre of the balance, and corresponds to the Greek ῥοπή. "

The existence of a *centre of heaviness* (centre of gravity) of a body was just as much an experimental fact to Galileo as it was to the Schoolmen. " Each body principally weighs through the centre in which it masses and unites all its impetuosity and weight. "

In turn, Galileo studied the lever, the steelyard, the lathe, the flywheel, the crane, the winch, the pulley and the screw.

The discussion of the screw entailed a study of the inclined plane and, in this connection, Galileo was able to do better than his predecessors had done.

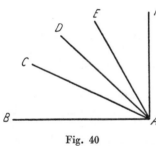

Fig. 40

He envisages a perfectly round and polished ball to be placed on a perfectly smooth surface. On the horizontal plane AB " the ball is indifferent to motion and rest, so that the wind or the smallest force can move it. " But a greater force is necessary in order to lift the ball on the inclined planes AC, AD, AE and finally, " it will only be possible to lift the ball on the perpendicular plane with a force equal to its whole weight. "

Galileo proceeds by considering two equal weight, A and C, in equilibrium on the lever ABC. In this way he is able to amend Pappus's demonstration, which he cites in this connection. If the arm BC falls to BF, the *moment* of the weight F becomes less than the moment of the equal weight A in the ratio $\dfrac{BK}{BF}$.

" When the weight is at F it is partly maintained by the *circular plane CI* and its slope—or the tendency which it has to the centre of the earth is diminished by the extent that BC exceeds BK. So that it is supported by this plane to the same extent as if it had been supported by the tangent GFH, more especially as the slope of the circumference at the point F only differs from the slope of the tangent GFH by the insensible angle of contact. "

By means of this remarkable artifice Galileo reduces the effect of the weight F on the inclined plane GFH to the effect of the same weight suspended as if from the arm of the lever BF. And he concludes that " the ratio of the total and absolute moment of the moving body, in the perpendicular to the horizon, to the moment which it has on the inclined plane HF is the same as the ratio of FH to FK. "

For $\dfrac{BF}{BK}$ is equal to $\dfrac{FH}{FK}$.

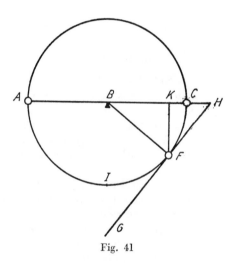

Fig. 41

Galileo also solves the problem of the inclined plane by appealing, this time, to the concept of virtual work.

" Imagine that in the triangle ABC the line AB represents the horizontal plane, the line AC the inclined plane whose height will be measured by CB. On the plane AC is placed a body E attached to the string EDF. At F the string carries a weight, or a force, which is related to the weight E in the ratio of the line BC to the line CA. If the weight F starts to fall, drawing the body E along the inclined plane, the body E will travel a path in the direction of AC which is equal to that which the heavy body F describes in its fall.   But the

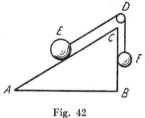

Fig. 42

following observations are necessary. It is true that the body E will have travelled all the line AC in the same time that the weight F will have taken to fall an equal distance. But during this time, the body E will not have been separated from the common centre of heavy things by a distance greater than the vertical BC, while the weight F, falling vertically, will have fallen a distance equal to the whole line AC. Now the bodies only resist an oblique motion to the extent to which they are taken away from the centre of the earth. . . . We can legitimately say that the path of the force F keeps the same ratio to the path of the force E as the ratio of the length AC to the length CB, and is therefore equal to the ratio of the weight E to the weight F. "

2. GALILEO AND THE FALL OF BODIES.

Thanks to a letter which Galileo wrote to Paolo Sarpi, dated October 16th, 1604, [1] we know that as early as this Galileo believed in the now classical law of distances $s = \text{constant} \times t^2$.

" The distances gone through in natural motion are in square ratio to the times of fall. Consequently the distances travelled in equal times are related to each other like the consecutive odd numbers starting from unity. "

Nevertheless, at first Galileo associated this law of distances with

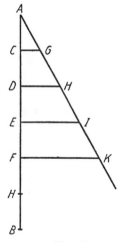

Fig. 43

an incorrect law of velocities, namely $v = k \cdot s$. We recall that as early as the XIVth Century Albert of Saxony had hesitated between this law and the correct one, $v = k \cdot t$.

To Galileo, the law $v = k \cdot s$ was explained in the following way.

" A body which moves naturally increases in velocity to the extent that it is separated from the source of its motion. "

The arguments by which Galileo sought to verify these two laws simultaneously are rather odd. Certainly they are incorrect, but they show us the development of his thought and deserve to be quoted as showing what detours he made before he became emancipated from them.

" If the heavy body starts from the point A and falls along the line AB, I suppose that the degree of velocity at the point D exceeds the degree of velocity at the point C in the ratio of DA to CA ; that in the same way, the degree of velocity at E is to the degree of velocity at D as EA is to DA. Thus, at every point of AB the body will have a velocity proportional to the

[1] *The Works of Galileo*, Italian National Edition, Vol. X, p. 115.

distance from this same point to the origin A. This principle appears to me to be very natural. It corresponds to all the observations that we make of machines whose purpose is hitting. Given this principle, I will demonstrate the rest.

" Let the line AK make any angle with the line AF and, through the points C, D, E, F let the parallels CG, DH, EI, FK be drawn. Since the lines FK, EI, DH, CG have the same relation to each other as the lines FA, EA, DA, CA, the velocities at the points F, E, D, C are therefore related to each other like the lines FK, EI, DH, CG. Therefore the degrees of velocity at all the points of the line AF are constantly increasing according to the increasing of the parallels drawn from these same points.

" Moreover, since the velocity with the moving body goes from A to D is made up of all the degrees of velocity acquired at all the points of the line AD, and since the velocity with which it has travelled the line AC is made up of all the degrees of velocity acquired at all the points of the line AC, the ratio of the velocity with which it has travelled the line AD to the velocity with which it has travelled the line AC is that between all the parallels drawn from all the points of the line AD to the line AH. This [ratio] is that of the triangle ADH to the triangle ACG, that is, the ratio of the square of AD to the square of AC. Therefore the relation of the velocity with which the moving body has travelled the line AD to the velocity with which it has run through the line AC is the square of the ratio of DA to CA.

" But the ratio of a velocity to a velocity is the inverse of the ratio of the corresponding times, for the time decreases when the velocity increases. The ratio of the duration of motion along AD to the duration of motion along AC is therefore the square root of the ratio of the distance AD to the distance AC. Therefore the distances from the starting point are as the squares of the times. However, the distances travelled in equal times are to each other as the consecutive odd numbers starting from unity. This is in accord with what I have always said and with observations made. All the verities are thus in accord. "[1]

Briefly, from the inexact hypothesis that $v = k \cdot s$, Galileo obtains the relation $\dfrac{v(D)}{v(C)} = \dfrac{DA}{CA}$.

Then by a consideration of a series of parallels erected from each point of AH he deduces, incorrectly, the relation $\dfrac{v(AD)}{v(AC)} = \left(\dfrac{DA}{CA}\right)^2$, where $v(AD)$ and $v(AC)$ are the mean velocities on AD and AC.

[1] *Complete Works of Galileo*, Italian Edition, Florence 1908, Vol. VIII, p. 373.

From the last relation, and again incorrectly, he concludes that the ratio

$$\frac{t(AD)}{t(AC)} = \sqrt{\frac{DA}{CA}}$$

and thus arrives at the correct law

$$s = \text{constant} \times t^2.$$

We shall now follow the treatise *Discorsi e dimostrazioni matematiche intorno a due nuove scienze attenanti alla Meccanica ed i movimenti locali.* The first edition of this work appeared in 1638 and was later supplemented by the author, although these additions only appear in an edition that was printed at Bologna in 1655.

In these *Discorsi* three characters, Salviati (Galileo), Sagredo (a Venetian senator and friend of Galileo) and Simplicio (who represents Scholasticism) discuss the work. This dialogue form has the obvious inconvenience of making the book difficult to read, but has the inestimable advantage of allowing the author to show the development of his thought.

The Text declares that the fall of bodies is *uniformly accelerated*, or that the " increase of the velocity is like that of the time. " Starting from rest, the moving body receives equal degrees of velocity. This the Text assumes *a priori*.

" Why indeed not believe that the increases in velocity follow the most simple and banal law ? "

Refusing to lose himself in the discussions which had occupied the Schoolmen, Salviati brushes aside all argument on the cause of the fall of bodies.

He recalls that he had, for some time, believed that the velocity could increase as the distance did. As we have just seen, this was his opinion in 1604. He now rejects this belief.

" If the velocities are proportional to the distances travelled, the distances will be travelled in equal times. Therefore, if the velocities with which the body travelled the 4 cubits were double those with which it travelled the first two cubits (as the distances are doubled) the durations of travel will be equal. But the same moving body can only travel the 4 or the 2 cubits in the same time if this motion is instantaneous. Now it is apparent that the motion of a heavy body lasts a certain time, and that it travels the first two cubits in less time than the four. Therefore it is not true that its velocity increases as the distance. "

We note here, with Jouguet,[1] that this argument of Galileo is not quite correct. The law $v = k \cdot s$ immediately leads to $s = s_o exp(kt)$. In order that there should be motion it is necessary that, contrary to the hypothesis, s_o should be different from zero when $t = 0$. Otherwise it is necessary to assume that in the first instant the body travels the distance s_o instantaneously.

Given this, Salviati makes the following postulate. " *I assume that the degrees of velocity acquired by the same moving body on differently inclined planes are equal whenever the heights of the planes are equal.* "

The moving body is assumed to be perfectly smooth and the planes to be perfectly polished.

In order to substantiate this principle Salviati starts from the following postulate, in which Galileo's physical intuition is very apparent.

" Imagine that this sheet of paper is a vertical wall, that a nail is fixed in it and that a ball of lead weighing an ounce or two is hung

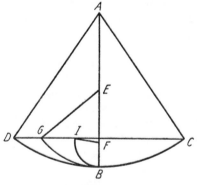

Fig. 44

from the nail by a thread AB. The thread is to be two or three cubits long, perpendicular to the horizon and at a distance of about two fingers from the wall. Draw a horizontal CD on the wall to cut the thread AB squarely. Draw aside the thread AB and the ball into the position AC. Then release the ball. We will see this descend, describing the arc CB, and pass the extremity B in such a way that it will go up again, along BD, almost to the line CD which has been drawn. Each time there will be a small deficiency, and this circumstance is precisely due to the resistance of the air and of the thread. From this we can conclude, in all truth, that the *impeto* at the point B which is acquired by the ball in its descent of the arc *CB* is such that it suffices to make it remount the identical arc *BD* to the same height. When this observation has been repeated again and again, fix in the wall a nail which projects about five or six fingers, exactly opposite the vertical *AB*—for example, at *E* or at *F*. The ball will describe the arc *CB*, the thread turning as before. When the ball comes to *B*, the thread will tangle in the nail *E* and the ball will be obliged to travel the circumference *BG* which has *E* as centre. Then we see that this can produce, at the extremity *B*, the very *impeto* that can make the moving body rise again along the arc *BD* until it almost reaches the horizontal *CD*. Now, gentlemen, you will see with pleasure that the ball attains the horizontal at the point *G*. The same thing would happen if the nail were fixed lower, at *F* for example. The ball will describe the arc *BI* and will always finish its ascent on the line *CD*. And, if the nail were too low for the ball to attain the height *CD* (this would happen if the nail were nearer *B* than *CD*) the thread would wrap itself round the nail. This observation prevents one from doubting the truth of the principle that has been supposed. Since the two arcs *CB*, *DB* are equal and similarly placed, the *momento* acquired at *B* along *CB* suffices to make the same body rise again along *BD*. Therefore the *momento* acquired along *DB* is equal to that which would make the same moving body rise again, along the same arc, from *B* to *D*. So that in general, the *momento* acquired along any arc is equal to that which can make the same body rebound along the same arc. But all the *momenti* which make the body rebound along the arcs *BD*, *BG*, *BI* are equal, since they are produced from the *momento* acquired in the descent *CB*, as observation shows. Therefore all the *momenti* acquired in descending the arcs *DB*, *GB*, *IB* are equal. "

Salviati goes on to consider motions along variously inclined planes.

" We cannot show with the same clarity that the same thing will happen when a perfect ball falls along inclined planes that are drawn along the chords of these same arcs. On the contrary, since the planes form an angle at the point *B*, it is plausible that the ball, having descended along the chord *CB* and meeting an obstacle at the bottom of the planes which mount along the chords *BD*, *BG*, *BI*, will lose a part of its *impeto* in rebounding, and will not be able to ascend again to the height of the line *CD*. But since the obstacle raised in this way prevents the observation, it seems to me that the mind will go on believing that the *impeto* (which contains, indeed, the force of the whole fall) will be able to make the body go up again to the same height. Therefore take this assertion as a postulate for the moment—its absolute truth will be established later when we shall see that the conclusions depending on this hypothesis are, in detail, in conformity with observation. "

Galileo then established the now classical laws of falling bodies. In particular, we shall describe how, going back on his opinion of 1604, he established the law of velocities.

" Since, in a accelerated motion, the velocity is continuously augmented, the degrees of the velocity cannot be divided into any determinate number. For since the velocity changes from moment to moment and increases continuously, they are of infinite number. However, we can represent our intention better by constructing a triangle ABC, taking as many equal parts AD, DE, EF, FG as we please on the side AC, and in drawing straight lines parallel to the base BC through the points D, E, F, G. Then, if the parts marked on the line AC are equal times, we assume that the parallels drawn through the points D, E, F represent the degrees of the accelerated velocity, degrees which increase equally in equal times. . . .

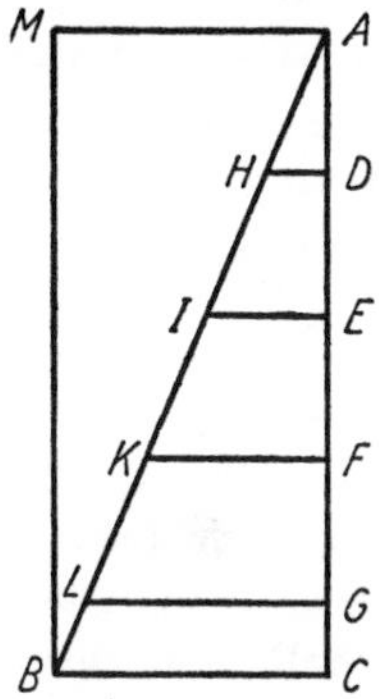

Fig. 45

" But because the acceleration is continuous from moment to moment and not of a discontinuous kind of this or that duration . . . before the moving body attains the degree of velocity DH that is acquired in the time AD, it has passed through an infinity of smaller and smaller degrees gained in the infinite number of instants that the time AD contains and which correspond to the infinity of points that lie on the line DA. However, in order to represent the infinity of degrees of velocity that precede the degree DH, it is necessary to imagine an infinity of lines, always smaller and smaller, which should be drawn from the various of the infinite number of points of the line DA. *In ultimo*, this infinity of lines will represent the surface of the triangle AHD.

" Complete the whole parallelogram $AMBC$ and produce as far as the side BM, not only the parallels which have been drawn in the triangle, but also the infinite number of parallels that was imagined to start from all the points of the side AD. The line BC, which is the longest parallel drawn in the triangle, represents the highest degree of the velocity acquired by the moving body in its accelerated motion. The total surface of the triangle is *the mass and the total* of all the velocity with which the body has travelled such a distance in the time AC. In the same way the parallelogram will be *the mass and the union* of degrees of velocity each of which is equal to the maximum degree BC. This latter mass of velocities will be twice the mass of the increasing velocities of the triangle, because the parallelogram is twice the triangle. *Conse-*

quently, if a moving body takes degrees of an accelerated velocity, in falling, which conform to the triangle ABC, and if it passes through such a distance in such a time, it will, when moving uniformly, travel twice the distance that it has travelled in the accelerated motion."

By an analogous argument whose detailed reproduction would serve no useful purpose, Galileo arrived at the following theorem.

" If a body starts from rest and moves with uniformly accelerated motion, the time that it takes to travel a certain distance is equal to the time that the same body would take to travel the same distance with a uniform motion whose degree of velocity was half of the greatest and final degree of the velocity of the uniformly accelerated motion. "

We know that the Schoolmen, thanks to the efforts of Oresme, Heytesbury and Soto, had already obtained this fundamental result. But Galileo did not confine himself to the *a priori* assertion that the fall of bodies was uniformly accelerated. He submitted the fall of a body on an inclined plane to an experiment which was, for the time, performed in a scrupulous manner and repeated a hundred times. We shall quote this essential passage of the *Discorsi*, noting its very marked difference from the tendencies of purely rationalist Scholasticism.

" In the thickness of a ruler, that is, of a strip of wood about twelve cubits long, half a cubit wide and three fingers thick, a channel, a little wider than one finger, was hollowed out. It was made quite straight and, in order that it should be polished and quite smooth, the inside was covered with a sheet of parchment as glazed as possible. A short ball of bronze that was very hard, quite round and well-polished, was allowed to move down the channel. The ruler, made as we have described, had one of its ends lifted to some height—of about one or two cubits—above the horizontal plane. As I have said, the ball was allowed to fall in the channel and the duration of its whole journey was observed in the way that I have explained. The same trial was repeated many times in order to be quite sure of the length of this time. In this repetition, no difference greater than a tenth of a pulse was ever found. When this observation had been repeated and established with precision, we made the ball fall through only a quarter of the length of the channel, and found that the measured duration of fall was always equal to half of the other. . . .

" When this observation had been repeated a hundred times, the distances travelled were always found to be in the ratio of the squares of the times, and this was true whatever the inclination of the plane, or that of the channel in which the ball fell, was made to be. We also observed that the durations of fall on differently inclined planes were in the proportion assigned to them [by our demonstrations].

" As for the measurement of the time, a large bucket filled with water was suspended in the air. A small hole in its base allowed a thin stream of water to escape, and this was caught in a small receptacle throughout the duration of the ball's descent of the channel, or of portions of it. The quantities of water caught in this way were weighed on a very accurate balance. The differences and relations of these weights gave the differences and relations of the times with such accuracy that, as I have said, these operations never gave a noticeable difference when repeated many times. "

Galileo then introduced the notion of *impeto*, which he also called *talento* and *momento del discendere*.

For a given body, this tendency to motion is greatest along the vertical *BA*. It is less on the planes *AD*, *AE*, *AF*. Finally, the *impeto* is completely reduced to nothing on the horizontal *CA*, where the body (as we have seen in reading the *Mechanics*) " is indifferent to motion or to rest, and does not of itself show any tendency to move in any direction or any resistance to being set in motion. " Salviati gives the following explanation of this fact.

" In the same way that it is impossible that a heavy body, or an ensemble of heavy bodies should, of its own accord, move upwards and thereby go further away from the common centre to which heavy things tend, so it is impossible that it should spontaneously move if its centre of gravity does not approach the common centre in its motion. Therefore the *impeto* of the moving body will be nothing on some chosen horizontal, or on a surface which is equidistant from the aforesaid centre and is without inclination. "

This is the Galilean form of the *principle of inertia*. Galileo arrived at it by a kind of limiting process, starting from the principle of virtual work.

Galileo then returned to the demonstration which he had given of the law of heaviness on an inclined plane, and in which he had appealed to the principle of virtual work. He completed it, however, in the following way.

" Manifestly, the resistance, or the smallest force which suffices to stop or prevent a heavy body in its descent, is as great as the *impeto* of that body. In order to measure this force, I shall make use of the gravity of another body. Imagine that a body *G* rests on the plane *FA* and that it is attached to a thread which passes over *F* and supports a weight *H*. . . . In the triangle *AFC*, the displacement of the body *G*, for example upwards from *A* to *F*, is made up of the transverse and horizontal motion *AC* and the vertical motion *CF*. Now the resistance to motion due to the horizontal displacement is zero. . . . Consequently the resistance is solely due to the fact that the body must climb the vertical *CF*. Therefore the body *G*, moving from *A* to *F*, only resists because of the vertical elevation *CF*. But the other body, *H*, necessarily descends the whole length *FA* in a vertical direction. . . . We can therefore say that when equilibrium is established the moments of the bodies, their velocities or tendencies to motion, that is, the distances which they would travel in the same time, will be in inverse ratio to their

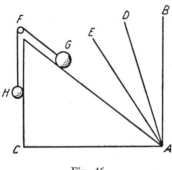

Fig. 46

gravities in accordance with the law which is true in every instance of motion in mechanics. It follows that to prevent the fall of *G*, it will suffice that *H* should be so much lighter with respect to *G* as the distance *CF* is less than *FA*. . . . *And since we have agreed that the impeto of a moving body, the energy, the moment or the tendency to motion has the same size as the force, or least resistance, which suffices to keep it still, we conclude that the body* H *is sufficient to prevent the motion of the body* G. . . . "

We note that Galileo's fundamental idea consists in measuring the *impeto*, or the tendency to motion, by means of the *static force* which can be opposed to it. This was an essentially original procedure which had escaped the notice of all the Schoolmen. As Jouguet[1] has legitimately remarked, the same word *impeto*, in Galileo's work, sometimes meant the *velocity acquired* by a body in a given time, and sometimes the *distances travelled* on differently inclined planes in a certain time, starting from rest.

By means of the preceding considerations Galileo verified that the postulate according to which the velocities of a body which starts from

[1] *L. M.*, Vol. I, p. 106.

rest and falls along the line of greatest slope on differently inclined planes of equal height are the same when it arrives at a given horizontal. He also showed that " if the same moving body falls, starting from rest, on an inclined plane and along the vertical with equal height, the durations of fall have the same relation as the lengths of the inclined plane and of the vertical. "

This demonstration was necessary in order to give full weight to his experimental verification of the law of falling bodies.

3. GALILEO AND THE MOTION OF PROJECTILES.

We have seen that the Schoolmen and those interested in mechanics in the XVIth Century had only been able to treat the motion of projectiles very imperfectly. Galileo solved this problem by means of a very remarkable analysis in which, together with the principle of inertia, there appears the principle of the composition of motions or of the independence of the effects of forces.

We shall quote from the text of the *Discorsi*.

" *The Text*. — I imagine a moving body thrown on a horizontal plane without any obstacle. *It is said that its motion on the plane will remain uniform indefinitely if the plane extends to infinity*. But if the plane is limited, and if it is set up in air, when the body, which we suppose to be under the influence of gravity, passes the end of the plane *it will add to the first uniform and indestructible motion, the downward propensity which it has because of its gravity*. From this will arise a *compound motion*, composed of the horizontal motion and the naturally accelerated motion of descent. I call this kind of motion, projection.

" *Animated by the motion composed of a uniform horizontal motion and a naturally accelerated falling motion, the projectile describes a parabola.*

" Let there be a horizontal or a horizontal plane, *AB*, which is placed in air and along which a body moves uniformly from *A* to *B*. At *B*, where its support is missing, the body, because of its weight, is forced by its gravity into a natural downward motion along the vertical *BN*. Produce *AB* into the line *BE*, which we shall use to measure the passage of time. Mark off equal lines *BC*, *CD*, *DE* on *BE*, and draw parallels to *BN* through the points *C, D, E*. On the first of these parallels take an arbitrary length *CI* ; on the next one, a length *DF* which is four times as great ; on the third, a length *EH* nine times greater ; and so on, the successive lengths increasing as the squares of *CB, DB, EB*. ... Imagine that the vertical descent along *CI* is added to the displacement of the body as it is carried from *B* to *C* in uniform motion.

At the time *BC* the body will be at *I*.   At the time *BD*, which is twice *BC*, its vertical distance of fall will be equal to 4*CI*.   For it has been proved that the distances are as the squares of the times in naturally accelerated motions.   In the same way, the distance *EH* that is travelled in the time *BE* will be nine times *CI*, so that the distances *EH*, *DF*, *CI* are related to each other as the squares of the lines *EB*, *DB*, *CB*. . . .   The points *I*, *F*, *H* therefore lie on a parabola. "

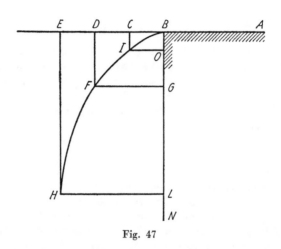

Fig. 47

The discussion between the three characters in the dialogue is of considerable interest.   Sagredo remarks that the argument supposes that the two motions combined in this way " neither alter each other, nor confuse each other, nor mutually hinder each other in mixing up. " He objects that, since the axis of the parabola is vertical and goes through the centre of the earth, the particle will be separated from this centre. . . .   Simplicio reproaches the text for, in the first place, neglecting the convergence of the verticals and, in the second, neglecting the resistance of the medium.

Salviati replies that, to a first approximation, these objections may be dismissed.   He has experimented on a ball of wood and one of lead which were arranged to fall from a height of 200 cubits.   The wooden ball, which was more sensitive to the resistance of the air, was not noticeably retarded.   Salviati recalls that the projectiles from firearms have such velocities that their trajectories can be modified by the resistance of the air.

4. GALILEO AND HYDROSTATICS.

Galileo took up the study of hydrostatics in a manuscript called *Discorso intorno alle cose che stanno in su l'acqua o che in quella si muovono.* This was published at Florence in 1612. Essentially, his hydrostatics was based on the principle of virtual velocities, which was directly inspired by Aristotle's mechanics. In this work, Galileo called the product of the force and the velocity, *momento.*

" I borrow two principles from the Science of mechanics. The first is this— two absolutely equal weights that are moved with equal velocities are of the same power, or the same *momento*, in all their doings.

" To students of mechanics, *momento* means that property, that action, that efficient power by which the motive agency moves and the body resists. This property does not only depend on the simple gravity, but also on the velocity of motion, the different inclinations and the different distances travelled. Indeed, a heavy body produces a greater *impeto* when it descends on a very steep surface than when it descends on a surface which is less steep. Whatever may be the ultimate cause of this property, it always keeps the name *momento.*

" The second principle is that the power of the gravitation increases with the velocity of the thing that is moved, so that absolutely equal weights that are animated with unequal velocities have unequal powers, strengths, unequal *momenti.* The more rapid is the more powerful, and this in the ratio of its own velocity to the velocity of the other weight. . . .

" Such a compensation between the gravity and the velocity is found in all machines. Aristotle has taken it as a principle in his *Problems of Mechanics.* Hence the assertion, that two weights of unequal size are in equilibrium with each other, and have equal *momenti*, whenever their gravities are in inverse ratio to the velocities of their motion, may be taken as wellestablished. "

In discussing the siphon, Galileo remarked that a small mass of water contained in a narrow vessel could maintain in equilibrium a large mass of water contained in a wide vessel, because a small lowering of the second entailed a great increase in the height of the first. In this respect Galileo preceded Pascal. If Duhem is to be believed, Galileo was guided by a tradition that went back to Leonardo da Vinci.[1]

The *Discorsi* were attacked by L. della Colombe and V. di Grazia, and defended by Benedetto Castelli (1577-1644), a faithful disciple of Galileo. The same Castelli was the author of a treatise on the measurement of running water (*Della misura dell'acque correnti*, 1628) which repeated Leonardo da Vinci's law of flow, $Sv =$ constant.

---

[1] *Études sur Léonard de Vinci*, Vol. II, p. 214.

Further, Galileo related the properties of the equilibrium of floating bodies to the principle of virtual velocities.

Like his contemporaries, Galileo also believed in the *horror vacui* (*resistenza del vacuo*). However, it is reported that he was very surprised to learn that a newly constructed pump, whose aspiration tube was very long, could not lift water higher than eighteen Italian ells. Therefore he believed that this height implied a kind of ceiling to the horror of the vacuum. In addition, Galileo attempted to determine the weight of air by weighing a balloon that was filled with air, then heated in order to partially expel the air, and weighed again. As Mach has remarked, it is very true that the heaviness of air and the *horror vacui* were quite separate concepts before Pascal's time.[1]

5. Galileo and the Copernican system.

We shall briefly summarise Galileo's astronomical work. By means of a lunette which he had had constructed at Venice in 1609, he discovered the satellites of Jupiter on Jan 7th, 1610 and observed that they accompanied the planet in its annual motion. This suggested the same possibility for the Moon in relation to the Earth. On the other hand, he noticed the phases of Venus and the sunspots, and thus obtained proof of the rotation of these two stars which was of first importance for supporting the hypothesis of the Earth's rotation. Finally he demonstrated a libration in the Moon's longitude. He was forced to retract his views on the Earth's rotation when he was first accused by the Inquisition in 1615. Nevertheless Galileo hastened to publish, at Florence in 1632, *Four Dialogues on the two principal systems of the World, those of Copernicus and Ptolemy*. (This in spite of the fact that he usually hesitated about printing his work because of his shortage of money; even though he was content to distribute a few copies of the *Discorsi* among his friends in 1636.)

The three speakers that will later appear in the *Discorsi*, Simplicio, Sagredo, and Salviati also appear in these dialogues. Galileo applied a searching dialectic to the scholastic arguments, here expressed by Simplicio. For example, in Dialogue II, Simplicio enumerates the scholastic axioms, such as the unity of the cause and the unity of the effect, the necessity of an extrinsic source for all motion, natural or otherwise. These axioms conflict with the triple motion of Earth which Copernicus has suggested. This triple motion comprises the diurnal motion, the annual motion and the displacement of the Earth's axis parallel to itself.

[1] Mach, *M.*, p. 106.

(Rather oddly, Copernicus had believed this to be one of the modes of the Earth's motion.) Salviati replies to this by assembling the experimental evidence. And if he dares to contradict Aristotle, it is because the telescope has made the eyes of the astronomer thirty times more powerful than those of the philosopher. " *Jam autem nos, beneficio Telescopii, tricies aut quadragies propius quam Aristoteles admovemur Caelo, sic ut in eo plurima possumus observare quae non potuit Aristoteles et, inter alia, maculas istas in Sole, quae prorsus ei fuerunt invisibiles. Ergo de Caelo, deque Sole, nos Aristotele certius tractare possumus.* " [1]

In his third dialogue Galileo concludes that though the copernican system may be difficult to visualise, it is simple in its effects. " *Systema Copernicanum intellectu difficile et effectu facile est.* "

It is reported that this work brought Galileo a denunciation from the Holy Office, which obliged him to renounce his copernican beliefs and to remain in compulsory residence at Arcetri, near Florence. Here he died, surrounded by a number of disciples. Among these was Torricelli, who had only belonged to the circle for a few months.

6. TORRICELLI'S PRINCIPLE.

We know that Galileo had already related the problem of the inclined plane to the principle of virtual work and that he had maintained, in his *Discorsi*, that an ensemble of heavy bodies could only start to move spontaneously if its centre of gravity came nearer to the common centre of heavy things.

Torricelli made this remark precise, and raised it to the status of a principle, in his treatise *De Motu gravium naturaliter descendentium et projectorum* (Florence, 1644).

" We shall lay down the principle that two bodies connected together cannot move spontaneously unless their common centre of gravity descends.

" Indeed, when two bodies are connected together in such a way that the motion of one determines that of the other, this connection being produced by means of a balance, a pulley or any other mechanism, the two bodies will behave as a single one formed of two parts. But such a body will never set itself in motion unless its centre of gravity falls. But if it is made in such a way that its centre of gravity cannot fall, the body will certainly remain at rest in the position that it occupies. From another point of view, it would move in vain because it would take a horizontal motion which did not tend downwards in any way. "

---

[1] We have quoted a Latin edition which appeared at Lyons in 1641.

Torricelli applied this principle to two bodies on differently inclined planes and attached to each other by a weightless thread. Similarly, he applied it to the balance. All these examples are instances of indifferent equilibrium. If ζ is the height of the centre of gravity, reckoned algebraically on an ascending vertical, Torricelli's principle may be written

$$\delta\zeta \leqq 0$$

for all virtual displacements compatible with the constraints. But the examples which he gave were all of the type

$$\delta\zeta = 0.$$

Torricelli's true merit is not so much that of having won this principle from Galileo's mechanics, but that of having specified that the verticals should be treated as parallel. At the same time he renounced the scholastic conception of a common centre of heavy bodies at a finite distance, where the verticals converged.

He writes, " This is an objection that is very common among the most thoughtful authors— Archimedes has made a false hypothesis in regarding the threads that support the two weights hung from a balance as being parallel to each other—in reality, the directions of these two threads meet at the centre of the Earth. . . .

" The foundation of mechanics which Archimedes adopted, namely the parallelism of the threads of a balance, may be deemed false when the masses hung from the balance are real physical masses, tending towards the centre of the Earth. It is not false when these masses, whether they be abstract or concrete, do not tend towards the centre of the Earth, or to any other point near the balance, but towards some point which is *infinitely distant*.

" We shall continue to call this point, towards which masses hung from the balance tend, the centre of the Earth. "

Beneath these verbal precautions, and in spite of the fact that he did not refer to the orders of magnitude, Torricelli's intention of treating the verticals as parallels is clear. In Torricelli's principle the word " descend " is intended to indicate a tendency towards a centre which is taken to infinity.

7. TORRICELLI AND THE MOTION OF PROJECTILES.

Galileo fully discussed the parabolic motion of a projectile which was thrown horizontally. Only in passing did he remark that, if a projectile was thrown obliquely from the point B with a velocity equal and opposite to that with which it arrived at B after having been thrown horizontally from A, it would describe the same parabola in the opposite direction. Galileo made use of this appeal to an inverse return without proof. Moreover, he announced that the greatest range for a given velocity, if the projectile was thrown from B, was obtained when the trajectory at B made half a right-angle with the horizontal.

In this matter too, Torricelli systematised Galileo's work. Thus, in Book II of his *De motu gravium*, he considered a body that was projected obliquely. He compounded the uniform velocity in the direction of the velocity of projection with the accelerated motion.

Fig. 48

Gassendi had studied the same problem as early as 1640, in a treatise *Tres Epistolae de motu impresso a motore translato*. He considered a body projected upwards from the deck of a ship in uniform motion, and showed that the trajectory was a parabola.

8. Torricelli's experiment.

No doubt inspired by Galileo's researches on the *resistenza del vacuo*, Torricelli was lead to make experiments on a column of mercury rather than a column of water. The classical experiment with which his name is still associated was, however, accomplished by Viviani in 1643.

9. Torricelli's law flow through an orifice.

Torricelli seems to have been the inventor of hydrodynamics. Thus he observed the flow of a liquid through a narrow orifice near the bottom of a vessel. Dividing the total duration of flow into equal parts, he established that the quantities of liquid caught by some suitable receptacle increased regularly, from the last interval of time to the first, and that they were proportional to the odd numbers taken consecutively. This analogy with the law of falling bodies induced him to investigate the height to which the water that flowed out of the orifice could rise, if suitably directed upwards. He established that this height was always less than that of the liquid in the vessel. Moreover, he supposed that the stream would attain this height if the resistances did not exist.   Torricelli then formulated the law that the velocity of the liquid flowing out of the orifice was proportional to the square root of the height of the liquid.   This statement was obtained by analogy with the motion of heavy bodies and was given without proof.   It attracted the attention of Newton and Varignon, and thus lies at the bottom of the first investigations in hydrodynamics.

CHAPTER THREE

MERSENNE (1588-1648)
AS AN INTERNATIONAL GO-BETWEEN
IN MECHANICS
ROBERVAL (1602-1675)

1. THE ARRIVAL OF FOREIGN THEORIES IN FRANCE. THE PART PLAYED BY MERSENNE.

In 1634 there appeared simultaneously, in French, the translation of Stevin's mathematical work, P. Hérigone's *Cours Mathématique* and Mersenne's translation of Galileo's *Mechanics*.

Hérigone's *Cours Mathématique* was inspired by Stevin's work, and took over the proof concerning equilibrium on an inclined plane. However, a column of liquid was unhappily substituted for Stevin's necklace of spheres. Hérigone also borrowed many things from Guido Ubaldo and the statics of Jordanus and his school—in particular, the solution of the problem of the inclined plane. In this he was helped by the Italian Renaissance and the tradition—which we have discussed in connection with Tartaglia—that honoured Jordanus' contribution to statics.

Though an omniverous reader, Father Mersenne (1588-1648) did not thereby arrive at a synthesis of his material. However, he established contact between the great students of mechanics, to whom he was continually posing questions, providing references and transmitting replies. His correspondence is like an international review of mechanics.

Mersenne's *Synopsis mathematica* (1626) reviewed the work of Archimedes, Luca Valerio, Stevin, Guido Ubaldo and many others. As early as 1634 he translated Galileo's *Mechanics*. He told of Galileo's first work on the fall of bodies in *Harmonicorum libri* (1636), and added to this a traetise on mechanics by Roberval. He also made the work of Benedetto and Bernardino Baldi known to french students of the subject. In 1644 he published another compilation under the title *Tractatus mechanicus*.

Much original work has only been preserved for us in the form of letters to Mersenne.   In a time when authors were not liberal of references, and disposed to pass of their writings as entirely original, Mersenne's self-imposed task of liason and dissemination was quite essential.

2. ROBERVAL AND COMPOUND MOTION.

We cannot describe the work in kinematics in the XVIIth Century, for an analysis of this would more properly belong to a history of geometry.   Nevertheless, Roberval's kinematic geometry deserves a special mention because he was able to solve the problem of drawing tangents to different curves—a preoccupation among geometers of the time—by means of the composition of velocities.

The *Treatise on compound motion* was only published by the *Académie des Sciences* in 1693.   It was edited by a gentleman of Bordeaux on the basis of Roberval's lectures, and the latter confined his own contribution to the addition of marginal notes.   The essential principle used in this treatise is the following one.

" Using the particular properties of the curved line that will be given to you, examine the different motions that a point which describes the line can have in the neighbourhood of the point at which you wish to find the tangent.   Of all these component motions, take the line of the direction of the compound motion.   You will have the tangent to the curved line. "

For example, in the curve described by a point M fixed on a circle which rolls without sliding on a straight line, Roberval compounds an elementary translation of the base with an elementary rotation of the circle.   This leads to a direction of compound motion which is perpendicular to the straight line joining M to the point of contact of the circle with the base.   Roberval treates fourteen examples in the same way—for instance the cycloid, the conchoid, Archimedes' spiral and the conics, and he succeeds in drawing their tangents correctly.   However, it turns out that he begs the question by giving the components of the velocity without precise justification.   Descartes was to replace this by a method that became that of the instantaneous centre of rotation.

3. ROBERVAL'S TREATISE ON MECHANICS.

In 1650 Roberval wrote to Hevelius, " We have constructed a new mechanics on the foundation already laid.   Except for a small number, the ancient stones with which it has previously been constructed have been completely rejected.   It consists of eight stages, corresponding to a similar number of books. "

The *Bibliothèque Nationale* (Paris) has a manuscript (No. 7226) which is undoubtedly an outline of this work. Even though he denies this, the mechanics which the author— " in the chair of Ramus " at the College of France —contemplates is most often inspired by Aristotle and the Italian Renaissance. Roberval claims to have only read Archimedes, Guido Ubaldo and Luca Valerio. But he is clearly subject to Baldi's influence. For example, this is what Roberval writes on the motion of projectiles.

" The violence of a cannon-shot is made up of two impressed [motions]. One is purely violent, arising from the cannon itself and from the powder which is ignited to drive the shot along. The other is natural, being caused by the shot's own weight. Of the first impression, the violence increases somewhat at some distance from the cannon because of the degrees acquired by the motion, which are added to the impression of the powder before this has decreased appreciably. It then happens that, since the impression decreases much more in itself than it is added to by the degrees of velocity acquired, it continually slows down and, after a certain time, finishes. Now at the beginning the line of the direction of this violent impression is directed towards the place at which the cannon points. Later it changes continually and the cause of this change is the natural impression, that is, the body's heaviness carrying it towards the centre of the Earth. For the mixture of these two impressions, violent and natural, means that the shot does not exactly proceed along one direction or the other. But at the beginning it almost entirely follows the violent one, which is, without comparison, much greater than the natural one. Later the violent one disappears bit by bit, and so the shot begins to descend a curved line, and this all the more as the violent impression decreases and the natural motion is added to by the degrees acquired. "

4. Roberval and the law of composition of forces.

Roberval's first claim to fame in statics is that of having justified the law of the parallelogram of forces. This he accomplished by starting from the condition for the equilibrium of the angular lever. We shall follow the treatise that Mersenne appended to his *Harmonicorum libri*. Roberval's work is more modest than the one to which we have just referred, and only occupies 36 pages.

We shall analyse Roberval's demonstrations instead of quoting them—their style is heavy and artificially complicated, while their basis is simple.

In the first place, Roberval considers a weight P suspended at B by two strings AB and BC. The string A passes through the fixed point A. Roberval sets out to determine the traction Q which must be applied to the string BC in order to support the weight P. He replaces the arm AB, whose length is fixed, by an *angular lever* Ap, Aq, where Ap

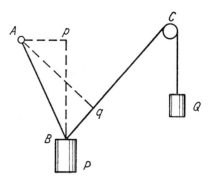

Fig. 49

is the perpendicular on the line of action of the weight P, and Aq is the perpendicular to the string BC. The equilibrium of an angular lever requires that

$$\frac{P}{Q} = \frac{Aq}{Ap}.$$

From this the value of Q is obtained.

Roberval then applies this result to the next diagram. Here QG and CB are, respectively, perpendicular to CA and QA. Further, CF and QD are perpendicular to the line of action of the weight A.

The weight A is suspended from the two strings CA, QA, to which are applied the powers K and E. The equilibrium of the lever CF, CB gives the ratio $\dfrac{A}{E} = \dfrac{CB}{CF}$. Similarily, the lever QD, QG gives the ratio $\dfrac{A}{K} = \dfrac{QG}{QD}$.

" Therefore it is observed that in both cases two perpendiculars are drawn from each power— one on the direction of the weight and the other on the string of the other power. Also that, in the ratios of the weight to the powers, the weight is homologous to the perpendiculars falling on the strings of the powers. Similarly the powers are homologous to the perpendiculars falling on the direction of the weight. "

By these purely geometrical considerations Roberval finally transforms the statement of the preceding rule and arrives at the decomposition of the weight into its two components in the directions of *CA* and *QA*.

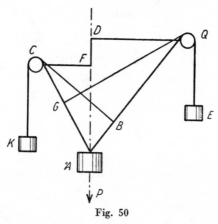

Fig. 50

" *If, from some point taken on the line of the direction of the weight, the line parallel to one of the strings is drawn to the other string, the sides of the triangle thus formed will be homologous to the weight and the two powers.* "

It is interesting to remark that Roberval attempted to relate the rule of the composition of forces to the principle of virtual work.

" In connection with a weight suspended by two strings, we have noticed a thing that has given us much pleasure. This is that when the weight is supported thus by two powers, it can neither rise nor fall without the reciprocal proportion of the paths with the weight and the two powers being changed, and this contrary to the common order. . . .

" If a line *AP* is taken underneath *A*, in the line of its direction, it turns out that if the weight *A* falls as far as *P*, drawing the strings with it and making the powers rise, the reciprocal ratio of the paths that the powers travel in rising and the path which the weight travels in falling will be greater than that of the same weight and the two powers taken together. Thus the powers will be raised further in the proportion that the weight descends in carrying them along, which is contrary to the common order. "

An analogous argument is applied to the rising of the weight, and this conclusion follows— " Consequently the weight *A*, in remaining in its place also remains in the common order. "

CHAPTER FOUR

DESCARTES' MECHANICS
PASCAL'S HYDROSTATICS

1. DESCARTES' STATICS.

Descartes' statics stems directly from the principle of *virtual work*, which he assumed *a priori*. We shall quote a letter from Descartes to Constantin Huyghens dated October 5th, 1637.

" The invention of all [simple machines] is only based on a single principle, which is that the same force that can lift a weight of, for example, a hundred pounds to a height of two feet, can also lift one of two hundred pounds to a height of one foot, or one of four hundred pounds to a height of half a foot, and so on, however this may be applied.

" And this principle cannot fail to be accepted if it is considered that the effect should always be proportional to the action which is needed to produce it. So that if it necessary to use the action by which a weight of a hundred pounds can be lifted to a height of two feet, in order to lift some weight to a height of one foot, this weight should weigh two hundred pounds. For it is the same to lift a hundred pounds to a height of one foot, and then again, to lift a hundred pounds to the same height of one foot as to lift two hundred pounds to a height of one foot and also the same as to lift one hundred pounds to a height of two feet.

" Now the machines which serve to make this application of a kind that acts on a weight over a great distance, and makes this rise by a smaller one, are the pulley, the inclined plane, the wedge, the lathe or turner, the screw, the lever and some others. For if it is not desired to relate some, they could be further enumerated. And if it is desired to relate them in such a way, there is no need to put down as many.

" If it is desired to lift a body F, of weight 200 pounds, to the height of the line BA, in spite of the fact that the *force* is is only sufficient to lift one hundred pounds, it is only necessary to drag or roll the body along the inclined plane *CA*, which I suppose to be twice as long as the line *AB*. For in order to bring it to the point *A* by this path, only the force which is necessary to make a hundred pounds rise twice as high would be used. . . .

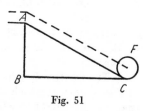

Fig. 51

" But to be set against this calculation is the difficulty there will be in moving the body *F* along the plane *AC* if this plane had been laid along the line *BC*, whose parts I assume to be equally distant from the centre of the Earth. Since this obstruction will be less as the plane is harder, more even and more polished, it is a fact that it can only be expressed approximately and is not very considerable. Further, there is no need to consider that the plane *AC* should be slightly curved on account of the fact that the line *BC* is a part of a circle which has the same centre as the Earth . . . for this is in no way appreciable. "

In Descartes' work on the lever, the resistance is always a weight hung from the lever, and the power is *constantly perpendicular to the arm of the lever.* Guido Ubaldo had made use of this practical observation and Descartes followed him. We shall return to the text.

" I have postponed speaking about levers until the end because, of all the machines, used to lift weights, this is the most difficult to explain.

" Consider this—that while the force which moves the lever descends along the whole semicircle *ABCDE*, although the weight also describes the semicircle *FGHIK* it is not lifted the whole length of the line *FGHIK*, but only the length of the straight line *FOK*. So that the proportion that the force which moves the weight must bear to the heaviness of the weight should not be measured by the proportion of the two diameters, but rather by the proportion of the greatest circumference to the smallest diameter.

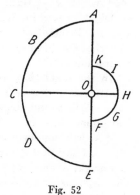

Fig. 52

" Moreover, consider that in order to turn the lever it is by no means necessary that the force should be as great when the lever is near *A* or near *E* as when it is near *B* or near *D*. The reason for this is that, there, the weight rises less, as it is easy to see. And to evaluate exactly what this force should be at each point of the curved line *ABCDE*, it is necessary to know that it acts in the same way as if drew the weight on a circular inclined plane.   Also that the inclination at each of these points on the circular plane should be measured by that of the straight line which touches the circle at that point. "

Not only did Descartes assert the principle of virtual work but—and in this regard his priority is certain—he indicated its infinitesimal character.

" The relative weight of each body should be measured by the *start* of the movement which the power that maintains it can produce, rather than by the height to which it can rise after it has fallen down.   Note that I say *start to fall* and not simply *fall*, because it is the start of the fall that must be taken care of. "

In passing, we recall Descartes' contempt of his contemporaries and predecessors.   Naturally Mersenne had drawn his attention to Galileo—here is Descartes reply.

" And in the first place, concerning Galileo, I will say to you that I have never seen him, nor have I had any communication with him, and that consequently I could not have borrowed anything from him. Also, I see nothing in his books that causes me envy, nor anything approaching what I would wish to call my own.

" It seems to me foolish to think of the screw as a lever—if my memory is correct, this is the fiction that Guido Ubaldo used. "

To assert his independence of Galileo he wrote to Mersenne in the following terms.

" As for what Galileo has written on the balance and the lever, it explains the *quod ita fit* rather well, but not the *cur ita fit* as I have done with my Principle. "   This shows that Descartes believed that a principle that had been set up overrode all other considerations, even experimental ones. . . .

In the texts that we have quoted, Descartes continually uses the word force to denote what we now call *work*.   Even in his own time some misunderstandings arose, and he was quick to take offence.   On November 15th, 1638, he wrote to Mersenne on this matter.

" At last you have understood the word force in the sense that I use it when I say that it takes as much force to lift a weight of 100 pounds to a height of one foot as to lift one of 50 pounds to a height of two feet. That is, that *as much action* or *as much* effort is needed. "   Descartes clarifies this later (September 12th, 1638).

" The force of which I have spoken *always has two dimensions* and is not the force which might be applied at some point to maintain a weight, which always has only one dimension. "

Force in Descartes sense is therefore expressed by the product pl of

a weight and a distance while the *momento* in Galileo's sense [1] is expressed by the product *pv* of a weight and a velocity. Descartes formally
claims to have excluded consideration of the velocity, " which would
make it necessary to attribute *three dimensions* to the force. " He adds—

" As for those who say that I should consider the velocity as Galileo
has done I believe, among ourselves, that they are people who only
talk nonsense and that they understand nothing in this matter. "

Writing to Boswell in 1646, Descartes returned to this theme, which
lay close to his heart.

" I do not deny the material truth of what the students of mechanics
are accustomed to say. Namely, that the greater the velocity at the
end of the long arm of a lever is, in relation to the velocity at the other
end, the less force it requires to be moved. But I deny that the velocity
or the slowness are the causes of this effect. "

Thus Descartes rejects all connection between statics and Aristotle's
dynamics—of which traces subsist even in some of Galileo's concepts.
Statics is made to depend on a single principle, which he asserts to be an
obvious reality. Writing to Mersenne on September 12th, 1638, he said,
" It is impossible to say anything good concerning the velocity without
having to explain what heaviness is, and, in the end, the whole system
of the World. "

With regard to Roberval, who had claimed Mersenne's recognition
of his own priority in connection with the postulate of statics that
Descartes used, the latter shows himself to be even more contemptuous.

" I have just read your Roberval's Treatise on Mechanics, in which
I learn that he is a professor—something I did not know. . . . As for
his Treatise, I would be able to find a large number of mistakes in it
if I wished to examine it carefully. But I will say to you that, on the
whole, he has taken a great deal of trouble to explain a thing that
is very easy, and that, by means of his explanation, he has made it
more difficult than it naturally is. Stevin showed the same things
before him, and in a much more facile and general way. It is true
that I do not know whether either of them is correct in his demonstrations,
for I cannot have the patience to read the whole of these books. When
he claims to have included something in a Corollary that is the same
as I have done in my Writing on Statics, *aberrat toto Caelo*, he is making
something that I made a principle, a conclusion, and he talks of time
and of velocity in places where I talk of distance. This is a very
serious mistake, as I have explained in my earlier letters. " [2]

This haughtiness had its inconveniences. For this refusal to

[1] *Cf.* above, p. 143.
[2] Letter to MERSENNE, October 11th, 1638.

read Roberval entailed Descartes' ignorance of the law of the composition of forces. In fact, the quantity *pl* had been considered as a measure of the work done by a weight by Jordanus and by Descartes contemporaries, Roberval and Hérigone.

We can agree with Duhem that " Descartes gave statics the order and the clarity which are the very essence of his method, but there is no truth in Descartes' statics that men had not know before. Blinded by his prodigious pride, he only saw the error in the work of his predecessors and contemporaries. "[1]

2. DESCARTES AND THE FALL OF HEAVY BODIES.

Descartes discussed the fall of bodies with Isaac Beeckman during his first stay in Holland (1617-1619). The fragment that we are going to analyse dates from this time, but Descartes returned to the subject in a letter to Mersenne dated November 16th, 1629.

Descartes starts by recalling that a body which falls from *A* to *B* and then from *B* to *C* travels much more quickly in *BC* than in *AB*,

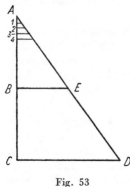

Fig. 53

" for it keeps all the *impetus* by means of which it moves along AB and besides, a new *impetus* which accumulates in it because of the effect of the gravity, which hurries it along anew at each instant. " This is the scholastic doctrine on the accumulation of *impetus*.

" The triangle *ABCDE* shows the proportion in which the velocity increases.

" The line 1 denotes the strength of the impressed velocity at the first moment, line 2, the strength of the velocity impressed at the second moment, etc. ... Thus the triangle *ABE* is formed and represents the *increase of the velocity* in the first half of the distance which the body travels. As the trapezium *BCDE* is three times greater than the triangle *ABE*, it follows that the weight falls three times more quickly from *B* to *C* than from *A* to *B*. That is, that if it falls from *A* to *B* in 3 moments, it will fall from *B* to *C* in a single moment. Thus in four moments its path will be twice as long as in three ; in twelve, twice as long as in nine ; and so on. "[2]

---

[1] *O. S.*, Vol. I, p. 351.
[2] *Œuvres complètes*, Vol. I, p. 69.

Therefore Descartes, like Galileo in 1604, assumed the law $v=ks$. Before he developed his own analytical geometry, he used the geometrical representation of uniformly varying quantities that was due to Oresme. Descartes called the *measure* of such a quantity the *augmentatio velocitatis*. But he confused the *augmentatio* along AB with the *mean velocity* along AB, which lead to a conclusion that was not only incorrect, but also in contradiction with the law from which he had started.

Isaac Beeckman's Journal, which was used by Adam and P. Tannery in their edition of Descartes' works, contains further details of these discussions.[1]

Beeckman assumes the correct law $v=kt$ and correctly deduces from it the law of distances. If AD represents a *duration* of one hour, the distance travelled is represented by the triangle ADE. In two hours the distance is represented by the triangle ADC. The ratio of the areas, and therefore that of the distances, is therefore the *squared ratio of the times.*

Beeckman makes use of the method of indivisibles in order to justify this result. " If, during the first moment of time, the body has travelled a *moment of distance AIRS*, during the first two moments of time it will have travelled 3 moments of distance, represented by the figure $AJTURS$. The distance travelled in any time whatever is therefore represented by the corresponding triangle supplemented by the small triangles ASR, RUT, etc. ... which are equal to each other. But these equal triangles added in this way are smaller as the moments of distance are smaller. Therefore these added areas will be of zero magnitude when it is supposed that the moment is of magnitude zero. It follows that the distance which the thing falls in one hour is to the distance through which it falls in two hours as the triangle ADE is to the triangle ACB. "

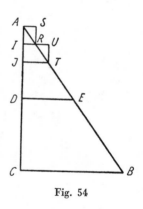

Fig. 54

The two propositions which Dominic Soto has stated are thus linked with each other by the bond of indivisibles.

Beeckman ascribed even this argument to Descartes. " *Haec ita demonstravit Mr. Peron.* "

Unfortunately Beeckman did not persist in this point of view. In another writing [2] he went back to the law $v = ks$ and repeated

[1] *Œuvres complètes*, Vol. X, p. 58.
[2] *Ibid.*, Vol. X, p. 75.

the very same error as Descartes in the evaluation of the mean velocity.

In spite of the efforts which Duhem made to elucidate them,[1] these essays remain somewhat confused. In this matter of the fall of heavy bodies, it remains that Descartes' contribution was not lasting and much less than that of Galileo, whose progress from 1604 to 1638 was continuous. Moreover, Galileo had a respect for observation, which Descartes eschewed.

3. DESCARTES AND THE CONSERVATION OF QUANTITIES OF MOTION.

As early as 1629 Descartes, writing to Mersenne, was categorical on the indestructibility of motion. " I suppose that the motion that is once impressed on a body remains there forever if it is not destroyed by some other means. In other words, that something which has started to move in the vacuum will move indefinitely and with the same velocity. "

In his *Dioptrics* Descartes fell back on a mechanical model to explain the laws of reflection. A ball impelled from A to B bounces off the earth CBE. He explicitly neglects " the heaviness, the size and

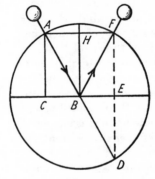

Fig. 55

the shape " of the ball, and supposed the earth to be " perfectly hard and flat. " He asserts that on meeting the earth the ball is reflected, and the " determination to tend to B which it had " is modified " *without there being any other alteration of the force of its motion than this.* " In this connection, but in passing, he denied the theory of *intermediate rest*, which was dear to the hearts of some of the Schoolmen. He deemed " the determination to move towards some direction, like the movement, to be divided into all the parts of which it can be imagined that it is composed. " The ball is thus animated by two " determinations. " One makes it descend and the other makes it travel horizontally. The impact with the ground can disturb the first but can have no effect on the second. Combining these principles with that of the conservation of the *force of the motion* of the ball, Descartes explained the laws of reflection.

In his *Principles* (1644), Descartes reasserts the conservation of motion in a very detailed way, making it part of a metaphysical system.

" God in his omnipotence has created matter together with the motion and the rest of its parts, and with his day-to-day interference, he keeps as much motion and rest in the Universe now as he put there when he created it. ... "

By motion, Descartes understands what we now call *quantity of motion*, however precise his ideas on mass may be.

" When a part of matter moves twice as quickly as another that is twice as large, we ought to think that there is as much motion in the smaller part as in the larger. And that each time the motion of one part decreases, that of some other part is increased proportionally. "

Further, Descartes asserts the relativity of motion.

" We would not be able to understand that the body *AB* is moved from the neighbourhood of the body *CD* if we did not also know that the body *CD* is moved from the neighbourhood of the body *AB*. No difficulty is created by saying that there is as much motion in one as in the other. "

Moreover, he distinguishes between the *proper* motion of a body and " the infinity of motions in which it can participate because it is part of other bodies which move differently. " In order to illustrate this, he gives the example of the watch of a sailor that takes part in the motion of his vessel.

Again, Descartes affirmed that the motion is conserved *in a straight line*:

" Each part of nature, in its detail, never tends to move along curved lines, but along straight lines. This rule ... results from the fact that God is immutable and conserves motion in nature by a very simple operation ; for he does not conserve it as it might have been some time previously, but as it is at the precise moment he conserves it. " And Descartes here recalls the motion of a stone in a sling ; he points out that we cannot " conceive any curvature in the stone, " of that we are " assured by experience " for the stone leaves " straight from the sling... ; which makes manifest to us that any body that is moved in a circle tends unceasingly to recede from the centre of the circle it describes ;

and we can even feel this with our hand while we turn the stone in the sling, because it pulls and makes the string taut in its effort to recede directly from our hand. "

4. DESCARTES AND THE IMPACT OF BODIES.

Descartes formulated the following rules for the impact of bodies.

1) If two equal bodies impinge on one another with equal velocity, they recoil, each with its own velocity.

2) If one of the two is greater than the other, and the velocities equal, the lesser alone will recoil, and both will move in the same direction with the velocity they possessed before impact.

3) If two equal bodies impinge on one another with unequal velocities, the slower will be carried along in such a way that their common velocity will be equal to half the sum of the velocities they possessed before impact.

4) If one of the two bodies is at rest and another impinges on it, this latter will recoil without communicating any motion to it.

5) If a body at rest is impinged on by a greater body, it will be carried along and both will move in the same direction with a velocity which will be to that of the impinging body as the mass of the latter is to the sum of the masses of each body.

6) If a body C is at rest and is hit by an equal body B, the latter will push C along and, at the same time, C will reflect B. If B has a velocity 4 it gives a velocity 1 to C and itself moves backwards with velocity 3.

This, as an example, is how Descartes justifies this rule.

" It is necessary that either B will push C along and not be reflected, and thus transfer 2 units of its own motion to C ; or that it will be reflected without pushing C along and, as a consequence, that it will retain these 2 units of velocity together with the 2 that cannot be lost to it ; or, further, that it will be reflected and retain a part of these two units and that it will push C along by transferring the other part. It is clear that, since the bodies are equal and, consequently, that there is no reason why B should be reflected rather than that it should push C along, these two effects will be equally divided. That is to say that B will transfer to C one of these 2 units of velocity and will be reflected with the other. "

7) Descartes also formulated a seventh rule relating to two unequal bodies travelling in the same direction.

We remark that Descartes' guiding idea was the conservation of the quantity of motion, $m|v|$, in absolute value. This idea was to persist among the Cartesians until the resolution of the controversy about living forces that we shall come to much later.

Nearly all Descartes' rules on impact are experimentally incorrect. We remark that he may have suspected this, without being very much disturbed, when he said—

" It often happens that, at first, the observations seem to be at variance with the rules I have just described. But the reason for this is clear. For the rules presuppose that the two bodies B and C are perfectly hard and so separated from all others that there is no other near them which can help or hinder their motion. We see nothing of this kind in the World. "

Jouguet has a very legitimate comment to make on the preceding declaration. " This passage is very characteristic of Descartes' thought. He could observe nature and argue accurately from his laws as well as any other. But he had the pretension of rebuilding everything in a rational way according to the principles of his philosophy. He considered that the source of certainty lay in thought alone. It is known that he did not wish to assume the principles that were accepted in geometry and physics—further, by an exaggeration of his system, he came to neglect observation. " [1]

5. The discussion between Descartes and Roberval on the centre of agitation.

In his *Exercitationes* we know that Bernardino Baldi had introduced a distinction between the centre of gravity and the *centre of violence*, or the centre of accidental gravity.[2]

Mersenne, who had read Baldi's work, suggested to geometers that they should search for a solid that would have the same period of oscillation as a simple pendulum of given length. Descartes replied to him in a letter dated March 2nd, 1646.[3]

" [The] point of your letter—which I do not wish to postpone answering—is the question concerning the size that each body that is hung in the air by one of its extremities should have, whatever its shape may be, in order that it should carry out its comings and goings equally with those of a lead hung by a thread of given length. . . . The general rule that I give in this connection is the following one. Just as there is a centre of gravity in all heavy bodies, so there is also a *centre of their agitation* in the same bodies, when these are hung from some

---

[1] *L. M.*, Vol. I, p. 90.
[2] See above, p. 106.
[3] *Œuvres complètes*, Vol. IV, pp. 362-364.

part and move. Also, that all those bodies in which this centre of oscillation is equally distant from the point from which they are suspended, execute their comings and goings in equal times, provided that the change of this proportion that can be produced by the air is excepted. "

He returned to this question on March 30th, 1646.[1]

" In the first place, since the centre of gravity is so situated in the middle of a heavy body, the action of any part of the body, which could, by its weight, divert this centre from the line along which it falls, is prevented by another part which is opposed to the first and which has just as much force. From this it follows that, in descent, the centre of gravity always moves along the same line as it would take if it were alone, and if all the other parts of which it is the centre were taken away. Similarly, what I call the *centre of agitation* of a suspended body is the point to which the different agitations of all the other parts of the body are related, so that the force which each part has to make itself move more or less quickly than it does is prevented by that of another which is opposed to it. From which it also follows, *ex definitione*, that this centre of agitation will move about the axle from which the body is suspended with the same velocity that it would have if the remainder of the body of which it is a part were taken away and, as a consequence, the same velocity as a lead hung from a thread at the same distance from the axle. "

Here Descartes raised an analogy to the status of a principle, and drew all the logical consequences from it with his accustomed vigour and clarity. But he did not confine himself to this. He tried to determine the position of the centre of agitation by forming the quantity mv for each element, or the proportional quantity mr (where r is the distance from the axis). But he took no account of the direction of these velocities—he always considered v, not $\vec{v}$—and related everything to a plane passing through the centre of gravity and the axis of rotation.

Roberval pointed out this error in a letter to Cavendish dated May, 1646.[2] " The defect of [Descartes'] argument is that he considers only the agitation of the parts of the agitated body, forgetting the direction of the agitation of each of those parts. For the centre of gravity is the cause of its reciprocation from left to right. "

Descartes replies [3] that Roberval was mistaken " in thinking that the centre of gravity contributes anything to the measure of its vibra-

---

[1] *Œuvres complètes*, Vol. IV, pp. 379-388.
[2] *Ibid.*, p. 400.
[3] *Ibid.*, p. 432.

tions beyond what the centre of agitation does. For the word centre of gravity is relative to bodies that move freely or which do not move in any way. For those which move about an axle to which they are attached, there is no centre of gravity with respect to that position and the motion has only a centre of agitation. "

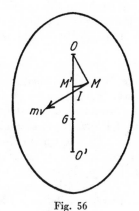

Fig. 56

Without delaying ourselves further with this controversy in which Descartes appears as his usual peremptory self, we shall confine ourselves to the question of a plane figure oscillating about an axis through O.[1] As we have already indicated, Descartes' calculation starts by bringing each element of the figure, M, to M' along the arc $\widehat{MM'}$ of centre O. The motion is then supposed to correspond to a quantity of motion applied perpendicularly to OG at M'. On the other hand, Roberval supposes that $\overrightarrow{mv}$ is applied at the point I on OG. Only the component normal to OG matters, since the other is nullified by the fixity of the support. Thus Roberval arrives at a correct determination of the centre of oscillation O' in the case of a sector of a circle oscillating about an axis passing through the centre.

Mersenne's correspondence has established that Huyghens was involved in the same question as early as 1646, at the age of seventeen. At the beginning, his attitude to the problem was determined by the Cartesian discipline and he did not emancipate himself from this until much later, when he solved the problem of the centre of oscillation in his *Horologium oscillatorium* (1673) by appealing to the principle of living forces.

[1] *Cf.* Jouguet, *L. M.*, Vol. I, p. 158.

6. The quarrel about geostatics.

We are now going to say a little about a controversy which has, at least, the interest of showing that traces of Scholasticism remained even in the most distinguished minds of the XVIIth Century.

In 1635 Jean de Beaugrand announced to Galileo, Cavalieri, Castelli and those interested in mechanics in France, that he had found the law which determined how the weight of a body varied with its distance from the centre of the Earth— the weight was proportional to the distance.

This result appeared in 1636 in his *Geostatics*.

In a letter to Mersenne, Descartes denied this in the following terms.

" Although I have seen many squarings of the circle, perpetual motions and many other would-be demonstrations which were false, I can nevertheless say that I have never seen so many errors united in one single proposition. . . . Thus I can say in conclusion that what this book on Geostatics contains is so irrelevant, so ridiculous and mistaken that I wonder that any honest man has ever deigned to take the trouble of reading it. I would be ashamed of that which I have taken in recording my feeling in this letter if I had not done so at your request. "

In May, 1636, Fermat formulated a proposition which he called *Propositio Geostatica* and which he expressed in the following way.

Let B be the centre of the Earth, BA a terrestrial radius and BC a part of the opposite radius. Consider two bodies A and C which are placed at A and C. If the weight A is to the weight C as BC is to BA, the two bodies are in equilibrium.

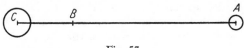

Fig. 57

He adds, " It is very easy to demonstrate this result by following in the steps of Archimedes. "

Given this, Fermat deduces the following result from his geostatic hypothesis.

Fig. 58

" Wherever a body N is placed between B and A, if the proportion of AB to BN is equal to the proportion of the weight N to the power R which is applied at A, the weight N will be kept in equilibrium by the power. Therefore the nearer a body approaches the centre of the Earth, the smaller is the power at A that is necessary to maintain it in equilibrium. This, errors apart, coincides with Beaugrand's geostatic proposition. " [1]

Fermat gave the following explanation to Mersenne, who had not indicated his accord with this strange proposition in which Fermat had applied the laws of a lever to longitudinal forces.[2]

" Every body, in whatever place except the centre of the Earth it may be, and taken by itself and absolutely, always weighs the same. . . . In my Proposition I never consider the body by itself, but only in relation to a lever, and thus there is nothing in the conclusions which is not included in the premises.

" Let A be the centre of the Earth, and let the body E be at the point E and the point N be in the surface or somewhere else that is further away from the centre than the body E. I do not say that E weighs less when it is at E than when it is at N. But I do say that if the body is suspended from the point N by the thread NE, this force at the point N will support it more easily than if it were nearer to the said force, and this in the proportion that I have indicated to you. "

Fig. 59

Fermat was legitimately attacked by Et. Pascal and Roberval. Descartes also condemned him, and in this connection, made clear his own ideas on heaviness in a Note of July 13th, 1638.

" It is necessary to decide what is meant by absolute heaviness. Most people understand it as an internal property or quality in each of those bodies that are called heavy, which makes these bodies tend towards the centre of the Earth. According to some, this property depends on the *shape*, according to others, only on the *material*. Now according to these two beliefs, of which the first is most common in the Schools and the second most often accepted by those who can understand something out of the ordinary, it is clear that the absolute heaviness of bodies is always the same and that it does not change in any way because of their different distances from the centre of the Earth.

" There is also a third belief—that of those who believe that there is no heaviness which is not relative, and that the force or property

[1] Œuvres complètes de Fermat, Vol. II, p. 6.
[2] Ibid., p. 17, Letter of Fermat to Mersenne, 24th June 1636.

which makes the bodies that we call heavy descend does not lie in the bodies themselves, but in the centre of the Earth or in all its mass, and that this attracts them towards itself as a magnet attracts iron. . . . And according to these, just as the magnet and all other natural agents which have a sphere of activity are always more effective at small than at great distances, so it should be said that the same body weighs more when it is closer to the centre of the Earth.

" For myself, I understand the true nature of heaviness in a sense that is very different from these three. . . . But all that I can say [here] is that by this I do not add anything to the clarification of the proposed question, [1] except that it is a *purely factual one*. That is to say that it can only be settled by a man in so far as he can make certain observations, and also that from the observations that are made here in our air it is not possible to know what there might be much lower down, towards the centre of the Earth, or much higher, among the clouds. Because if there is a decrease or increase of heaviness, it is not obvious that it follows the same proportion throughout. "

Descartes was of the opinion that observation seemed to show that heaviness decreased as a weight was separated from the centre of the Earth. He gave strange evidence for this, such as " the flight of birds, the paper dragons that children fly and the balls of pieces of artillery that are fired directly towards the zenith and appear not to fall down again. " Another piece of evidence was that " since the planets which do not have light inside themselves, like the Moon, Venus and Mercury, are probably bodies of the same kind as the Earth, and since the skies are liquid as nearly all Astronomers of this Century believe, it seems that these planets would be heavy things and would fall towards the Earth if their great separation from it had not removed this inclination. "

Returning to observation in the neighbourhood of the Earth, Descartes considered the *absolute* heaviness as being practically constant. " If this equality in the absolute heaviness is supposed, it can be shown that the *relative* heaviness of all hard bodies, considered in the free air without any support, is somewhat less when they are near the centre of the Earth than when they are separated from it, though it may not be the same for liquid bodies. On the contrary, if it is supposed that two equal weights are opposed to each other on a perfectly accurate balance, when the arms of the balance are not parallel to the horizon, that one of the two bodies which is nearer the centre of the Earth will weigh more precisely to the extent that it is closer to it. To leave the example of the balance, it also follows that of the equal parts of the

---

[1] The question is that of knowing whether a body weighs more or less when it is near the centre of the Earth than when it is further away.

same body, the highest parts weigh more than than the lowest ones to the extent that they are further separated from the centre of the Earth, so that the centre of gravity cannot be fixed in any body, even a spherical one. "

As a general rule the convergence of the verticals renders consideration of the centre of gravity illusory. Such was Descartes' conclusion. He also assumed, as Guido Ubaldo had already done, the law of attraction $\frac{k}{r}$ (inversely proportional to the distance).

This conclusion was important and, if one is to believe Duhem,[1] it stemmed from the Italian School—from Torricelli through the agency of Castelli. However, the Beaugrand-Fermat law of attraction kr (proportional to the distance) allowed the existence of a fixed centre of gravity in a body. This had been shown by P. Saccheri in his *Neo Stattica*.

7. PASCAL'S HYDROSTATICS.

Mersenne had advertised Torricelli's experiment in France as early as 1644. Pascal set about repeating the experiment with the collaboration of Petit and convinced himself, by this means, of the possibility of a vacuum, " which Nature does not avoid with as much horror as many imagine. " In 1647 Pascal published *New Experiments concerning the Vacuum*. He gave full details of his plan for his great experiment in a letter to Perier dated November 15th, 1647, and the experiment itself was completed at the Puy de Dôme—a mountain in Central France—on September 19th, 1648. The account of the experiment was published in October of the same year. The principal result was that the difference of level of mercury columns separated by a height of 500 toises was 3 pouces, one and a half lines.

The *Treatise on the Equilibrium of Liquids and the Heaviness of the Mass of Air* appeared in 1663.

In this work Pascal established that liquids " weigh " according to their height, and that in this respect a vessel of ten pounds capacity was equivalent to a vessel of one ounce capacity if both heights were the same. From this Pascal directly obtained the principle of the hydraulic press.

" A vessel full of water is a new principle in mechanics, and a new machine for multiplying forces to whatever degree might be desired. "

Pascal immediately relates this principle to that of *virtual work*.

[1] *O. S.*, Vol. II, p. 183.

" And it is wonderful that in this new machine there is encountered the same constant order that is found in all the old ones, namely the lever, the windlass, the endless screw, . . . which is that the path increases in the same proportion as the force. . . . It is clear that it is the same thing to make a hundred pounds travel a path of one pouce as to make one pound travel a path of a hundred pouces. "

It is in this connection that Pascal defines the *pressure*—the water beneath both pistons of a hydraulic press is equally compressed.

In Chapter II of the *Treatise on the Equilibrium of Liquids* there is mention of a " small treatise on mechanics, " now lost, in which Pascal had given " the reason for all the multiplications of forces which are found in all the instruments of mechanics so far invented. "

This principle does not seem to have differed from Torricelli's principle which Pascal used, without quotation, in hydrostatics.

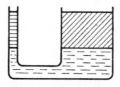

Fig. 60

" I take it as a principle that a body never moves because of its weight unless the centre of gravity descends. From this I prove that the two pistons in the diagram are in equilibrium of this kind. For their common centre of gravity is at a point which divides the line of their respective centres of gravity in the proportion of their weights. Now if they move, if this is possible, their paths will be related to each other as their reciprocal weights, as we have shown. Now if their common centre of gravity is taken in this second situation, it will be found in exactly the same position as previously. Therefore the two pistons, considered as one and the same body, move in such a way that their centre of gravity does not descend, which is contrary to the principle. Therefore they are in equilibrium. Q. E. D. "

The reason for the equilibrium in all Pascal's examples lies in the fact that " the material which is extended over the base of the vessels, from one opening to the other, is liquid. " In mechanics this property which belongs to incompressible fluids of wholly transmitting a pressure has been called *Pascal's Principle*.

We shall not detain ourselves further with Pascal's *Treatise*—which has become classical—nor with the suggestive forms that he gave the *hydrostatic paradox*. Stevin had anticipated some of these ideas.

Pascal's contribution to physics was in marked contrast with Descartes' fragile conceptions in the same field. However, it had its contemporary opponents—thus Pascal was obliged to contend with Aristotelians like Father Noël. At first Pascal took various verbal precautions when he argued the futility of the horror of the vacuum. But eventually he became forthright in his conviction that this scholastic prejudice was absurd. " Let all the disciples of Aristotle gather together all the strength in the writings of their master and his commentators in order, if they can, to make these things reasonable by means of the horror of the vacuum. Except that they know that experiments are the true masters that must be followed in physics. And that what has been accomplished in the mountains reverses the common belief of the world that Nature abhors a vacuum ; it has also established the knowledge —which will never die—that Nature has no horror of a vacuum, and that the heaviness of the mass of air is the true cause of all the effects which have previously been attributed to this imaginary cause. "

CHAPTER FIVE

THE LAWS OF IMPACT
(WALLIS, WREN, HUYGHENS, MARIOTTE)
THE MECHANICS OF HUYGHENS
(1629-1697)

1. THE MECHANICS OF WALLIS (1616-1703).

In 1668 the Royal Society of London took the initiative of choosing for discussion the subject of the laws of impacting bodies. Wallis (November 26th, 1668) discussed the impact of inelastic bodies, while Wren (December 17th, 1668) and Huyghens (January 4th, 1669) discussed the impact of elastic bodies.

Wallis' memoir should be set against the background of his other work in mechanics, which was the subject of a treatise *Mechanica, sive de Motu* (London, 1669-1671). In this treatise Wallis considered the *vis motrix* and the *resistantia*, which were opposed to each other in every machine. He hesitated between the concept of *momento* in Galileo's sense, which could be expressed by the product pv_z of the weight and the vertical component of the velocity, and the notion of *moment*, or the product ph of the weight and the height of fall. In the first Chapter, Wallis even went as far as to consider a mixed solution by the product pv_z as a measure of the *momentum* of the motive force and the product ph as a measure of the *impedimentum* of the resistance. Very fortunately he did not adhere to this peculiar choice and declared himself in favour of the product ph in the second Chapter of his Treatise, which was concerned with the fall of bodies. He went further and generalised the principle that Descartes had laid at the foundation of statics by extending it to forces *other than the heaviness.*

" In an absolutely general way, the progress effected by a motive force is measured by the movement effected in the direction of this force, the recoil by the movement in the opposite direction. . . . The progress or recoil effected under the action of any force is obtained by taking the products of the forces by the lengths of the progress or recoil reckoned in the line of direction of the force. "

Here Wallis supposes the displacements to be finite and rectilinear and the forces to be constant in magnitude and direction. But he assumes that a curvilinear trajectory can be represented as a limit of its tangents. It is said that Wallis also exploited the notion of *work* in all its generality, in so far as displacements due to a constant force were concerned.

Without delaying ourselves over-much with this treatise, we shall discuss Wallis's treatment of the laws of impact. Wallis called a body *perfectly hard* if it did not yield in any way in impact. This is a category that must be distinguished from *soft* and *elastic*.

" A soft body is one that yields at impact in such a way as to lose its original shape, like clay, wax, lead. . . . For these bodies part of the force is used to deform them—the whole force is not expended in the obstacle. It is necessary to take account of this part. "

Like Jouguet,[1] we remark that from the point of view of energy, Wallis had good reason to make this distinction. According to whether the internal energy of a body depends on its deformation or not, the living force lost in the impact is not equivalent, or is equivalent, to the heating of the body. But from the point of view of the quantity of motion, which was the one that Wallis took, this energetic distinction is irrelevant. Mariotte abandoned it, and was followed by others.

Wallis called a body *elastic* if it yielded in impact, but then spontaneously regained its original shape, like a steel spring.

Finally, he defined the *direct impact* of two bodies in the way that is now accepted.

We now come to Wallis's proposition relating to *soft impact* (Chapter XI, Proposition II).

" If a body in motion collides with a body at rest, and the latter is such that it is not moving nor prevented from moving by any external cause, after the impact the two bodies will go together with a velocity which is given by the following calculation.

" Divide the *momentum* furnished by the product of the weight and the velocity of the body which is moving by the weight of the two bodies taken together. You will have the velocity after the impact.

" Indeed, let a body A be in motion along the line AA that passes through its centre of gravity and through that of the body B which is at rest. Let p and v be the weight and the velocity of A. The impelling force *(vis impellens)* will be pv. Let p' be the weight of the body B

[1] *L. M.*, Vol. I, p. 123.

whose velocity is nothing. The weight of the two bodies is $p + p'$. After the impact the two bodies will move with the same velocity. Indeed, B cannot go more slowly than A, since A follows it. Neither can it go more quickly, for it is supposed that there is no other cause of motion than that which arises from the impulsion of A. (If there is another force, like the elastic force, which can impell B more quickly, the problem is of another order, to which we shall return.) Therefore the weight $p + p'$ is moved by a force pv and its velocity is $\dfrac{pv}{p + p'}$. "

This demonstration is based on the conservation of the total *quantity of motion* of the system. It has already been remarked that Wallis did not distinguish between the weight and the mass.

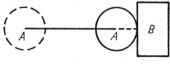

Fig. 61

Wallis generalised this argument to the situation in which B is in motion with a smaller velocity than that of A, but in the same direction. If v' is the velocity of B, the common velocity of the two bodies A and B after the impact is

$$\frac{pv + p'v'}{p + p'}.$$

This is always obtained by dividing the sum of the moments by the sum of the weights.

If the velocity of B is in the opposite direction to the velocity v of A, and is denoted by $-v'$, the common velocity of the two bodies A and B after the impact is

$$\frac{pv - p'v'}{p + p'}.$$

It is seen that Wallis, unlike Descartes, was careful to take account of the signs of the quantities of motion. That is why he arrived at rules which were, apart from the confusion of the weight and the mass, correct. Finally, Wallis remarked that " the magnitude of the impact is equal to twice the decrease that is experienced by the greatest moment in direct impact. "

Indeed, " Consider the body which has the greatest moment as hitting, and the other body as being hit. The body which is hit receives as much moment as is lost by the body which hits it. These moments that are gained or lost are both the result of the impact. The impact is therefore equal to their sum, that is, to twice the decrease experienced by the greater moment. "

Wallis was also concerned with elastic impact (in Chapter XIII of his Treatise). He related this to the theory of soft impact.

He introduced an elastic force *(vis elastica)* whose nature he did not specify, confining himself to an appeal to experimental facts. He stated the following proposition.

" If a body hits an obstacle directly, and if the two bodies—or only one of them—are elastic, the first body will rebound with a velocity equal to that which it had before the impact, and will follow the same direction. "

Indeed, " if the *elasticity* were nothing, the body A would come to rest. "—(This conclusion is obtained by applying the theory of soft impact to an immovable obstacle B. Moreover, we add that Wallis extended this result to an obstacle whose *force of resistance* was limited by comparing this force with a moment greater than that of the body A.) Wallis continues—

" Therefore all motion remaining after the impact is the result of the *vis elastica*. Now this is always equal to the force of the impact. . . . Indeed, the elasticity does not resist as a simple *impedimentum*, but rather as a contrary force acting by reaction and with the same energy that the compression requires. Now what the elasticity suffers during the compression is equal to the impact. The restitutive force is therefore equal to the impact. . . . Now, in particular, since the body A has weight p and velocity v, the magnitude of the impact is $2pv$. This is also that of the elastic force. Since it is developed equally in the two parts half of this force, that is pv, acts on the obstacle and is dissipated there, while the other half repels the body A with velocity v. "

Wallis also treats, for example, the elastic collision of two equal bodies which have equal and opposite velocities and—borrowing this from Wren—the collision of unequal bodies with velocities inversely proportional to their weights. Each body rebounds with the velocity that it has before the impact. These demonstrations are analogous to the preceding one and we shall not describe them here.

2. Wren (1632-1723) and the laws of elastic impact.

We shall describe the paper presented by Wren at the meeting in 1668 and which is included in the *Philosophical Transactions* of 1669.

Wren starts from the concept of *proper velocity* which, for any body, is inversely proportional to the weight. The impact of two bodies R and S which travel with their proper velocities results in the conservation of these velocities. If the velocities differ from their proper values, the bodies R and S " are brought back to equilibrium by the impact. " This is to say that if, before the impact, the velocity of R is greater than its proper velocity by a certain amount and that of S is less than its proper velocity by the same amount, as a result of the impact this amount is added to the proper velocity of S and subtracted from that of R.

Wren seems to have regarded the impact of two bodies with their proper velocities as equivalent to a balance oscillating about its centre of gravity. Wren expressed this analogy in the diagrams that he used to represent the effect of the impact. Strictly speaking, he did not justify his results in a logical and satisfactory way, but he had the merit of making experiments and of embodying his conclusions in a clear and precise law.

3. HUYGHENS (1629-1697) AND THE LAWS OF IMPACT.

Following Wren's example, Huyghens confined himself to elastic impact. His researches were collected in a posthumous volume *De Motu corporum ex percussione* (1700).

Huyghens' investigation was based on the following three hypotheses.

1) The first is the *principle of inertia*. " Any body in motion tends to move in a straight line with the same velocity as long as it does not meet an obstacle. "

2) The second is the following principle. *Two equal bodies which are in direct impact with each other and have equal and opposite velocities before the impact, rebound with velocities that are, apart from sign, the same.*

3) The third hypothesis asserts the *relativity of motion*. Huyghens shows himself to be a Cartesian in this matter.

" The expressions ' motion of bodies ' and ' equal or unequal velocities ' should be understood relatively to other bodies that are considered as at rest, although it may be that the second and the first both participate in a common motion. And when two bodies collide, even if both are subject to a uniform motion as well, to an observer who has this common motion they will repel each other just as if this parasitical motion did not exist.

" Thus let an experimenter be carried by a ship in uniform motion and let him make two equal spheres, that have equal and opposite velocities with respect to him and the ship, collide. We say [hypothesis 2] that the two bodies will rebound with velocities that are equal with respect to the ship, just as if the impact were produced in a ship at rest or on *terra firma.* "

Huyghens appealed to this relativity in order to justify the following proposition.

" *Proposition I.* — *If a body is at rest and an equal body collides with it, after the impact the second body will be at rest and the first will have acquired the velocity that the other had before the impact.*

" Imagine that a ship is carried alongside the bank by the current of a river and that it is so close to the edge that a passenger on the ship can hold the hands of an assistant on the bank. In his two hands, A and B, the passenger holds two equal bodies E and F which are hung from threads. Let the distance EF be divided into two equal parts by the point G. By displacing his two hands equally towards each other, the passenger will make the two spheres E and F collide with equal velocities with respect to himself and the ship (hypotheses 2 and 3). But during this time the ship is supposed to be carried to the right with a velocity GE equal to that with which the right hand of the passenger is moved towards the left.

" Consequently the right hand, A, is motionless with respect to the bank and the assistant who is placed there, while the passenger's left hand, B, is displaced with a velocity EF—twice of GE or FG—with respect to the assistant. Suppose that the assistant placed on the bank grasps the passenger's hand A, as well as the end of the thread which supports the globe E, with his own hand C. Also, that with his hand D he grasps the passenger's hand B, which is the one that holds the thread from which the sphere F hangs. It is seen that when the passenger makes the spheres E and F meet each other with velocities that are equal with respect to himself and the ship, at the same time his assistant makes the sphere E—motionless with respect to himself and the bank— collide with the sphere F whose velocity is FE. And it is certain that, if the passenger displaces the spheres in the way that has been described, there is nothing to prevent his assistant on the bank from seizing his hands and the ends of the threads, provided only that he accompanies the motion and does not oppose any hindrance to them. In the same way, when the assistant on the bank is directing the sphere F against the motionless sphere E, there is no obstacle to the passenger grasping his hands, even though the hands A and C are at rest with respect to the bank while the hands D and B move with the same velocity EF.

" As we have seen, the spheres E and F rebound after the impact with velocities that are equal with respect to the passenger and the ship—that is, the sphere E with the velocity GE and the sphere F with the velocity GF. During this time the ship moves towards the right

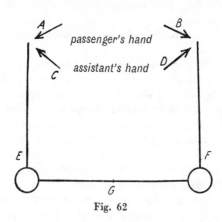

Fig. 62

with the velocity GE or FG. Therefore, with respect to the bank and the assistant on it, the sphere F remains motionless after the impact and the sphere E moves to the left with a velocity twice GE—that is, with the velocity FE with which F has hit E. We therefore show that, to an observer on the earth, when a motionless body is hit by an equal one, after the impact the second one loses all its motion which is, on the other hand, completely taken over by the other. "

With the help of this remarkable artifice Huyghens treats all instances of the impact of two equal bodies by starting from the symmetrical case, whose solution he assumes *a priori*. Thus he shows that two equal bodies that have unequal velocities exchange these velocities in a direct impact.

He then passes to a consideration of unequal bodies and establishes the *conservation of the relative velocity* in the impact of two elastic bodies. It is of some interest to remark here that Huyghens showed, by examples, that the *quantity of motion* was not always conserved. In this context he was concerned with the quantity of motion $m|v|$, in Descartes' sense.

Finally, he demonstrates Wren's rule on the conservation of proper velocities.

" *Proposition VIII.* — *If two bodies moving in opposite directions and with velocities inversely proportional to their magnitudes collide with each other, each one rebounds with the velocity that it had before the impact.*

" Let two bodies A and B collide with each other. $(A > B)$. Suppose that the velocity BC of the body B is to the velocity AC of the body A as the magnitude A is to the magnitude B. We wish to show that after the impact, A will be reflected with a velocity CA and B with a velocity CD. If the former is true for A, the latter is true for B (conservation of relative velocity). Suppose that A is reflected with

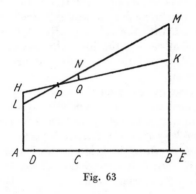

Fig. 63

a velocity $CD < CA$. Then B will rebound with a velocity $CE > CB$ and $DE = AB$. Imagine that A has acquired its original velocity AC by falling from the height HA and that its vertical motion has then been changed into the horizontal motion with velocity AC. In the same way, suppose that B has acquired its velocity by falling from the height KB. These heights are in the square ratio of the velocities. That is, $\dfrac{HA}{KB} = \left(\dfrac{CA}{CB}\right)^2$. Then suppose that after the impact the bodies A and B change their horizontal motions—whose velocities are CD and CE—into vertical upwards motions and thus arrive at the points L and M such that $\dfrac{AL}{BM} = \left(\dfrac{CD}{CE}\right)^2$.

" When the centre of gravity of A is at H and that of B at K, their common centre of gravity is at Q. After the impact this centre of gravity is at the point N. Now it can be shown that N is above Q. "[1]

In this matter Huyghens invokes a principle which we find developed in some detail in the *Horologium oscillatorium*,[2] and reduces, in fact, to the *principle of the conservation of living forces*.

[1] This is a question of pure geometry.
[2] See below, p. 187.

" It is a well-established principle of Mechanics that, in the motion of several bodies under the influence of their centre of gravity alone, the common centre of gravity of these bodies cannot be raised. "

If this principle is assumed, the supposition which has been made about the velocity with which A rebounds ($CD < CA$) implies a contradiction.   Huyghens dismisses the hypothesis that A is reflected with a velocity $CD > CA$ in an analogous way.   Therefore A rebounds with the velocity CA and B with the velocity CB.   Q. E. D.

Huyghens related all cases of the direct elastic impact of two unequal bodies to the preceding situation by using the artifice of the moving ship on every occasion.

We now know that by invoking the *relativity of impact phenomena*, Huyghens carried the discussion into a privileged field.   It follows from the rule of the composition of velocities that a percussion remains the same when a " fixed " system of reference is replaced by a " moving " one, from the moment when the relative motion of the two systems becomes *continuous*.   With this restriction alone, the relative motion of the two systems can be accelerated in any way.   Huyghens, however, restricted himself to a uniform and rectilinear relative motion—namely, that of the ship with respect to the river-bank.

We add that in using the principle of inertia, on the other hand, Huyghens confined himself to an infinite number of reference points, to day termed *absolutes*.   In fact, the principle of inertia is irrelevant to impact phenomena because of their instantaneous character.

In all his writings, Huyghens took care to explain his hypotheses clearly and to deduce his propositions from them logically.   His style is similar to that of Archimedes.   By this means a perfect clarity is achieved at the price of some tedium.   However, the rigour of this work is sometimes only an apparent one.

Jouguet has come to the following conclusion after an exhaustive study of the interdependence of certain of Huyghens' hypotheses. [1] It was sufficient for Huyghens to assume the conservation of the total living force of the system in *every* system of reference, or its conservation in *two arbitrary* systems in *continuous* (and not zero) relative motion with respect to each other.   Such a hypothesis is itself equivalent to the twin hypothesis of the conservation of living force in *one* arbitrary system of reference and the simultaneous conservation of the total quantity of motion in direction and sign as Wallis intended—not in Descartes' sense of absolute value.

[1] *L. M.*, Vol. I, p. 151.

4. THE PLAN OF HUYGHENS' FUNDAMENTAL TREATISE.

We now come to Huyghens' major work in dynamics—the treatise *Horologium oscillatorium sive de motu pendulorum ad horologia aptato demonstrationes geometricae* (Paris, 1673).[1]

This work consists of five parts— a description of the clock; on the fall and motion of bodies on a cycloid; the evolution and dimensions of curved lines; on the centre of oscillation or agitation; and finally, on the construction of a new clock with a circular pendulum, and theorems on the centrifugal force.

Huyghens had constructed cycloidal clocks at The Hague since 1657. He planned this treatise over a long period of time and only completed it in 1669. In 1665 he was invited by Colbert to visit Paris and to work at the Academy of Sciences, where he obtained a royal monopoly for the reproduction of his clocks. As early as 1667 there is mention of " three clocks made at Paris, at the expense of the King, to be used on the voyage to Madagascar. " Huyghens had already tried out his marine clock aboard an English vessel in 1664. Aware of the defect of the isochronism of the finite oscillations of a circular pendulum, Huyghens strove to find a pendulum which might be theoretically isochronous for all amplitudes. " It is the oscillations of marine clocks that most noticeably become unequal, because of the ship's continual shaking. So that it is necessary to take care that oscillations of large and small amplitudes should be isochronous. " Huyghens also made astronomical clocks, both at Leyden and Paris, that were correct to one second a day.

" We have, " he wrote, " regarded the cyloid as the cause of this property of isochronism that we have found, without having the least understanding of anything except that it is consistent with the rules of the craft. " In theory, " it has been necessary to corroborate and to extend the doctrine of the great Galileo on the fall of bodies. The most desired result and, so to speak, the greatest, is precisely this property of the cyloid that we have discovered. . . . To be able to relate this property to the use of pendulums, we have had to study a new theory of curved lines that produce others in their own evolution. " The question here is that of the theory of developable curves, and Huyghens established that the development of a cycloid is an equal cycloid.

Huyghens describes his " automatic " at great length. In this clock the motion of the pendulum is determined by pulleys that are actuated

[1] *The complete Works of Christiaan Huyghens*, published by the Dutch Society of the Sciences, The Hague, 1934, Vol. XVII.

by the driving weights—" the oscillations of the pendulum impose the law and the rule of motion on the whole clock. " He also evolved a new and ingenious suspension which assured the continuity of the clock's motion when the driving weights were removed and replaced.

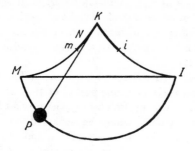

Fig. 64

We return to the question of the cycloidal pendulum. The pendulum oscillates between two thin plates whose function is to assure the constancy of the period in spite of variations in the amplitude. These plates, Km and Ki, are cut from two half-cycloids, KM and KI. The pendulum KMP has a length equal to twice the diameter of the generating circle. Huyghens wrote, " I do not know whether any other line has this remarkable property, namely that of describing itself in its evolution "—in other words, of being identical with its development. This mathematical problem was to be taken up by Euler.

5. Huyghens and the Fall of Bodies.

Though, as we have remarked in connection with the laws of impact, Huyghens was a Cartesian to the extent that he used the relativity of motion, he was much more a direct successor of Galileo and Torricelli, and provided a link between them and Newton. To use his words, he " corroborated and extended Galileo " in the matter of the fall of bodies.

He clearly stated the principle of inertia and the principle of the composition of motions, and applied these principles to the fall of bodies and to rectilinear uniform motion in any direction. " Each of these motions can be considered separately. One does not disturb the other. "   He accepted Galileo's laws on the rectilinear fall of bodies and improved the associated demonstrations. For example, he established the following proposition.

"*Proposition I.* — *In equal times the increases of the velocity of a body which starts from rest and falls vertically are always equal, and the distances travelled in equal times form a series in which successive differences are constant.*

"Suppose that a body, starting from rest at A, falls through the distance AB in the first time and has acquired a velocity at B which would allow it to travel the distance BD during the second time.

"We know then that the distance travelled in the second time will be greater than BD, since the distance BD would be travelled even if all the action of the weight ceased at B. In fact, the body will be animated by a compound motion consisting of the uniform motion which would allow it to travel the distance BD and the motion of the fall of bodies, by means of which it must necessarily fall a distance AB. Therefore at the end of the second time, the body will arrive at the point E which is obtained by adding to BD a length DE which is equal to AB. ..."

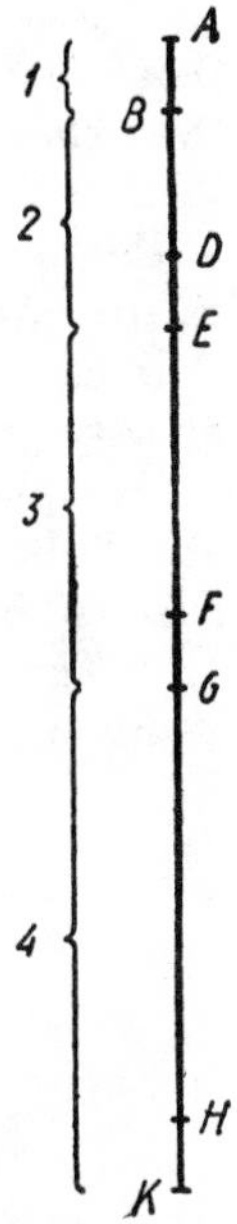

Huyghens showed in the same way that, at the end of the third time, the body will arrive at G—a point such that $EF = 2BD$ and $FG = AB$. The same procedure was repeated.

Thus Huyghens arrived at the following proposition.

Fig. 65

"*Proposition II.* — *The distance that a body starting from rest travels in a certain time is half the distance it would travel in uniform motion with the velocity acquired, in falling, at the end of the time considered.*"

To show this, Huyghens considers the distances AB, BE, EG and GK travelled in the first four intervals of time. He doubles the value of these times so that the body travels along AE in the first instant and along EK in the second. Necessarily

$$\frac{AB}{BE} = \frac{AE}{EK} \quad \text{thence} \quad \frac{EK}{AE} = \frac{BE}{AB} = \frac{AD}{AB}.$$

Now $KE = 2AB + 5BD$ and $EA = 2AB + BD.$

Therefore $KE - EA = 4BD$

and consequently $\dfrac{4BD}{AE} = \dfrac{AD - AB}{AB} = \dfrac{BD}{AB}.$

Therefore $AE = 4AB$ and $BD = 2AB.$ Q. E. D.

These demonstrations have been quoted because they differ from those of Galileo. In particular, they make use of a composition of the velocity acquired and the new fall of the body at each instant. With a little good-will it is possible to regard them as the expression of the ideas of the Schoolmen of the XIVth Century in a more sophisticated mathematical language. Thus Buridan, in particular, believed that heaviness continually caused a new *impetus* to the one that was already present.

Huyghens then sets out to establish a hypothesis " that Galileo asked should be granted to him as obvious. " Thus Salviati, in the *Discorsi*, had been obliged to take the following principle as a postulate.[1]

" *The velocities acquired by a body in falling on differently inclined planes are equal when the heights of the planes are so.* "

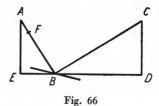

Fig. 66

" . . . Let there be two inclined planes whose sections by a vertical plane are *AB* and *CB*. Their heights, *AE* and *CD*, are equal. I maintain that in these two circumstances the velocity acquired at *B* is the same. Indeed, if in falling along *CB* the body acquires a velocity that is smaller than in falling along *AB*, this velocity will be equal to that which would be acquired in some descent *FB* < *AB*. But along *CB* the body acquires a velocity that allows it to rise again along the whole length of *BC*. [This in virtue of proposition IV, which we have not quoted and which demonstrates this fact for a rectilinear rise.] Therefore it will acquire, along *FB*, a velocity which can make it rise again along *BC*—this can be achieved by reflection at an oblique surface. It will therefore rise as far as *C*, or to a height greater than that from which it fell, which is absurd. "

[1] See above, p. 135.

In the same way it is shown that in descending along AB the body cannot acquire a smaller velocity than in falling along CB. This establishes the proposition.

Huyghens shows that the durations of fall have the same relation to each other as the lengths of the planes. He also shows that when the body falls in a *continuous* motion from a given height along any number of differently inclined planes, it always acquires the same velocity as that obtaining at the end of a vertical fall from the same height. Conversely, in rising again along a trajectory formed of contiguous and differently inclined planes, the body will achieve its original height (Proposition IX). A passage to the limit then allows the question of the motion of a body on a curve contained in a vertical plane to be considered.

6. The isochronism of the cycloidal pendulum.

Huyghens arrived at a proof of the isochronism of a cycloidal pendulum by means of an argument in infinitesimal geometry. Admirable though this was, it required no less than a dozen propositions, and we cannot reproduce it here.[1] The principal result is stated in the following terms.

" *Proposition XXV.* — *In a cycloid whose axis is vertical and whose summit is placed below, the times of descent in which a particle starts from any point on the curve and reaches the lowest point are equal to each other ; their ratio with the time of vertical fall along the whole axis of the cycloid is equal to the ratio of half the circumference of a circle to its diameter.* "

This result may be obtained easily by means of a well-known analysis. Thus the motion of a heavy particle on a cycloid is defined by the differential equation

$$\frac{d^2s}{dt^2} + \frac{g}{4R}\, s = 0.$$

Here s denotes the distance of the particle from A as measured along the arc. If the particle starts from rest at the point B, and if the arc AB is equal to s_0, it follows that $s = s_0 \cos \sqrt{\frac{g}{4R}} \cdot t$. The summit, A, is attained in a time $T = \frac{\pi}{2} \sqrt{\frac{4R}{g}}$. Now if the particle were

[1] *Cf. The Complete Works of Christiaan Huyghens*, Vol. XVIII, pp. 152 to 184.

allowed to fall freely along DA, it would reach A after a time T' given by $2R = \dfrac{1}{2} gT'^2$, from which it follows that $T' = \sqrt{\dfrac{4R}{g}}$. Therefore it must be that $\dfrac{T'}{T} = \dfrac{\pi}{2}$, which is the statement that Huyghens makes.[1]

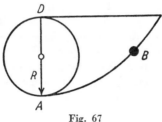

Fig. 67

We shall pass over the third part of Huyghens' treatise, which is devoted to curved lines. The question is that of the search for developments. Huyghens called the development of a curve, the *evoluta*, and the development the *descripta ex evolutione*.

Notable among this work is the development of the cycloid, which rationally justified the use of the cycloidal shape in his clock. He also studied the developments of conics—in particular, those of the parabola, which he called *paraboloides*.[2]

7. THE THEORY OF THE CENTRE OF OSCILLATION.

We now come to the fourth part of the *Horologium oscillatorium*, which is devoted to an investigation of the centre of oscillation.

" A long time ago, when I was still almost a child, the very wise Mersenne suggested to me, and to many others, the investigation of centres of oscillation or agitation. " Thus Huyghens expresses himself at the beginning of the fourth part of his major work. At first he found nothing "which might open the way to this contemplation. " However, he returned to the question in order to improve the pendulums of his " automatic, " to which he had been led to add moveable weights above the principal fixed weight. Huyghens completely resolved

[1] It is of some interest to remark that this study of the cycloid was very popular among XVIIth Century geometers. WREN had calculated its length, ROBERVAL had defined the tangents while PASCAL determined the centre of gravity and calculated the area. WALLIS too, had made analogous investigations.

[2] HUYGHENS knew the development of a parabola as early as 1659, as a result of his work with Jean VAN HEURAET of Harlem.

this question by appealing to a kind of generalisation of Torricelli's principle that depended on the *principle of living forces.*

First he defines the compound pendulum and the centre of oscillation. The latter is the point on the perpendicular to the axis of oscillation through the centre of gravity which is separated from the axis by a distance equal to the length of the simple isochronous pendulum.

Huyghens starts from the following fundamental hypothesis.

" *We suppose that when any number of weights starts to fall, the common centre of gravity cannot rise to a height greater than that from which it starts.* "

In the commentary which accompanies this hypothesis, Huyghens specifies that verticals should be considered as parallels if the consideration of a centre of heaviness is to have any meaning. His hypothesis reduces to the following— no heavy body can rise by the sole agency of its own gravity ; what is true for a single body is also true for bodies which are attached to each other by rigid rods.

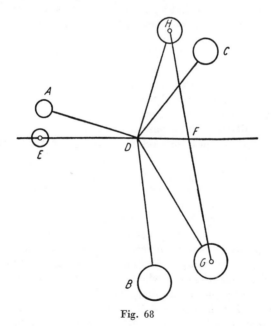

Fig. 68

If, now, the bodies considered are no longer connected to each other, they nevertheless have a common centre of gravity, and it is this which cannot rise spontaneously.

" Let there be weights A, B, C and let D be their common centre of gravity. Suppose the horizontal plane is drawn and that EDF is a right section of it. Let DA, DB, DC be the rigid lines joining the points to each other in a rigid way. Now set the weights in motion so that A comes to E in the plane EF. Since all the rods are turned through the same angle, B will now be at G and C at H.

" Finally suppose that B and C are joined by the rod HG which cuts the plane EF in F. The point F must also be the centre of gravity of these two weights taken together, since D is the centre of gravity of the three weights at E, G, H and that of the body E is also in the plane DEF. The weights H and G are once more set in motion about the point F as about an axis and, without any force, simultaneously arrive in the plane EF. Thus it appears that the three weights, which were originally at A, B and C, have been carried exactly to the height of their centre of gravity by their own equilibrium, Q. E. D. The demonstration is the same for any other number of weights.

" Now the hypothesis that we have made is also applicable to liquid bodies. By its means, not only may all that Archimedes has said about floating bodies be demonstrated, but also many other theorems in mechanics. And truly, if the inventors of new machines who strive in vain to obtain perpetual motion were able to make use of this hypothesis, they would easily discover their errors for themselves and would understand that this motion cannot be obtained by any mechanical means. "

Huyghens' second hypothesis consists of the neglect of the resistance of the air and all other disturbances of motion.

His first three propositions relate to the geometry of masses. We come to the fourth.

" *Proposition IV.* — *If a pendulum composed of several weights, and starting from rest, has executed some part of its whole oscillation, and it is imagined that, from that moment on, the common bond of the weights is broken and that each of the weights directs its acquired velocity upwards and rises to the greatest height possible, then by this means the common centre of gravity will rise to the height it had at the start of the oscillation.*

" Let a pendulum composed of any number of weights A, B, C be connected by a weightless rod which is suspended from an axis D perpendicular to the plane of the diagram. The centre of gravity, E, of the weights A, B, C is supposed to be in this plane. The line of the centre, DE, makes an angle $\widehat{EDF}$ with DF and the pendulum is drawn aside as far as this. Suppose that it is released in this position and that it executes a part of its oscillation in such a way that the weights A, B, C come to G, H, K. Suppose that each of these weights directs its velocity upwards when the bond is broken (this can be arranged by the adjunction of certain inclined planes) and rises to the greatest possible height, as far as *L*, *M*, *N*. Let *P* be the centre of gravity of the weights when they have attained these positions. *I maintain that this point is at the same height as E.*

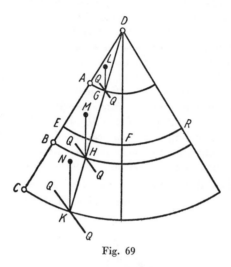

Fig. 69

" First, it is certain that *P* is not higher than *E* (hypothesis I). But neither is it at a lesser height. Indeed, if this is possible, let *P* be lower than *E*. Suppose that the weights fall down again through the same heights that they travelled in mounting—namely *LG*, *MH*, *NK*. It is clear that they will attain the same velocity as they had at the beginning of their climb—that is to say the velocity they acquired in the motion of the pendulum from *CBAD* to *KHGD*. Consequently, if they are simultaneously attached again to the rod which supported them, they will continue their motion along the arcs which they had started along. (This will happen if, before coming to the rod, they rebound on the planes *QQ*.) The pendulum reconstituted in this way will effect the rest of its motion without any interruption. So that the centre of gravity *E* travels, in rising and falling, along the equal arcs *EF* and *FR*, and finds itself at *R*—at the same height as at *E*. But we have supposed that *R* is higher than *P*, the centre of gravity of the weights when they are at *L*, *M*, *N*. Therefore *R* will also be higher than *P*. The centre of gravity of the weights which have fallen from *L*, *M*, *N* will therefore have risen by a height greater than that from which they fell, which is absurd. The centre of gravity *P* is not, therefore, lower than *E*.

No more is it at a greater height.   It must therefore be that it is at the
same height. Q. E. D. "

" *Proposition V. — Being given a pendulum composed of any number
of weights, if each of these is multiplied by the square of the distance from
the axis of oscillation, and the sum of these products is divided by the
product of the sum of the weights with the distance of their centre of gravity
from the same axis of oscillation, there will be obtained the length of the
simple pendulum which is isochronous with the compound pendulum—that
is, the distance between the axis and the centre of oscillation of the compound pendulum.* "

We shall analyse this demonstration instead of reproducing it.   Let
A, B, C be the material points which constitute the compound and
a, b, c be their weights.   Suppose that $DA = e$, $DB = f$, $DC = g$ and
$ED = d$.   Also suppose that E is the centre of gravity of the weights.
Initially the compound pendulum is released from rest in the position
$DABC$.   Let FG be a simple pendulum isochronous with the compound
pendulum and placed, initially, at FG.   Let the angle $\widehat{FGH}$ be equal
to the angle $\widehat{EDF}$.   On DE, mark off the length of the simple pendulum,

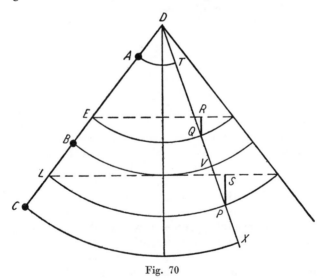

Fig. 70

$x = FG$.   There is isochronism between the simple pendulum and the
compound pendulum if, at corresponding points of the two oscillations,
O and P—such that the arc $\widehat{GO}$ is equal to the arc $\widehat{LP}$—the velocities
of G and L are equal.

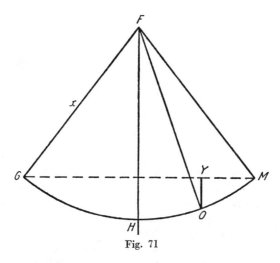

Fig. 71

We shall show that this equality holds for

$$x = \frac{ae^2 + bf^2 + cg^2}{(a + b + c)d}.$$

Indeed, suppose that the velocity from L to P is greater than that from G to O.

Let SP, RQ, YO be the descents from the points L, E, G to the corresponding points P, Q, O.

If $SP = y$, then $RQ = y\,\dfrac{d}{x}$.

The simple pendulum G has a velocity at O which is sufficient to enable it to return to the height of M, either along $\widehat{OM}$ or along OY by means of a suitably chosen elastic impact. Then the point L has, at P, a velocity greater than that which would enable it to return along $PS = OY$.

Let h_L be greater than y, the height to which L can return.

The points A, B, C travel the arcs $\widehat{AT}$, $\widehat{BV}$, $\widehat{CX}$ while L travels $\widehat{LP}$. Thus for A there obtains the relation

$$\frac{v(A) \text{ to } T}{v(L) \text{ to } P} = \frac{DA}{DL} = \frac{e}{x}.$$

Now the heights of return are proportional to the squares of the velocities. Therefore the height of return of A, say h_A, is greater than $\dfrac{e^2}{x^2}y$, from the moment that L exceeds y.

The same is true for B and C, so that

$$ah_A + bh_B + ch_C > \frac{ae^2 + bf^2 + cg^2}{x^2}\, y = \frac{(a + b + c)dy}{x} = (a + b + c)\, RQ.$$

Now this inequality expresses the fact that the centre of gravity E can return to a greater height than the one—RQ—from which it fell, for $h_E = \dfrac{ah_A + bh_B + ch_C}{a + b + c}$. This result is in contradiction with proposition IV and therefore impossible. In the same way the hypothesis that the velocity from L to P will be less than the velocity from G to O implies a contradiction with the same proposition. Therefore the pendulum FG of length x is synchronous with the compound pendulum. This establishes the required result.

We shall not discuss the applications of these propositions here, but shall indicate how Huyghens was able to demonstrate the reciprocity between the axis of suspension and the axis of oscillation. Huyghens stated the following proposition.

" *Proposition XVIII. — If the plane space, whose product with the number of particles of the suspended magnitude is equal to the sum of the squares of their distances from the axis of gravity, is divided by the distance* between the two axes, *the result obtained is the distance from the centre of gravity to the centre of oscillation.* "

In this enunciation, the axis of gravity is the axis through the centre of gravity and parallel to the axis of suspension.

Huyghens' *plane space* has the value $\dfrac{\sum r'^2}{n}$, where r' is the distance from the axis of gravity of one of the n equal particles constituting the suspended magnitude. It is therefore identical with the square of the radius of gyration of the pendulum, ϱ^2, about this axis. If x denotes the length of the simple isochronous pendulum and d the distance between the centre of gravity and the axis of suspension, Huyghens' statement may be written

$$\frac{\sum r'^2}{nd} = \frac{\varrho^2}{d} = x - d.$$

Now, because of proposition V, the length x is equal to $\dfrac{\sum r^2}{nd}$, where r is the distance from one of the n particles to the axis of suspension. This is equal to $\dfrac{K^2}{d}$, where K is the radius of gyration of the pendulum about this axis. Huyghens' long demonstration of proposition XVIII reduces, then, to the verification of the equality

$$\frac{K^2}{d} - d = \frac{\varrho^2}{d} \quad \text{or} \quad \varrho^2 = K^2 - d^2.$$

This is an immediate consequence of the very definition of the moment of inertia, which was to be introduced by Euler.

Huyghens then states the following proposition.

" *Proposition XIX. — When the same magnitude oscillates, the suspension being sometimes shorter, sometimes longer, the distances from the centre of oscillation to the centre of gravity are inversely proportional to the distances from the axes of suspension to the centre of gravity.* "

This statement is equivalent to the equation $\dfrac{x-d}{x'-d'} = \dfrac{d'}{d}$ and is a direct consequence of proposition XVIII.

Finally, Huyghens was able to state the reciprocity of the two axes.

" *Proposition XX. — The centre of oscillation and the point of suspension are reciprocal.* "

This reciprocity is a direct consequence of Proposition XIX and the constancy of the product $d\,(x-d)$.

In Huyghens' work the whole theory of the centre of oscillation rests on the fundamental hypothesis described on page 187. This is equivalent to the *a priori* assumption of the conservation of living forces. In Lagrange's opinion, he thus sets out from an " indirect precept. " Huyghens' theory produced its critics, like Roberval, Catelan, Jacques Bernoulli and others—but these had nothing more than " evil objections. " [1] However, criticism had the value of attracting the attention of geometers—as did the efforts of Descartes and Roberval to solve the same problem—to the investigation of the velocities lost or gained in the *constrained motion* of the elementary weights that constitute a compound pendulum. Jacques Bernoulli was at first mistaken in this investigation, in that he considered the velocities acquired in a finite time. The Marquis de l'Hospital drew his attention to the fact that an *infinitesimal* motion of the system should be considered. It was due to this remark, made in 1680, that Jacques Bernoulli arrived at a new solution of the problem of the centre of oscillation (1703). We shall return to a discussion of this solution, which prepared the way for d'Alembert's principle.

---

[1] *Mécanique analytique*, Part II, Section I.

8. THE THEORY OF CENTRIFUGAL FORCE.

The *Horologium oscillatorium* finishes with thirteen unproved propositions on centrifugal force and the conical pendulum. These propositions were the subject of *De vi centrifuga*. This manuscript was written in 1659 but did not appear until 1703, in a form that was edited by de Volder and Fullenius and published posthumously with the remainder of Huyghens work.

In this treatise Huyghens regards gravity as a tendency *(conatus)* to fall. This tendency is made apparent by the tension of the thread which supports a body. To measure it, it is necessary to consider the *first motion* of the body after the thread has been broken. In this way the *conatus* is caught in life, before there has been time for it to have been destroyed.

Given this, Huyghens sets out to determine the *conatus* of a body attached to a revolving wheel. By an artifice whose object is clearly that of introducing a *reference system bound to the wheel*, he assumes that the wheel is sufficiently large to carry a man who is attached to it. This man holds a thread, supporting a ball of lead, in his hand. Because of the rotation the thread is stretched with the same force as if it were fixed at the centre of the wheel. In equal times the man travels the

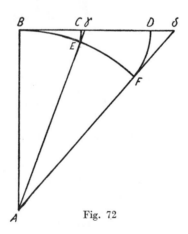

Fig. 72

very small arcs BE and BF. If it is released at B, the lead will travel along the rectilinear paths BC and CD which are equal to these arcs. The points C and D do not fall on the radii AE and EF, but very slightly behind them.

If the points C and D coincide with γ and δ, points on the radii *AE*, *AF*, the lead will tend to move away from the man along the radius. The distances *Eγ*, *Fδ*, . . . increase as the square numbers 1, 4, 9, 16, . . . and this becomes more accurate as the arcs *BE*, *EF*, . . . become smaller. Now, according to Galileo's laws, the distances travelled by a body that starts its fall from rest increase as the successive square numbers 1, 4, 9, 16, . . . The *conatus* which is sought will therefore be, on this hypothesis, the same as that of a heavy body suspended by a thread.

In fact, however, the points *C* and *D* lie behind γ and δ. Therefore, with respect to the radius on which it is placed, the weight tends to describe a path which is tangential to the radius. But at the moment of the separation of the lead and the wheel, these curves can be regarded as being the same as their tangents *Eγ*, *Fδ*, . . . with the consequence that the distances *EC*, *FD*, . . . must be considered as increasing as the series 1, 4, 9, 16, . . . And here is Huyghens' conclusion.

" The *conatus* of a sphere attached to a revolving wheel is the same as if the sphere tended to advance along the radius with a uniformly accelerated motion. . . . It is sufficient, indeed, that this motion should be observed *at the beginning*. Afterwards, the motion can follow every other law. This cannot affect the *conatus* that exists at the beginning of the motion in any way. This *conatus* is entirely similar to that of a body hung by a thread. From which we conclude that the centrifugal forces of unequal particles that move with equal velocities on equal circles have the same relation to each other as their gravities, that is, as their *quantities of solid (quantitates solidae)*. [Here we catch a fleeting glimpse of the concept of *mass*.] Indeed, all bodies tend to fall with the same velocity in the same uniformly accelerated motion. But their *conatus* has a moment *(momentum)* that is greater as the bodies themselves are greater. It must be the same for bodies that tend to move away from a centre, since their *conatus* is similar to that which arises from their gravity. But while the same sphere always has the same tendency to fall whenever it is hung from a thread, the *conatus* of a sphere attached to a revolving wheel depends on the velocity of rotation of the wheel. It remains to us to find the magnitude or the quantity of *conatus* for different velocities of the wheel. " [1]

So much for the principle of centrifugal force that Huyghens developed in his preamble. We shall now state the propositions that he established in a much abbreviated form.

1) For a given period of rotation, the centrifugal force is proportional to the diameter.

---

[1] *The Complete Works of Christiaan Huyghens*, Vol. XVI, p. 266.

2) For a given velocity on the circumference, it is inversely proportional to the diameter.

3) For a given radius, it is proportional to the square of the velocity on the circumference.

4) For a given centrifugal force, the period of revolution is proportional to the square root of the radius.

5) *" When a particle moves on the circumference with the velocity that it would have acquired in falling from a height equal to a quarter of the diameter, its centrifugal force is equal to its gravity. In other words, it will stretch the cord to which it is attached with the same force as if it were suspended. "*

We shall summarise Huyghens' proof of this proposition.

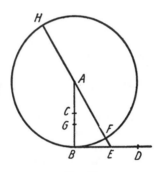

Fig. 73

The particle is supposed to describe the circumference of a circle with uniform motion and the velocity $\left(\sqrt{Rg}\right)$ which it would have acquired in falling from the height $CB = \dfrac{R}{2}$. If it is detached at B, it travels along the tangent uniformly, covering a distance $BD = R$ in the time $\left(\sqrt{\dfrac{R}{g}}\right)$ which it would have spent in falling along CB. We consider a *very small fraction* of BD—namely BE—and draw the straight line $EFAH$. We also suppose that $\dfrac{CG}{CB} = \left(\dfrac{BE}{BD}\right)^2$. Then BE, or $\left(\sqrt{2R \cdot CG}\right)$, is proportional to the time of free fall along CG, which is equal to $\left(\sqrt{\dfrac{2CG}{g}}\right)$.

The particle detached at B travels the distance BE uniformly in the time it would have spent in falling freely from the height CG. Now BE can be approximated to by the arc $\overset{\frown}{BF}$. If it is shown that $CG=FE$, it will have been proved that the *conatus* of the centrifugal force is equal to the *conatus* of the gravity, for the particle considered.

Now $FE = \dfrac{\overline{BE}^2}{2R} = \dfrac{R}{2}\left(\dfrac{BE}{R}\right)^2 = CG$. The proposition is etablished.

It must be pointed out that for Huyghens centrifugal force is in no way a fictitious force. On the contrary, he attributed to it both measurement and special privilege by identifying it with gravity in the particular case we have just seen.

Let us continue our examination of the propositions that end the *Horologium oscillatorium* :

6) A conical pendulum is isochronous with a simple pendulum having as length half the *latus rectum* (parameter) of the paraboloid of which it describes a parallel.

7) The period of a conical pendulum depends only on the height of the cone.

8) It is proportional to the square root of this height.

9) The period of motion of a conical pendulum " on extremely small circumferences " is equal to π T, T being the time it falls freely from a height twice the length of the pendulum.

10) If a mobile runs along a circumference and if the period of its uniform motion is equal to the time in which a conical pendulum, whose length is equivalent to the

Fig. 74

radius, describes an extremely small circumference, its centrifugal force is equal to its gravity.

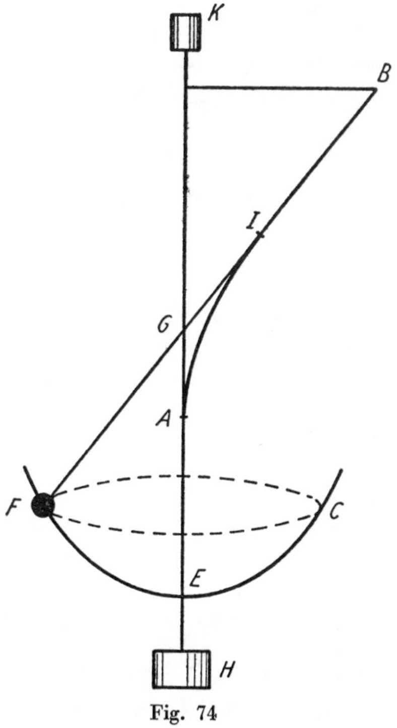

11) The period of revolution of a conical pendulum is equal to the time it takes to fall freely from a height equal to its length, if the string

forms with the horizontal plane an angle the sine of which is equal to $\dfrac{1}{2\pi^2}$.

12) The tension of the string of a conical pendulum of given height is proportional to the length of the pendulum.

13) When a simple pendulum performs a maximum lateral motion, that is, when it descends by a whole quarter of the circumference, the tension of the string at the lowest point is three times the weight of the pendulum.[1]

Without spending time over the demonstrations of these propositions, we may mention, for the sake of curiosity, a clock constructed by Huyghens, which illustrates this theory.

The axis *KH* is vertical, the curved line *AI* is the evolute of the parabola *FEC*. In the rotation of the axis, the pendulum *BF*, which escapes tangentially from the evolute, describes a parallel, a straight section of the paraboloid engendered by *FEC*.

9. HUYGHENS AND THE PRINCIPLE OF RELATIVITY.

In Huyghens' sense, the principle of relativity is that it is impossible that an observer in uniform rectilinear motion can discover his own translation. We have seen how Huyghens exploited this principle in his study of the laws of impact.

Huyghens appears to have assumed, however, in his work on the centrifugal force, the tangible reality of uniform circular motion, which he called *motus verus*. Nevertheless, he went back on this opinion after the appearence of Newton's *Principia*. In the last analysis, he rejected the concept of *absolute motion* and remained a Cartesian. Indeed, in a fragment of his writing that must be placed later than 1688, he wrote, " In circular motion as well as in straight and free motion there is nothing that is not relative, in the sense that this is all there is to know of motion. "

We do not have the time to deal with Huyghens' contribution to physics. We only recall that he outlined an *undulatory theory of light* in 1673—he had learnt of Erasme Bartholin's experimental discovery of the birefringence of Iceland Spar in 1670, and sought to find a rational explanation of this phenomenon. Huyghens read his *Treatise on Light* before the *Académie des Sciences* at Paris in 1690.

---

[1] *Œuvres complètes de Huyghens*, vol. XVI, p. 280 sq.

10. MARIOTTE AND THE LAWS OF IMPACT.

Mariotte dealt with the theory of impact in his *Treatise on the percussion or impact of bodies.*

This work added nothing essentially novel to the work of Wallis, Wren and Huyghens, but was distinguished by its much more experimental approach to the subject. Mariotte rejected the concept of *perfectly hard* bodies in the sense that Wallis had used the term. He confined himself to bodies that were *flexible and resilient* (that is, perfectly elastic) and bodies that were *flexible and not resilient* (that is, perfectly soft).

In order to obtain velocities of direct impact that were in any given relationship, Mariotte described an apparatus consisting of two equal pendulums which were allowed to fall from positions DI', EL' that could be chosen at will.

Mariotte had the merit of recognising the part played by the mass in the laws of impact. Thus he wrote—

" The weight of a body is not understood, here, to be the tendency which makes it move towards the centre of the Earth, but rather to be its volume together with a certain solidity or condensation of the parts of its material which is probably the cause of its heaviness. "

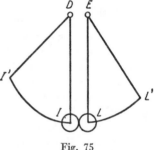

Fig. 75

Mariotte was also concerned with the investigation of *centres of percussion.* He combined the laws of statics with those of impact in this investigation. He investigated the percussions exerted on a balance by jets of water of known amount and came to the following conclusion. " Two bodies that fall on a balance, on one side and the other, are in equilibrium at the moment of impact if the distances [from the centre] of the points where they fall are in reciprocal proportion to their quantities of motion. "

CHAPTER SIX

NEWTON
(1642-1727)

1. THE NEWTONIAN METHOD.

Thanks to Galileo and Huyghens, mechanics had been emancipated from the scholastic discipline. Essential problems like the motion of projectiles in the vacuum and the oscillations of a compound pendulum had been solved. Nevertheless, the task of constructing an organised corpus of principles in dynamics remained. This was the work of Newton, who set his seal on the foundation of classical mechanics at the same time that he extended its field of application to celestial phenomena.

Newton's work in mechanics is called *Philosophiae naturalis principia mathematica* (1687).[1] It proceeds by a method that is at once rational and experimental, to which the author himself gave us the key.

A first rule of the newtonian method consists in not assuming other causes than those which are necessary to explain the phenomena. A second is to relate as completely as possible analogous effects to the same cause. A third, to extend to all bodies the properties which are associated with those on which it is possible to make experiments. A fourth, to consider every proposition obtained by induction from observed phenomena to be valid until a new phenomenon occurs and contradicts the proposition or limits its validity.

It was by relying on the third of these rules that Newton was able to formulate the law of universal gravitation. In expressing this law, Newton had no intention of assigning a cause to gravitation. " But hitherto I have not been able to discover the cause of those properties of gravity from phenomena, and I frame no hypotheses *(hypo-*

---

[1] The manuscript of the *Principia* was deposited with the Royal Society on April 28th, 1686. It was published for the first time in 1687, on the intervention of HALLEY.

theses non fingo). For whatever is not deduced from the phenomena, is to be called an hypothesis ; and hypotheses, whether metaphysical or physical, whether of occult qualities or mechanical, have no place in experimental philosophy. In this philosophy particular propositions are inferred from the phenomena, and afterwards rendered general by induction. " [1]

It is not surprising that Newton himself should have departed from this rule and that he should have introduced purely abstract entities into some of his arguments. But on the whole, his work is a practical expression of the natural philosophy whose foundation he laid.

2. THE NEWTONIAN CONCEPTS.

Newton introduced the notion of *mass* into mechanics. This notion had appeared in Huyghen's work, but only in an impermanent form.[2]

" *Definition I.* — *The Quantity of Matter is the measure of the same, arising from its density and bulk conjunctly.*

" . . . I have no regard in this place to a medium, if such there is, that freely pervades the interstices between the parts of bodies. It is this quantity that I mean hereafter every where under the name of Body or Mass. And the same is known by the weight of each body— For it is proportional to the weight, as I have found by experiments on pendulums, very accurately made. . . . "

In these experiments Newton worked with pendulums made of different materials and of the same length, and established that their acceleration did not depend on the nature of the material. He eliminated the variations of the resistance of the air by using pendulums formed of spheres of the same diameter, suitably hollowed-out to ensure equality of the weights.

When Newton declared that the mass was known by the weight, he contemplated the weight *in a given place.* For he was well aware of the fact that the weight of a body varied with its distance from the centre of the Earth, while its mass remained constant.

This Newtonian definition of the mass has been often and justly criticised. Thus Mach wrote, " The vicious circle is clear, since the density can only be defined as the mass of unit volume. Newton clearly believed that to each body was associated a characteristic determinant

---

[1] In the English translation of the present book quotations from the work of NEWTON are taken from Andrew MOTTE's translation of the *Principia* (1724).

[2] See above, p. 195.

of its motion, which was different from its weight and which we, with him, call *mass*. But he did not succeed in expressing this idea correctly." [1]

" *Definition II. — The Quantity of Motion is the measure of the same, arising from the velocity and quantity of matter conjunctly.*

" *Definition III. — The vis insita, or Innate Force of Matter, is a power of resisting, by which every body, as much as in it lies, endeavours to persevere in its present state, whether it be of rest, or of moving uniformly forward in a right line.* "

To Newton, this *vis insita* was always proportional to the quantity of matter. He also gave it—with a meaning different from that which is accepted now—the name of *force of inertia*. This force is resistive when it is desired to change a body's state of motion, and impulsive to the extent that a body in motion acts on an obstacle.

" *Definition IV. — An impressed force* (vis impressa) *is an action exerted upon a body, in order to change its state, either of rest, or of moving uniformly forward in a right line.* "

Therefore, to Newton, the *vis impressa* is the *determinant of the acceleration*. " This force consists in the action only ; and remains no longer in the body, when the action is over. For a body maintains every new state it acquires, by its *vis insita* only. " The impressed force acts by impact, pressure or at a distance.

" *Definition V. — A Centripetal force is that by which bodies are drawn or impelled, or any way tend, towards a point as to a centre.* "

As examples of centripetal force, Newton cites the gravity which makes bodies tend towards the centre of the Earth, the magnetic force that attracts iron towards a magnet, and that force—whatever its nature might be—that makes each planet describe a curved orbit.

The force exerted by a hand that whirls a stone in a sling is also a centripetal force. Newton adds, " And the same thing is to be understood of all bodies, revolved in any orbits ; and were it not for the opposition of a contrary force which restrains them to, and detains them in their orbits, which I therefore call Centripetal, would fly off in right lines, with an uniform motion. "

Newton then distinguishes the *absolute* quantity, the *accelerative* quantity and the *motive* quantity of the centripetal force (Definitions VI, VII and VIII).

The absolute quantity depends on the efficacy of the cause that propagates the centripetal force—for example, the size of a stone or the strength of a magnet.

The accelerative quantity is measured by the velocity produced in

[1] *M.*, p. 190.

a given time. Therefore, in modern language, it is the acceleration produced by the force.

Newton takes the value of the motive quantity to be the *quantity of motion produced in a given time*. Therefore it is the motive quantity which satisfies the law that is now written—

(1) $$\vec{F} = m\vec{\gamma}.$$

For heavy bodies, the motive quantity becomes identified with the weight.

In this way Newton multiplied the definitions and concepts. Instead of deducing the concept of motive force from the concepts of mass and acceleration by using the law (1), he consciously regarded the mass and the force as two primarily distinct notions.

Newton also took certain precautions in order to anticipate the objections of the cartesian philosophy, and to make the notion of action at a distance acceptable. " I likewise call Attractions and Impulses, in the same sense, Accelerative and Motive ; and use the words Attraction, Impulse or Propensity of any sort towards a centre, promiscuously, and indifferently, one for another ; considering those forces not Physically but Mathematically— Wherefore the reader is not to imagine, that by those words I any where take upon me to define the kind, or the manner of any Action, the causes and the physical reason thereof, or that I attribute Forces, in a true and Physical sense, to certain centres which are only mathematical points. "

Newton then proceeds to discuss the currently used concepts of time, space, place and motion. He introduces a distinction between the relative, apparent and common senses of the words and the absolute, true and mathematical senses.

" I. Absolute, true and mathematical time, of itself, and from its own nature flows equably without regard to any thing external and by another name is called duration— relative, apparent and common time is some sensible and external (whether accurate or unequable) measure of duration by the means of motion, which is commonly used instead of true time ; such as an hour, a day, a month, a year.

" II. Absolute space, in its own nature, without regard to anything external, remains always familar and immoveable. Relative space is some moveable dimension or measure of the absolute spaces ; which our senses determine, by its position to bodies ; and which is vulgarly taken for immoveable space.... "

Or again, " It may be, that there is no such thing as an equable motion, whereby time may be accurately measured. All motions may be accelerated and retarded, but the true, or equable progress, of absolute time is liable to no change. . . .

" For times and spaces are, as it were, the places as well of themselves as of all other things. All things are placed in time as to order of succession ; and in space, as to order of situation. It is from their essence or nature that they are places ; and translations out of those places, are the only absolute motions. "

Newton concerns himself with distinguishing absolute and relative motions by their causes and their effects.

" The causes by which true and relative motions are distinguished, one from the other, are the forces impressed upon bodies to generate motion. True motion is neither generated nor altered, but by some force impressed on the body moved— but the relative motion may be generated or altered without any force impressed upon the body. . . .

" The effects which distinguish absolute from relative motion, are the force of receding from the axe of circular motion. For there are no such forces in a circular motion, purely relative, but in a true and absolute circular motion, they are greater or less, according to the quantity of motion.

" If a vessel, hung by a long cord, is so turned about that the cord is strongly twisted, then filled with water, and held at rest together with the water ; after by the sudden action of another force, it is whirled about the contrary way, and while the cord is untwisting itself, the vessel continues for some time in this motion ; the surface of the water will at first be plain, as before the vessel began to move— but the vessel, by gradually communicating its motion to the water, will make it begin sensibly to revolve, and recede by little and little from the middle, and ascend to the sides of the vessel, forming itself into a concave figure (as I have experienced) and the swifter the motion becomes, the higher will the water rise, till at last, performing its revolutions in the same times with the vessel, it becomes relatively at rest in it. This ascent of the water shows its endeavour to recede from the axe of its motion ; and the true and absolute circular motion of the water, *which is here directly contrary to the relative*, discovers itself, and may be measured by this endeavour. At first, when the relative motion of the water in the vessel was greatest it produced no endeavour to recede from the axe— the water shewed no tendency to the circumference, nor any ascent towards the sides of the vessel, but remained of a plain surface, and therefore its true circular motion had not yet begun. But afterwards, when the relative motion of the water had decreased, the ascent thereof towards the sides of the vessel, proved its endeavour to recede from the axe ; and this endeavour shewed the real circular motion of the water perpetually increasing, till it had acquired its greatest quantity, when the water rested relatively in the vessel. . . . "

Again Newton stresses the distinction between absolute and relative quantities. " And if the meaning of the words is to be determined by their use ; then by the names time, space, place and motion, their measures are properly to be understood ; and the expression will be unusual, *and purely mathematical, if the measured quantities themselves are meant.* Upon this account, they do strain the Sacred Writings, who there interpret those words for the measured quantities. Nor do those less defile the purity of Mathematical and Philosophical truths, who confound real quantities themselves with their relations and vulgar measures. "

Newton did not conceal the difficulty of distinguishing true from apparent motions, because " the parts of that immoveable space in which those motions are performed, do by no means come under the observation of our senses. " In order to accomplish this it is necessary, according to him, to make use simultaneously of the apparent motions, " which are the differences of the true motions " and the forces, " which are the causes and the effects of the true motions. "

As an example he cites the motion of two spheres attached by an inflexible thread and turning about their centre of gravity. The tension of the thread allows " the quantity of circular motion " to be measured.

To Newton, force therefore appears as a true or absolute element and is opposed to motion, which only has a relative character with respect to a suitably chosen reference system. Certain modern critics—notably Mach—reproach the Newtonian philosophy for its metaphysical character in this connection. Absolute space and time appear to them as purely abstract entities which can only be deduced from observation. More correctly, theoretical physics is based on the introduction of pure *unobservables* as intermediaries in the calculation. Under a cloak of metaphysical appearance it contains a profound physical truth. It " explicitly proclaims to the student of mechanics the necessity of considering the privileged reference frames in time and space and of thus avoiding the confusion that is so apparent in the ideas of Descartes and Huyghens. "[1]

3. THE NEWTONIAN LAWS OF MOTION.

Newton stated the principle of inertia at the beginning. This had already been discovered by Galileo and reformulated by Huyghens.

" *Law I. — Every body perseveres in its state of rest, or of uniform*

[1] JOUGUET, *L. M.*, Vol. II, p. 11, note 9.

motion in a right line, unless it is compelled to change that state by forces impressed thereon. "

He next repeats the idea that the motive force is the determinant of acceleration.

" *Law II. — The alteration of* [*the quantity of*] *motion is ever proportional to the motive force impressed ; and is made in the direction of the right line in which that force is impressed.* "

The third law constitutes the principle of the equality of the action and the reaction.

" *Law III. — To every action there is always opposed an equal reaction— or the mutual actions of the two bodies upon each other are always equal, and directed to contrary parts.* "

" Whatever draws or presses another is as much drawn or pressed by that other. If you press a stone with your finger, the finger is also pressed by the stone. If a horse draws a stone tyed to a rope, the horse (if I may so say) will be equally drawn back towards the stone— For the distended rope, by the same endeavour to relax or unbend itself, will draw the horse as much towards the stone, as it does the stone towards the horse, and will obstruct the progress of the one as much as it advances that of the other.

" If a body impinge on another, and by its force change the motion of the other, that body also (because of the equality of the mutual pressure) will undergo an equal change, in its own motion, towards the contrary part.

" The changes made by these actions are equal, not in the velocities, but in the [quantities of] motion of the bodies ; that is to say, if the bodies are not hindered by any other impediments. For because the motions are equally changed, the changes of velocity made towards contrary parts, are reciprocally proportional to the bodies. This law also takes place in attractions. . . . "

It is interesting to see how Newton—contrary to the custom of the time—pays homage to his predecessors.

" Hitherto I have laid down such principles as have been received by all mathematicians, and are confirmed by abundance of experiments. By the two first Laws and the first two Corollaries, Galileo discovered that the descent of bodies observed the duplicate ratio of the time, and that the motion of projectiles was in the curve of a parabola ; experience agreeing with both, unless so far as these motions are a little retarded by the resistance of the air. . . .

" On these same laws and corollaries depend those things which have been demonstrated concerning the times of vibration of pendulums, and are confirmed by the daily experiments of pendulum clocks. By the same together with the third Law Sir Christopher Wren, Dr. Wallis and Mr. Huyghens, the greatest geometers of our times, did severally determine the rules of the congress and reflection of hard bodies, and much about the same time communicated their discoveries to the Royal Society, exactly agreeing among themselves, as to those rules. Dr. Wallis indeed was something more early in the publication ; then followed Sir Christopher Wren, and lastly, Mr. Huyghens. But Sir Christopher Wren confirmed the truth of the things before the Royal Society by the experiment of pendulums, which Mr. Mariotte soon after thought fit to explain in a treatise entirely upon that subject. "

For his part, Newton repeated the experiments with great care. From them he concluded that " the quantity of motion, collected from the sum of the motions directed towards the same way, or from the difference of those that were directed towards contrary ways, was never changed, " whether the bodies were hard or soft, elastic or not.

In order to justify the equality of action and reaction in the case of attractions, Newton argued in the following way.

" Suppose an obstacle is interposed to hinder the congress of any two bodies A, B, mutually attracting one the other— then if either body as A, is more attracted towards the other body B, than that other body B is towards the first body A, the obstacle will be more strongly urged by the pressure of the body A than by the pressure of the body B ; and therefore will not remain in aequilibrio— but the stronger pressure will prevail, and will make the system of the two bodies, together with the obstacle, to move directly towards the parts on which B lies ; and in free spaces, to go forward *in infinitum* with a motion perpetually accelerated. Which is absurd, and contrary to the first law. For by the first law, the system ought to persevere in it's state of rest, or of moving uniformly forward in a right line ; and therefore the bodies must equally press the obstacle, and be equally attracted one by the other.

" I made the experiment on the loadstone and iron. If these placed apart in proper vessels, are made to float by one another in standing water ; neither of them will propel the other, but by being equally attracted, they will sustain each others pressure, and rest at last in equilibrium. "

4. NEWTON AND THE DYNAMICAL LAW OF COMPOSITION OF FORCES.

We have seen how Stevin and Roberval had established the rule of the composition of forces in statics. Newton arrived at the law of the parallelogram of forces by purely dynamical considerations.

" *Corollary I* (to the second law). — *A body by two forces conjoined will describe the diagonal of a parallelogram, in the same time that it would describe the sides, by those forces apart.*

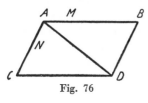

Fig. 76

" If a body in a given time, by the force M impressed apart in the place A, should with an uniform motion be carried from A to B ; and by the force N impressed apart in the same place, should be carried from A to C— compleat the parallelogram ABCD, and by both forces acting together, it will in the same time be carried in the diagonal from A to D. For since the force N acts in the direction of the line AC, parallel to BD, this force (by the second law) will not at all alter the velocity generated by the other force M, by which the body is carried towards the line BD. The body therefore will arrive at the line BD in the same time, whether the force N be impressed or not ; and therefore at the end of that time, it will be found somewhere in the line BD. By the same argument, at the end of the same time it will be found somewhere in the line CD. Therefore it will be found in the point D, where both lines meet. But it will move in a right line from A to D by Law I. "

Newton's demonstration is clearly based on the postulate of the independence of forces. The words " with an uniform motion " show that he considered the impulsion of the force M or N during an infinitely short time. This force acts instantaneously, like a percussion—this explains the importance of the laws of impact in Newton's thought. The students of mechanics in the XVIIth Century had all perceived that the phenomenon of impact was a means of crystallising the effect of a force into the velocity acquired in a first instant.

" *Corollary II.* — *And hence is explained the composition of any one direct force AD, out of any two oblique forces AB and BD ; and, on the contrary the resolution of any one direct force AD into two oblique forces AB and BD— which composition and resolution are abundantly confirmed from Mechanics.* "

Newton deduces the condition of equilibrium for simple machines (the balance, inclined plane and wedge) from this proposition.

We have already seen that Aristotle compounded motions according to the rule of the parallelogram.[1] Since the force was the determinant of the velocity in his belief, it may be held, as Duhem has done, that Aristotle compounded forces in the same way.

[1] See above, p. 21.

For Newton too, the composition of forces according to the rule of the parallelogram had an origin in dynamics. But, to him, the force was the generator of a quantity of motion in a given *elementary* time (Definition VIII, para. 2 above).

5. The motion of a point under the action of a central force.

By a very simple and direct geometrical argument, Newton established that the motion of a material point that is subject to a central force was contained in a plane, and followed the law of areas which Kepler had formulated in a semi-empirical way (the radius vector sweeps through equal areas in equal times). Here is Newton's argument.

" Suppose the time to be divided into equal parts, and in the first part of that time, let the body by its innate force describe the right line *AB*. In the second part of that time, the same would (by Law I), if not hindered, proceed directly to *c*, along the line *Bc* equal to *AB* ;

so that by the radii *AS*, *BS*, *cS* drawn to the centre, the equal areas *ASB*, *BSc*, would be described. But when the body arrived at *B*, suppose that a centripetal force acts at once with a great impulse, and turning aside the body from the right line *Bc*, compells it afterwards to continue its motion along the right line *BC*. Draw *cC* parallel to *BS* meeting *BC* in *C* ; and at the second part of the time,

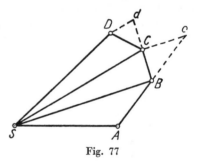

Fig. 77

the body (by Cor. I of the laws) will be found in *C*, in the same plane with the triangle *ASB*.[1]   Join *SC*, and, because *SB* and *Cc* are parallel, the triangle *SBC* will be equal to the triangle *SBc*, and therefore also to the triangle *SAB*.

" By the like argument, if the centripetal force acts successively in *C*, *D*, *E*, and *c* and makes the body in each single particle of time, to describe the right lines *CD*, *DE*, *EF*, and *c* they will all lye in the same plane ; and the triangle *SCD* will be equal to the triangle *SBC*, and *SDE* to *SCD*, and *SEF* to *SDE*. And therefore in equal times, equal areas are described in one immoveable plane— and, by composition, any sums *SADS*, *SAFS*, of those areas, are one to the other, as the times

---

[1] We shall encounter this method of argument again in the next paragraph. It is equivalent to making use of the deviation (in the kinematic sense) produced by the force during an infinitely short time.

in which they are described.   Now let the number of those triangles be augmented, and their breadth diminished *in infinitum* ; and their ultimate perimeter *ADF* will be a curve line. . . . "

Newton establishes the converse of this proposition.   He then examines the circular trajectory of a body gravitating about the centre of this trajectory—its gravity is equal to the centripetal force.   Therefore the gravity can be evaluated by using the propositions given by Huyghens in his *Horologium oscillatorium*.[1]

Newton then studies a particle which describes a circular orbit under the action of a force emanating from any point in the plane of the circle. Given this, he comes to the fundamental problem of the motion of the planets.

6. NEWTON'S EXPLANATION OF THE MOTION OF THE PLANETS.

We shall quote the original text of the *Principia (De motu corporum, Liber I, Prop. VI, cor. 5)*.

" *Si corpus P revolvendo circa centrum S describat lineam curvam APQ ; tangat vero recta ZPR curvam illam in puncto quovis P et ad*

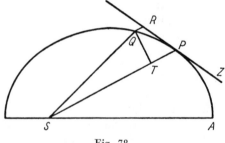

Fig. 78

tangentem ab alio quovis curvae puncto Q agatur QR distantiae SP parallela, ac demitatur QT perpendicularis ad distantiam illam SP : vis centripeta erit reciproce ut solidum $\dfrac{SP \, quad \times QT \, quad}{QR}$ *si modo solidi illius ea semper sumatur quantitas quae ultimo sit, ubi coeunt puncta P et Q.*

" *Nam QR aequalis est sagittae dupli arcus QP in cujus medio est P ; et duplum trianguli SPQ, sive SP $\times$ QT, tempori quo arcus iste duplus describitur, proportionale est ; ideoque pro temporis exponente scribi potest.*"

---

[1] These are the propositions which HUYGHENS included, without proof, at the end of his treatise.

That is, " If a body P revolving about the centre S, describes a curve line APQ, which a right line ZPR touches in any point P ; and from any other point Q of the curve, QR is drawn parallel to the distance SP, meeting the tangent in R ; and QT is drawn perpendicular to the distance SP— the centripetal force will be reciprocally as the solid $\dfrac{\overline{SP^2} \cdot \overline{QT^2}}{QR}$, if the solid be taken of that magnitude which it ultimately acquires when the points P and Q coincide.

" For QR is equal to the versed sine (sagitta) of double the arc QP, whose middle is P— and double the triangle SQP, or $SP \times QT$ is proportional to the time, in which that double arc is described ; and therefore may be used for the exponent of the time. "

No purpose is served in indicating the generality and the remarkably direct character of this argument. The quantity QR is now called, in kinematics, the deviation.

This deviation has the value $\vec{\gamma}\,\dfrac{dt^2}{2}$, where $\vec{\gamma}$ is the acceleration. Now since, here, the acceleration is central, like the force, and since it passes through the pole S, it is seen that QR is parallel to SP. Since the law of areas is applicable, the area of the triangle SPQ is proportional to dt. Since the force is itself proportional to the acceleration it is therefore, in the last analysis, inversely proportional to the expression $\dfrac{\overline{SP^2} \cdot \overline{QT^2}}{QR}$.

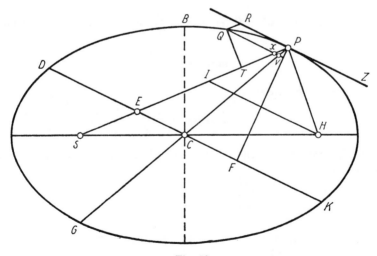

Fig. 79

Newton applies this general law to trajectories which are conic sections. We shall confine ourselves here to the elliptic trajectory and shall summarise the solution of Problem VI, proposition XI— " *Revolvatur corpus in ellipsi : requiritur lex vis centripetae tendentis ad umbilicum ellipseos.* "

The original solution has, to some extent, the character of a rebus— we shall attempt to distinguish the essential steps and to express them in a way that will make this argument clearer.

If DK is the diameter conjugate to CP, Newton first verifies that $PE = a$. Then drawing the line Qxv parallel to the tangent, cutting SP and PC in x and v, he verifies that $\dfrac{QR}{Pv} = \dfrac{PE}{PC} = \dfrac{a}{PC}$ because of the similar triangles. In the same way, if QT is perpendicular to SP and PF is perpendicular to the tangent,

$$\frac{Qx}{QT} = \frac{a}{PF}.$$

Further, by Apollonius' Theorem

$$\frac{a}{PF} = \frac{CD}{b}.$$

Hence

$$QR = \frac{a \cdot Pv}{PC} \quad \text{and} \quad QT = \frac{PF \cdot Qx}{a} = \frac{b \cdot Qx}{CD}.$$

Now in the limit when Q tends to P, $\dfrac{Qv}{Qx}$ tends to unity. Therefore

$$\lim QT = \frac{b \cdot Qv}{CD}.$$

We form the expression

$$\lim \frac{\overline{SP^2} \cdot \overline{QT^2}}{QR} = \lim \overline{SP^2} \cdot \frac{\overline{Qv^2} \cdot b^2 \cdot PC}{\overline{CD^2} \cdot a \cdot Pv}.$$

By the equation of the ellipse referred to oblique conjugate axes CD and CP

$$\frac{\overline{Qv^2}}{\overline{CD^2}} + \frac{\overline{Cv^2}}{\overline{CP^2}} - 1 = 0$$

whence

$$\frac{\overline{Qv^2}}{\overline{CD^2}} = \frac{\overline{CP^2} - \overline{Cv^2}}{\overline{CP^2}} = \frac{(CP + Cv)\,Pv}{\overline{CP^2}}.$$

Therefore

$$\lim \frac{\overline{SP^2}\cdot\overline{QT^2}}{QR} = \lim \overline{SP^2}\cdot\frac{(CP+Cv)Pv\cdot b^2\cdot PC}{\overline{CP^2}\cdot a\cdot Pv} = \frac{2\overline{PC^2}\cdot b^2}{a\overline{PC^2}}\cdot\overline{SP^2} = \frac{2b^2}{a}\overline{SP^2}.$$

Whence the conclusion—

" *Vis centripeta reciproce est ut* $\dfrac{2b^2}{a}\cdot\overline{SP^2}$, *id est reciproce in ratione*

duplicata distantiae SP." The law of force is inversely proportional to the distance.

In short, this proof rests on the newtonian definition of force ; on the use of the kinematic idea of deviation ; and on a direct argument of infinitesimal geometry making use of the classical properties of conics. Except for the finite properties of conics, all its steps were unknown to Newton's predecessors, and were indispensible for the justification of Kepler' semi-empirical laws and for the fashioning of celestial mechanics into a chapter of dynamics.

7. THE UNIVERSAL ATTRACTION.

The scope of this work does not allow us to deal with the numerous problems that are treated in the *Principia*. We shall only describe the path that was travelled by Newton's predecessors, and by Newton himself, and which ended in the law of universal attraction.[1]

Only an excessive schematisation can make the spontaneous blossoming of a physical theory credible. The fall of an apple did not suffice to give Newton the idea of universal gravitation—rather, this was the product of a long development.

As Early as the XIIIth Century Pierre de Maricourt, in a letter written *in castris* in 1269, analysed the polarities of a magnet in a very detailed way—the magnetic property tends to conserve the integrity of the magnet by binding its parts together.

We have seen how the Schools of the XIVth Century, in the persons of Jean de Jandun, William of Ockham and Albert of Saxony, discussed the possibility of action at a distance.[2] We have seen how Copernicus maintained that gravity was only a " natural desire " given to the parts of the Earth in order that their integrity might result.[3]

In *De sympathia et antipathia rerum* (1555), Frascator held that when two parts of the same whole were separated from each other, each of

---

[1] For further details the reader should consult DUHEM's *Théorie physique* (Paris, 1906), pp. 364 *et seq.*
[2] See above, pp. 57-58.
[3] *Ibid.*, p. 85.

them emitted a *species* which was propagated in the intermediate space.

In *De Magnete* (London, 1600), Gilbert argued that the rectilinear motion of heavy bodies was the motion of the reunion of separated parts. He added that " this motion, which is only the inclination towards its source, does not only belong to the parts of the Earth, but also to the parts of the Sun, the Moon, and to those of the other celestial orbs. " Here Gilbert enters on the metaphysical plane. " We give the cause of this coming together and this motion which touches all nature. . . . It is a substantial form which is special, particular, belonging to primary and principal spheres ; it is a proper entity and an essence of their homogeneous and their uncorrupted parts, which we call a primary, radical and astral form ; it is not Aristotles' first form, but that special form by which the orb conserves and disposes what is its own. . . . It constitutes that true magnetic form that we call the primary energy. "[1]

This *animist philosophy*, as Duhem has called it, was adopted by Francis Bacon. Kepler himself was stimulated by it, but substituted in it the idea on one single property belonging to any part of any star. We have already remarked [2] that Kepler regarded gravity as a " mutual affection between parent bodies that tends to unite them. "

As far as the tides are concerned, Ptolemy had already produced an explanation by invoking a special influence of the Moon on the seas. In order to get rid of what seemed to them an occult quality, Averroës, Albertus Magnus and Roger Bacon attributed this action to the heat of the light from the Moon. Albert of Saxony championed an animist theory of the tides. Cardan, followed by Scaliger, believed only in an obedience of the waters to the Moon.

Kepler himself wrote, " Observation proves that everything that contains humidity swells when the Moon waxes and shrinks when the Moon wanes. " [3] But later he corrected this opinion, and thereby anticipated the Newtonian thesis.

" The Moon acts not as a moist or damp star, but as a mass similar to the mass of the Earth. It attracts the waters of the sea, not because they are fluids but because they are gifted with terrestrial substance, to which they also owe their gravity. " [4]

This attraction is reciprocal. " If the Moon and the Earth were in no way held by a sensual force or by some equivalent force, each in its orbit, the Earth would rise towards the Moon and the Moon would descend towards the Earth until these two stars joined together. If the

---

[1] Translated into French by DUHEM.
[2] See above, p. 118.
[3] *Opera omnia*, Vol. I, p. 422.
[4] *Ibid.*, Vol. VII, p. 118.

Earth ceases to attract the waters that cover it to itself, the waves of
the sea would all rise and run towards the body of the Moon. " [1]

Returning to a thesis which had already been put forward by
Calcagnini, Galileo held that the ebb and the flow of the sea was explained by the following relative motion. The Earth turns from East to
West at the same time that it is animated by a translational velocity v.
At a the two motions add together—at b,
they tend to cancel out. Because of their
inertia, the waters of the sea do not follow
this motion exactly. The ebb and flow,
thanks to the delay, is produced twice a
day although, if the composition of the
motions were perfect, they would have the
period of the rotation of the Earth.
Therefore Galileo interpreted the tidal
phenomena as proof of the motion of the

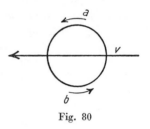

Fig. 80

Earth, while the opponents of the copernican system held to a lunar
attraction.

The astrologers of the XVIth Century, following Grisogone, were
inspired to separate the whole tide into a solar tide and a lunar tide.
In 1528 Grisogone wrote, " The Sun and the Moon attract the swelling
of the sea towards themselves, so that the maximum swelling is perpendicularly beneath each of them. Therefore, for each of them, there are
two maxima of swelling, one beneath the star and one on the opposite
side, which is called the nadir of the star. "

Ideas on the law of attraction itself were yet more vague and chan-
geable. To Roger Bacon, all actions at a distance were propagated in
straight rays, like light. Kepler took up this analogy—now, it has been
known since the time of Euclid that the intensity of the light emitted
by a source varies in inverse ratio of the square of the distance from the
source. In this optical analogy, the *virtus movens* emanating from the
Sun and acting on the planets must follow the same law. But in dynamics Kepler remained an Aristotelian—force was, to him, proportional
to velocity. Therefore Kepler deduced the following result from the
law of areas rv = constant. The *virtus movens* of the Sun on the planets
is inversely proportional to the distance from the Sun. In order to
reconcile this law with the optical analogy Kepler held that light spread
out in all directions in space, while the *virtus movens* was only effective
in the plane of the solar equator.

Boulliau, writing *Astronomia Philolaïca* in 1645, carried the optical

[1] *Opera omnia*, Vol. III, p. 151.

analogy to its limit and supported the law of attraction inversely proportional to the square of the distance. But it should be remarked that this attraction was *normal* to the radius vector, and not central as the Newtonian theory demanded.

Descartes confined himself to replacing Kepler's *virtus movens* by a vortical ether.

Borelli has the merit of having invoked the example of the sling in order to explain why the planets did not fall on the Sun—he sets the instinct by which the planet carries itself towards the Sun against the tendency of all bodies in rotation to move away from their centre—this *vis repellens* is inversely proportional to the radius of the orbit.

In a paper called *An Attempt to prove the annual Motion of the Earth* (1674) Hooke, curator of the Royal Society, clearly formulated the principle of universal gravitation. " All celestial bodies without exception exert a power of attraction or heaviness which is directed towards their centre ; in virtue of which they not only retain their own parts and prevent them from escaping, as we see to be the case on the Earth, but also they attract all the celestial bodies that happen to be within the sphere of their activity. Whence, for example, not only do the Sun and the Moon act on the progress and motion of the Earth in the same way that the Earth acts on them, but also Mercury, Venus, Mars, Jupiter and Saturn have, because of their attractive power, a considerable influence on the motion of the Earth in the same way that the Earth has an influence on the motion of these bodies. " Hooke assumed that the attraction decreased with the distance and, in 1672, declared himself for the inverse square law. No doubt he was guided by the optical analogy.

In order to justify this result it was necessary to know the laws of centrifugal force. Now we know that although Huyghens had written his treatise *De vi centrifuga* as early as 1659, only the statements of the thirteen propositions that conclude the *Horologium oscillatorium* were published during his lifetime.

Halley appears to have applied Huyghens' theorems to Hooke's hypothesis. By assuming Kepler's third law $\left(\dfrac{a^3}{T^2} = \text{constant}\right)$ he discovered the law of the inverse square.

This whole development, that we have only been able to summarise, shows that one cannot talk of the spontaneous generation of the theory of gravitation.

For his part, Newton was in possession of the laws of uniform circular motion in 1666. By an analysis analogous to that which Halley had made, and starting from Kepler's third law, he formulated the law of an attraction inversely proportional to the square of the distance. But more careful than his predecessors, Newton sought experimental verification for this law. He tried to discover whether the attraction exerted by the Earth on the Moon corresponded to this law, and whether this attraction could be identified with terrestrial heaviness.

Since the radius of the Earth's orbit is of the order of 60 terrestrial radii, the force that maintains the Moon in its orbit is 3600 times weaker than the heaviness at the centre of the Earth. Now a body falling freely in the neighbourhood of the Earth falls a distance of 15 Paris feet [1] in the first second. The Moon would therefore fall a distance of $\frac{1}{20}$ pouce in the first second. Knowing the period of the Moon's motion and the radius of its orbit, it is easy to calculate this fall of the Moon. With the data on the Earth's radius that were accepted in England, Newton obtained a fall of only $\frac{1}{23}$ pouce.

Faced with this divergence, he gave up his idea. It was only 16 years later (1682) that he learnt of the measurement of the terrestrial meridian that had been made by Picard. (This happened at a meeting of the Royal Society.) By assuming the value given by this determination, Newton obtained the expected value of $\frac{1}{20}$ pouce. He was then able to declare, " *Lunam gravitare in Terram et vi gravitatis retrahi semper a motu rectilineo et in orbe suo retineri* "; and, by an induction conforming to the very principles of his philosophy, to affirm the doctrine of universal gravitation.

The theory of the attraction of spheres allowed him to concentrate at their centres the masses of stars that were supposed to be formed of homogeneous concentric layers, and thus to reduce them to material points whose mutual attractions could be studied.

Newton evaluated the masses and densities of the Sun and the planets that were surrounded by satellites. He also calculated the heaviness at a point on their surface. He showed that the comets described very elongated elliptical trajectories and replaced these by parabolas whose elements he calculated. In this way he was able to connect the segments of trajectory of a comet that had appeared on each side of the Sun in 1680. Halley then showed that the appearances in 1531, 1607 and 1682 were those of this same comet.

Newton also showed that the rotation of the Earth must entail its flattening at the two poles, and calculated the variation of gravity

---

[1] In these discussions in the *Principia*, the distances are given in French units.

along a meridian. He related the theory of tides to the combined attraction of the Moon and the Sun and thus justified the anticipations of the astrologers of the XVIth Century. Finally, calculating the actions of the Moon and the Sun on the equatorial bulge, he arrived at a theory of the precession of the equinoxes.

CHAPTER SEVEN

LEIBNIZ AND LIVING FORCE

1. THE " VIS MOTRIX " IN THE SENSE OF LEIBNIZ.

Leibniz protested against the cartesian mechanics in a memoir which appeared in 1686 in the *Acta eruditorum* at Leipzig, under the title *A short demonstration of a famous error of Descartes and other learned men, concerning the claimed natural law according to which God always preserves the same quantity of motion ; a law which they use incorrectly, even in mechanics.*

Leibniz set out to show that the *vis motrix* (or, in the words of the XVIIIth Century, the *force of bodies in motion*), was distinct from the *quantity of motion* in Descartes' sense.

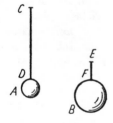

Fig. 81

Like Huyghens, Leibniz assumes that a body falling freely from a given height will acquire the " force " necessary to rise again to the same height, if the resistance of the medium is neglected and no external inelastic obstacle is encountered. On the other hand, like Descartes, he assumes that the same " force " (in the modern sense of work) is needed to lift a body A, whose weight is one pound, to a height DC of four ells as to lift a body B, whose weight is four pounds, to a height of one ell.

In falling freely from the height CD the body A acquires the same " force " as the body B acquires in falling from the height EF. For when it has arrived at D, the body A has acquired the force that it needs to climb again to C, and the body B, when it has come to F, has acquired the force needed to climb to E. By hypothesis, these two forces are equal. Now the quantities of motion of A and B are far from being equal.

Indeed, Galileo's laws show that the velocity acquired in the free fall CD is twice the velocity acquired in the free fall EF. The quantity of motion of A is then proportional to 1×2, while that of B is proportional to 4×1, and is therefore twice that of A. This contradicts the cartesian thesis in which the quantity of motion is used to evaluate the " force. "

Leibniz recognised that in simple machines (the lever, the windlass, the pulley, the wedge and the screw) the same quantity of motion tended to be produced, in one part and the other, when equilibrium obtained. " Thus it happens by accident that the force can be reckoned as the quantity of motion. But there are other instances in which this coincidence no longer exists. "

And Leibniz concludes, " It should be said, therefore, that the forces are in compound proportion to the bodies (of the same specific weight or density) and the generating heights of the velocities—that is, the heights from which the bodies are able to acquire their velocities in falling, or more generally (since often no velocity has been produced at this point), the heights that will be generated. " [1]

2. LEIBNIZ AND THE LAWS OF IMPACT.

Writing to the Abbé de Conti in 1687, Leibniz suggested that for the cartesian principle of the conservation of the quantity of motion should be substituted a natural law which he took as universal and inviolate. This was, " that there is always a perfect equality between the complete cause and the whole effect. "

In this connection he went on to discuss Descartes' third rule on the impact of bodies.[2]

" Suppose that two bodies, B and C, each weighing one pound and travelling in the same direction, collide with each other. The velocity of B is 100 units and that of C, 1 unit. Their total quantity of motion will be 101. But if C, with its velocity, can rise to a height of one pouce, the velocity of B will enable it to rise to a height of 10,000 pouces. Thus the force of the two united bodies will be that of lifting one pound to 10,001 pouces. Now according to Descartes third rule, after the impact the bodies will go together in company with a common velocity of 50 and a half. . . . But then these 2 pounds are only able to lift themselves to a height of 2550 pouces and a quarter, which is equivalent to lifting one pound to 5100 pouces and a half. Thus almost half the force will be lost according to this rule, without there being any reason and without its having been used for anything. "

---

[1] Translated into French by JOUGUET.
[2] See above, p. 162.

In this discussion lies the germ of the controversy about living forces that was to divide the geometers at the beginning of the XVIIIth Century, and to which we shall return. We know now that Descartes third rule is correct and is applicable to perfectly soft bodies (soft, in order that they should travel together after the impact). The total quantity of motion is conserved (no difficulty of sign occurs here) and a part of the living force is transformed into heat.

3. LIVING AND DEAD FORCES.

Leibniz showed himself to be even more systematic in his *Specimen dynamicum* (1695).

We shall pass over the several quantities that he introduced and only concern ourselves with the distinction between *living forces* and *dead forces*.

" Force is twin. The elementary force, which I call *dead* because motion does not yet exist in it, but only a solicitation to motion, is like that of a sphere in a rotating tube or a stone in a sling.

" The other is the ordinary force associated with actual motion, and I call it *living*.

" Examples of dead force are provided by centrifugal force, by gravity or centripetal force, and by the force with which a stretched spring starts to contract.

" But in percussion that is produced by a body which has been falling for some time, or by an arc which has been unbending for some time, or by any other means, *the force is living and born of an infinity of continued impressions of the dead force.* "

Leibniz reproached the Ancients " for having had exclusively an understanding of dead forces, and for only having studied the first *conatus* [in Huyghens sense] of bodies to each other, even though the latter had not acquired an *impetus* [in the sense of quantity of motion] by the action of the forces.

In modern language, Leibniz's assertion that the living force is born of an infinity of impressions of the dead force may be expressed by

$$d\left(m\,\frac{v^2}{2}\right) = F \cdot ds.$$

This leads to the fundamental law $m\,\dfrac{dv}{dt} = F$ and identifies the dead force as the static force.

CHAPTER EIGHT

THE FRENCH - ITALIAN SCHOOL OF ZACCHI AND VARIGNON

1. ZACCHI AND SACCHERI. LAMY AND THE COMPOSITION OF FORCES.

We shall devote this chapter to a brief analysis of some works which can only appear as miniatures in comparison with those of Galileo, Huyghens and Newton. But in leaving the peaks on which the work of the creators of dynamics lies, we shall have a better appreciation of the extent to which those dominated their own century.

In a *Nova de machinis philosophia* (Roma, 1649) Zacchi was concerned with an attempt to isolate the principles implicit in Aristotle's statics. Under the term *virtus* he confused the concepts of force and work, and thus misunderstood Descartes' principle.

Father Fabri (1606-1688) was a teacher at the Jesuit College at Lyons and a friend of Mersenne. His *Tractatus physicus de motu locali* (1646) was a work on dynamics which took over the ideas of Jordanus and Albert of Saxony. Among the moderns it only makes mention of Galileo's statics—and this, as we have seen, was impregnated with the ideas of Aristotle.

Father Lamy attacked Descartes in his *Treatise on Mechanics* (1679) and contested Stevin's argument on the inclined plane. He claimed that nothing proved that the lower part of the chain of balls hung symmetrically.

We now know that this criticism is not justified and that if the number of balls is infinite the necklace outlines a perfectly symmetrical catenary underneath the plane.

In order to solve the problem of the equilibrium of a body on an inclined plane, Lamy preferred to return to the arguments of Bernardino Baldi and da Vinci.

To set against this is a letter addressed by Lamy to M. de Dieulamant, an engineer at Grenoble, which is concerned with the law of the composition of forces and deserves a little of our attention.

" 1. When two forces draw the body Z along the lines AC and BC, which are called the *lines of direction* of the forces, it is clear that the body Z will not travel on the line AC or on the line BC, but on another line between AC and BC, say X.

" 2. If the path X were closed then Z, which is forced to travel by this path, would remain motionless, so that the forces would be in equilibrium.

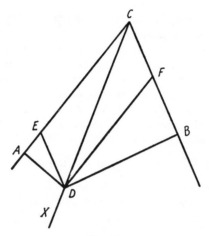

Fig. 82

" 3. Force is that which can move things. Motions are only measured by the distances which they travel. Suppose then that the force A is to the force B as 6 is to 2. Then if A, *in a first instant*, draws the body Z as far as the point E on its own, in the same instant B would only draw it as far as $F\left(CF = \dfrac{1}{3}CE\right)$. We have seen that Z cannot go along AC or along BC. Thus it is necessary that in the first instant it should come to D, where it corresponds to E and to F—that is to say, where it has travelled the value of CE and of FC. . . .

" This line X is related to the lines of direction of the two forces A and B in such a way that at any point from which two perpendiculars on the two lines are drawn, their relation to each other will be the reciprocal of that of the forces, or the relation of DE to DF. "

Lamy's demonstration is very similar to that on Newton. The simultaneity (1687) of the two demonstrations makes it seem however, that they were independent of each other. On the other hand, Lamy was accused of plagiarism from Varignon, who published his plan for a new mechanics at the same time. Lamy vigorously defended himself against this charge. If, like Duhem,[1] we put the emphasis on the words " *in a first instant,* " it is reasonable to believe that Lamy used an argument which would have been acceptable in modern mechanics. On the other hand, Varignon—who only cared for statics, a branch of the subject in which he showed great skill—did not progress beyond Aristotle's dynamics.

We must also say a little about *Néostatique* (1703), a rather original work due to Father Saccheri.

Saccheri regarded the *vis motrix* as proportional to the *impetus*, the term which he used to denote the absolute value of the velocity. As he was not concerned with the *impetus* of a body, starting from rest, in the first instant, this rule is equivalent to that of Aristotle. However, Saccheri arrived at an accord with Newton's dynamics. Thus he called the oriented velocity the *impetus vivus*, and used the term *impetus subnascens* for a quantity which, for a body of weight p, reduced to the

projection of the acceleration $\dfrac{p}{m}$ on the tangent. In identifying the

impetus subnascens as the *incrementum* of the *impetus vivus*, he was able to write down the Newtonian law of motion. This illustrates the extent to which the language and the ideas of the XVIIth Century were confused.

Father Ceva had drawn Saccheri's attention to the law of Beaugrand and Fermat which we have mentioned in connection with the controversy on geostatics.[2]

This is the law of an attraction which is proportional to the distance.

Saccheri had the merit of showing that, according to this law, the heaviness passed through a centre of gravity that was fixed in the body. Also, that a body falling freely from rest and subject to this law, arrived at the common centre of heavy bodies in a time which did not depend on its distance from the centre.

2. The statics of Varignon (1654-1722).

Varignon produced his *Project for a New Mechanics* in 1687, and the *New Mechanics or Statics* only appeared posthumously in 1725. At the beginning of the *Project* Varignon acknowledged the influence of Wallis and that of Descartes. The latter had declared that it was " a ridiculous thing to wish to use the argument of the lever in the pulley " ; Varignon persuaded himself that it was equally useless to treat the

---

[1] *O. S.*, Vol. II, p. 259.
[2] See above, p. 166.

inclined plane by starting from the lever. Of a more practical mind than his predecessors, he attached more weight to a study of the modes of equilibrium than to its necessity, and reduced everything to the principle of compound motions. " It seems to me that the physical reason for the effects that are most admired in machines is exactly that of compound motion. "

It is important to remark that Varignon interpreted the composition of forces and of motions in Aristotle's sense, for he remained consciously faithful to aristotelian dynamics. Indeed, in his *New Mechanics* he wrote—

" *Axiom VI.* — The velocities of a single body or of bodies of equal mass are as all the motive forces which are there used, or which cause these velocities ; conversely, when the velocities are in this ratio they are those of a single body or of bodies of equal masses. "

To Varignon, all force is analogous to the tension of a thread. In the diagrams which appear in his books, the hands holding the threads materialise the powers. He neglects all friction and even heaviness, which he identifies with a tension.

" *Requirement II.* — That it may be permissible to neglect the heaviness of a body and to consider it as if it had none ; but to regard it as a power which may be applied to the weight ; when it will be considered as weight, notice will be given. . . . "

Varignon starts from a general principle which he expresses in the following way.

" Whatever may be the number of forces or powers, directed as may be chosen, that act at once on the same body, either this body will not be displaced at all ; or it will travel along one path and along a line which will be the same as if, instead of being pushed in this way, compressed or drawn by all these powers at once, the body was only following the same line in the same direction by means of a single force or power equivalent or equal to the resultant of the meeting of all those forces. "

Therefore everything reduces to the determination of this resultant. And it is here that Varignon affirms his allegiance to the Ancients.

" It is what we are going to find by means of compound motions known to the ancients and the moderns— Aristotle treats them in the problems of mechanics ; Archimedes, Nicodemus, Dinostratus, Diocles, etc. . . . have used them for the description of the spiral, the conchoid, the cissoid, etc. . . . ; Descartes used them to explain the reflection and refraction of light ; in one word, all mathematicians use compound motions for the generation of an infinity of curved lines, and all correct physicists for determining the forces of impact or of oblique percussions, etc. . . .   Thus I claim nothing but the principle I indicated nearly forty years ago, and that I use once more for the explanation of machines. "

Given this, it is easy to see how Varignon reduced the composition of forces to that of velocities.   The superiority of Varignon's work in statics is a didactic one.   He treats all simple machines in detail by means of the composition of forces alone—this by ingenious procedures that are still commonly used.

In Duhem's opinion, it does not seem that the geometers of the XVIIth Century, and even of the XVIIIth Century, had attached any importance to the distinction that can now be made between the method of Newton and of Lamy on the one hand, and of Varignon on the other, in the matter of the proof of the rule of the parallelogram.   " The propositions that aristotelian dynamics, over a period of two thousand years, had made customary in physics were also familar to all minds. They were still invoked naturally on all occasions when conscience did not too violently conflict with the truths of the new Dynamics.   When Varignon, in 1687, produced his *Project of a New Mechanics*, he took as his starting point axioms which were said to have been borrowed from *Physica auscultatio* or *De Caelo* ; but at the same time Newton and Lamy showed that the same consequences could be obtained from an accurate dynamics. " [1]

3. Varignon and Torricelli's law of flow.

In the second Book of the *Principia* Newton had undertaken a proof of Torricelli's law of flow.   He remarked that a column of liquid falling freely in a vacuum assumed the shape of a solid of revolution whose meridian was a curve of the fourth degree.   Indeed, the velocity of each horizontal slice is proportional to the square root of the height from which it has fallen.   On the other hand, this same velocity is inversely proportional to the section of the column, and consequently to the square of the radius.   In a vessel having this shape and kept filled with water, it is clear that each particle of the fluid has its velocity of free fall and that, in consequence, Torricelli's law is justified.

Newton then imagines that in a cylindrical vessel whose base is pierced with a hole, the fluid separates into two parts.   One, *the cataract*, takes the shape of free fall of which we have spoken.   The other remains motionless.   It is easy to see that this solution contradicts the principles of hydrostatics.

Varignon had the merit of giving Torricelli's law a more natural

[1] *O. S.*, Vol. II, p. 260.

explanation. He assumed that the water remained sensibly immobile up to the immediate neighbourhood of the hole. At that point each particle instantaneously received, in the form of a finite velocity, the effect of the weight of the fluid that was above it. It is easy to see, taking account of the quantity of water flowing out, that the quantity of motion thus created in each particle is proportional to the square of the velocity. If the weight of the column of water above it is proportional to the height h, Varignon can retrieve Torricelli's law, $h = kv^2$.

Lagrange[1] criticised this argument by observing that the pressure cannot suddenly produce a finite velocity. This was a difficulty that could not detain Varignon, to whom all force was generated by velocity. But one can, like Lagrange, assume that the weight of the column acts on the particle throughout the time that it is leaving the vessel. If it is then assumed that the fluid remains sensibly stagnant in the very interior of the vessel, Torricelli's law can be verified.

[1] *Mécanique analytique*, Section VI, part I — *Sur les principes de l'hydrostatique*.

THE ORGANISATION AND DEVELOPMENT OF THE PRINCIPLES OF CLASSICAL MECHANICS IN THE XVIIIth CENTURY

CHAPTER ONE

JEAN BERNOULLI
AND THE PRINCIPLE OF VIRTUAL WORK
(1717)
DANIEL BERNOULLI
AND THE COMPOSITION OF FORCES
(1726)

1. Jean Bernoulli and the principle of virtual work.

Classical mechanics was born in the XVIIth Century. The organisation and development of the general principles had still to be accomplished—this was to be the work of the XVIIIth Century.

The achievements of Galileo, Huyghens and Newton appear rather as disjointed parts than as the continuous development of a single discipline. Their successors, on the other hand, were to participate in a collective labour which, in the hands of Lagrange, was to end in an ordered science whose form approached perfection.

In the preceding parts of this book we have treated each author in isolation from his contemporaries, and have attempted to follow the chronological order. In order to analyse the collective work of the XVIIIth Century, it will be more satisfactory if we devote each chapter to an attempt to collect together the work of different men that was relevant to one single topic.

Although, at the end of the XVIIth Century, Varignon had tried to found statics on the one law of the composition of forces, we see Jean Bernoulli, in a letter to Varignon himself (January 26th, 1717), taking up the generalisation of what was really the principle of virtual work. We have seen that this principle had been used implicitly as early as the XIIIth Century, by the School of Jordanus, and that later it had been affirmed by Descartes and Wallis.

Jean Bernoulli wrote, in the letter to Varignon, " Imagine several different forces which act according to different tendencies or in different directions in order to hold a point, a line, a surface or a body in equilibrium. Also, imagine that a small motion is impressed on the whole system of these forces. Let this motion be parallel to itself in any direction, or let it be about any fixed point. It will be easy for you to understand that, by this motion, each of the forces will advance or recoil in its direction ; at least if one or several of the forces do not have their tendency perpendicular to that of the small motion, in which case that force or those forces will neither advance nor recoil. For these advances or recoils, which are what I call *virtual velocities*, are nothing else than that by which each line of tendency increases or decreases because of the small motion. And these increases or decreases are found if a perpendicular is drawn to the extremity of the line of tendency of any force. This perpendicular will cut off a small part from the same line of tendency, in the neighbourhood of the small motion, which will be a measure of the virtual velocity of that force.

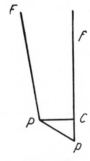

Fig. 83

" For example, let P be any point in a system which maintains itself in equilibrium. Let F be one of the forces, which would push or draw the point P in the direction FP or PF. Let Pp be a short straight line which the point P describes in a small motion, by which the tendency FP assumes the position fp. Either this will be exactly parallel to FP, if the small motion is, at every point, parallel to a straight line whose position is given ; or it will make an infinitely small angle with FP when this is produced, and if the small motion of the system is made around a fixed point. Therefore draw PC perpendicular to fp and you will have Cp for the *virtual velocity* of the force F, so that $F \times Cp$ is what I call the *energy*.

" Observe that Cp is either *positive* or *negative*. The point P is pushed by the force F. It is *positive* if the angle FPp is obtuse and *negative* if the angle *FPp* is acute. But on the contrary, if the point *P* is pulled, *Cp* will be *negative* when the angle *FPp* is obtuse and *positive* when it is acute. All this being understood, I form the following general proposition.

" In all equilibrium of any forces, in whatever way they may be applied and in whatever direction they may act—through intermediaries or directly—the sum of the positive energies will be equal to the sum of the negative energies taken positively. "

Jean Bernoulli's statement is much more general than those of his predecessors. Nevertheless, it must be remarked that the virtual displacements that are contemplated reduce to translations or rotations, to displacements in which the system behaves as a solid. Displacements of this kind are not necessarily compatible with the constraints of the system—they do not necessarily include the most general virtual displacement which is compatible with the constraints.

Jean Bernoulli's principle does not seem to have accomplished a modification of Varignon's point of view. The latter was content to verify the principle in a large number of examples, which he treated with the methods to which he was accustomed.

2. DANIEL BERNOULLI AND THE COMPOSITION OF FORCES.

In a memoir which appeared in 1726, called *Examen principiorum mechanicae et demonstrationes geometricae de compositione et resolutione virium*, Daniel Bernoulli set out to show that the law of the composition of forces was of *necessary*, and not of *contingent*, truth. We shall find that Euler and d'Alembert had similar preoccupations in other fields. The search for such a separation of purely rational truths from those which are subject to the uncertainties of, and correction by, experiment, was ever present in learned minds throughout the XVIIIth Century.

The question, by its nature, is illusory. But the influence of Bernoulli's demonstration remained alive, and even Poisson was subject to it in 1833.

Bernoulli regarded the hypothesis of the composition of motions on which Varignon had based his statics to be of a *contingent* kind. But a necessary truth can arise from *two* contingent hypotheses. In particular, the necessary law of the composition of forces depends, not only on the contingent hypothesis of the proportionality of the forces to the velocities that they produce but also, on the following hypothesis— A force which acts on a body that is already moved by another force impresses the same velocity on the body as if the latter were at rest.

Basically, the development of Bernoulli's demonstration is the following [1]—

Hypothesis I. — The composition of forces is associative.

Hypothesis II. — The composition of two forces in the same direction reduces to algebraic addition.

Hypothesis III. — The resultant of two equal forces is directed along their internal bisector— " a metaphysical axiom that must be regarded as a necessary truth. "

With this basis, Bernoulli shows that if three forces are in equilibrium, so too are three forces which are the multiples of the first by the same number. He then establishes that the resultant of two equal forces at right angles is the diagonal of the square that has these two forces as sides.

He continues with a consideration of two unequal rectangular forces and finds that the resultant is equal to the diagonal of the rectangle of which these two components form two sides. He also discusses the direction of the resultant.

Bernoulli then treats pairs of components forming a rhombus whose angle is equal to $\left(\dfrac{1}{2^n}\right)\left(\dfrac{\pi}{2}\right)$. Then, in order, components forming any rhombus, a rectangle, and a parallelogram.

---

[1] For further details, *cf.* JOUGUET, *L. M.*, Vol. II, p. 58.

CHAPTER TWO

THE CONTROVERSY ABOUT LIVING FORCES

We know that, as early as 1686,[1] in criticising the Cartesian notion of the conservation of quantities of motion, Leibniz had suggested that the " force " acquired by a body falling freely should be evaluated by the height to which this body could rise. Thus a body whose velocity is twice that of another is endowed with a force that is four times a great.

The Abbé de Catelan protested that the body effected this ascent in twice the time. To produce a quadruple effect in twice the time is not to have a quadruple force, but only one which is twice as great. A child, in time, and bit by bit, will carry a sack of corn weighing 240 pounds. All force will be infinite if no regard is paid to time.

After much hesitation, Jean Bernoulli came round to the opinion of Leibniz. In 1724 the *Académie des Sciences*, without using the words *living force*, chose the subject of the *communication of motion* for competition.

Father Mazière, an adversary of the doctrine of living forces, was the successful competitor, in spite of a contribution from Jean Bernoulli that defended Leibniz. In this debate MacLaurin, Stirling and Clarke were opposed by the supporters of Leibniz—s'Gravesande, Wolf and Bulfinger.

Bernoulli believed that the law $v = k\sqrt{h}$ was related to that of gravity and that it was not an independent *a priori* law. Bodies would rise to infinity if no cause prevented them. The limitation is due to gravity, whose reiterated obstacles consumed a body's force of ascent. Bernoulli made use of other examples, of which the following is typical.

If a perfectly elastic sphere A, moving with the velocity AC, collides obliquely with an identical sphere which it projects in the direction CD, the body C will be displaced on CD with the velocity $CD = BC$, while the body A will continue its journey with the velocity $CE = CB$.

---

[1] See above, p. 219.

Now the sum of the forces after the impact must be the same as the sum of the forces before the impact. This would be impossible if the force were proportional to the velocity, for $CE + CD > AC$. On the other hand, this relation is verified if the force is proportional to the square of the velocity, for $\overline{AC}^2 = \overline{CD}^2 + \overline{CE}^2$.

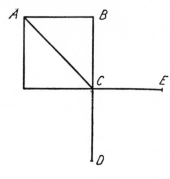

Fig. 84

In a *Dissertation of the Estimation and the Measurement of the Motive Forces of Bodies* (1728) de Mairan, like the Abbé de Catelan, opposed the evaluation of the force that the followers of Leibniz had suggested. His premises were simple.

" As soon as I conceive that a body may be in motion, I conceive of a force that makes it move [to be understood as the *vis motrix* or the force of a body in motion, and not the corresponding *dead force*, which is zero for uniform motion]. A uniform motion can never indicate to us another measure of the force than the product of the simple velocity and the mass. "

Here is the argument—

" A massive body having two units of velocity is in such a state that it can mount to a height that is four times as great as that to which a body with only one unit of velocity would mount.

" This proportion implies common measure. This common measure is the time ; at least I can take the time or the times to be equal. . . .

" Now given this, in the effects of a body which has twice as much velocity, I only find an effect which is double and not quadruple— a distance travelled which is double, and a displacement of matter which is double, in equal times. From which I conclude, by the very principle of the proportionality of causes to their effects, that the Motive

Force is not quadruple but only double, as the simple velocity and not the square of the velocity. "

And de Mairan adds, " Strictly speaking, the concept of motion only includes uniformity. All motion should, on its own, be uniform, just as it should be effected in a straight line ; the acceleration and retardation are limitations which are *foreign to its nature*, as the curve that it is made to describe is to its proper direction. . . .

" It is not the distances travelled by the body in retarded motion that give the evaluation and the measure of the motive force, but rather, the distances which are *not travelled*, and which should be travelled, in each instant by uniform motion. These distances which are not travelled are proportional to the simple velocities. And therefore the distances which correspond to a retarded or decreasing motive force, in so much as it is consumed in its action, are always proportional to this force and to the motion of the body, just as much in retarded motions as in uniform motions. "

To explain this " kind of paradox, " de Mairan considers the example of two bodies, A and B, which ascend along AD and $B\delta$. The body A has two units of velocity and B has only one.

" If nothing opposes its *motive force*, in the first time B will travel the two toises $B\delta$ without losing any part of this force or any part of the unit of velocity which gives rise to it. But because the contrary impulsions of the heaviness, which are continually applied to it succeed in consuming this force and its velocity, and in completely stopping it, the body will only travel one toise in its retarded motion.

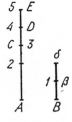

Fig. 85

" In the same way, A would travel four toises in the first instant. The impulsions of the heaviness make it fall back through one toise, so that it only travels three. These impulsions have consumed one unit of force and one unit of velocity, as for B. But A remains with one unit and, at C, it finds itself in the initial case of B. It therefore has what it needs to travel the two toises CE. But the impulsions of the heaviness oppose it and it only travels CD, being pulled back through the one toise ED."

Thus the distance which is not travelled by B in the first instant is $\beta\delta$. In the first instant the distance not travelled by A is CD, and in the second, is DE.

This discussion is interesting—its metaphysical content is so apparent that we shall not emphasise it.

Supporters and adversaries of the doctrine of living forces opposed each other with examples of impacting bodies.

Thus Herman considers a perfectly elastic body M, of mass 1 and velocity 2, colliding with a motionless sphere N of mass 3. The body N will take, after the impact, the velocity 1 while the body M will be thrown back with the velocity 1. If M then meets a motionless body O of mass 1, it can communicate its velocity to the latter and remain at rest. Therefore the force of M, which has mass 1 and velocity 2, is equivalent to four times the force of a body of mass 1 and velocity 1, which verifies the law of living forces and contradicts that of quantities of motion.

De Mairan observed that this coincidence was accidental and stemmed from the equality $2 + 2 = 2 \times 2$. For his part, he considered a body M of mass 1 and velocity 4 which he arranged to collide with a body N of mass 3 which was initially at rest. If M communicates a velocity 2 to N, the force of N is as 6. The body M, which keeps the velocity 2, can transfer this to a body O of mass 1, initially at rest. The total force of M is therefore as $6 + 2 = 8$, and not as 16 as the law of living forces would require.

The Marchioness of Châtelet came round to the doctrine of living forces rather late in the day, and added an *erratum* to her book on the nature of fire (1740). While Koenig was a supporter of Leibniz, Maupertuis and Clairaut remained indifferent to this controversy. In the meantime, de Mairan tried to convince the Marchioness of Châtelet and, in 1741, Voltaire himself proclaimed his doubts about the measure of living forces.

The error of the Cartesians, which was corrected in the course of the controversy by de Mairan, was that of reckoning the quantity of motion as $m|v|$, without regard to the direction of the velocities. The reader will easily verify, in all the examples which have been cited—which are examples of elastic impact—that if the direction is introduced, that is, if quantities of motion $m\vec{v}$ are considered, then the quantities $\sum m\vec{v}$ and $\sum mv^2$ are both conserved. Therefore the controversy of living forces was based on a mis-statement of the doctrine. It rested on a misunderstanding concerning the definition of quantity of motion which, as d'Alembert observed, divided the geometers for more than thirty years.

CHAPTER THREE

EULER AND THE MECHANICS OF A PARTICLE
(1736)

Euler (1707-1783) was concerned with all branches of dynamics, and we shall have occasion to return to his work in different connections. For the moment, we shall confine ourselves to the basic ideas of his treatment of the dynamics of a particle. This is found in *Mechanica, sive motus scientia analytice exposita* which was published in 1736.

The very title is a programme. Euler had read the great creators of mechanics, especially Huyghens and Newton, and he set out to fashion mechanics into a rational science by starting from definitions and logically ordered propositions. He tried to demonstrate the laws of mechanics in such a way that it would be clear that they were not only correct, but also *necessary truths*.

To Euler, *power* or *force* is characterised by the modification of the motion of a particle that is produced by it. A power acts along a definite direction at each instant. This is what Euler expresses in the following definitions.

" *Potentia est vis corpus vel ex quiete in motum perducens, vel motum ejus alterans.* "

" *Directio potentiae est linea recta secundum quam ea corpus movere conatur.* "

In passing we remark that, in Euler's work, the term " *corpus* " denotes a particle.

In the absence of force a particle either remains at rest, or is animated with a rectilinear and uniform motion. Euler expresses this principle with the help of the concept of " force of inertia. "

" *Vis inertiae est illa in omnibus corporibus insita facultas vel in quiete permanendi vel motum uniformiter in directum continuendi.* "

Euler believes that " the comparison and the measurement of different powers should be the task of Statics. " Euler's dynamics is therefore primarily based on the notion of force, which he borrows directly from statics in accordance with Galileo's procedure.

Euler attempted to show that the composition or the equivalence of forces in statics could be extended to their dynamical effects. In fact, he was here concerned with a postulate. He also distinguished between *absolute* powers, such as gravity, that acted indifferently on a body at rest or in motion, and *relative* powers, whose effects depended on the velocity of the body. As an example of such a power, he cited the force exerted by a river on a body—this force disappears when the velocity of the body is the same as that of the river.

In order to determine the effect of a relative power, an absolute power is associated with it, at least when the body has a known velocity.

We return to the *vis inertiae* in the sense that Euler used it. For any body, this is proportional to the quantity of matter that the body contains.

" The force of inertia is the force that exists in every body by means of which that body persists in its state of rest or of uniform motion in a straight line. It should therefore be reckoned by the force or power that is necessary to take the body out of its state. Now different bodies are taken out of their state to similar extents by powers which are proportional to the quantities of matter that they contain. Therefore their forces of inertia are proportional to these powers, and consequently, to the quantities of matter. " Euler assigns the same *vis inertiae* to one body, whether it is at rest or in motion. For in both cases the body is subject to the same action and the same absolute power.

Here we see a systematisation of Newtonian ideas. Basically Euler introduces the mass—in the guise of a logical deduction—by means of the physical assertion of proportionality between the powers necessary to produce a given effect and the quantities of matter.

As an example of Euler's analysis, we shall give his treatment of the following problem.

" *Proposition XIV. — Problem. — Being given the effect of an absolute power on a particle at rest, to find the effect of the same power on the same particle when the latter is moving in any way.* "

The absolute power which is given will make a body A, initially at rest, travel the path $dz = AC$ in the time dt.

If A has the velocity c, in the absence of any power it will travel the path $AB = cdt$ in the time dt.

But the given power, being absolute, acts on A in motion in the

same way as it acts on A at rest. Therefore the effect of the power
is compounded with that of the velocity, and the body A comes to
D, where $BD = AC$.

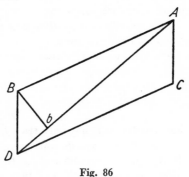

Fig. 86

Under the effect of the given power, the velocity of A will become

$$c + dc = \frac{AD}{dt}.$$

A simple geometrical argument shows that

$$dc = \frac{dz}{dt} \cos \widehat{BAC}.$$

Strictly speaking it would be more natural to regard the effect
of the power as being the increase of the velocity between the time
t and the time $t + dt$; that is, to consider the quantity $2dc$ instead
of the difference between the initial velocity c and the mean velocity
of A during the time dt.[1]

Euler then studies the effect of a power B on a body when the
effect of a power A on the same body is given. He concludes—

" If a body is affected by many powers, at first it may be thought
of as divided into as many parts, on each of which one of the powers
acts. Then, when the different parts have been drawn by their respective powers for an element of time, it is imagined that they suddenly
unite. When this is accomplished, the position of their reunion will
be that at which the whole body would have arrived in the same time
by the simultaneous action of all the powers. The truth of this principle can be illustrated by remarking that the parts of a body can be
held together by very strong springs which though they act in an
undefined manner, can be supposed to relax completely in the interval

---

[1] *Cf.* JOUGUET, *L. M.*, Vol. II, p. 43.

of time, and to contract suddenly with an infinite force, afterwards, in such a way that the conjunction of the separated parts takes no time. "

Thus Euler's law of dynamics takes the form—

The increase, dc, of the velocity is proportional to pdt, where p is the power acting on the body during the time dt. This applies to a single body ; if several bodies are considered simultaneously, their masses must be introduced.

Therefore this law emphasises the *impulse of the force* during an elementary time, or the impulse that gives rise to an increase of momentum.

Euler declared that this law was not only true, but also a *necessary truth*, and that a law identifying mdc as p^2dt or as p^3dt would imply a contradiction. Clearly this is an illusion of the author.

Euler's treatise then continues with a study of a large number of problems. First he treats a free particle, and concludes with a particle bound on a curve or a surface, either in a vacuum or a resisting medium. His work was the first to merit, for the order and the precision of its demonstrations, the name of a treatise of rational mechanics.

CHAPTER FOUR

JACQUES BERNOULLI
AND THE CENTRE OF OSCILLATION
(1703)
D'ALEMBERT'S TREATISE ON DYNAMICS
(1743)

1. Jacques Bernoulli and the centre of oscillation.

In 1703 Jacques Bernoulli returned to the famous problem of the search for a centre of oscillation, and gave a solution of it which contained the germ of d'Alembert's principle. Jacques Bernoulli's paper was called " *General demonstration of the centre of balancing and of oscillation deduced from the nature of the lever.* "

He considers a lever which is free to turn about a point A and whose different arms carry weights or powers which act perpendicularly to the arms. If the powers are divided into two groups that act on the lever in opposite senses, and if the sum of the products of the arms of the lever and the powers has the same absolute value for each group, then the lever will remain in equilibrium. This had been shown by Mariotte in the *Treatise on the percussion of bodies.*

Given this, let A represent the axis of suspension, and let AC and AD join A to two arbitrary elements of a compound pendulum (for simplicity assumed to be plane). Then let AM be the simple pendulum isochronous with the compound pendulum.

Consider the motion of the elements C, D and M of the compound pendulum. Their velocities are proportional to AC, AD and AM. At each instant the gravity adds an impact or an impulse which is represented by MN, CO, DP, " short vertical and equal lines. " Take NK, OT and PV perpendicular to the arcs MK, CT, DV.

Bernoulli considers the " motions " MN, CO, DP as being decomposed into motions MK and KN ; CT and TO ; DV and VP. The motions KN, TO, VP " distribute themselves over the whole axis A

and there lose themselves completely. " Because of the isochronism of the points C, D and M, the motions MK, CT and VD suffer " some change. " If, for example, M comes to K (without alteration), then C comes to R and D to S, and the arcs MK, CR and DS will be similar. The effort of gravity acting on the point C is not exhausted at R and " the remainder, RT, must be used to push the body D along VS. " But D itself resists as much as it is pushed, and everything happens as if D travelled to S—as if there were a force " which tries to repel it from S to V. "

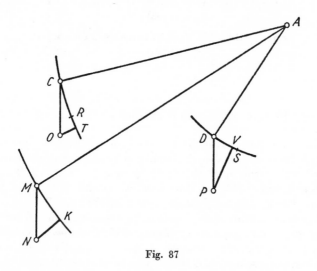

Fig. 87

To sum up, the lever CAD is in equilibrium under the action of weights like C, " pulling or pushing from one side with forces or velocities RT," and weights like D, pulling or pushing in the opposite sense.

Therefore Bernoulli writes $\Sigma (C \times CA \times RT) = \Sigma (D \times AD \times VS)$ and, from this, deduces the solution of the problem of the centre of oscillation.

2. The introductory argument of d'Alembert's Treatise on dynamics.

The first edition of d'Alembert's *Traité de dynamique* is dated 1743. Here we shall follow an edition of 1758, which was corrected and added to by the author.

In an introductory discussion, d'Alembert explains his philosophy of mechanics. The Sciences are divided into two groups—those which are based on principles which are necessarily true and clear in themselves; and those which are based on physical principles, experimental truths, or simply on hypotheses. Mechanics belongs to the first category of purely rational sciences, although it appears to us as less direct than Geometry and Algebra. It has failed to clarify the mystery of impenetrability, the enigma of the nature of motion, and the metaphysical principle of the laws of impact. . . .

The best method of discussing any part of mathematics " is to regard the particular subject of that science in the most abstract and direct way possible ; to suppose nothing, and to assume nothing about that subject, that the properties of the science itself does not suppose. "

D'Alembert sets out " to throw back the boundaries of mechanics and to smooth out the approach to it . . . and, in some way, to achieve one of these objects by means of the other. That is, not only to deduce the principles of mechanics from the clearest concepts, but also to apply them to new ends. " He strives " to make everything clear at once ; both the futility of most of the principles that have so far been used in mechanics, and the advantage that can be obtained from the combination of others, for the progress of that Science. In a word, *to extend the principles and reduce them in number.* "

The nature of motion has been much discussed. " Nothing would seem more natural than to conceive of it as the successive application, of the moving body to the different parts of infinite space. " But the Cartesians, " a faction that, in truth, now barely exists, " refuse to distinguish space from matter. In order to counter their objections, d'Alembert makes a distinction between *impenetrable* space, provided by what are properly called bodies, and space pure and simple, penetrable or not, which can be used to measure distances and to observe the motion of bodies.

" The nature of time is to run uniformly, and mechanics supposes this uniformity. " This is Newtonian.

" A body cannot impart motion to itself." There must be an external cause in order to move it from rest. But " if the existence of motion is once supposed, without any other particular hypothesis, the most simple law that a moving body can observe in its motion is the law of uniformity, and consequently, this is that which it must conform to. . . . Therefore motion is inherently uniform. "

D'Alembert defines the *force of inertia* as the property of bodies of remaining in their state of rest or motion. Among the means that can alter the motion of a body, apart from constraints, he only allows two—impact (or impulse) and gravity (or, more generally, attraction).

In this connection it seems that d'Alembert criticises the very principle of Euler's mechanics.

" Why have we gone back to the principle, which the whole world now uses, that the *accelerating or retarding force is proportional to the element of the velocity* ?   A principle supported on that single vague and obscure axiom that the effect is proportional to its cause.

" We shall in no way examine whether this principle is a necessary truth or not.   We only say that the evidence that has so far been produced on this matter is irrelevant.   Neither shall we accept it, as some geometers have done, *as being of purely contingent truth, which would destroy the exactness of mechanics and reduce it to being no more than an experimental science.*   We shall be content to remark that, true or false, clear or obscure, it is useless to mechanics and that, consequently, it should be abolished. "

This shows in what sense d'Alembert interpreted the task of making mechanics into a rational science, and the extent to which he valued his own principle.

D'Alembert made appeal to a principle of the *composition of motions*, of which he intended to give simple evidence.

When a body changes in direction, its motion is made up of the *initial* motion and an *acquired* motion.   Conversely, the initial motion can be compounded of a motion which is *assumed* and a motion which is *lost*.

D'Alembert established the laws of motion in the presence of any obstacle in the following way.   The motion of the body *before meeting the obstacle* is decomposed into two motions— one which is *unchanged*, and another which is *annihilated* by the obstacle.

If the obstacle is *insurmountable*, the laws of equilibrium are used. These laws are expressed by a relation of the kind

$$\frac{m}{m'} = -\frac{v'}{v}$$

where v, v' are the velocities with which the masses m, m' *tend to move*.

Only when there is perfect symmetry, or when

$$m = m' \qquad\qquad v' = -v$$

does the problem appear inherently clear and simple to d'Alembert, and he tries to reduce all other situations to this one.   We have seen that this was an illusion which Archimedes had in his investigation of the equilibrium of the lever.

And d'Alembert concludes, " The principle of equilibrium, together with the principles of the force of inertia and of compound motion, therefore leads us to the solution of all problems which concern the motion of a body in so far as it can be stopped by an impenetrable and immovable obstacle—that is, in general, by another body to which it must necessarily impart motion in order to keep at least a part of its own. From these principles together can easily be deduced the laws of the motion of bodies that collide in any manner whatever, or which affect each other by means of some body placed in between them and to which they are attached. "

Lagrange said, and it is often repeated, that d'Alembert had reduced dynamics to statics by means of his principle. The last quotation shows clearly that d'Alembert himself did not accept such a simple interpretation. On the contrary, he stressed the fact that " the three principles of the force of inertia, of compound motion and of equilibrium are essentially different from each other. "

D'Alembert's beliefs are thus clearly expressed in the first pages on his introduction. But he also made clear his view on the problems which were popular in his time. Above all, he intended to take account of motion without being concerned with *motive causes* ; he completely banished the forces inherent to bodies in motion, " as being obscure and metaphysical, and which are only able to cover with obscurity a subject that is clear in itself. "

This is why d'Alembert refused " to start an examination of the celebrated question of *living forces*, which has divided the geometers for thirty years. " To him, this question was only a dispute about words, for the two opposing sides were entirely in agreement of the fundamental principles of equilibrium and of motion. Their solutions of the same problem coincided, " if they were sound. "

D'Alembert also discussed the question of knowing *whether the laws of mechanics are of necessary or contingent truth.* This question had been formulated by the Academy of Berlin.

In order that this question may have a meaning, it is necessary to dispense with " every sentient being capable of acting on matter, every will of intellectual origin. " It is said that d'Alembert rejected every finalist explanation involving the wisdom of the Creator—we shall return to this in connection with the principle of least action.

To d'Alembert, the principles of mechanics are of necessary truth. " We believe that we have shown that a body left to itself must remain forever in its state of rest or of uniform motion ; that if it tends to move along the two sides of a parallelogram at once, the diagonal is the direction that it must take ; that is, that it must select from all the others. Finally, we have shown that all the laws of the communication of motion between bodies reduce to the laws of equilibrium, and that the laws of equilibrium themselves reduce to those of the equilibrium of two equal bodies which are animated in different senses by equal virtual velocities. In the latter instance, the motions of the two bodies evidently cancel each other out ; and by a geometrical consequence, there will also be equilibrium when the masses are inversely proportional to the velocities. It only remains to know whether the case of equilibrium is unique—that is, whether one of the bodies will necessarily force the other to move when the masses are no longer inversely proportional to the velocities. Now it is easy to believe that as soon as there is one possible and necessary case of equilibrium, it will not be possible for others to exist without the laws of impact —which necessarily reduce to those of equilibrium—becoming indeterminate. And this cannot be, since, if one body collides with another, the result must necessarily be unique, the inevitable consequence of the existence and the impenetrability of bodies. ''

3. D'ALEMBERT AND THE CONCEPT OF ACCELERATING FORCE.

Of all the causes that could influence a body, d'Alembert was of the opinion that only impulse (that is, impact) was perfectly determinate. All other causes are entirely unknown to us and can only be distinguished by the variation of motion which they produce. The '' accelerating force '' φ is introduced by the relation $\varphi dt = du$, a relation between the time t and the velocity u—the only observable kinematic quantities. This relation is the *definition* of φ.

Therefore, to d'Alembert, this force was a derived concept, though to Daniel Bernoulli and Euler it constituted a primary concept.

To Daniel Bernoulli, the law $\varphi dt = du$ was a contingent truth ; to Euler, a necessary truth.

D'Alembert wrote, '' for us, without wishing to discuss here whether this principle is a necessary or a contingent truth, we shall be content to take it as a definition, and to understand by the phrase ' accelerating force ', merely the quantity to which the increase in velocity is proportional. '' [1]

4. D'ALEMBERT'S PRINCIPLE.

D'Alembert's principle was made the subject of a letter to the *Académie des Sciences* as early as 1742. In this book, we shall follow the presentation of the principle which appears in the 1758 edition of the *Traité de Dynamique* (2nd Part, Chapt. I, p. 72).

[1] *Traité de Dynamique*, cor. VI, p. 25 (1758 edition).

PRESENTATION OF THE PRINCIPLE

" Bodies only act on each other in three ways that are known to us— either by immediate impulse as in ordinary impact; or by means of some body interposed between them and to which they are attached ; or finally, by a reciprocal property of attraction, as they do in the Newtonian system of the Sun and the Planets.   Since the effects of this last mode of action have been sufficiently investigated, I shall confine myself to a treatment of bodies which collide in any manner whatever, and of those which are acted upon be means of threads or rigid rods. I shall dwell on this subject even more readily because the greatest geometers have only so far (1742) solved a small number of problems of this kind, and because I hope, by means of the general method which I am going to present, to equip all those who are familar with the calculations and principles of mechanics so that they can solve the most difficult problems of this kind.

DEFINITION

" In what follows, I shall call *motion* of a body the velocity of this same body and shall take account of its direction.   And by *quantity of motion*, I shall understand, as is customary, the product of the mass and the velocity.

GENERAL PROBLEM

" *Let there be given a system of bodies arranged in any way with respect to each other ; and suppose that a particular motion is imparted to each of these bodies, which it cannot follow because of the action of the other bodies— to find the motion that each body must take.*

SOLUTION

" Let A, B, C, etc. . . . be the bodies that constitute the system and suppose that the motions a, b, c, etc. . . . are impressed on them ; let there be forces, arising from their mutual action, which change these into the motions a, $\overline{b}$, c, etc. . . .   It is clear that the motion a impressed on the body A can be compounded of the motion a which it acquires and another motion α.   In the same way the motions b, c, etc. . . . can be regarded as compounded of the motions $\overline{b}$ and β, c and $\varkappa$, etc. . . . From this it follows that the motions of the bodies A, B, C, etc. . . . would be the same, among themselves, if instead of their having been given the impulses a, b, c, etc. . . . they had been simultaneously given the twin impulsions a and α, $\overline{b}$ and β, c and $\varkappa$, etc. . . . Now, by supposition, the bodies A, B, C, etc. . . . have assumed, by their own action, the motions a, $\overline{b}$, c, etc. . . . Therefore the motions α, β, $\varkappa$, etc. . . . must be such that they do not disturb the motions a, $\overline{b}$, c, etc. . . . in any way. That is to say, that if the bodies had only received the motions α, β, $\varkappa$, etc. . . . these motions would have been cancelled out among themselves, and the system would have remained at rest.

" From this results the following principle for finding the motion of several bodies which act upon each other. *Decompose each of the motions a, b, c, etc. . . . which are impressed on the bodies into two others, a and α, $\overline{b}$ and β, c and $\varkappa$, etc. . . . which are such that if the motions a, $\overline{b}$, c, etc. . . . had been impressed on the bodies, they would have been retained unchanged ; and if the motions α, β, $\varkappa$, etc. . . . alone had been impressed on the bodies, the system would have remained at rest. It is clear that a, $\overline{b}$, c, etc. . . . will be the motions that the bodies will take because of their mutual action. This is what it was necessary to find.* "

5. D'ALEMBERT'S SOLUTION OF THE PROBLEM OF THE CENTRE OF OSCILLATION.

Although d'Alembert's principle is perfectly clear, its application is difficult, and the *Traité de Dynamique* remains a difficult book to read.

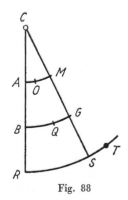

Fig. 88

As a concrete example of its application, we shall give d'Alembert's solution of the celebrated problem of the centre of oscillation.[1]

" *Problem.* — *To find the velocity of a rod CR fixed at C, and loaded with as many weights as may be desired, under the supposition that these bodies, if the rod had not prevented them, would have described infinitely short lines AO, BQ, RT, perpendicular to the rod, in equal times.*

" All the difficulty reduces to finding the line RS travelled by one of the bodies, R, in the time that it would have travelled RT. For then the velocities BG, AM, of all the other bodies are known.

" Now regard the impressed velocities, RT, BQ, AO as being composed of the velocities RS and ST; BG and $-GQ$; AM and $-MO$. By our principle, the lever CAR would have

---

[1] *Traité de Dynamique*, p. 96.

remained in equilibrium if the bodies R, B, A had received the motions ST, $-GQ$, $-MO$ alone.

" Therefore

$$A \cdot MO \cdot AC + B \cdot QG \cdot BC = R \cdot ST \cdot CR.$$

" Denoting AO by a, BQ by b, RT by c, CA by r, CB by r', CR by ϱ and RS by z, we will have

$$R(c\text{-}z)\varrho = Ar\left(\frac{zr}{\varrho} - a\right) + Br'\left(\frac{zr'}{\varrho} - b\right).$$

" Consequently

$$z = \frac{Aar\varrho + Bbr'\varrho + Rc\varrho^2}{Ar^2 + Br'^2 + R\varrho^2}.$$

" *Corollary.* — Let F, f, φ be the motive forces of the bodies A, B, R. The accelerating force will be found to be

$$\frac{Fr + fr' + \varphi\varrho}{Ar^2 + Br^2 + R\varrho^2} \times \varrho$$

on giving a, b, c, their values $\dfrac{F}{A}, \dfrac{f}{b}, \dfrac{\varphi}{R}$. Therefore, if the element of arc described by the radius CR is taken to be ds and the velocity of R to be u, then, in general,

$$\frac{Fr + fr' + \varphi\varrho}{Ar^2 + Br^2 + R\varrho^2} \varrho ds = u du$$

whatever the forces F, f, φ may be. It is easy, by this means, to solve the problem of centres of oscillation under any hypothesis.

6. THE PRIORITY OF HERMAN AND EULER IN THE MATTER OF D'ALEMBERT'S PRINCIPLE.

After recalling Jacques Bernoulli's solution of the problem of the centre of oscillation, d'Alembert remarks that Euler, in Volume III of the old Commentaries of the Academy of Petersbourg (1740), had used the principle according to which the powers $R \cdot RS$, $B \cdot BG$, $A \cdot AM$ must be equivalent to the powers $R \cdot RT$, $B \cdot BQ$, $A \cdot AO$. " But M. Euler has in no way demonstrated this principle and this, it seems to me, can only be done by means of ours. Moreover, the author has only applied this principle to the solution of a small number of problems concerning the oscillation of flexible or inflexible bodies, and the solution that he has given to one of these problems is not correct. [This was the problem of the oscillation of a solid body on a plane.] This shows to what extent our principle is preferable for solving not only problems of that kind, but in general, all questions of dynamics. " [1]

Lagrange had the following comment to make on this matter.

" If it is desired to avoid the decompositions of motions that d'Alembert's principle demands, it is only necessary to establish immediately the equilibrium between the forces and the motions they generate, but taken in the opposite directions. For if it is imagined that there is impressed on each body the motion that it must take, in the opposite sense, it is clear that the system will be reduced to rest. Consequently, it is necessary that these motions should destroy those that the body had received and which they would have followed without this interaction. Thus there must be equilibrium between all these motions or between the forces which can produce them.

" This method of recalling to mind the laws of Dynamics is certainly less direct than that which follows from d'Alembert's principle, but it offers greater simplicity in applications. It reduces to that of Herman [2] and of Euler,[3] who used it in the solution of many problems in Mechanics, and which is found in many treatises on mechanics under the name of *d'Alembert's Principle.* "

However clear these priorities may be, they do not detract from the originality of d'Alembert's conceptions. His work stands out because of its philosophic breadth of view, because of its property of unifying and generalising, and its equal is not found among the work of his immediate predecessors.

7. D'ALEMBERT AND THE LAWS OF IMPACT.

D'Alembert systematically applied his principle to the solution of all the problems which appear in his *Traité,* whether they concern bodies which are supported by threads or rods, bodies which oscillate on planes, bodies which interact by means of threads on which they can run freely, or different modes of impact.

In the problems of impact d'Alembert, at first, only considers " hard bodies " (that is, bodies deprived of their elasticity). Thus, if a body of mass M and velocity U collides directly with a body of mass m and velocity u, d'Alembert writes the following relations between the velocities.

$$u = v + u - v$$
$$U = V + U - V$$

---

[1] *Traité de Dynamique*, p. 101.

[2] *Phoronomia, sive De viribus et motibus corporum solidorum et fluidorum*, Amsterdam, 1716.

[3] The paper cited by d'ALEMBERT (see the beginning of this §).

Here v and V are the velocities of the first and the second bodies after the impact.

After the impact $V = v$ and because of the principle, $m(u—v) + M(U—V) = 0$. Therefore V and $v = \dfrac{mu + MU}{M + m}$.

D'Alembert next deduces the laws of the impact of elastic bodies from those of the impact of hard bodies by the following procedure. " *If as many bodies as may be desired collide with each other so that when it is supposed that they are perfectly hard and without elasticity, they all remain at rest after the impact ; I say that if they are of perfect elasticity each one will rebound after the impact with the velocity it had before the impact.* For the effect of the elasticity is to give back to each body the velocity which it has lost because of the action of the others. " [1]

Thus d'Alembert separated the theory of impact from all appeal to the conservation of living forces.

8. D'ALEMBERT AND THE PRINCIPLE OF LIVING FORCES.

D'Alembert prepared the way for Lagrange by setting out to show that the principle of the conservation of living forces was a *consequence* of the laws of dynamics for systems with restraints composed of threads and inflexible rods, just as the laws of impact were a consequence of this same principle. Without giving a general demonstration of this fact, d'Alembert gave " the principles sufficient for obtaining the demonstration in every particular case. "

We shall confine ourselves here to a very simple case.

" Imagine two bodies, A and B, of an infinitely small extension, to be attached to an inflexible rod AB. And suppose that any directions and velocities are imparted to these bodies, and that these velocities are represented by the infinitely short lines AK and BD. According to our principle, it is necessary to construct the parallelograms MC and NL, in which $LC = AB$ and $B \times BM = A \times AN$. The velocities and the directions of the bodies B and A will be BC and AL. Now $\overline{BC^2} = \overline{BD^2} — 2CE \cdot CD—\overline{CD^2}$ and $\overline{AL^2} = \overline{AK^2} + 2PL \cdot KL — \overline{KL^2}$. Therefore $B \cdot \overline{BC^2} + A \cdot \overline{AL^2} = A \cdot \overline{AK^2} + B \cdot \overline{BD^2} + A(2PL \cdot KL — \overline{KL^2}) — B(2CE \cdot CD + \overline{CD^2})$, which reduces to $A \cdot \overline{AK^2} + B \cdot \overline{BD^2} — A \cdot \overline{KL^2} — B \cdot \overline{CD^2}$, since $CE = PL$ and $A \cdot KL = B \cdot CD$.

" Therefore
$$B \cdot \overline{BC^2} + A \cdot \overline{AL^2} = A \cdot \overline{AK^2} + B \cdot \overline{BD^2} — A \cdot \overline{KL^2} — B \cdot \overline{CD^2} \text{ " [2]}$$

---

[1] *Traité de Dynamique*, p. 218.
[2] *Ibid.*, p 253.

CHAPTER FIVE

THE PRINCIPLE OF LEAST ACTION

1. RETURN TO FERMAT.

On January 1st, 1662, Fermat wrote to C. de la Chambre concerning refraction.

" M. Descartes has never demonstrated his principle. For apart from the fact that comparisons are not of much use as foundations for demonstrations, he uses his own in the wrong way and even supposes that the passage of light is easier in dense bodies than in rare ones, which is clearly false. "

In his investigation of the refraction of light, Fermat starts " *from the principle, so common and so well-established, that Nature always acts in the shortest ways.* "

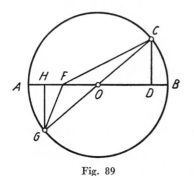

Fig. 89

He first shows that in a particular numerical example, the rectilinear path is not the most rapid for the traversal of two media by light.

If the medium AGB is supposed to be more dense than the medium ACB, " so that the passage through the rarer medium is twice as easy as that through the denser one, " the time taken by the light in going from C to G by the straight line COG can be represented " by the sum of half CO and the whole of OG. "

Taking $CO = 10$, $HO = OD = 8$ and $OF = 1$, Fermat shows that

$$\frac{CO}{2} + OG = 15 \qquad CF = \sqrt{117} \qquad FG = \sqrt{85}$$

and that, consequently, $\dfrac{CF}{2} + FG$ is less than $\dfrac{59}{4}$, and therefore less than 15.

Fermat adds, " I arrived at this point without much trouble, but it was necessary to carry the investigation further ; and because, in order that my conscience might be satisfied, it was not sufficient to have found a point such as F through which the natural motion was accomplished more quickly, more easily and in less time than by the straight line COG, it was also necessary to find the point which allowed the passage from one side to the other in less time than any other there might be. In this connection, it was necessary to use my method of *maximis* and *minimis*, which is rather successful for expediting this kind of problem. "

Fermat did not doubt the truth of his principle, but he had been warned from all sides that experiments confirmed Descartes' law. Therefore it was dangerous to try to introduce a " proportion different from those which M. Descartes has given to refractions. " Moreover, it was necessary to " overcome the length and the difficulty of the calculation, which at first presented four lines by their fourth roots and accordingly became entangled in assymmetries. . . . " However, his " passionate desire " to succeed fortunately inspired him to find a method which shortened his work by a half, in reducing these four asymmetries to only two.

Fermat's calculation is found in his paper *Synthesis ad refractiones*, probably written in February, 1662.

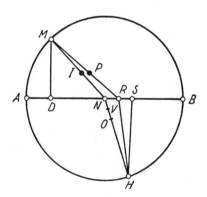

Fig. 90

Let there be a circle of diameter ANB, an incident ray MN and a refracted ray NH. Let MRH be another trajectory passing through any point of AB, chosen, for example, on the right of N.

Fermat introduces the ratios

(1) $$\frac{\text{velocity on } MN}{\text{velocity on } NH} = \frac{MN}{IN} = \frac{\text{velocity on } MR}{\text{velocity on } RH} = \frac{MR}{PR} > 1$$

whence

$$\frac{\text{time on } MN}{\text{time on } NH} = \frac{MN}{NH} \cdot \frac{IN}{MN} = \frac{IN}{NH}$$

and, similarly,

$$\frac{\text{time on } MR}{\text{time on } RH} = \frac{PR}{RH};$$

so that

$$\frac{\text{time on } MNH}{\text{time on } MRH} = \frac{IN + NH}{PR + RH}.$$

The point H, " at which Nature herself takes aim " corresponds to a projection on AB such that

(2) $$\frac{DN}{NS} = \frac{MN}{NI} > 1.$$

It is necessary to show that

$$PR + RH > IN + NH.$$

Immediately

$$\frac{DN}{NS} = \frac{MR}{PR} \text{ by (1) and (2).}$$

Putting

(3) $$\frac{MN}{DN} = \frac{RN}{NO} \quad \text{and} \quad \frac{DN}{NS} = \frac{NO}{NV}.$$

(4) $\begin{cases} \text{As} \quad\quad DN < MN, \quad \text{therefore} \quad NO < RN \\ \text{As} \quad\quad NS < DN, \quad \text{therefore} \quad NV < NO. \end{cases}$

Now, by (3),

$$MR^2 = MN^2 + NR^2 + 2DN \cdot NR = MN^2 + NR^2 + 2MN \cdot NO.$$

Therefore, by (4)

(5) $$MR > MN + NO.$$

Now, by (1), (2), (3),

$$\frac{DN}{NS} = \frac{MN}{IN} = \frac{NO}{NV} = \frac{NO + MN}{NV + IN} = \frac{MR}{RP}.$$

Therefore, by (5),

$$RP > NI + NV.$$

It remains to prove that

$$RH > HV$$

for then it is clear that

$$PR + RH > NI + NH.$$

Now

$$RH^2 = NH^2 + NR^2 - 2SN \cdot NR$$

and by (3)

$$\frac{HN}{DN} = \frac{MN}{DN} = \frac{NR}{NO} \qquad \frac{DN}{NS} = \frac{NO}{NV}$$

Therefore

$$\frac{HN}{NS} = \frac{NR}{NV} \quad \text{or} \quad SN \cdot NR = HN \cdot NV$$

since, by (4), $NR > NV$ it follows that $RH > NH - NV = HV$, which completes the proof.

We return to the letter which we quoted at the beginning of this section. Fermat concludes, " The reward of my work has been most extraordinary, most unexpected, and the most fortunate that I have ever obtained. For after having gone through all the equations, multiplications, antitheses and other operations of my method, and finally having settled the problem. . . , I found that my principle gave exactly the same proportion of the refractions that M. Descartes has established. I was so surprised by a happening that was so little expected that I only recovered from my astonishment with difficulty. I repeated my algebraic operations several times and the result was always the same, though my demonstration supposes that the passage of light through dense bodies is more difficult than through rare ones—something I believe to be very true and necessary, and something which M. Descartes believes to be the contrary.

" What must we conclude from this ? Is it not sufficient, Sir, that as friends of M. Descartes, I might allow him free possession of his theorem ? Is it not rather glorious to have learned the ways of Nature in one glance, and without the help of any demonstration ? I therefore cede to him the victory and the field of battle. . . . "

2. Cartesian objections to Fermat's principle.

Although his demonstration was mathematically incontestible, Fermat was not successful in convincing the Cartesians, who opposed it with metaphysical objections—which, at that time, took place over pure and simple reason.

These facts emerge from the correspondence between Fermat and Clerselier. Thus Clerselier, writing to Fermat on May 6th, 1662, declares that Fermat's principle is, in his eyes, " a principle which is moral and in no way physical ; which is not, and which cannot be, the cause of any effect of Nature. " To Clerselier, the straight line is the only *determinate*—" this is the only thing that Nature tends to in all her motions. " And he explains—

" The shortness of the time ? Never. For when the radius MN has come to the point N, according to this principle it must there be indifferent to going to all parts of the circumference BHA, since it takes as much time to travel to one as to the other. And since this reason of the shortness of time will not, then, be able to direct it towards one place rather than towards another, there will be good reason that it must follow the straight line. For in order that it might select the point H rather than any other, it is necessary to suppose that this ray MN, which Nature cannot send out without an indefinite tendency towards a straight line, remembers that it has started from the point M with the order to discover, at the meeting between the two media, the path that it must then travel in order to arrive at H in the shortest time. This is certainly imaginary, and in no way founded on physics.

" Therefore what will make the direction of the ray MN (when it has come to N) change at the meeting with the other medium, if not that which M. Descartes urges ? Which is that the same force that acts on and moves the ray MN, finding a different natural arrangement for receiving its action in this medium than in the other, one which changes its own in this respect, makes the direction of the ray conform to the disposition that it has at the time. "

And Clerselier concludes—

" That path, which you reckon the shortest because it is the quickest, is only a path of error and bewilderment, which Nature in no way follows and cannot intend to follow. For, as Nature is determinate in everything she does, she will only and always tend to conduct her works in a straight line. "

As for the velocity of light in dense and rare bodies, Clerselier believed that it would be " clearly more reasonable " to accept Fermat's thesis.

But, with a fine assurance, he writes, " M. Descartes—in the 23rd page of his *Dioptrique*—proves and does not simply suppose, that light moves more easily through dense bodies than through rare ones. "

A letter from Fermat to Clerselier, dated May 21st, 1662, contains the following bitter ironical reply.

" I have often said to M. de la Chambre and yourself that I do not claim and that I have never claimed, to be in the private confidence of Nature. She has obscure and hidden ways that I have never had the initiative to penetrate ; I have merely offered her a small geometrical assistance in the matter of refraction, supposing that she has need of it. But since you, Sir, assure me that she can conduct her affairs without this, and that she is satisfied with the order that M. Descartes has prescribed for her, I willingly relinquish my pretended conquest of physics and shall be content if you will leave me with a geometrical problem, quite pure and *in abstracto*, by means of which there can be found the path of a particle which travels through two different media and seeks to accomplish its motion as quickly as it can. "

Thus the problem was taken back on to the mathematical plane, the only profitable one.

In a letter written in 1664 to an unknown person, Fermat returns to " the intrigue of our dioptrics and our refractions. " If one is to judge from the text, the Cartesians had not confessed themselves beaten.

" The Cartesian gentlemen turned my demonstration, which was communicated to them by M. de la Chambre, upside down. At first they were of the opinion that it must be rejected, and although I represented to them very sweetly that they might be content that the field of battle should remain with M. Descartes, since his opinion was justified and confirmed, albeit by reasons different from his own ; that the most famous conquerors did not regard themselves less fortunate when their victory was won with auxiliary troops than if it was won by their own. At first they had no wish to listen to raillery. They determined that my demonstration was faulty because it could not exist without destroying that of M. Descartes, which they always understood to have no equal. . . . Eventually they congratulated me, by means of a letter from M. Clerselier. . . . They acclaimed as a miracle the fact that the same truth had been found at the ends of two such completely opposed paths and announced that they would prefer to leave the matter undecided, saying that they did not know, in this connection, whether to value M. Descartes' demonstration more highly than my own, and that posterity would be the judge. "

3. Leibniz and the path of " least resistance " for light.

In a paper in the *Acta* of Leipzig for 1682, Leibniz rejected Fermat's principle.   Light chooses the easiest path, which must not be confused with the shortest path or with that which takes the shortest time.

Leibniz contemplated a path of least resistance or, more accurately, a path for which the product of the path and the " resistance " might be a minimum.   Leibniz also supported Descartes' opinion on the relative velocity of light in rare and dense bodies with the aid of the following arguments.   Although glass " resists " more than air, light proceeds more quickly in glass than in air because the greater resistance prevents the diffusion of the rays, which are confined in the passage after the manner of a river which flows in a narrow bed and thus acquires a greater velocity.

4. Maupertuis' law of rest.

Before coming to Maupertuis' dynamics, we shall devote a little attention to a law of minimum and maximum which was put forward by this author in the *Mémoires de l'Académie des Sciences* for 1740, and in which the concept of potential makes its appearance.

" Let there be a system of bodies which gravitate, or which are attracted towards centres by the forces that act on each one, as the nth power of their distances from the centre.   In order that all these bodies should remain at rest, it is necessary that the sum of the products of each mass with the intensity of the force [1] and with the $(n + 1)^{th}$ power of its distance from the centre of its force (which may be called the sum of the forces at rest) should be a maximum or a minimum. "

By means of this *law of rest* Maupertuis rediscovered the essential theorems of elementary statics (the rule of the parallelogram, the equilibrium of an angular lever).

5. The principle of least action in Maupertuis' sense (1744).

The debate between Fermat and the Cartesians, and Leibniz's objections to Fermat's principle, prepared the way for Maupertuis' intervention.   The latter stated the principle of least action in a paper read to the *Académie des Sciences* on April 15th, 1744.   The paper is entitled *The agreement between the different laws of Nature that had, until now, seemed incompatible.*

---

[1] The force is here of the form kmr^n.

Maupertuis starts by recalling the laws which light must obey—rectilinear propagation in a uniform medium, the law of reflection and the law of refraction. He seeks simple mechanical analogies.

" The first of the laws is common to light and to all bodies. They move in a straight line unless some outside force deflects them.

" The second is also the same as that followed by an elastic ball which is thrown at an immoveable surface.

" But it is also very necessary that the third law should be explained as satisfactorily. When light passes from one medium into another, the phenomena are quite different from those which occur when a ball is reflected from a surface which does not yield to it in any way ; or those which occur when a ball, on meeting one that does yield to it, continues its progress, only changing the direction of its path. . . . Several mathematicians have extracted some fallacy which had escaped the notice of Descartes, and have made the error of his explanation clear.

" Newton gave up the attempt to deduce the phenomena of refraction from those which occur when a body encounters an obstacle, or when it is forced along in media that resist differently, and fell back on his attraction. Once this force, which is distributed through all bodies in proportion to the quantity of matter, is assumed, the phenomena of refraction are explained in the most correct and rigorous way. . . .

" M. Clairaut, who assumes that light has a tendency towards transparent bodies, and who considers this to be caused by some atmosphere which could produce the same effects as the attraction, has deduced the phenomena of refraction. . . .

" Fermat was the first to become aware of the error of Descartes' explanation. . . . He did not rely on atmospheres about the bodies, or on attraction, although it is known that the latter principle was neither unknown nor disagreeable to him.[1] He sought the explanation of these phenomena in a principle that was quite different and purely metaphysical.

" This principle was ' *that Nature, in the production of her effects, always acts in the most simple ways.* ' Therefore Fermat believed that,

---

[1] MAUPERTUIS is here referring to a passage from FERMAT's work (*var. oper. math.*, p. 114) and which he cited elsewhere with the intention of showing that FERMAT had anticipated NEWTON. This does not seem very convincing, for FERMAT's attraction remained metaphysical in essence. Here is this passage. " The common opinion is that gravity is a quality which resides in the falling body itself. Others are of the opinion that the descent of bodies is due to the attraction of another body, like the Earth, which draws those that descend towards itself. There is a third possibility—that it is a mutual attraction between the bodies which is caused by the mutual attraction that bodies have for each other, as is apparent for iron and a magnet. "

in all circumstances, light followed at once the shortest path in the shortest time.[1] This led him to assume that light moved more easily and more quickly in the rarer media than in those in which there is a greater quantity of matter. "

When Maupertuis wrote it was generally agreed that light moved more quickly in denser media, in the manner specified by the newtonian law of the proportionality of the indices of refraction to the velocities of propagation.

"All the structure that Fermat has built up is therefore destroyed. . . . In the paper that M. de Mayran has given on the reflection and refraction, there can be found the history of the dispute between Fermat and Descartes, and the difficulty and inability there has so far been to reconcile the law of refraction with the metaphysical principle. "

Therefore, unlike Fermat, Maupertuis sought a minimum principle that might be compatible with the newtonian law, and not with the now generally accepted law which goes back to Huyghens. The strange thing is not that he succeeded in finding it. Rather it is that, in boldly extending—one is even tempted to say gratuitously extending—this minimum principle into the field of dynamics, he was led to a law which was truly sufficient, and which he successfully opposed to the thesis of Descartes on the conservation of momentum and of Leibniz on the conservation of kinetic energy.

Up to this point, our author has only criticised the different interpretations of the laws of refraction that had been put forward. We shall now look at his achievement. The relevant passage merits quotation in its entirely—on the rational plane, it would be impossible to conceal its extreme weakness.

We now enter the metaphysical plane in the most complete sense of the word.

" In meditating deeply on this matter, I thought that, since light has already forsaken the shortest path when it goes from one medium to another—the path which is a straight line—it could just as well not follow that of the shortest time. Indeed, what preference can there be in this matter for time or distance ? Light cannot at once travel along the shortest path and along that of the shortest time—why should it go by one of these paths rather than by the other ? Further, why should it follow either of these two ? It chooses a path which has a very real advantage— *the path which it takes is that by which the quantity of action is the least.*

---

[1] As far as it concerns the path, this is incorrect. What is a minimum, to FERMAT, is the sum $ln + l'n'$, the sum of the products of each trajectory with the corresponding refractive index in SNELL's sense.

" It must now be explained what I mean by the quantity of action. When a body is carried from one point to another a certain action is necessary. This action depends on the velocity that the body has and the distance that it travels, but it is neither the velocity nor the distance taken separately. The quantity of action is the greater as the velocity is the greater and the path which it travels is the longer. It is proportional to the sum of the distances, each one multiplied by the velocity with which the body travels along it.[1]

" *It is the quantity of action which is Nature's true storehouse, and which it economises as much as possible in the motion of light.* "

Maupertuis' demonstration follows.

" Let there be two different media, separated by a surface which is represented by the line *CD*, such that the velocity in the upper medium is proportional to *m* and the velocity in the lower medium is proportional to *n*. Let there be a ray of light, starting from the given point *A*, which must pass through the given point *B*. In order to find the point *R* at which it must break through, I seek the point at which, if the ray breaks through, *the quantity of action is least.* I have $m \cdot AR + n \cdot RB$, which must be a minimum.

" Or, having drawn the perpendiculars *AC*, *BD*, to the common surface of the two media, I have

$$m \sqrt{\overline{AC^2} + \overline{CR^2}} + n \sqrt{\overline{BD^2} + \overline{DR^2}} = \text{Min.}$$

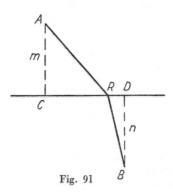

Fig. 91

or, since *AC* and *BD* are constants

$$\frac{mCR \cdot dCR}{\sqrt{\overline{AC^2} + \overline{CR^2}}} + \frac{nDR \cdot dDR}{\sqrt{\overline{BD^2} + \overline{DR^2}}} = 0.$$

[1] A footnote adds the following detail— " As there is only one body, the mass is neglected. "

" But, since CD is a constant, there obtains

$$dCR = - dDR.$$

" Therefore

$$\frac{mCR}{AR} = \frac{nDR}{BR} \quad \text{and} \quad \frac{CR}{AR} \cdot \frac{RD}{BR} \cdot \cdot \frac{n}{m},$$

or, in words, *the sine of the incidence, or the sine of the refraction, are in inverse proportion to the velocity which the light has in each medium.*

" All the phenomena of refraction now agree with the great principle that *Nature in the production of her works, always acts in the most simple ways.* "

Maupertuis then shows without difficulty that " this basis, this quantity of action that nature economises in the motion of light through different media, she also saves in the reflection and the linear propagation. In both these circumstances, the least action reduces to the shortest path and the shortest time. And it is *this consequence that Fermat took as a principle.* "

Maupertuis concludes, " I know of the repugnance that several mathematicians have for *final causes* when applied to physics, and to a certain extent I am in accord with them. I believe that they are not introduced without risk. The error, which men like Fermat and those that followed him, have committed, only shows that, too often, their use is dangerous. It can be said, however, that it is not the principle which has betrayed them, but rather, the haste with which they have taken for the principle what is merely one of the consequences of it. It cannot be doubted that all things are regulated by a Supreme Being who, when he impressed on matter the forces which denote his power, destined it to effect the doings which indicate his wisdom. "

6. The application of the principle of least action to the direct impact of two bodies.

In a paper published by the Royal Academy of Berlin in 1747, and called *On the laws of motion and of rest*, Maupertuis applied the principle of least action to the direct impact of two bodies.

He only considered the effect of the direct impact of two homogeneous spheres, and started from the hypothesis that " *the magnitude of the impact of two given bodies depends uniquely on their respective velocity,* " that is, on their relative velocity. He distinguished between—

" *Perfectly hard bodies.* These are those whose parts are inseparable and inflexible, and whose shape is consequently unalterable.

" *Perfectly elastic bodies.* These are those whose parts, after being deformed, right themselves again, taking up their original situation and restoring to the body its original shape. "

Modern language would call the first category *completely devoid of elasticity* or *perfectly soft.* But the important matter here is that of the experimental laws which Maupertuis stated.

"*After the impact, hard bodies travel together with a common velocity. . . . The respective velocity of elastic bodies after the impact is the same as that before.* "

Maupertuis did not treat the intermediate case, " that of soft or fluid bodies, which are merely aggregates of hard or elastic ones. "

He started from the principle that " when any change takes place in Nature, the quantity of action necessary for this change is the smallest possible. "

We shall quote (notation apart) Maupertuis' argument on the impact of hard bodies.

" Let there be two hard bodies A and B, whose masses are m and m', which move in the same direction with velocities v and v_0 ; but A more quickly than B, so that it overtakes B and collides with it. Let the common velocity of the two bodies after the impact $= v_1 < v_0$ and $> v_0'$. The change which occurs in the Universe consists in that the body A, which used to move with a velocity v_0 and which, in a certain time, used to travel a distance $= v_0$, now moves with the velocity v_1 and travels no more than a distance $= v_1$. The body B, which only used to move with a velocity v_0' and travelled a distance $= v_0'$, moves with the velocity v_1 and travels a distance $= v_1$.

" This change is therefore the same as would occur if, while the body A moved with the velocity v_0 and travelled a distance $= v_0$, it were carried backwards on an immaterial plane, which was made to move with a velocity v_0-v_1, through a distance $= v_0-v_1$; and that while the body B moved with the velocity v_0' and travelled a distance $= v_0'$, it were carried forwards by an immaterial plane, which was made to move with a velocity v_1-v_0' through a distance v_1-v_0'.

" Now whether the bodies A and B move with their appropriate velocities on the moving planes, or whether they are immobile there, the motion of the planes loaded with these bodies being the same, the quantities of action produced in Nature will be $m(v_0-v_1)^2$ and $m'(v_1-v_0')^2$, and it is the sum of these which must be as small as possible. Therefore it must be that

$$mv_0^2 - 2mv_0v_1 + mv_1^2 + m'v_1^2 - 2m'v_1v_0' + m'v_0'^2 = \text{Min.}$$

or $\qquad - 2mv_0dv_1 + 2mv_1dv_1 + 2m'v_1dv_1 - 2m'v_0'dv_1 = 0$

whence the common velocity

$$v_1 = \frac{mv_0 + m'v_0'}{m + m'}$$

is obtained. "

No purpose would be served by reproducing the argument relevant to two bodies moving towards each other. Here too the condition of least action reduces to the conservation of the total momentum.

Next treating the impact of elastic bodies, Maupertuis used an argument which was completely analogous to that which we have reproduced. Apart from sign, the " respective " velocity is conserved after the impact, or

$$v_0 - v_0' = v_1' - v_1.$$

The quantity of action involved has the value

$$m (v_0 - v_1)^2 + m' (v_1' - v_0')^2$$

and it follows from the condition of least action that

$$v_1 = \frac{mv_0 - m'v_0 + 2m'v_0'}{m + m'} \qquad v_1' = \frac{m'v_0' - mv_0' + 2mv_0}{m + m'}.$$

On this occasion the living forces are conserved, " but this conservation only takes place for elastic bodies, not for hard ones. *The general principle, which applies to the first and to the others, is that the quantity of action necessary to produce some change in Nature is the smallest that is possible.* "

At the end of his paper, Maupertuis dealt with the principle of the lever, and deduced it from the principle of least action.

" Let c be the length of the lever, which I suppose to be immaterial, and at whose ends are placed two bodies whose masses are A and B. Let z be the distance of the body A from the point of support which is sought, and $c - z$ be the distance of the body B. It is clear that, if the lever has some small motion, the bodies A and B will describe small arcs which are similar to each other and proportional to the distances of the bodies from the point which is sought. Therefore these arcs will be the distances travelled by the bodies, and at the same time will represent their velocities. The quantity of action will therefore be proportional to the product of each body by the square of its arc. Or (since the arcs are similar) to the product of each body by the square of its distance from the point about which the lever turns, that is, to

Az^2 and $B(c-z)^2$, and it is the sum of these which must be the smallest possible. Therefore

$$Az^2 + B(c - z)^2 = \text{Min.}$$

or

$$2Azdz + 2Bzdz - 2Bcdz = 0$$

from which it is deduced that

$$z = \frac{Bc}{A + B}$$

which is the fundamental proposition of statics. ''

7. THE PRINCIPLE OF LEAST ACTION IN MAUPERTUIS' WORK.

Maupertuis, who had been a musketeer, had a great liking for geometry. He was a surveyor and, in an amateur way, a geographer, astronomer, biologist, moralist and linguist.... And to crown and grace it all, Maupertuis was a metaphysician. Although he had a systematic mind, because of a trait rather common to men of his province he was not free from fantasy. From this fantasy, or perhaps from his temperament, sprang naivety.

We shall therefore turn over the pages of Maupertuis' work, seeking an explanation of the principle of least action.[1]

Here we shall only dwell on the *Essai de Cosmologie*. In this document Maupertuis contrasted the rationalist school, " wishing to submit Nature to a purely material regime and to ban *final causes* entirely, " with the school which, on the contrary, " makes continual use of these causes and discovers the intentions of the Creator in every part of Nature.... According to the first, the Universe could dispense with God ; according to the second, the tiniest parts of the Universe are as much demonstrations " of the existence of God.

He declared, " I have been attacked by both these factions of philosophy.... Reason defends me from the first, an enlightened century has not allowed the other to oppress me. "

Thus Maupertuis flattered himself with having found a happy mean between these two extreme attitudes. " Those who make immoderate use (of final causes) have wished to persuade me that I seek to deny the evidence of the existence of God—which the Universe everywhere presents to the eyes of all men—in order to substitute for it one which has only been given to a few. "

[1] We have referred both to the Dresden edition (Walter, 1752) and the Lyons edition (Bruyset, 1756).

Among the evidence of the existence of God, Maupertuis intended to dispense with all that was provided by metaphysics. He also took no account of that which sprang from the structure of animals and plants, such as the proof—to cite only one—offered by the folds in the skin of a rhinoceros, who would not be able to move without them.

" Philosophers who have assigned the cause of motion to God have been reduced to this because they did not know where else to place it. Not being able to conceive that matter had any ability to produce, distribute and destroy motion, they have resorted to an *Immaterial Being*. But when it is known that all the laws of motion are based on the principle of *better*, it cannot be doubted that these have their foundation in an *omnipotent and omniscient Being*, whether he gave bodies the power to interact with each other, or whether he used some other means which is still less understood by us. "

He was not concerned, like Fermat, with assuming that Nature acts in the most simple ways. He was not concerned, as Descartes was, with assuming that the same quantity of motion was always conserved in Nature— " He deduced his laws of motion from this ; observation belied them, for the principle was not true. " Finally he was not concerned, like Leibniz, in assuming that the living force was always conserved.   Huyghens and Wren had discovered the laws of the impact of elastic bodies simultaneously, but Huyghens had not taken these laws onto the plane of a universal principle.   The conservation of living forces does not apply to the impact of *hard bodies* and, on this occasion, Maupertuis accused the followers of Leibniz " *of preferring to say that there are no hard bodies in Nature* " than to give up their principle. " This has been reduced to the strangest paradox that love of a system could produce— for the primitive bodies, the bodies that are the elements of all others, what can they be but hard bodies ? "

Therefore Maupertuis denied all general principles that were not final.   " In vain did Descartes imagine a world which could arise from the hand of the Creator. (Strictly speaking, Descartes' system supposes the initial intervention of the Creator, and the continuance of his " customary assistance. ")   In vain did Leibniz, on another principle, devise the same plan. "

And he concludes, " After so many great men have worked on this matter, I hardly dare say that I have discovered the principle on which all the laws of motion are founded ; a principle which applies equally to hard bodies and elastic bodies; from which the motions of all corporeal substances follow. . . .   Our principle, more in conformity with the ideas of things that we should have, leaves the world in its natural need of the power of the Creator, and is a necessary result of the wisest doing of that same power. . . . What satisfaction for the human mind, in contemplating these laws—so beautiful and so simple—that they may be the only ones that the Creator and the Director of things has established in matter in order to accomplish all the phenomena of the visible world. "

8. D'Alembert's condemnation of final causes.

D'Alembert himself was not directly involved in the polemic on the principle of least action that we shall describe in the next section. But he did completely condemn the intervention of final causes in the principles of mechanics.

Indeed, he wrote,[1] " The laws of equilibrium and of motion are necessary truths. A metaphysician would perhaps be satisfied to prove this by saying that it was the wisdom of the Creator and the simplicity of his intentions never to establish other laws of equilibrium and of motion than those which follow from the very existence of bodies and their mutual impenetrability. But we have considered it our duty to abstain from this kind of argument, because it has seemed to us that it is based on too vague a principle. The nature of the Supreme Being is too well concealed for us to be able to know directly what is, or is not, in conformity with his wisdom.[2] We can only discover the effect of his wisdom by the observation of the laws of nature, since mathematical reasoning has made the simplicity of these laws evident to us, and experiment has shown us their application and their scope.

" It seems to me that this consideration can be used to judge the value of the demonstrations of the laws of motion which have been given by several philosophers, in accordance with the principle of final causes ; that is, according to the intentions that the Author of nature might have formulated in establishing these laws. Such demonstrations cannot have as much force as those which are preceded and supported by direct demonstrations, and which are deduced from principles that are more within our grasp. Otherwise, it often happens that they lead us into error. It is because he followed this method, and because he believed that it was the Creator's wisdom to conserve the same quantity of motion in the Universe always, that Descartes has been misled about the laws of impact.[3] Those who imitate him

---

[1] *Traité de Dynamique, Discours préliminaire*, 1758 edition, p. 29.

[2] Clearly an allusion to Maupertuis.

[3] The reader knows that Descartes' error is not, in fact, that of having asserted the conservation of momentum, but of having considered $m|v|$ instead of $m\vec{v}$.

run the risk of being similarly deceived ; or of giving as a principle, something that is only true in certain circumstances ; or finally, of regarding something which is only a mathematical consequence of certain formulae as a fundamental law of nature. "

9. THE POLEMIC ON THE PRINCIPLE OF LEAST ACTION.

In the *Acta* of Leipzig for 1751 Koenig, Professor at The Hague, reproduced part of a letter which he alleged had been written by Leibniz to Herman in 1707, and which contained the following passage.

"Force is therefore as the product of the mass and the square of the velocity, and the time plays no part, as the demonstration which you use shows clearly. But action is in no way what you think. There the consideration of the time enters as the product of the mass by the distance and the velocity, or of the time by the living force. I have pointed out that in the variations of motions, it usually becomes a minimum or a maximum. From this can be deduced several important propositions. It can be used to determine the curves described by bodies that are attached to one or several centres. I wished to treat these things in the second part of my Dynamics but I suppressed them, because the hostile reception with which prejudice, from the first, accorded them, disgusted me. "

Maupertuis, for his part, represented the affair in the following way.[1]

" Koenig, Professor at The Hague, took it into his head to insert in the proceedings of Leipzig a dissertation in which he had two ends in view—rather contradictory ones for such a zealous partisan of M. de Leibniz, but which he found it possible to unite. . . . He attacked my principle as strongly as possible. And, for those that he was unable to persuade of its falsehood, he quoted a fragment of a letter from Leibniz from which it could be inferred that the principle belonged to that one. "

Summoned by Maupertuis to produce the letter, Koenig referred him to " a man whose head has been cut off " (Henzi, of Berne). No trace of this letter was found in spite of all the searches ordered by the King at the request of the *Académie*. The matter became a very acrimonious one. " It was no longer a matter of reasons. M. Koenig and his supporters only replied with abuse. Finally they resorted to libel. . . ."

At the time, Maupertuis presided over the Academy of Berlin on the appointment of Frederic II. Koenig returned his diploma to the Academy and published an *Appeal to the public* from the judgement

---

[1] *Œuvres complètes*, 1756 edition, Letter XI. — *Sur ce qui s'est passé à l'occasion du principe de moindre action.*

that the Academy of Berlin had pronounced in this matter. In 1753 he emphasised this with a *Defence of the Appeal to the public* which he addressed to Maupertuis and which he claimed not only the priority of Leibniz, but also that of Malebranche, Wolf, s'Gravesande and Engelhardt.

Voltaire took part in the controversy. Maupertuis wrote,[1] " The strangest thing was to see appear as an auxiliary in this dispute a man who had no claim to take part. Not satisfied with deciding at random on this matter—which demanded much knowledge which he lacked—he took this opportunity to hurl the grossest insults at me, and was soon to cap them with his *Diatribe*.[2] I allowed this torrent of gall and filth to run on, when I saw myself defended by the pen and the sceptre. Although the most eloquent pen of all had uttered these libels, justice made his work burn on the gibbets and in the public places of Berlin. "

" My only fault, " declared Maupertuis, " was that of having discovered a principle that created something of a sensation. " Euler, director of the Academy of Berlin, presented the following report. " This great geometer has not only established the principle more firmly than I had done but his method, more ubiquitous and penetrating than mine, has discovered consequences that I had not obtained. After so many vested interests in the principle itself, he has shown, with the same evidence, that I was the only one to whom the discovery could be attributed. "

---

[1] *Ibid.*

[2] *La Diatribe du Dr Akakia, médecin du Pape*, is too well known to need emphasis here. We confine ourselves to the extraction of what is directly relevant to our subject.

At the beginning VOLTAIRE writes, " We ask forgiveness of God for having pretended that there is only proof of his existence in $A + B$ divided by Z, etc. . . . " This is both a reference to the demonstration of the equilibrium of the lever by means of the principle of least action and to MAUPERTUIS' rejection, in his *Essai de Cosmologie*, of metaphysical proofs of the existence of God.

Then, in the guise of a *Decision of the professors of the College of Wisdom*, VOLTAIRE makes, in spite of the malicious terms in which it is couched, an accurate criticism. " The assertion that the product of the distance and the velocity is always a minimum seems to us to be false, for this product is sometimes a maximum, as Leibniz believed and as he has shown. It seems that the young author has only taken half of Leibniz's idea ; and, in this, we vindicate him of ever having had an idea of Leibniz in its entirety."

And finally, concerning the part played by EULER, which MAUPERTUIS had not thought of concealing, the same *Decision* declares, " We say that the Copernicus's, the Kepler's, the Leibniz's . . . are something, and that we have studied under the Bernoulli's, and shall study again ; and that, finally, Professor Euler, who was very anxious to serve us as a lieutenant, is a very great geometer who has supported our principle with formulae which we have been quite unable to understand, but which those who do understand have assured us they are full of genius, like the published works of the professor referred to, our lieutenant. . . . "

We must also add that MAUPERTUIS is caricatured in a consistently malicious way in *Micromégas, Candide,* and in *L'Homme aux quarante écus.*

10. EULER'S JUDGEMENT ON THE CONTROVERSY ON LEAST ACTION.

Traces of Euler's opinion about the controversy on least action can be found in a *Dissertation on the principle of least action, with an examination of the objections to this principle made by Professor Koenig.* This was printed at Berlin, in Latin and French, in 1753.

Euler discloses a great respect for Maupertuis, our " illustrious President. " He pays homage to Maupertuis *law of rest* in the following terms. This principle indicates " the marvellous accord of the equilibrium of bodies, whether rigid, flexible, elastic or fluid. From each attraction can be deduced the *Efficacy* of each force, and there is equilibrium when the sum of all the efficacies is least. "

Euler remarks, " Professor Koenig places us under the twin obligation of proving that the principle of least action is true, and that it does not belong to Leibniz. "

To Koenig, all instances of equilibrium can be deduced successfully from the principle of living forces.

The " Koenigian principle " consists of " the annihilation of the living force if there were no equilibrium. "

It can be seen, " more clearly than the day, " that where the applied forces produce no living force, there is equilibrium. In short, in stating the principle of *the nullity of the living force* Professor Koenig is " concealing that which he found first, ' *that in the state of equilibrium there is neither motion nor living force* '. " In this form, the principle of Koenig may appear a truism but, to be accurate, his method proceeds in the following way. First, the system is displaced from its equilibrium position and the living force calculated. Then this is cancelled out and the conditions of equilibrium deduced. This method searches the difficulty, for the calculation of motion is, in general, more difficult than that of equilibrium. And Euler concludes, " Koenig's principle usually leads to great circumlocutions and is, often, incapable of application. "

To Koenig, action does not differ from living force. He considers himself able to assert that " It is clearly seen that all equilibrium arises from the nullity of the living force or from the nullity of the action, taken correctly, and in no way from their Min. of Max. " Euler forthrightly condemned this thesis and, in passing, made the following observation.

" Professor Koenig seems too attached to metaphysical speculations to be able successfully to withdraw his mind from those subtle abstractions and to apply it to the ordinary and material ideas such as those which are the subject of mechanics. "

In the next section we shall study Euler's personal contribution to the extremum principle in dynamics. In the document which concerns us here, he only made the following allusion to this matter.

" I am not in any way concerned, here, with the observation which I have made in the motion of the celestial bodies and, general, of those attracted to fixed centres of force, that if the mass of the body is multiplied by the distance travelled, at each instant, and by the velocity, then the sum of all these products is always the least. " To Euler, the question is one of an *a posteriori* verification and not of an *a priori* deduction.

Further, Euler acknowledged Maupertuis' priority in the principle of least action. " Since this remark was only made after M. Maupertuis had presented his principle, it should not imply any prejudice against his originality. "

11. EULER AND THE LAW OF THE EXTREMUM OF $\int mvds$.

As early as 1744 Euler published a work called *Methodus inveniendi lineas curvas maximi minimive proprietate gaudentes* (Bousquet, Lausanne). Here we are concerned with his Appendix II— *De motu projectorum in medio non resistente per methodum maximorum ac minimorum determinando.*

Euler starts from the following principle. " Since all the effects of Nature obey some law of maximum or minimum, it cannot be denied that the curves described by projectiles under the influence of some forces will enjoy the same property of maximum or minimum. It seems less easy to define, *a priori*, using metaphysical principles, what this property is. But since, with the necessary application, it is possible to determine these curves by the direct method, it may be decided which is a maximum or a minimum. " [1]

Euler emphasised, in the *Dissertation* which we have analysed in § 10, that the matter was, to him, one of the *a posteriori* verification of the existence of an extremum in particular examples of the dynamics of a particle.

The quantity which Euler considered was, at first, $Mds\sqrt{v}$. Here M is the mass of the particle, ds the element of distance travelled and v the height of fall. Since the velocity is $\sqrt{v}$, $dt = ds : \sqrt{v}$, and $\int ds \sqrt{v} = \int vdt$. The first integral refers to momenta and the second to living forces. This duality enabled Euler to emphasise that he did not offend the feelings of any party to the controversy on living forces.

[1] Translated into French by JOUGUET.

Euler verified that the integral $\int ds \sqrt{v} = \int v\, dt$ is an extremum in the parabolic motion of a particle subject to a central force. He then generalised this result to a particle attracted by any number of fixed centres.

Mach remarks in this connection, " Euler, a truly great man, lent his reputation to the principle of least action and the glory of his invention to Maupertuis ; but he made a new thing of the principle, practicable and useful. " One should observe, however, that Euler did not condemn the doctrine of final causes as d'Alembert had done. On the contrary, the true significance of an extremum principle should be, in his opinion, sought in a sound metaphysics. Indeed, he concludes in the following terms.

„ Since bodies, because of their inertia, resist all changes of state, they will obey forces which act on them as little as possible if they are free. Therefore, in the motion generated the effect produced by the forces will be less than if the bodies were moved in some other way. The strength of this argument may not be sufficiently clear. If, however, it is in accord with the truth I have no doubt that a sounder metaphysics will enable it to be demonstrated clearly. I leave this task to others, who make a profession of metaphysics *(quod negotium aliis, qui metaphysicam profitentur, relinquo).* "

12. Final remark.

To recapitulate, Fermat, in geometrical optics, stated the first minimum principle that was not trivial. He was not able to convince the Cartesians although he eventually accepted a reduction of his principle to the rank of a " small geometrical assistance " offered to Nature without any pretension to dictate her doings.

No one accepted Fermat's conclusion, however plausible it might have been, on the relative velocity of light in dense and rare media. Maupertuis cannot be reproached for having shared the errors of his time, reinforced as they were by the double authority of Descartes and Newton.

By means of a very simple differential argument, Maupertuis succeeded in making both the newtonian law of propagation and that of refraction amenable to an extremum law.

Was the development of his thought as was said at the time, of an exclusively metaphysical kind ? Yes, if the explanation of his motives with which he prefaced his analysis is considered on its own. I am reluctant to suggest a more natural, but much more worldly, explanation—that Maupertuis had, in his presentation, reversed the order of the arguments ; that he first discovered the differential argument which we have reproduced and then presented it, *a posteriori*, as the consequence of an economic principle which indicated both the power and the wisdom of the creator.

If this had been the whole of Maupertuis' contribution, his name would have fallen into oblivion, at least as far as the invention of principle is concerned. For in optics only Fermat's principle, which Maupertuis had set out to demolish, has survived.

Maupertuis' extension of the principle of least action to dynamics appears rather gratuitous, for it rests on a fragile analogy—yet it is this principle which has survived and assured the fame of its author. Certainly, as early as 1744, Euler gave the exact mathematical justification of the principle in the special but important case of the mechanics of a particle. Following Euler's example, Lagrange stated the principle *of the greatest or the least living force* without Maupertuis. But Euler himself was determined to leave the honour of having to discovered the principle of least action to Maupertuis ; and on this fact, he knew the evidence.

The term " least " is only justifiable on the metaphysical plane, where every maximum would be evidence of the imperfection of the Creator's wisdom. Despite the criticism of Lagrange and, later, that of Hamilton, the name has survived and even now it is encountered in all the books.

In the domain of the laws of impact Maupertuis' contribution was most constructive. His principle enabled him to encompass the cases of elastic bodies and hard bodies which had previously appeared separate, if not contradictory. A trace of this disjunction was still apparent in Lagrange's work.

CHAPTER SIX

EULER AND THE MECHANICS OF SOLID BODIES
(1760)

In 1760 Euler published a *Theoria motus corporum solidorum seu rigidorum*. This was eventually amended and added to by his son, in a new edition which appeared in 1790.

The treatise starts with an introduction in which Euler confirms the principles of his *Mechanica* (1736).[1]

In connection with the mechanics of solids, Euler states that he will consider the characteristic property of a solid to be the conservation of the mutual distances of its elements. For every solid he defines a *centrum massae* or *centrum inertiae*, remarking that the term " centre of gravity " implies the more restricted concept of a solid that is only heavy, while the centre of mass of inertia is defined by means of the inertia alone *(per solam inertiam determinari)*, the forces to which the solid is subject being neglected. This apt comment has not prevailed against usage.

Euler also defines the *moments of inertia*—a concept which Huyghens lacked and which considerably simplifies the language—and calculates these moments for homogeneous bodies.

He systematically studies the motion of a solid body about a fixed axis, the given forces being at first zero and then being equated to the gravity alone. He demonstrates the existence of spontaneous or permanent axes of rotation for a solid body and thus clarifies the notion of the principle axes of inertia.

He then investigates the motion of a free solid by decomposing it into the motion of the centre of inertia, and the motion of the solid about the centre of inertia. Euler clearly distinguishes—

1) the variation of the velocity of the centre of inertia I;

2) the variation of the direction of the point I;

3) the variation of the rotation of the solid about an axis passing through I.

4) the variation of this axis of rotation.

[1] See above, p. 239.

We shall make this clear by an analysis of problems which Euler himself treated.

" *Problem 86.* — *Being given a solid body actuated by a given angular velocity about some axis passing through its centre of inertia, to find the elementary forces which must act on the elements of the solid in order that the axis of rotation and the angular velocity should undergo given variations in the time dt.*" [1]

Let I be the centre of inertia ; IA, IB, IC the principal [or central] axes of the solid ; α, β, γ the angles between the axis of rotation and IA, IB, and IC ; ω the angular velocity of rotation of the solid ; x, y, z the coordinates of some element of the solid with respect to the principal axes ; u, v, w the components of the velocity of this element along the same axes ; X, Y, Z the unknown force applied to the particular element considered, whose mass is dM.

The data of the problem are $d\alpha$, $d\beta$, $d\gamma$ and $d\omega$ and the unknowns, X, Y, Z. According to the fundamental law of Euler's dynamics, du, dv and dw are proportional to $\dfrac{Xdt}{dM}$, $\dfrac{Ydt}{dM}$ and $\dfrac{Zdt}{dM}$. Therefore the problem reduces to the calculation of du, dv and dw. Now

$$\begin{cases} u = \omega\,(z\cos\beta - y\cos\gamma) \\ v = \omega\,(x\cos\gamma - z\cos\alpha) \\ w = \omega\,(y\cos\alpha - x\cos\beta) \end{cases} \qquad \begin{cases} dx = udt = \omega dt\,(z\cos\beta - y\cos\gamma \\ dy = \dots. \\ dz = \dots. \end{cases}$$

A simple differentiation gives

$$\begin{cases} du = d\omega\,(z\cos\beta - y\cos\gamma) - \omega z d\beta\,\sin\beta + \omega y d\gamma\,\sin\gamma \\ \qquad\qquad\qquad + \omega^2 dt\,(y\cos\alpha\cos\beta + z\cos\alpha\cos\gamma - x\sin^2\alpha) \\ dv = \dots \\ dw = \dots \end{cases}$$

and the unknown forces, X, Y, Z, applied to the element $dM\,(x, y, z)$ are deduced from the fundamental law.

In the next problem (No. 87) Euler calculates the moments P, Q, R, with respect to the principal axes, of the forces applied in the conditions specified. By the definition of moments,

$$dP = \frac{dM}{dt}\,(ydw - zdv) \qquad dQ = \dots \qquad dR = \dots$$

for the element dM.

---

[1] Page 337 of the new latin edition of EULER's work, which we follow here.

By summing over all the solid body, after replacing du, dv and dw by their appropriate value, Euler finds

$$(E) \begin{cases} P = \dfrac{1}{dt} \ (Ad\omega \cos \alpha - \omega A d\alpha \sin \alpha + \omega^2 \ (C - B) \ dt \cos \beta \cos \gamma). \\ Q = \ldots \\ R = \ldots \end{cases}$$

(A, B, C, central moments of inertia of the solid.)

Having obtained this result, Euler poses the following problem.

" *Problem 88.* — *If a solid body, turning about an axis passing through its centre of inertia with angular velocity ω, is acted upon by some forces, to find the variation of the axis of rotation and the angular velocity at the end of a time dt.* "

A linear combination of the equations (E) gives the result in the form

$$d\omega = \frac{(C - B) \ (A - C) \ (B - A)}{ABC} \ \omega^2 dt \cos \alpha \cos \beta \cos \gamma$$
$$+ \ dt \left(\frac{P \cos \alpha}{A} + \frac{Q \cos \beta}{B} + \frac{R \cos \gamma}{C} \right).$$

But Euler discovered that the equations can be cast into a much more simple form by introducing the components of the angular velocity of instantaneous rotation about the central axes of inertia ; that is, by introducing the quantities $p = \omega \cos \alpha$, $q = \omega \cos \beta$ and $r = \omega \cos \gamma$.

Under these conditions α, β and γ no longer appear in the equations (E), which take the form

$$\begin{cases} P = A \dfrac{dp}{dt} + (C - B) \ qr \\ \\ Q = B \dfrac{dq}{dt} + (A - C) \ rp \\ \\ R = C \dfrac{dr}{dt} + (B - A) \ pq \end{cases}$$

Euler immediately saw the importance of these equations—" *summa totius Theoriae motus corporum rigidorum his tribus formulis satis simplicibus continebitur.* "

Thus Euler's mathematical talent enabled him to discover the equations which express the general motion of a solid body under the influence of arbitrary forces by means of the decomposition of this motion into the motion of the centre of inertia and the rotation about the centre of inertia. An essential part of this process was the consideration of the central axes of inertia—that is, the consideration of a real *moving reference frame* fixed in the solid.

CHAPTER SEVEN

CLAIRAUT AND THE FUNDAMENTAL LAW OF HYDROSTATICS

1. CLAIRAUT'S PRINCIPLE OF THE DUCT.

Clairaut (1713-1765) was led to formulate the general law of the equilibrium of a fluid mass by the contemporary investigations of the figure of the Earth.[1]

Clairaut did not openly take sides in the conflict between the doctrine of vortices and that of attraction. However, he remarked that the Neo-Cartesians, while recognising part of the newtonian system, assumed *a priori*, whatever the shape of the Earth might be, that the gravity was inversely proportional to the square of the distance. They then compounded this gravitational force with the centrifugal force calculated for a given shape of the Earth. In the procedure of the attraction, on the other hand, the law of gravity depended on the shape of the Earth itself. " The Newtonians must find a spheroid such that a corpuscle, placed at an point on its surface and which is subject to both the centrifugal force and the attractions of all the parts of the spheroid, will take a direction perpendicular to that surface. " [2]

Huyghens assumed that the gravity must be normal at each point of the surface of a fluid mass. Newton supposed the equality of the weights of two liquid columns ending at the centre of mass. Bouguer had the merit of observing, as early as 1734 [3], that these two hypotheses were incompatible for certain laws of gravity. Whence the theme of Clairaut's investigation—

" *To find the laws of hydrostatics which agree equally with all kinds of hypotheses about gravity.* " [4]

---

[1] *Théorie de la figure de la Terre tirée des principes de l'hydrodynamique* (Durand, Paris, 1743).

[2] *Loc. cit.*, p. xxj.

[3] *Comparaison des deux lois que la Terre et les autres Planètes doivent observer dans la figure que la Pesanteur leur fait prendre, Mémoires de l'Académie des Sciences*, 1734.

[4] CLAIRAUT, *loc. cit.*, p. xxxij.

Clairaut states the following principle at the beginning of his paper.

" *A mass of fluid cannot be in equilibrium unless the efforts of all the parts which are contained in a duct of any shape, which is imagined to traverse the whole mass, cancel each other out.* " [1]

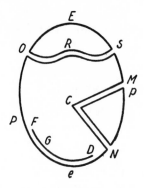

Fig. 92

He justifies this in the following way.

" Since the whole mass *PEpe* is supposed to be in equilibrium, any part of the fluid could become solid without the remainder changing its condition. Suppose that all the mass is solidified except for what is necessary to form the duct *ORS*. The duct will therefore be in equilibrium. Now this can only occur if the efforts of *OR* to leave the duct through *S* are equal to those of *SR* to leave through *O*. "

This principle includes Newton's hypothesis, as may be verified by the consideration of a duct *MCN* passing through the centre *C*. It also includes Huyghens' hypothesis— it suffices, indeed, to consider a duct *FGD* lying along the surface. This duct may be in equilibrium in two ways— the first, from the fact of Huyghens' hypothesis itself; the second, because of the fact that a part *FG* thrusting towards *D* is compensated by a part *GD* thrusting towards *F*. But since the length of the duct is arbitrary, a small piece *FG* should be in equilibrium just as much as the whole duct, which excludes the preceding compensation. Therefore it is necessary to return to Huyghens' hypothesis.

But the most valuable form in which Clairaut stated his principle is the following one.

[1] CLAIRAUT, *loc. cit.*, p. 1.

" In order that a mass of fluid may be in equilibrium, it is necessary that the efforts of all the parts of the fluid which are contained in a duct which is re-entrant upon itself should cancel each other out. " [1]

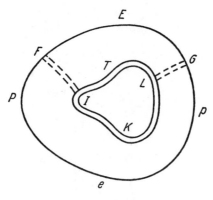

Fig. 93

This proposition can be justified immediately by solidifying all the fluid except that in the duct *IKLT*. The efforts of the parts *IKL*, *ITL* must be equivalent to each other, " or else there would be a perpetual current in the duct. " It can also be deduced from the preceding proposition by the consideration of two ducts *FIKLG* and *FITLG*.

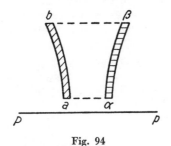

Fig. 94

Clairaut then observes that in the consideration of two ducts *ab*, *αβ*, which are filled with liquid and rotate about an axis *Pp*, the total effort of the centrifugal force on the duct *ab* will be the same as that

[1] CLAIRAUT, *loc. cit.*, p. 5.

on the duct $\alpha\beta$ if a and α, b and β are respectively at the same distance from the axis. It follows that " when it is desired to investigate whether a law of gravity is such that a mass of fluid turning about an axis can preserve a constant shape, no purpose it served by paying attention to the centrifugal force. That is, that if the mass of fluid can have a constant shape when not rotating, it will also be able to have one when rotating. "

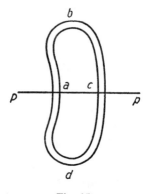

Fig. 95

If a duct $abcd$ is considered in the mass, this must be in equilibrium in order that the mass should have a constant shape. Now it is seen that the sum of the effects of the centrifugal force on $abcd$ is nothing ; for ab and cb will thrust on b to the same extent, just as ad and cd will thrust equally on d. Therefore the rotation will not prevent the equilibrium of a duct which is re-entrant upon itself. Accordingly, " if the duct is in equilibrium when only gravity alone is considered, it will still be in equilibrium if it is supposed that, instead of gravity, the actual weight, composed of gravity and the centrifugal force, is considered. "

2. The condition to be satisfied by the law of gravity to assure the conservation of the shape of a rotating fluid mass.

Clairaut supposes that gravity has two components, P and Q, which are parallel to the axes CP and CE. He considers an arbitrary duct ON which ends on the surface, and an element of this duct, Ss, which has $Sr = dx$ and $sr = dy$.

The force Q acts along SH. The projection of Q on the direction of the duct has the value $Q \cdot \dfrac{rs}{Ss}$. By multiplying by the mass of the element, it is found that " the *effort* which the force Q will cause the cylinder to exert on the point O " has the value Qdy. Similarly, the force P gives the *effort* Pdx. Whence the total *effort* of gravity is

$$Pdx + Qdy.$$

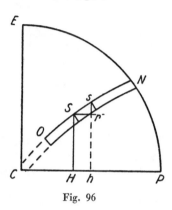

Fig. 96

It is necessary that the sum of the efforts of gravity on any duct ON should have the same value as if any other duct passing through the points O, N had been taken.   " The equilibrium of the fluid requires that the weight of ON does not depend on the curvature of OSN or, that is, on the value of y as a function of x.   Therefore it is necessary that $Pdx + Qdy$ can be integrated without knowing the value of x ; that is, it is necessary that $Pdx + Qdy$ should be a complete differential, or that

$$\frac{\partial P}{\partial y} = \frac{\partial Q}{\partial x}.\text{ " [1]}$$

3. CONDITION FOR THE EQUILIBRIUM OF A FLUID MASS.

Clairaut set himself ¦the following problem.[2] " Supposing that the force which actuates the particles of a fluid had been decomposed into three others, of which the first acts perpendicularly to the plane

---

[1] CLAIRAUT gave this last condition in the *Mémoires de l'Académie des Sciences* for 1740, p. 294.

[2] CLAIRAUT, *loc. cit.*, p. 96.

QAP and the second and third along two perpendicular directions, QA and AP, in the plane QAP, it is required to find the relation there must be between these three forces in order that the equilibrium of the fluid may be possible. "

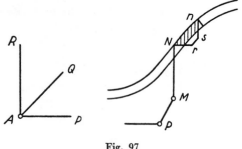

Fig. 97

If P, Q and R are the components of the force parallel to AP, AQ and AR, and Nr, rs and sn are represented by dx, dy and dz, then an argument quite analogous tŏ that of the preceding § gives the condition

$$Pdx + Qdy + Rdz = \text{complete differential}$$

or

$$\frac{\partial P}{\partial y} = \frac{\partial Q}{\partial x}, \ \frac{\partial P}{\partial z} = \frac{\partial R}{\partial x}, \ \frac{\partial Q}{\partial z} = \frac{\partial R}{\partial y}.$$

Apart from this, Clairaut also defines the lines and surfaces of intensity of gravity. He verifies that a fluid mass that is imagined to be divided into an infinite number of layers which are defined by the surfaces of intensity, will be in equilibrium if, at each point of one of the surfaces, the weight is inversely proportional to the thickness of the layer.

This remarkable analysis, which can be regarded as the introduction of the concept of potential, enabled Clairaut to make an important contribution to the theory of the figure of the Earth. Newton had assumed, *a priori*, the shape of an elliptic spheroid. He considered two columns—one connecting the centre to a Pole and the other, the centre to the Equator—and equated the difference in their weight to the sum of the centrifugal force on the different portions of the Equitorial column. The ratio of the axes obtained in this way was $\frac{229}{230}$.

The Neo-Cartesians, on their side, announced the ratio $\frac{576}{577}$.

As early as 1737 Clairaut was able to show that the elliptic spheroid was an equilibrium figure.

Newton's value—or that of MacLaurin $\frac{230}{231}$ —supposed the Earth to be *homogeneous*. This value was disproved by experiments made in Lapland by an expedition sent at the command of the King. This expedition consisted of four members of the *Académie des Sciences*— Clairaut, Camus, Le Monnier and Maupertuis—who were joined by two others—Celsius and the Abbé Outhier. It embarked at Dunkirk on May 2nd, 1736.[1]

The relative magnitudes of degrees of meridian obtained in this way indicated a flattening about $\frac{1}{300}$, less than that which Newton had announced. It was therefore necessary to give up the hypothesis of the homogeneity of the Earth. If the Earth were formed of similar layers, it shape would not conform to the fundamental law of hydrostatics. And Clairaut decided on the existence of layers which were *flatter as they were further from the centre*, the flattening following a law that depended on the decrease of the density between the centre and the surface.

[1] For the account of this mission, see MAUPERTUIS' *Discours lu dans l'Assemblée publique de l'Académie Royale des Sciences sur la mesure de la Terre au Cercle polaire* (*Œuvres complètes*, Vol. III, 1756, p. 89).

CHAPTER EIGHT

DANIEL BERNOULLI'S HYDRODYNAMICS
D'ALEMBERT AND THE RESISTANCE OF FLUIDS
EULER'S HYDRODYNAMICAL EQUATIONS
BORDA AND THE LOSSES OF KINETIC ENERGY IN FLUIDS

1. RETURN TO THE HYDRODYNAMICS OF THE XVIIth CENTURY.

We have already described the attempts of Newton and Varignon to explain the law which Torricelli had formulated. We must now return to the work of Mariotte, who emerges as the forerunner of the important XVIIIth Century school of hydrodynamics.

As early as 1668 a Committee of the *Académie des Sciences* was formed and instructed to verify Torricelli's law experimentally. The Committee's members were Huyghens, Picard, Mariotte and Cassini. It extended its investigations to the determination of the effect of the impact of a fluid stream on a plane surface.

Influenced by these investigations, Mariotte published, in 1684, the *Traité du mouvement des eaux* in which he carried the subject further. He verified Torricelli's law without observing the contraction of the stream. Newton corrected this error in the second edition of the *Principia*.

In the matter of the impact of a fluid stream on a surface, Mariotte had the merit of demonstrating the importance of the deviation from the momentum of the fluid. But he compared this problem with the action of a fluid current on a completely immersed solid, thus disregarding the reconstitution of the stream-lines behind the obstacle.

Mariotte was also the first to introduce considerations of similitude in the resistance of fluids, and the first after Huyghens to state that the resistance of a fluid was proportional to the square of the velocity.

Finally, Mariotte turned his attention to hydraulics, and studied the velocity of flow in rivers or canals, and the friction of water in pipes.

2. Daniel Bernoulli's hydrodynamics.

The beginning of the XVIIIth Century was a period of extraordinary development for both theoretical and applied hydrodynamics. In the compass of the present history—limited to a study of the evolution of the principles of dynamics—it would be impossible to analyse the complex development of the mechanics of fluids in any detail. This investigation rapidly became an independent branch of science, both theoretical and experimental.

However, we consider it valuable to deal with a few typical achievements of the XVIIIth Century in this field.

In 1738 Daniel Bernoulli published *Hydrodynamica, sive de viribus et motibus fluidorum commentarii*. This was a most remarkable work which has only aged a little in the last two centuries.

In D. Bernoulli's sense, the term *hydrodynamica* includes hydrostatics—the science of equilibrium—and hydraulics—the science of fluid motion. " My theory is novel, " he declared, " because it considers both the pressure and the motion of fluids. It might be called *hydraulico-stattica.* "

D. Bernoulli's guiding principle was that of the *conservation of living forces* or, more accurately, that of the equality between the actual descent and the potential ascent *(aequalitas inter descensum actualem ascensumque potentialem)*.

As the controversy on living forces was then at its height, D. Bernoulli took certain precautions in this respect. Quite legitimately, he emphasised that the doctrine of Leibniz stemmed, in fact, from a principle that Huyghens had formulated. (No body freely can rise to a height greater than that from which it has fallen.) [1]

Moreover inelastic impact, which entails a loss of kinetic energy, has its analogy in hydrodynamics, where it appears as a reduction of the *ascensus potentialis*. D. Bernoulli excluded this occurrence from the theory, adding that this was a reason for applying the theory with care.

Apart from the hypothesis of the conservation of living forces, D. Bernoulli also assumed that all the particles of a slice of fluid which was perpendicular to the direction of motion moved with the same velocity, which was inversely proportional to the cross-section of the slice. Further, he only studied stationary states *(fingenda est uniformitas in motu aquarum)*.

We shall give a concrete example of one of the numerous problems which D. Bernoulli solved.

[1] See above, p. 187.

" *Let there be a vessel of very large cross-section, ACEB, which is kept full of fluid, and let it be pierced with a horizontal cylindrical tube ED. At the end of the tube is an orifice O through which the fluid escapes with constant velocity. It is required to find the pressure exerted on the walls of ED.*

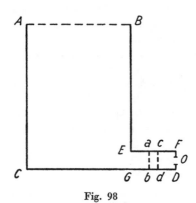

Fig. 98

" Let a be the height of the surface AB above the orifice o.   When the steady state is established *(si prima fluxus momenta excipias)* the velocity with which the water leaves through o is constant and equal to $\sqrt{a}$, since we suppose that the vessel remains full.   If n is the ratio of the section of the tube to that of the orifice, the velocity of the water in the tube will be $\dfrac{\sqrt{a}}{n}$.   If the whole end FD were missing the velocity of the water in the tube would be $\sqrt{a}$, which is greater than $\dfrac{\sqrt{a}}{n}$.   Therefore the water in the tube strives towards a greater motion, to which the end FD presents an obstacle.   From this there results an over-pressure which is transmitted to the boundary walls.   The pressure on the boundary walls is thus proportional to the acceleration which the fluid would take if the obstacle were instantaneously taken away and the fluid were allowed to escape into the atmosphere.

" All this happens as if, during the escape through the orifice o, the tube FD were suddenly broken off at cd and the acceleration of the small portion of fluid $abcd$ were sought. . . .   Thus we must consider the vessel $ABEcdC$ and, with its help, try to find the acceleration which the particle, of velocity $\dfrac{\sqrt{a}}{n}$, takes on escaping.

"Let v be the variable velocity of the water in the tube Ed; n be the cross-section of the tube; c its length, equal to Ec; and let dx be the length ac. A portion of the fluid enters the tube at E at the same time that $abcd$ escapes from it. The portion at E, whose mass is ndx, acquires the velocity v or the living force nv^2dx, which is generated suddenly. Indeed, since the section of the vessel Ae is infinite, the portion of the liquid at E does not have any velocity before entering the tube. To the living force nv^2dx will be added the increase of the living force which the water receives in Eb when the portion ad escapes— say $2ncvdv$. These two quantities together are due to the *real descent of the portion of the fluid from* the height BE, or a. Therefore

$$nv^2dx + 2ncvdv = nadx$$

or

$$\frac{vdv}{dx} = \frac{a - v^2}{2c}.$$

"Throughout the motion the increase dv of the velocity is proportional to the pressure produced in the time $\frac{dx}{v}$. Therefore the pressure which is exerted on the portion ad is proportional to the quantity $v\frac{dv}{dx}$; that is, to $\frac{a - v^2}{2c}$.

"At the instant when the tube is broken, $v = \frac{\sqrt{a}}{n}$ or $v^2 = \frac{a}{n^2}$. This expression is to be substituted in $\frac{a - v^2}{2c}$, which becomes $a\frac{n^2 - 1}{2n^2c}$.

"And it is this quantity which is proportional to the pressure of the water on the portion ac of the tube, whatever the section of the tube may be and whatever the orifice in the end might be....

"If the orifice is infinitely small, or n is infinitely large compared with unity, it is clear that the water exerts the whole pressure corresponding to the height a. This pressure we call a. But, then, unity is vanishingly small compared with n^2 and the quantity to which the pressure is proportional has the value $\frac{a}{2c}$.... If the quantity $\frac{a}{2c}$ corresponds to the pressure a, the pressure corresponding to the quantity $a\frac{n^2 - 1}{2n^2c}$ will be $a\frac{n^2 - 1}{n^2}$, which is independent of c, Q.E.D."

There is no need to quote further from among the demonstrations which D. Bernoulli developed in a similar way. They are remarkable for their ingenuity and all start from the single hypothesis which we have recorded.   We notice that in the same treatise D. Bernoulli enunciated the theorem with which his name is still associated, and which appears, in a different form, in all the classical treatises on hydrodynamics. D. Bernoulli also treated the impact of a fluid stream on a plane in a way which was superior to that used by Mariotte.   He showed that this problem was distinct from the one which concerns the effect of a fluid current on a completely immersed solid.   Together with the theoretical solutions of the problems which he treated, D. Bernoulli recorded much experimental material in support of his demonstrations.

We also remark, without being able to devote a discussion to it, the fact that MacLaurin (in his *Treatise on Fluxions*) and Jean Bernoulli (in his *Nouvelle hydraulique*) had sought to dispense with the assumption of the conservation of living forces which D. Bernoulli had made.

But, in the words of Lagrange, " MacLaurin's theory is not very rigorous and seems to be contrived in advance to agree with the results that he wished to obtain. "   As for that of Jean Bernoulli, it leaves much to be desired in clarity and accuracy and, in d'Alembert's opinion, " its general principle is deduced so easily from that of the conservation of living forces that it appears to be nothing else than that same principle presented in another form ; and again, Jean Bernoulli seeks to confirm his method by means of indirect solutions supported by the laws of the conservation of living forces. "

3. D'Alembert and the motion of fluids.

D'Alembert's contribution to the theory of fluids was both extensive and important.   In the confines of this history we must restrict ourselves to the indication of the principles of this work, and to the citation of some typical examples.

Generally speaking d'Alembert remained faithful, in his treatment of the mechanics of fluids, to the *Discours Préliminaire* of his *Traité de Dynamique*.   Nevertheless he did not carry into this field the conviction that science could be of a purely rational origin.

He wrote, " The mechanics of solid bodies depends only on metaphysical principles which are independent of experiment.   Those principles which must be used as the foundation for others can be determined exactly.   The foundation of the theory of fluids, on the other hand, must be experiment, from which we receive only a very little enlightenment. "

After having devoted long and laborious efforts to an attempt at elucidating the motion and resistance of fluids, d'Alembert remained optimistic about the future of this theory.   " When I speak of the limitations by which the theory must be prescribed, " he wrote, " I only contemplate the theory with such assistance as it can now obtain, and not the theory as it may be in the future, aided by what is yet to be discovered. For, in whatever subject one might be, one should not be too ready to erect a wall of separation between nature and the human mind. " [1]

However, d'Alembert did not conceal the great difficulties that arose in the translation of such complicated phenomena as those of the motion and the resistance of fluids into a rational language. In this period of enlightenment, when all men, including d'Alembert, were willing to trust—perhaps too readily—in the universal validity of science, he made the reservation—" not to exalt the algebraic formulae into physical truths or propositions too readily. . . . Perhaps the spirit of calculation which has displaced the spirit of system rules, in its turn, a little too strongly. For in each century there is a dominant style of philosophy. This style almost always entails some prejudice, and the best philosophy is the one which has least of this consequence. " [2]

In 1744, one year after the publication of his *Traité de Dynamique*, d'Alembert wrote a *Traité de l'Équilibre et du Mouvement des Fluides*. In it he refused to start from the principle of the conservation of living forces, as Daniel Bernoulli had done in his *Hydrodynamica*. For this principle, as we know from the discussion in the *Discours Préliminaire*, d'Alembert regarded as not being sufficiently well-established to be used at the basis of hydrodynamics. " One of the greatest advantages that follow from our theory is that of being able to show that the well-known law of mechanics called the conservation of living forces is as appropriate in the motion of fluids as in that of solid bodies. "

D'Alembert, a keen critic of his predecessors, reproached Daniel Bernoulli for not having brought forward, in Volume II of the *Mémoires de Petersbourg* (1727), " other evidence for the conservation of living forces in the fluids than that a fluid could be regarded as aggregate of fluid particles which press on each other, and that the conservation of living forces is generally accepted to be applicable to the impact of a system of bodies of this kind. . . . Therefore it seemed to me that it is necessary to prove, more clearly and exactly, the question of whether the principle is applicable to fluids. I had tried to demonstrate this, in a few words at the end of my *Traité de Dynamique*.[3] Here will be found a more detailed and more extended proof. " [4]

[1] *Essai d'une nouvelle théorie de la résistance des fluides* (David, Paris, 1752), p. xxxiv.
[2] *Ibid.*, p. xlj.
[3] *Traité de Dynamique* (1758 ed.), p. 269.
[4] *Traité de l'Équilibre et du Mouvement des Fluides* (1744), Preface.

At the beginning of the *Traité des Fluides* d'Alembert explicitly stated the hypothesis of flow by parallel slices of fluid, whose parallelism was conserved throughout the motion. Moreover, he assumed that all the points of the same slice had the same velocity. Necessarily this hypothesis restricted the generality of his analysis.

Given these hypotheses, d'Alembert extended to fluids the principle that he had used as a basis for his dynamics.

" In general, let the velocities of the different slices of the fluid, at the same instant, be represented by the variable v. Imagine that dv is the increment of the velocity in the next instant, the quantities dv being different for the different slices, positive for some and negative for others. Or, briefly, imagine that $v \mp dv$ expresses the velocity of each layer when it takes the place of that which is immediately below. I say that if each layer is supposed to tend to move with an infinitely small velocity $\pm dv$, the fluid will remain in equilibrium.

" For since $v = v \mp dv \pm dv$, and the velocity of each slice is supposed not to change in direction, each layer can be regarded, at the instant that v changes to $v \mp dv$, as if it had both the velocity $v \mp dv$ and the velocity $\pm dv$. Now, since it only retains the first of these velocities, it follows that the velocity $\pm dv$ must be such that it does not affect the first and is reduced to nothing. Therefore if each slice were actuated by the velocity $\pm dv$ the fluid would remain at rest. " [1]

D'Alembert accompanied this theorem with the following observation.

" If it is supposed that the particles of the fluid are subject to an accelerating force φ, different for each slice if so desired, then it is clear that at the end of the instant dt the velocity v will be $v + \varphi dt$ if the slices do not interact in any way. Therefore, if at the end of the instant dt the velocity v becomes $v \mp dv$ because of the interaction of the slices, it would be necessary to suppose that

$$v + \varphi dt = v \mp dv + \varphi dt \pm dv$$

and it is clear that the fluid would remain in equilibrium if only the slice were actuated with the velocity $\varphi dt \pm dv$. " [2]

Starting from this principle, d'Alembert went back to the problems which had been studied by Daniel Bernoulli and treated them anew. We shall pause on a single example, the first and the most simple which d'Alembert treated.

The question is that of the flow in parallel slices of a fluid which

---

[1] *Traité de l'Équilibre et du Mouvement des Fluides* (1744), p.70.
[2] *Ibid.*, p. 71.

is supposed incompressible in a vessel of any shape. The liquid is homogeneous, has no weight, and is set in motion by an unspecified means ; for example, by the impulsion of a piston.

Let u be the velocity of the layer GH. That of the layer CD will be $\dfrac{uGH}{CD}$.

If y is the width, and x the side, of one of the slices, then it is possible to write $ydx = $ constant.

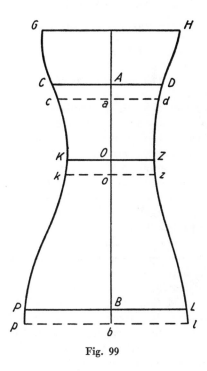

Fig. 99

Between the instant t and the instant $t + dt$ the fluid moves from the position $CDLP$ to the position $cdlp$. If v is the variable which represents the velocity of a slice in the position $CDLP$ and if $v - dv$ is (algebraically) the value of the same variable at $cdlp$, the fluid will remain in equilibrium if each slice tends to move with the velocity dv. D'Alembert expresses this by

$$\int dvdx = 0$$

or

$$\int \frac{ydx \cdot dv}{y} = 0.$$

But $v = \dfrac{uGH}{y}$.   Therefore

$$\int \frac{ydx \cdot vdv}{uGH} = 0.$$

Now " GH is constant as well as u, with respect to the variables v and dv " and ydx is constant. It follows therefore that

$$\int ydx \cdot V^2 = \int ydx \cdot v^2.$$

Here V is the velocity of each slice in the instant which follows that at which its velocity is v.

" *Therefore it is seen that the principle of conservation of living forces also applies to fluids.* "

To d'Alembert however, this principle was a corollary which had to be verified in each instance. And indeed, he was not surprised to retrieve, by means of his principle, the solutions of the problems which had been treated by D. Bernoulli. In this connection, d'Alembert makes the following observation in the preface to his *Traité des Fluides.*

" And indeed I am forced to confess that the results of my solutions always agree with those of M. Daniel Bernoulli. Nevertheless it is necessary to except a small number of these problems. These are problems in which that skilful geometer used the principle of the conservation of living forces to determine the motion of a fluid in which there is a portion whose velocity increases suddenly by a finite quantity. Such, for example, is the problem in which the question is to find the velocity of a fluid leaving a vessel which is kept filled to the same height. It is supposed that the small sheet of fluid which is added at each instant receives its motion from the fluid below, by which it is drawn along. It is clear that in a similar hypothesis this sheet of fluid, which had no velocity at the instant that is was added to the surface, in the next instant receives a finite velocity equal to that of the surface which draws it along. Now, without wishing to ask whether or not this hypothesis is in conformity with nature, it is certain that the principle of living forces should not be used to investigate the motion of any system when it is supposed that there is some body in this system whose velocity varies in an instant by a finite quantity. "

The *Traité des Fluides* also contains an attempt to investigate the resistance of fluids. D'Alembert used rather a flimsy mechanical model, analogous to that which he had reproached Daniel Bernoulli for using, in order to substantiate the conservation of living forces.

" First I determined the motion that a solid body must communicate to an infinity of small balls with which it was supposed to be covered. Then I showed that the motion lost by this body in a given time was the same whether it collided with a certain number of balls at once, or whether it merely collided with them in succession. Further, I showed that the resistance would be the same when the corpuscles had some other shape than the spherical one, and when they were arranged in any manner that might be desired, provided that the total mass of these small bodies which was contained in a given space remained the same. By this means I arrived at the general formulae for the resistance, in which there only appeared the relationships of the densities of the fluid and the body moving in it. " [1]

Finally, d'Alembert devoted a chapter to " fluids that move in vortices and to the motion of bodies which are immersed in them. " He did not make this study to bolster up " a cause as desperate as that of Descartes' vortices, " but because the subject seemed in itself to be " rather curious, independently of any application to the case of planets that one might desire to make. "

4. D'ALEMBERT AND THE RESISTANCE OF FLUIDS — HIS PARADOX.

The *Essai d'une nouvelle théorie de la résistance des fluides* (Paris, David, 1752) has its origin in d'Alembert's participation in the competition held by the Academy of Berlin in 1750. The subject chosen concerned the theory of the resistance of fluids. The prize was not awarded, the Academy having required the authors to give proof of the agreement of their calculations with experiment by means of a supplement to their work.

D'Alembert, who had competed, seems to have become rather bitter at this decision, for the observations available at that time were often contradictory and the theory was sufficiently difficult to require a worker's undivided attention. His *Essai* is, apart from some details, his contribution to the Berlin competition.

In his Essai d'Alembert obtains, at least for plane motions, the general equations of the motion of fluids. [2] However, his analysis is so long and tortuous that we cannot attempt to summarise it. In this achievement d'Alembert preceded Euler by some years—but he did not succeed in presenting the equations of hydrodynamics in the " direct and luminous " [3] way that Euler was able to discover.

---

[1] *Traité de l'Équilibre et du Mouvement des Fluides*, Preface.

[2] See also *Opuscules mathématiques* by d'ALEMBERT, especially Vol. VI, p. 379 (1778 ed.).

[3] LAGRANGE, *Mécanique analytique*, Part II, Section X.

Faithful to his principles, d'Alembert reduced the search for the resistance to the laws of equilibrium between the fluid and the body. The resistance was given by the momentum lost by the fluid.

Rather than attempt to analyse the difficult *Essai sur la résistance des fluides*, we consider it more valuable to reproduce a portion of the work in which d'Alembert, guided, it seems, by his logical insight alone, discovers the hydrodynamical paradox with which his name is still associated.

" *Paradoxe proposé aux géomètres sur la résistance des fluides.*[1]

" Suppose that a body is composed of four equal and similar parts and placed in an indefinite fluid contained in a rectilinear vessel.

" Imagine that the body is fixed and immoveable, and that the parts of the fluid all receive an equal impulsion parallel to the sides of the vessel and the axis of the body. It is clear that the parts of the fluid at the front of the body must be deflected and must slide along the body, thus describing curves which are more like straight lines as they are further from the body ; this up to a certain distance, which will be at least that of the boundary walls of the vessel, where the parts of the fluid will move in straight lines.

" If the solid is not terminated in a very sharp point, in such a way that the derivative $\dfrac{dy}{dx}$ may be either finite or infinite at the origin, there will be or there may be a small portion of the liquid which is stagnant at the front of the body.[2] But in order to avoid this diffi-

---

[1] *Opuscules mathématiques*, Vol. V, p. 132.
[2] We reproduce here the diagram used by d'ALEM-BERT. The "stagnant part, " which, in d'ALEMBERT's opinion, will exist in general, is referred to in the *Essai sur la résistance des fluides*, in the following terms—

" All moving bodies which change direction only do so by imperceptible degrees. The particles which move along TF do not travel as far as A because of the right angle $TA\alpha$— they leave TF at F, for example. Therefore, in front of and behind the solid, there are spaces in which the fluid is necessarily stagnant. "

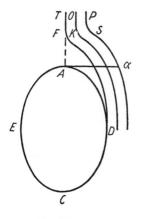

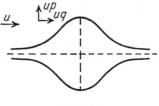

Fig. 100   Fig. 101

culty, I suppose that $\dfrac{dy}{dx} = 0$ at the origin, so that the point of the body is infinitely sharp. Then there will no longer be stagnant fluid, and the fluid will run along the forward surface as far as the point at which dy again becomes parallel to the axis.

" With regard to the backward part, it seems at first sight that the motion must be different from that at the forward part. For at the forward part the fluid is not free to follow its original direction, while it is so free at the backward part. However, for the moment suppose that the particles of the fluid do have the same motion at this backward part as at the front. It is easily discovered, by our theory of the motion of fluids, that under this supposition the laws of the equilibrium and the incompressibility of the fluid will be fully satisfied. For since the back is similar to the front (hyp.), it is easy to see that the same values of p and q [1] that will yield the equilibrium and the incompressibility of the forward part will yield the same results for the backward part. Therefore there is nothing but justification for the supposition concerned. Therefore the fluid can move in this way at the back. Now if it can, it must. For there is only one possible way in which the fluid can be moved by the body.

" In this condition, the fluid will exert a pressure on the body, though this pressure will not have the effect of separating the body from its position because the body is immoveable and fixed at rest in the middle of the fluid (hyp.). Let u be the velocity imparted to the fluid. The pressure, if it exists at the first instant, will be exerted on the front of the body and will be ku, where k is a quantity which depends on the shape of the body.

" Let uq be the velocity of the fluid parallel to the axis and let up be the velocity of the fluid perpendicular to the axis. Since it is supposed that, at the first instant, with velocity u parallel to the axis and that it changes this velocity into the velocities uq and up, the fluid will exert a pressure on the body which will be the same as if the fluid were at rest and the body were moved with velocities $u-uq$ and $u-up$. Now because of the velocity u, the pressure exerted will be, by the principles of hydrostatics, equal to Mu and in the opposite direction to u, if M is the mass of the body. And because of the velocities $-uq$ and $-up$, the pressure exerted will be in a direction opposite to that of Mu and will be equal to $4u \int dy \int ds \sqrt{(p^2 + q^2)}$. It is easy to see this by supposing that $\int ds \sqrt{(p^2 + q^2)} = 0$ and $\int dy \int ds \sqrt{(p^2 + q^2)} = 0$ at the point at which $dy = 0$ and which

---

[1] Quantities proportional to the components of the velocity in d'ALEMBERT's theory.

is not the summit of the body—that is, at the point at which the tangent is parallel to the axis. For it is clear that, since $dy = 0$ (hyp.) for $x = 0$ and the body is composed of four equal and similar parts, there must be a point on each side of the axis, and in the centre of the axis, where $dy = 0$. Therefore, if it is supposed that $\int dy \int ds \sqrt{(p^2 + q^2)} = R$, in its totality, then $k = 4R - M$.

" In the following instants parts of the fluid evidently retain the velocities uq and up, from which it is easy to see, by means of our theory on the resistance, that the pressure of the fluid on the body will be absolutely nothing. For the pressure on the forward surface is equal and opposite to the pressure on the backward surface.

" And if it were supposed that a force γ, constant or variable from one instant to another but always the same for all parts of the fluid at the same instant, acted on all these parts, they would nevertheless continue to describe the same lines with a velocity that would be increased in the ratio of γdz to u. And from this it would only result that a new pressure equal to $k\gamma$ was exerted on the surface of the body.

" Suppose now that the fluid is at rest and the body is moved along in it with the velocity v, of which it only retains a part u. Give the whole system—fluid and body—a motion u in the opposite direction. The body will be at rest in fixed space. In front there will be the force $M(v - u)$, while the fluid will exert a pressure ku on the body. The latter will nullify the former.

" Therefore

$$ku = (v - u) M \qquad u = \frac{Mv}{k + M} = \frac{Mv}{4R}.$$

" Now suppose that the body continues to move in the fluid with a velocity which decreases by an amount γdt at each instant so that $du = -\gamma dt$. Also suppose that the system of body and fluid moves on the opposite direction with this decreasing velocity. It is apparent that the body will be at rest and that the pressure at each instant will be $k\gamma$, or kdu, which must counterbalance Mdu. Therefore

$$Mdu = du(4R - M) \quad \text{or} \quad Mdu = 2Rdu.$$

" Therefore either $du = 0$ or $2R = M$.

" If $du = 0$, the body will move uniformly and it will be true that $u = \frac{Mv}{4R}$.

" If $2R = M$, then $u = \frac{v}{2}$ and the quantity du remains indeterminate. It could be supposed to be zero, and even must be supposed to be zero, since there is no other equation to determine it than $Mdu = 2Rdu$.

" Thus the greatest alteration that could occur in the original motion of the body is that the velocity v—which is supposed to have been impressed on it—should be changed to $\dfrac{Mv}{4R}$ in the first instant, and after this the body will move without suffering any resistance due to the fluid.

" If the shape of the body is such that $\int dy \int ds \sqrt{(p^2 + q^2)} = R = \dfrac{M}{4}$ then $u = v$. Whence $Mdu = du \,(4R - M) = 0$ and therefore the body will not lose any velocity in the first instant. This seems also borne out by experiment.

" I do not ask whether the quantities p and q which are obtained by the theory are such that $4R = M$ for any shape of the body—it appears rather doubtful that this should be so. Neither do I ask whether $4R$ could be greater than M for some shapes and less than M for others. These conditions would imply $u < v$ (contrary to experiment) and $u > v$ (contrary to common sense). Therefore we will have $du = 0$ and, from our theory, it will follow that the body, supposed to be of four equal and similar parts, will suffer no resistance from the fluid.

" And whatever relation there is supposed to be between $4R$ and M, it is apparent that the velocity v will, at the most, only experience an alteration in the first instant, and that it will then remain uniform. This would be much worse if $4R < M$ for then the initial velocity would first increase and afterwards remain uniform.

" I must therefore confess that I do not know how the resistance of fluids can be explained by the theory in a satisfactory way. On the contrary, it seems to me that this theory, handled with all possible rigour, yields a resistance which is absolutely nothing in at least several situations. I bequeath this strange paradox to the geometers, that they may explain it. "

5. EULER AND THE EQUILIBRIUM OF FLUIDS.

In a paper given to the Academy of Berlin in 1755,[1] Euler directed his attention to the equilibrium of fluids.

He considered a fluid, either compressible or not, which was subjected to any given forces. " The generality that I include ," he declared, " instead of dazzling us, will rather discover the true laws of Nature

---

[1] *Principes généraux de l'état d'équilibre des fluides*, *Mémoires de l'Académie de Berlin*, 1755, p. 217.

in all their splendour, and there will be found yet stronger reasons for wondering at their beauty and simplicity. "

The general problem which Euler poses is the following one.

" *The forces which act on all the elements of the fluid being given, together with the relation which exists at each point between the density and the elasticity of the fluid, to find the pressures that there must be, at all points of the fluid mass, in order that it may remain in equilibrium.* "

In the fluid mass Euler considers an elementary rectangular parallelipiped with one corner at the point Z, of coordinates x, y, z and with sides dx, dy and dz.

The components of the " accelerative force " applied to each element are called P, Q and R, and q is the density of fluid at Z.

Then the element of volume $dxdydz$ is subject to the " motive force " whose components are

$$Pqdxdydz \qquad Qqdxdydz \qquad Rqdxdydz.$$

If p is the unknown pressure at the point Z, then

$$dp = Ldx + Mdy + Ndz.$$

By a very simple geometrical argument—which has become classical—Euler deduces the general conditions of equilibrium

$$L = Pq \qquad M = Qp \qquad N = Rq.$$

If L, M and N are the partial derivatives of a function $p\,(x, y, z)$, equilibrium requires the conditions

$$\frac{\partial\,(Pq)}{\partial y} = \frac{\partial\,(Qq)}{\partial x} \quad \frac{\partial\,(Qq)}{\partial z} = \frac{\partial\,(Rq)}{\partial y} \quad \frac{\partial\,(Rq)}{\partial x} = \frac{\partial\,(Pq)}{\partial z}.$$

If p is a given function of q at each point of the fluid, the relation

$$dp = q\,(Pdx + Qdy + Rdz)$$

shows that $Pdx + Qdy + Rdz$ is the total differential of the function $\dfrac{dp}{q}$.

This differential represents the " effort " or the " efficacy " of the given force—this was a notion which Euler used in the case of central forces.

In a second paper, which will be discussed in the next §, Euler deduced a general conclusion from the equation of equilibrium

$$\frac{dp}{q} = Pdx + Qdy + Rdz.$$

"The forces P, Q, R must be such that the differential form $Pdx +$ $Qdy + Rdz$ either becomes integrable when the density q is constant or uniquely dependent on the elasticity p, or becomes integrable when multiplied by some function."

Euler did not refer to Clairaut in this connection. Although he had the merit of introducing the pressure and relating it to the accelerative force at each point it must be observed, with Lagrange, that Euler's achievement was that of applying, by generalising it, the principle of Clairaut.

6. EULER AND THE GENERAL EQUATIONS OF HYDRODYNAMICS.

We now come to a fundamental paper of Euler on the equations of hydrodynamics.[1] So perfect is this paper that not a line has aged.

In assuming this difficult task, Euler declared, "I hope to emerge successful at the end, so that if difficulties remain they will not be in the field of mechanics, but entirely in the field of analysis."

Euler considers a fluid which is compressible or incompressible, homogeneous or inhomogeneous. Its *original state*—that is, the arrangement of the particles and their velocities—is supposed known at a given instant, as are the external forces acting on the fluid.

It is necessary to determine, at all times, the pressure at each point of the fluid together with the density and the velocity of the element passing through that point.

In order to study the *present state* of the fluid, Euler uses the components of the accelerative force, P, Q, R which are known functions of x, y, z and t.[2]

The density q, the pressure p, and the components u, v, w of the velocity of the element of the fluid which is at the point Z at the time t are unknown.

During the time dt the element of fluid at Z will be carried to the point Z', whose coordinates will be

$$x + udt \qquad y + vdt \qquad z + wdt.$$

The element of fluid at z, of coordinates

$$x + dx \qquad y + dy \qquad z + dz$$

[1] *Principes généraux du mouvement des fluides*, *Mémoires de l'Académie de Berlin*, 1755, p. 274.

[2] EULER also refers to a variable r, the "heat at the point Z, or that other property which, apart from the density, affects the elasticity."

has a velocity whose components are

$$\begin{cases} u + \dfrac{\partial u}{\partial x}\,dx + \dfrac{\partial u}{\partial y}\,dy + \dfrac{\partial u}{\partial z}\,dz \\ v + \ldots \\ w + \ldots \end{cases}$$

and, during the time dt, is carried to the point z'. In order to perform the calculation, Euler first considers a segment Zz which is parallel to the axis of x. During the time dt this segment will turn through an infinitely small angle, and its length will become

$$Z'z' = dx\left(1 + dt\,\frac{\partial u}{\partial x}\right) + \ldots$$

to the second order.

In a latin paper, *Principia motus fluidorum*, Euler elucidates the problem—then entirely novel—of the kinematics of continuous media. He calculates the form which the elementary parallelipiped, whose origin is Z and sides are dx, dy and dz, will assume at the time $t + dt$ because of the motion of the fluid. He finds that the volume becomes

$$dxdydz\left(1 + dt\,\frac{\partial u}{\partial x} + dt\,\frac{\partial v}{\partial y} + dt\,\frac{\partial w}{\partial z}\right).$$

Similarly the density, q, of the fluid at Z becomes, at Z',

$$q + dt\,\frac{\partial q}{\partial t} + udt\,\frac{\partial q}{\partial x} + vdt\,\frac{\partial q}{\partial y} + wdt\,\frac{\partial q}{\partial z}.$$

At this point Euler expresse the *conservation of the mass* in the course of the motion. " As the density is reciprocally proportional to the volume, the quantity q' will be related to q as $dxdydz$ is related to

$$dxdydz\left(1 + dt\,\frac{\partial u}{\partial x} + dt\,\frac{\partial v}{\partial y} + dt\,\frac{\partial w}{\partial z}\right);$$

whence, by carrying out the division, the *very remarkable condition which results from the continuity of the fluid*,

$$\frac{\partial q}{\partial t} + u\,\frac{\partial q}{\partial x} + v\,\frac{\partial q}{\partial y} + w\,\frac{\partial q}{\partial z} + q\,\frac{\partial u}{\partial x} + q\,\frac{\partial v}{\partial y} + q\,\frac{\partial w}{\partial z} = 0.$$

This may be written more simply as

$$\frac{\partial q}{\partial t} + \frac{\partial}{\partial x}\,(qu) + \frac{\partial}{\partial y}\,(qv) + \frac{\partial}{\partial z}\,(qw) = 0$$

and, for an incompressible fluid, it reduces to

$$\frac{\partial u}{\partial x} + \frac{\partial v}{\partial y} + \frac{\partial w}{\partial z} = 0 \cdot "$$

Euler then calculates the acceleration of the element of fluid which is at Z at the instant t. He first writes the components of the velocity at the point Z', to which the point Z is carried at the end of the time dt, in the following form

$$\begin{cases} u + dt\,\frac{\partial u}{\partial t} + udt\,\frac{\partial u}{\partial x} + vdt\,\frac{\partial u}{\partial y} + wdt\,\frac{\partial u}{\partial z} \\ v + \dots \\ w + \dots \end{cases}$$

whence the acceleration or the increment of the velocity

$$\begin{cases} \dfrac{\partial u}{\partial t} + u\,\dfrac{\partial u}{\partial x} + v\,\dfrac{\partial u}{\partial y} + w\,\dfrac{\partial u}{\partial z} \\ \dfrac{\partial v}{\partial t} + \dots \\ \dfrac{\partial w}{\partial t} + \dots \end{cases}$$

The pressure exerts the " accelerative force, " whose components are

$$-\frac{1}{q}\frac{\partial p}{\partial x} \qquad -\frac{1}{q}\frac{\partial p}{\partial y} \qquad -\frac{1}{q}\frac{\partial p}{\partial z}$$

on the elementary mass of the parallelepiped. Thus the equations of motion of the fluid, to be joined to the equation of continuity, are

$$\begin{cases} P - \dfrac{1}{q}\dfrac{\partial p}{\partial x} = \dfrac{\partial u}{\partial t} + u\,\dfrac{\partial u}{\partial x} + v\,\dfrac{\partial u}{\partial y} + w\,\dfrac{\partial u}{\partial z} \\ \dots \\ \dots \end{cases}$$

Euler was too aware to misunderstand the difficulty of the study of these equations of motion. Thus he wrote—

" If it does not allow us to penetrate to a complete knowledge of the motion of fluids, the reason for this must not be attributed to mechanics and the inadequacy of the known principles, for analysis itself deserts us here. . . . "

Lagrange, in this connection, wrote—

" By the discovery of Euler the whole mechanics of fluids was reduced to a matter of analysis alone, and if the equations which contain it were integrable, in all cases the circumstances of the motion and behaviour of a fluid moved by any forces could be determined. Unfortunately, they are so difficult that, up to the present, it has only been possible to succeed in very special cases. "

Without concerning ourselves with the particular problems which he treats, we remark that Euler indicated the simplicity that results if

$$udx + vdy + wdz$$

is a complete differential. Much later this was distinguished as the case in which a velocity potential existed, or the case of irrotational motion.

In a third paper on the motion of fluids [1] Euler draws attention to a plane irrotational motion of an incompressible fluid, which is characterised by the two conditions

$$\frac{\partial u}{\partial x} + \frac{\partial v}{\partial y} = 0 \qquad \frac{\partial v}{\partial x} = \frac{\partial u}{\partial y}.$$

In this connection, Euler acknowledges a debt to d'Alembert for having conceived the device of considering $u - iv$ as a function of $x + iy$, and $u + iv$ as a function of $x - iy$.[2] (This was before Cauchy had systematised the notion of analytic function, and long before the modern school of hydrodynamics existed.)

Euler also writes, with some hint of sarcasm, " However sublime may be the investigations on fluids for which we are indebted to MM. Bernoulli, Clairaut and d'Alembert, they stem so naturally from our two general formulae that one cannot but admire this agreement of their profound meditations with the simplicity of the principles from which I have deduced my two equations, and to which I was directly led by the first axioms of mechanics. "

Just because of the analytical difficulties of the general problem, Euler did not misunderstand the importance of the part-experimental, part-theoretical considerations that were used in hydraulics. On the contrary, in that he was personally concerned with the Segner water wheel, had analysed the working of turbines and had himself designed a reaction turbine, he was a pioneer of modern technics.

---

[1] *Continuation des recherches sur la théorie du mouvement des fluides*, *Mémoires de l'Académie de Berlin*, 1755, p. 316.

[2] This device is used by d'ALEMBERT in his *Essai sur la résistance des fluides*.

7. BORDA AND THE LOSSES OF KINETIC ENERGY IN FLUIDS.

In this paragraph we shall follow a work of Chevalier de Borda (1733-1799) called *Mémoire sur l'écoulement des fluides par les orifices des vases*.[1]

Borda's analysis is based on both Daniel Bernoulli's hydrodynamics and the mechanics of fluids which d'Alembert had related to his own principle. At first Borda discusses problems of flow, and on each occasion his analysis owes something to Bernoulli and d'Alembert. Notable among these problems is the determination of the *contracted section*—in this connection he considers a *re-entrant nozzle*, where the contracted section can be calculated and turns out to be equal to half of that of the orifice.

But the essential interest of Borda's study is that he drew attention to "*hydrodynamical questions in which a loss of living force must be assumed.*" Such losses appear in a tube which is abruptly enlarged or contracted. With a bold insight, Borda compared the phenomenon which occurs in the fluid to an impact in which a loss of kinetic energy was involved— that is, in the language of the time, to an impact of *hard bodies*.

First Borda establishes the following Lemma, and thus anticipates Carnot's theorem in a special case.

"*Lemma.* — Let there be a hard body a, whose velocity is u, which hits another hard body A whose velocity is V. It is required to find the loss of living force which occurs in the impact.

"Before the impact the sum of the living forces was $\dfrac{au^2 + AV^2}{2g}$. After the impact this sum has the value

$$\frac{a + A}{2g}\left(\frac{au + AV}{a + A}\right)^2$$

whence, by difference,

$$\frac{aA}{a + A}\frac{(u - V)^2}{2g} \qquad\qquad \text{Q. E. D.}"$$

Borda considers (see figure) the immersion of a cylindrical vessel into an indefinite fluid $OPQR$, and seeks the motion which the fluid will have on entering the vessel. He starts from the following consideration.

"The motion of the water in the vessel can be regarded as that of a system of hard bodies that interact in some way. Now we know that the principle of living forces only applies to the motion of such

[1] *Mémoires de l'Académie des Sciences*, 1766, p. 579.

bodies when they act on each other by imperceptible degrees, and that there is necessarily a loss of living force as soon as one of the bodies collides with another. "

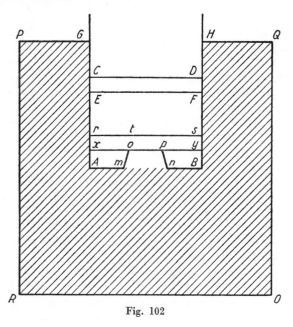

Fig. 102

In the example with which we are concerned, " the slice *mopn* which enters the vessel at one instant, occupies the position *rsqy* at the next instant. It is clear that before it occupies this position the small slice will have lost a part of its motion against the fluid above, as if it had been an isolated mass which had been hit by another isolated mass. But in the case of these two isolated masses there would have been a loss of living force. Therefore there will also be such a loss in the case that we are discussing. "

And here is Borda's solution, which follows Daniel Bernoulli's method. Suppose that the fluid has travelled to *EF*, and that in the next instant it travel to *CD*. Put $AE = x$, $Ag = a$ and $AB = b$. Let u be the velocity of the fluid at E. Assume that the living force of the fluid in the indefinite vessel *ROPQ* remains zero. Under these conditions, the living force of the fluid in the inner vessel can be written

$$\frac{u^2bx}{2g}$$

if the living force of the slice which enters the vessel is neglected.

" Thus the difference of the living force of all the fluid contained in the vessel will be

$$\frac{u^2bdx + 2bxudu}{2g}.$$

Now while the fluid acquires this increment of living force the slice $DCFE$, or bdx, is supposed to descend from the height GE, or $a - x$. Therefore, if the principle of living forces applied without restriction, it would be true that

$$(a - x)\,bdx = \frac{u^2bdx + 2bxudu}{2g}.$$

" But there is a loss of living force in the whole of the fluid, which arises from the action of the small slice $rsqy$ on the fluid $rCDs$ which is above it. It is easy to see, by the lemma, that if the velocity of the slice $opmn$ is denoted by V, then this loss of living force is

$$\frac{badx}{a + dx}\frac{(V - u)^2}{2g} = bdx\,\frac{(V - u)^2}{2g}.$$

" Therefore, adding this quantity to the second term of the equation above, the correct solution of the problem is obtained—

$$u^2bdx + 2bxudu + bdx\,(V - u)^2 = 2g\,(a - x)\,bdx.$$

" It only remains to determine V. For this purpose it is sufficient to observe that the stream of fluid which enters the vessel contracts in the same way as if it left the vessel by the same orifice and entered free space. This must be since, in both cases, the fluid which arrives at the orifice is travelling in the same directions. Now the loss of living force must be distributed from the slice that has the greatest velocity—that is, from that which is at the point of greatest contraction.

" Therefore suppose that this point is at o and that m is the ratio of EF to op. Then $V = mu$, whence

$$u^2dx + 2xudu + u^2dx\,(m - 1)^2 = 2g\,(a - x)\,dx.$$

This equation is integrated by supposing that $x = e$ and $u = o$ at the beginning of the motion. "

Borda then repeats his argument and, this time, follows d'Alembert's method.

" What we have just said of the principle of conservation of living forces is also applicable to M. d'Alembert's principle. Not that the latter principle is always true, for there are some instances in which the way it is applied to the motion of fluids must be somewhat modified. Indeed, we have seen that the slice *rsxy* only acts on the fluid above in the way that an isolated mass would lose a part of its motion to another mass with which it collided. Whence it follows that in the equation of equilibrium, the accelerating force must not be multiplied by the volume $\frac{1}{2} mt$ which the slice occupies at the middle to the time interval, but by the volume *ot* which it occupies at the *end of this interval.* For the volume *ot* represents the mass of the small slice and *rC* represents that of the fluid *rCDs.* " No purpose would be served by reproducing the calculation which follows, which leads to the same result as the analysis reproduced above.

However bold it may have been, Borda's hypothesis is discovered to be in satisfactory agreement with experiment.

" A tube 18 lines in diameter and one foot long was made of very uniform tinplate whose edges were tapered. Then, closing the upper orifice with the hand, the tube was plunged into a vessel filled with water. It was assured that the air contained in the tube did not allow the water to enter to the same extent as if both openings had been free. Then the upper orifice of the tube was opened and the water mounted inside the tube to a height greater than its level outside. The experiment was repeated several times and the water rose to its peak which was 4 pouces above the outside level. According to the calculation of M. Bernoulli, it should have risen to 8 pouces. " [1]

The ascent calculated by Borda was 49 ½ lines. He observed an ascent of 47½ lines and attributed the difference to the friction of the fluid on the walls.

[1] *Mémoires de l'Académie des Sciences,* 1766, p. 147.

CHAPTER NINE

EXPERIMENTS ON THE RESISTANCE OF FLUIDS
(BORDA, BOSSUT, DU BUAT)
COULOMB AND THE LAWS OF FRICTION

1. BORDA'S EXPERIMENTS AND NEWTONIAN THEORIES.

During the same time that the principles of dynamics were being organised and the foundations of hydrodynamics were being developed, there grew up a complete experimental approach that was determined by requirements of an essentially practical kind. To pause on this remarkable movement is not to move away from the principles of mechanics, for here it can be seen how experiment is dominant in fields where the theory is impotent before the very complexity of even the most tangible phenomena.

We shall only deal with some examples of this experimental work in mechanics during the XVIIIth Century. Besides being characteristic, these examples are ones in which the origins of modern research should be sought, and in which the modest methods deployed (for example, the motive agencies were invariably provided by falling weights) were no obstacle to the application of a rigorous experimental method.

But before coming to these examples, it is necessary that we should describe some essays of the theoreticians, who had, indeed, preceded the experimentalists by several years.

Newton had developed a schematic theory of fluids, which he considered to be formed of an aggregate of elastic particles which repelled each other, were arranged at equal distances from each other, and were free. If the density of this aggregate was very small, Newton assumed that if a solid moved in the fluid then the parts of the fluid which were driven along by the solid were displaced freely, and did not communicate the motion which they received to neighbouring parts.

In this framework, Newton calculated the resistance of a fluid to the translation of a cylinder. He found that this resistance was equal to the weight of a cylinder of fluid of the same base as the solid, and whose height was twice that from which a heavy body would have to fall in order to acquire the velocity with which the solid moved.

The resistance offered to the translation of a sphere, according to the same newtonian theory, is half the resistance which the cylinder encounters under the same conditions.

Jean Bernoulli adopted these laws in the discussion of the communication of motion which he gave in connection with the controversy on living forces.

Newton also formulated a second theory on the resistance of fluids, and applied this to water, oil and mercury. His first theory was only applied to the resistance of air.[1]

In this second theory, particles of the fluid are contiguous. Newton compares the resistance to the effect of the impact of a stream of fluid on a circular surface, the stream being imagined to leave a cylindrical vessel through a horizontal orifice. He passes to the limit by infinitely increasing the capacity of the vessel, and also the dimensions of the orifice, in order to simulate the conditions of an indefinite fluid. He then substitutes the motion of the circular surface for that of the fluid in the first model of impact.

Given this, Newton calculates the resistance offered to the translation of a cylinder and finds that the resistance is equal to the weight of a cylinder of fluid whose base is the same as that of the solid and whose height is half that from which a heavy body would have to fall in order to acquire the velocity with which the solid moves in the fluid. This resistance is four times smaller than that provided by the first theory.

Further, in the second theory the length of the moving cylinder does not affect the result, for only its base is exposed to the impact of the fluid. Under these circumstances the resistance offered to the translation of a sphere is equal to that which would be offered to the translation of a cylinder circumscribed about the sphere. This result is half that provided by the first theory.

The second newtonian theory is applicable to the *oblique impact* of a stream of fluid on a plane wall. Under these conditions, it leads to a resistance which is proportional to both the square of the velocity and the *square of the sine* of the angle of incidence. These were the proportions which the experimenters tried to verify.

We also add that Daniel Bernoulli, although he did not offer an alternative theory, had already remarked on considerable differences

---

[1] In fact this theory goes back to HUYGHENS (1669), MARIOTTE (1684) and PARDIES (1671).

between the newtonian laws and experiment.[1] Moreover, he legitimately emphasised that it was necessary to distinguish between the
impact of a fluid on a wall and the impact of a fluid on a *completely
immersed* plane.[2] Finally, we recall that d'Alembert, in his early
work on the resistance of fluids, also calculated the impact of a moveable
surface on an infinity of small elastic balls which represented a fluid.[3]

With this in mind, we come to the experiments of Chevalier de Borda.

In the first place, Borda studied the resistance of air.[4] By means
of a driving weight he made a flywheel rotate and attached plane
surfaces of different shapes to the circumference. He took care to
correct the results for the friction of the flywheel and to confine the
observations to a period of uniform motion, when a steady state had
been established.

These are Borda's conclusions.

1) The total resistance of the air cannot be calculated as the sum
of the partial resistances of each of its elements. For example, the
resistance of a circle is not the sum of the resistances of two semicircles.
This conclusion is very important—it shoes that the resistance is a
phenomenon which behaves integrally, and also makes it clear that
the resistance cannot be obtained by an integration which depends
on a simple elementary law.

2) The aggregate resistance is proportional to the square of the
velocity and the sine of the angle of incidence (not to the square of
this sine).

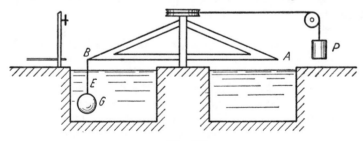

Fig. 103

As far as the resistance of water is concerned, Borda confines himself, in this first paper, to the verification of the proportionality to
the square of the velocity.

[1] *Mémoires de Petersbourg*, Vol. II, 1727.
[2] *Ibid.*, Vol. VIII, 1741.
[3] See above, p. 295.
[4] *Mémoires de l'Académie des Sciences*, 1763, p. 358.

Borda returned to the resistance of fluids in a paper dated 1767.[1]

He worked with a circular vessel 12 feet in diameter. By means of driving weights varying from 4 ounces to 8 pounds, he made a sphere of 59 lines diameter move through the water. The sphere was made of two equal parts, which could be joined together or separated as desired.

When working with one hemisphere, Borda allowed it to present either the section of a great circle or the convex part, to the fluid.

Borda took care to allow for " the friction and the impact of the air on the flywheel " by making the apparatus rotate freely without the sphere.

He verified that the resistance was very accurately proportional to the square of the velocity. In addition, he established that the resistance of the hemisphere was nearly independent of the surface that was presented to the fluid. From this he concluded that " *at these small velocities, the forward part of the body is the only one which has resistance.* "

Borda next turned his attention to the absolute magnitude of the resistance, and compared the values observed with those calculated from what we have called the second newtonian theory (*Principia*, Book II, Proposition XXXVIII). He found that the resistance of the hemisphere when it offered a section of a great circle to the fluid was $2\frac{1}{2}$ times as great as the resistance of the whole sphere, itself accurately equal to the resistance of the hemisphere when this offered its convex side to the fluid. Now, according to the newtonian theory, the first resistance is twice the second. The disagreement is evident.

Similarly, Borda determined the *oblique resistance*. He established, exactly as in his experiments in air, that the law of the square of the sine was not true, and even declared " that when the angles of incidence are small the resistance does not decrease as much as the simple sine. "

Borda also studied the influence of the depth on the resistance in water. He established that the resistance decreased with the depth, and that, at the surface, it increased more rapidly than the square of the velocity. In this connection, he attempted an explanation which was only half convincing, by falling back on his own theory of the losses of living force in fluids.[2]

" It is clear that when the sphere is only 6 pouces below the surface it does not impart such great velocities to the neighbouring parts as when it moves in the surface of the water. For in the first case the fluid is free to run round the whole circumference of the sphere

[1] *Mémoires de l'Académie des Sciences*, 1767, p. 495.
[2] See above, p. 305.

while in the second, it cannot escape along the upper part of the sphere. From which it follows that in the first instance the fluid neither gains nor loses as great a quantity of living forces as in the second. "

Borda then worked with a model $ABCD$ in which $AH = HD = 6$ pouces and $BC = 4$ pouces.

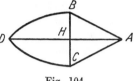

Fig. 104

The difference between the resistance when the side A (angle BAC), and then the side D (two arcs of circles, BD and DC, with centres on BC), were offered to the fluid was negligible—the newtonian theory predicted a ratio of 28 to 15 for these resistances.

Borda's general conclusion was that the newtonian theory could not account for the resistances of fluids. "The ordinary theory of the impact of fluids only gives relationships which are absolutely false and, consequently, it would be useless and even dangerous to wish to apply this theory to the craft of the construction of ships. "

2. THE ABBÉ BOSSUT'S EXPERIMENTS.

In 1775 Turgot asked the *Académie des Sciences* "to examine means of improving navigation in the Realm. " A committee consisting of d'Alembert, Condorcet and the Abbé Bossut (as secretary) immediately took up the investigation and, between July and September, 1775, conducted numerous experiments "on a large stretch of water in the grounds of the Military College. " They secured the cooperation of the mathematicians attached to that College, including Legendre and Monge.

The committee reported to the *Académie des Sciences* on April 17th, 1776, and this report, *Nouvelles expériences sur la résistance des fluides* was published at Paris (Jombert) in 1777, under the names of the three members of the committee.

The experimental method is referred to in the following terms. "To ask questions of nature by doing experiments is a very delicate matter. In vain do you assemble the facts if these have no relation to each other; if they appear in an equivocal form; if, when they are produced by different causes, you are unable to assign and distinguish the particular effects of these causes with a certain precision. . . . Do not heed the limited experimenter, the one who lacks principles; guided by an unreasoning method, he often shows us the same fact in different guises—of necessity, and perhaps without recognising this himself; or he gathers at random several facts whose differences he is unable to explain.   A science without reasoning does not exist or, what comes to the same thing, a science without theory does not exist. "[1]

Bossut explicitly distinguished between the resistance of fluids that were indefinitely extended (a ship on the sea or on wide and deep rivers) and the resistance in narrow channels (shallow or narrow rivers and canals).

Borda's experiments were conducted in fluids that were, for practical purposes, indefinitely extended.   On the other hand, in order to study the effect of the depth of immersion, Franklin had worked on a small scale with a canal and a model of a ship which was 6 pouces long, and $2\frac{1}{2}$ pouces wide.[2]

The basin at the Military College was 100 feet long and 53 feet wide at the centre, its maximum depth being $6\frac{1}{2}$ feet.   A weight hung over a pulley assured the traction of the model, which was equiped with a rudder in order that its motion might be determinate.

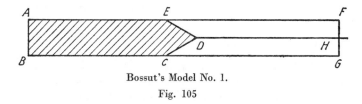

Bossut's Model No. 1.

Fig. 105

Bossut used twelve different models of ships and carried out a total of about 300 trials, of which about 200 were in an effectively indefinite fluid and the remainder in an artificially constructed channel whose depth and width were variable at will.

When he compared the experimental results with the second newtonian theory, Bossut came to the following conclusions.

1) On a given surface, and at different velocities, the resistance is " approximately in the square ratio, just as much for oblique impacts as for direct impacts.   More accurately, the resistance increases in a greater ratio than the square. "   He gives the following explanation of this fact.   " The fluid has greater difficulty in deflecting itself when the velocity increases—it piles up in front of the prow and is lowered near the stern. "[3]

2) " For surfaces which are equally immersed in the fluid and only different in respect of their width, the resistance *sensibly* follows

[1] *Nouvelles expériences sur la résistance des fluides*, p. 5.
[2] *Œuvres complètes de Franklin*, Vol. II, p. 237.
[3] BOSSUT, *op. cit.*, p. 147.

the ratio of the surfaces. . . . More precisely, it increases in a ratio which is a little greater than that of the extent of the surfaces. " [1]

" The resistance of bodies which are entirely submerged is a little less than that of bodies which are only partly submerged. " [2]

3) " The law of the square of the sine is less justified when the angles are very small. " [3] In order to express the results of these experiments, the Abbé Bossut chose a provisional law of the form

$$\sin^n i$$

where i is the angle of incidence. He found that the exponent n varied from 0.66 to 1.79, according to the model studied. This led him to conclude—

" The resistances which occur in oblique impacts cannot be explained by the theory of resistances by introducing, instead of the square, some other power of the sine of the angle of incidence in the expression for the resistance. " [4]

In order to determine the magnitude of the resistance of water, Bossut made two corrections. The first depended on the friction of the pulley which supported the cable and the motive weight—he measured this friction by varying the motive weight. The second correction arose because of " the impact of the air " on the model. Indeed, to the author, resistance was an impact phenomenon. He was guided throughout by the second newtonian theory. In order to eliminate the impact of the air, Bossut measured the surface of the model which was offered to the impact of air, and assumed that the impacts of the water and the air on the model were respectively " in compound proportion to the impacted surfaces and the densities of the two fluids."

Having made these two corrections, Bossut concluded—

" The resistance perpendicular to a plane surface in an indefinite fluid is equal to the weight of a column of fluid having the impacted surface as its base and whose height is that which corresponds to the velocity with which the percussion occurs. " [5]

Bossut tried to analyse further the phenomenon of resistance ; he sought to emphasize the part played by the " tenacity " of the fluid and the " friction caused along the length of the boat by the water. " From this somewhat arbitrary decomposition, he felt justified in drawing the following conclusions :

" We have observed that as soon as the friction is overcome, the

[1] BOSSUT, *op. cit.*, p. 152.
[2] *Ibid.*, p. 157.
[3] *Ibid.*, p. 163.
[4] *Ibid.*, p. 164.
[5] *Ibid.*, p. 173.

slightest force sets the boat in motion. From which we have concluded that the tenacity of the water is extremely small and that this resistance must be considered absolutely nil in comparison with that caused by inertia. The same applies to the friction of the water along the sides and bottom of the boat. This friction is very slight and its effect cannot be distinguished from that of the pulleys or of the resistance of the air. " [1]

Again Bossut noted the resistance in a narrow canal, superior to the resistance in an unlimited fluid, and he underlined the influence of the transversal dimensions and of the form of the vessel used for comparison. For the construction itself of the canals, his paper is limited to cautious generalities : the canal should be as large and as deep as possible, " without nevertheless going to superfluous expense " ; subterranean canals should be avoided unless local circumstances make their use indispensable. Indeed, concludes this sagacious rapporteur, " a canal is an object of utility and not an instrument for ostentation. "

3. Du Buat (1734-1809). Hydraulics and the resistance of fluids.

Du Buat began by directing the construction of fortifications and on this occasion was the promotor of ' géométrie cotée '. He later devoted himself to hydraulics, as " Captain of the infantry, engineer to the King."

The *Principes d'hydraulique*, the first edition of which is dated 1779, deals with " the motion of water in rivers, canals and conduits ; the origin of rivers and the formation of their beds ; the effect of locks, bridges and reservoirs [2] ; of the impact of water ; and of navigation on rivers as well as on narrow canals. "

Du Buat wrote, " there is no argument which can be used to apply the formulae for flow through orifices to the uniform flow of a river, which can only owe the velocity with which it moves to the slope of its bed, taken at the surface of the current. "

Gravity is, on both cases, certainly the cause of the motion. " I therefore set out to consider whether, if water was perfectly fluid and ran in the part of a bed which provided no resistance, it would accelerate its motion like bodies which slide on an inclined plane. . . . Since it is not so, there exists some obstacle which prevents the accelerating force from imparting fresh degrees of velocity to it. Now, of what can this obstacle consist, except the friction of the water against the walls of the bed and the viscosity of the fluid ? "

And Du Buat stated this principle— " When water runs uniformly in some bed, the force which is necessary to make it run is equal to

[1] Bossut, *op. cit.*, p. 173.
[2] Read " weirs. "

the sum of the resistances to which it is subject, whether they are due to its own viscosity or to the friction of the bed. "

In the 1786 edition of his *Traité*, Du Buat amends this statement and no longer speaks of the viscosity, but only of the resistance of the bed or the containing walls. The viscosity only enters indirectly, " in order to communicate the retardation due to the walls, step by step, to those parts of the fluid which are not in contact with them. " This effect only influences " the relation between the mean velocity and that which is possessed by the fluid against the walls. "

In canals of circular or rectangular section, Du Buat introduced the notion of *mean radius* (the ratio of the area of the cross-section to the length of the perimeter in contact with the fluid) and evaluated the resistance of the walls to unit length of the current by the product of this radius and the friction on unit surface. He assumed that the resistance of the bottom was proportional to the square of the velocity of the current, and likened it to the impact of the water on the irregularities on the bottom.

Du Buat did not confine himself to this theoretical outline but, like the Abbé Bossut, he sought experimental confirmation. The second edition of his *Traité* is concerned with these experiments.

Du Buat verified that the friction of fluids was independent of their pressure. This he did by making water oscillate in two siphons of very different depth. He investigated the friction of fluids on different materials (glass, lead and tin) and, having observed that this friction was always the same, he assumed that the water " itself prepares the surface on which it runs " by wetting the pores and cavities as a varnish does. He went further, and even held that the resistance of the walls did not depend on their roughness—a conclusion that was very far from being correct.

In order to obtain the resistance of the walls, Du Buat worked with an artificial canal of oak planks, whose section could be varied in shape and size. He also worked with pipes of tinplate or glass of very different diameters. He discovered that the resistance of the walls was *in a smaller ratio* than the square of the velocity [1] and

---

[1] DE PRONY advocated a formula for the resistance which had the form $av + bv^2$. Much later, after having observed the oscillations of a circular plate in a fluid medium, COULOMB was to say, " There must be two kinds of resistance. One, due to the coherence of the molecules which are separated from each other in a given time, is proportional to the number of these molecules and, consequently, to the velocity. The other, due to the inertia of the molecules which are stopped by the roughnesses with which they collide, is proportional to both their number and their velocity and, consequently, to the square of their velocity. " COULOMB was, before STOKES, the first to hold that the velocity of a viscous fluid relative to a solid was *nothing at the surface of contact*, and that it then varied continuously in the fluid.

gave an empirical formula for this resistance which was only surpassed in accuracy by those of Darcy (1857) and Bazin (1869).

Du Buat then turned to the empirical relationships between the mean velocity, the velocity at the centre of the surface and the velocity at the centre of the bed. To account for the resistance due to bends, Du Buat assumed a series of impacts on the banks, and expressed the resistance as a number proportional to the square of the mean velocity, the square of the sine of the angle of incidence and the number of " ricochets. " He applied his empirical formula to the eddies and local variations of level which are found upstream from barrages and narrows by considering small consecutive lengths of the current.

Du Buat also treated the decrease of the slope, and the increase of the depth, from the source of a river to its mouth. He took account of tributaries, temporary and periodic floods, changes of course, the retarding effect of the wind and even the influence of the weeds which grew in the bottom.

In order to ascertain *the resistance of fluids to the translation of a solid*, Du Buat exposed a tinplate box to the current. The box was either cylindrical or in the form of a parallelipiped whose edges were parallel to the flow lines. The boxes were provided with holes which could be opened or closed at will. A float allowed the difference of the levels, outside and inside the box, to be measured, and thus the pressures at different points of the surface of the box to be estimated. In this way Du Buat showed the existence of an *over-pressure* at the front (with respect to the previously existing state, in which the level was uniform) and a " *non-pressure,* " or suction " at the back acting in the same direction as the over-pressure. "

The total observable resistance corresponds to the sum of these two effects. Du Buat measured them separately, and showed that the over-pressure was approximately the same for a thin plate, for a cube and for a parallelipiped. On the other hand, the " non-pressure " decreased rapidly when the solid became relatively longer.

Du Buat found that the resisting force of a fluid mass to a solid in translation was less than the resistance of the solid at rest to the moving fluid, if the relative velocity was the same in both cases. This is explained by the fact that he worked on a limited fluid mass. Du Buat then set out to measure the amount of the fluid which accompanied a solid in its motion through a practically indefinite fluid. He made a solid oscillate, like a pendulum in the fluid, and studied the variation of the amplitude of small oscillations—a consequence of the decrease of the weight of the solid body due to the upthrust of the fluid, and the increase of its mass due to the mass a fluid carried along. If p

is the weight of the oscillating body (weighed in the fluid), P the weight of fluid displaced, nP the sum of the weights of the fluid displaced and the fluid carried along, l the length of the pendulum and a the length of an isochronous pendulum in the vacuum, then

$$\frac{l}{a} = \frac{p}{p + nP} \quad \text{whence} \quad n = \frac{p}{P}\left(\frac{a}{l} - 1\right).$$

Indeed, in the fluid

$$\frac{p + nP}{g}\gamma = p \quad \text{or} \quad \gamma = \frac{pg}{p + nP}$$

and also

$$\frac{l}{a} = \frac{\gamma}{g}.$$

Du Buat established, by working in water with metallic bodies and in air with distended balloons, that the amount of fluid carried along by solids was approximately proportional to the resistance obtained by other methods. Further, he suggested extending the measurement of oscillations in order to determine the resistance of fluids, by working with pendulums consisting of long columns, so that the curvature of the trajectory might be a minimum.

From all these investigations, which place Du Buat among the greatest experimenters of his time, the author concludes that he has " not done much more than destroy the old theoretical structure," and he appealed for more experiments, in the hope that a more correct theory might emerge from them.

4. Coulomb's work on friction.

Coulomb was not the first to make experiments on the friction of sliding and the stiffness of ropes.

Amontons, in 1699[1], had stated that the friction was proportional to the mutual pressure of the parts in contact. Muschenbroek introduced the amount of the area of contact. De Camus, in a *Traité des forces mouvantes*, and Désaguillers in a *Cours de physique*, remarked that the friction at rest was much greater than the friction in motion.

In connection with the stiffness of ropes, Amontons showed that the force necessary to bend a rope round a cylinder was inversely proportional to the radius of the cylinder and directly proportional to the tension and the diameter of the rope.

[1] *Mémoires de l'Académie des Sciences*, 1699.

In 1781 the *Académie des Sciences* chose the subject of the laws of friction and the stiffness of ropes for a competition, asking for a return " to new experiments, made on a large scale and applicable to machines valuable to the Navy, such as the pulley, the capstan and the inclined plane. "

Coulomb, who was then senior captain of the Royal Corps of Engineers, won the prize with his *Théorie des machines simples en ayant égard au frottement et à la roideur des cordages.*[1]

In the frontispiece of his paper, Coulomb quotes this saying of Montaigne— " Reason has so many forms that we do not know which to choose—Experiment has no fewer " (*Essais*, Book III, Chapter XIII). In fact Coulomb's work is a model of experimental analysis, carried out with precision and exemplary detail, and from which he obtained a theory applicable to machines.

The parameters which Coulomb used in his study of friction were the following— the nature of the surfaces in contact and of their coatings ; the pressure to which the surfaces are subject ; the extent of the surfaces ; the time that has passed since the surfaces were placed in contact ; the greater or lesser velocity of the planes in contact ; and, incidentally, the humid or dry condition of the atmosphere.

He described his apparatus in great detail and, for example, mentioned " a plank of oak, finished with a trying-plane and polished with seal-skin. " He studied the friction of oak on oak, " seasoned, along the grain of the wood, with as high a degree of polish as skill could achieve. " All the result obtained were recorded, experiment by experiment, with the rigor of an official report.

He first studied the friction of sliding between two pieces of seasoned wood (oak on oak, oak on fir, fir on fir, elm on elm). He then studied the friction between wood and metals, between metals with or without coatings, etc. . . .

By way of an example, here is a summary of some of his conclusions.

" 1. The friction of wood sliding, in the dry state, on wood opposes a resistance proportional to the pressures after a sufficient period of rest; in the first moments of rest this resistance increases appreciably, but after some minutes it usually reaches its maximum and its limit.

" 2. When wood slides, in the dry state, on wood, with any velocity, the friction is once more proportional to the pressures but its intensity is much less that which is discovered on detaching the surfaces after some moments of rest.

" 3. The friction of metals sliding on metals, without coatings,

---

[1] *Mémoires des Savants étrangers*, Vol. X.

is similarly proportional to the pressures but its intensity is the same whether the surfaces are detached after some moments of rest, or whether they are forced into some uniform velocity.

" 4. Heterogeneous surfaces, such as wood or metals, sliding upon each other without coatings, provide, in their friction, very different results from the preceding ones. For the intensity of their friction, relatively to a time of rest, increases slowly and only reaches its limit after four or five days, or even more. . . . Here the friction increases very appreciably as the velocities are increased, so that the friction increases approximately in an arithmetic progression when the velocities increase according to a geometric progression. "

The most debatable part of Coulomb's paper is that in which he attemps to construct a model of the production of friction.

" The friction can only arise from the engaging of the projections from the two surfaces, and coherence can only affect it a little. . . . The fibres of wood engage in each other as the hairs of two brushes do ; they bend until they are touching without, however, disengaging ; in this position the fibres which are touching each other cannot bed themselves down further, and the angle of their inclination, depending on the thickness of the fibres, will be the same under all degrees of pressure. Therefore a force proportional to the pressure will be necessary for the fibres to be able to disengage. "

At first Coulomb used the same arrangement as Amontons for the investigation of the stiffness of ropes. Later he developed a new one which allowed him to work with more industrial cables, namely, " ropes of three untarred strands. " He summarised the effect of the stiffness of ropes by means of the formula

$$\frac{A + BT}{R}$$

where $A = hr^q$, $B = h'r^\mu$ where R is the radius of the pulley, r the radius and T the tension of the rope. The exponents q and μ are approximately equal.

The mechanics of friction was still a very skeletal one in Coulomb's paper. Coulomb assumes that, in order to draw a weight P along a horizontal plane, it is necessary to deploy a force

$$T = A + \frac{P}{\mu}$$

In this formula, A is a small constant depending on the " coherence " of the surfaces and μ is a coefficient (the reciprocal of the coefficient of friction which is now commonly used) depending on the nature of the surfaces.

Turning his attention to the observations made of the launching of ships at the port of Rochefort in 1779, Coulomb calculated the

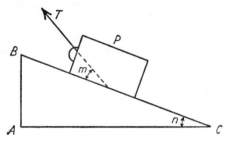

Fig. 106

force necessary to hold a body on an inclined plane. He obtained the result that

$$T = \frac{A\mu + P\,(\cos n + \mu \sin n)}{\mu \cos m + \sin m}$$

where n is the inclination of the plane and m the angle between the force T and the plane BC. From this he easily deduced that T is a minimum for $\mu = \dfrac{\cos m}{\sin m}$.

The mechanics of friction was born of some experiments in physics in the XVIIth Century and then, for an essentially practical purpose, was systematised by Coulomb. But, at the time, it remained linked to the common practice of engineering, while rational mechanics developed, without regard to friction, in the mathematical field.

LAZARE CARNOT'S MECHANICS

1. CARNOT AND THE EXPERIMENTAL CHARACTER OF MECHANICS.

In 1783 Lazare Carnot (1753-1823) published an *Essai sur les machines en général.* He later extended this under the title of *Principes généraux de l'équilibre et du mouvement* (1803). In this interval Lagrange published the first edition of his *Mécanique analytique* (1788). But Carnot's ideas varied so little from the *Essai* to the *Principes* that it can be maintained that Lagrange had no influence on Carnot. Further, it is natural to think of Carnot as a predecessor of Lagrange, in spite of details of simple chronology.[1]

In the field of principles, we are indebted to Carnot because he was the first to assert the experimental character of mechanics— universally accepted now. This is quite in contrast with the ideas professed by Euler, and more often, by d'Alembert. The declarations which follow are taken from the *Principes* and are to be contrasted, in particular, with the introduction to d'Alembert's treatise.

" The Ancients established the axiom that all our ideas come from our senses ; and this great truth is, today, no longer a subject of controversy. . . . [Here Carnot is invoking Locke's *Essay on Human Understanding.*]

" However, all the sciences do not draw on the same experimental foundation. Pure mathematics requires less than all the others ; next come the physico-mathematical sciences ; then the physical sciences. . . .

" Certainly it would be satisfactory to be able to indicate exactly

---

[1] CARNOT himself wrote, in the preface to the *Principes*, " Since the first edition of this work in 1783, under the title of *Essai sur les machines*, there have appeared, in all branches of mechanics, works of such beauty and of such scope that there hardly remains room for some remembrance of mine. However, as it contained some ideas that were new at the time it appeared, and as it is always valuable to contemplate the fundamental truths of science from the various points of view that can be chosen, a new edition has been asked of me. . . . "

the point at which each science ceased to be experimental and became entirely rational [read, in order to develop rationally, starting from principles obtained from experiment] ; that is, to be able to reduce to the smallest number the truths that it is necessary to infer from experiment and which, once established, suffice to embrace all the ramifications of the science, being combined by reason alone. But this seems to be very difficult. In the desire to penetrate more deeply by reason alone, it is tempting to give obscure definitions, vague and inaccurate demonstrations. It is less inconvenient to take more information from experiment than would strictly be necessary. The development may seem less elegant. But it will be more complete and more secure. . . .

" It is therefore from observation that men derived the first concepts of mechanics. However, the fundamental laws of equilibrium and motion which serve as its foundation offer themselves so naturally to reason on the one hand, and on the other, show themselves so clearly in the most common facts, that it is difficult to say whether it is from the one rather than from the other that we derive our perfect conviction of these laws; and whether this conviction would exist without the concurrence of these laws with the first. These facts seem too familiar for us to be able to know at what point, without them, reason alone would be able to establish definitions. And, on the other hand, if reason is unable to connect these facts by analogy, they appear too isolated for us to be able to weld them into principles. " [1]

2. THE CONCEPTS AND POSTULATES OF CARNOT'S MECHANICS.

Carnot had certainly studied Euler and d'Alembert, and thus knew of the theory of *forces* and also of that of *motions* (in the purely kinematic sense). He reproaches the first for " being founded on a metaphysical and obscure notion of forces. " If, on the contrary, the word *force* is understood to be the momentum impressed on a system, the first theory reduces to the second and requires an appeal to experiment.

At least in principle, Carnot adopts the second attitude and seeks to reduce mechanics to the study of the communication of motion. He applies the laws of mechanics to the reasoned observation of problems of impact. He then reduces the action of a continuous force to that of a series of infinitely small impacts.

" Weight and all forces of the same kind act in imperceptible degrees and produce no sudden changes. However, it seems rather

[1] *Principes généraux de l'équilibre et du mouvement,* p. 2.

natural to consider them as dealing infinitely small blows, at infinitely short intervals, to the bodies which they actuate. "

Thus the fundamental law of Carnot's mechanics is written, apart from notation, in the form

$$\vec{F}dt = d(\vec{mv}).$$

But Carnot accompanied this fundamental law with the following commentary.

" At first I shall repeat that the question here is not that of the original causes which create motion in bodies, but only that of the motion already produced and inherent in each of them. The quantity of motion already produced in a body is called its *force* or its *power*. Thus the forces which are considered in mechanics are not metaphysical or abstract entities. Each of them resides in a determinate mass. *The force is the product of this mass and the velocity which the body takes if it is not obstructed by the motions of other bodies which are incompatible with its own.* Such incompatibility makes some bodies lose a part of their quantity of motion ; it makes others add to it, and creates it in those which had none. Each body assumes a kind of combined velocity, in between the one which it must have already had and those which are newly impressed on all its parts. Now it is this compound velocity that it is necessary to determine, at each instant and for each point of the system, when the shapes of the different parts which compose it, their masses and the velocities which they are supposed to have received previously—whether by earlier impacts or by external agencies of any kind—are known. Thus, in a word, we do not seek the laws of motion in general, but rather the laws of the communication of motion between the different material parts of a single system. "[1]

In fact, Carnot did not rigorously dispense with the concept of force. It may even be said that he multiplied the names for it, as we shall see. Moreover, this conforms with his general attitude—his mechanics did not depend on a closed set of axioms.

Carnot variously called the product of a body's mass and the *accelerating force* [read " acceleration "] its *motive force, force of pressure* or *dead force*. Thus gravity or heaviness is an accelerating force and weight, a motive force.

By *moving force* Carnot understood " the motive force applied to a machine in order to overcome the resistances, or to produce any motion at all. " If the *living force* is expressed by the product mv^2, the *latent living force* is expressed by the product PH of a weight and

---

[1] *Principes généraux de l'équilibre et du mouvement*, p. 47.

a height.   The elementary work of a force is called, by Carnot, the *moment of activity achieved by a motive force.*

As for the *moment of absolute activity of a moving body*, this can be expressed in modern language by the product

$$\overrightarrow{mv}\,\overrightarrow{(v + dv)}$$

where $\overrightarrow{v + dv}$ is the velocity of the body at the time $t + dt$ (if the motion is continuous).   In impact, the same moment of activity would be written

$$\overrightarrow{mv}\,\overrightarrow{(v + \Delta v)}$$

where Δv is a finite increment.

Carnot next introduces the *force of inertia* by means of the following definition— " The resistance offered by a body to a change of state " or the " reactions opposed to a system of bodies which make it pass from rest to motion. "   For example, in an impact (the external actions being supposed negligable) the force of inertia of a body of mass m whose velocity changes from $\vec{v}_0$ to $\vec{v}_1$ would be, in Carnot's sense, $m\,(\vec{v}_0 - \vec{v}_1)$.   Here the force of inertia coincides with the *quantity of motion lost*.   But, in general, the quantity of motion lost is the "resultant of the quantity of motion produced by the motive force and the quantity of motion produced by the force of inertia. "   Finally, Carnot understands the *force exerted* on a body of the system to be the resultant of the motive force and the force of inertia.

In passing, we note a curious discussion on this subject.   In his *Sixty-Sixth Letter to a German Princess*, Euler had criticised the expression " force of inertia " as uniting the concept of force (capable of changing the state of a body) and the word inertia (expressing the property of a body that tends to preserve it in its state).

Carnot objected that " the inertia is merely a property which may not be introduced in the calculations, while the *force of inertia* is a real measurable property ; it is the quantity of motion, which this body imparts to any other body, that displaces it from its state. " [1]

Carnot assumed the following postulates as a foundation for his mechanics.

1) The principle of inertia.

2) A system in equilibrium remains in equilibrium under the application of forces which are in equilibrium among themselves.

3) In a system of forces in equilibrium, each force is equal and opposed to the geometric sum of all the others.

4) " The quantities of motion of motive forces which, in a system of bodies, destroy each other at all times, can always be decomposed

---

[1] *Principes généraux de l'équilibre et du mouvement*, p. 73.

into other forces which are, taken in pairs, equal and directly opposed along the direction of the straight line which connects the two bodies to which they belong. And, in each of these bodies, each force can be regarded as nullified by the action of the other. "

5) The action of one body on another by impact, traction or pressure, only depends on the relative velocity of the bodies.

6) " The quantities of motion or the dead forces which the bodies impress on each other through threads or rods are directed along these threads or rods ; and those which they impress on each other by impact or pressure are directed along the perpendicular erected at their common surface at the point of contact. "

7) Hypotheses expressing the laws of inelastic, elastic and partially-elastic impact.

Given these definitions, Carnot introduced the concept of *geometrical motion* into mechanics in the following way.

" Every motion which is imparted to a system of bodies and which does not alter the intensity of the action which they exert or could exert on each other when any other motions whatever are imparted to them, will be called a *geometrical motion*. Then the velocity which each body assumes will be called its *geometrical velocity*. "[1]

Carnot has the following comment to make about this concept.

" This denomination of geometrical motion is based on the fact that the motions concerned have no effect on the action which can be exerted between the bodies of the system, and that they are independent of the rules of dynamics. . . . They only depend on the conditions of constraint between the parts of the system and, consequently, can be determined by geometry alone.

" The theory of geometrical motions is, in a sense, a science intermediate between geometry and mechanics. It is the theory of the motions that a system of bodies can assume without the bodies hindering each other, or exerting any action or reaction on each other. "[2]

In modern language, Carnot's *geometrical motions* are *virtual displacements* (finite or infinitely small) compatible with the constraints between the bodies of the system.

3. CARNOT'S THEOREM.

In the second part of his *Principes fondamentaux* Carnot studied the motion of systems, taking as his basis the problems of impact between " hard bodies "—that is, bodies devoid of elasticity.

[1] *Principes généraux de l'équilibre et du mouvement*, p. 108.
[2] *Ibid.*, p. 106.

Carnot first shows that " if a system of hard bodies suffers an impact or any instantaneous action, either directly or by means of some mechanism without elasticity, the motion taken by the system is necessarily a geometrical one. "

Indeed, if the bodies contiguous with the system by which the action is propagated are considered in pairs, after the impact they have no relative velocity in the line of their reciprocal action. Their real motions after the impact cannot therefore produce any action between them. It follows that the motion of the system after the impact is necessarily a geometrical one. Moreover, it is easy to see that every geometrical motion which is imparted to any system is received by the system without alteration.

Turning to the consideration of a system of hard bodies which sustains an impact, Carnot decomposes (after the manner of d'Alembert) the motion of the system before the impact into two others. The first of these is that which remains after the impact and the second is, consequently, necessarily destroyed by the impact.

If only the first motion is imparted to the system, it will necessarily be received without alteration.

Under the influence of the second motion, also considered in isolation, the system remains in equilibrium.

Carnot writes, " This is what constitutes d'Alembert's famous principle. But it must be recalled that it is only applicable to perfectly hard bodies and to mechanisms without elasticity—this, I think, has not been observed explicitly before. If the bodies were elastic, the motion before the impact would decompose into two in the same way as for hard bodies. One of these motions would be the motion that remains after the impact and the other would be destroyed. But the independence of these motions would not subsist ; for if the first alone were suppressed, there would not be equilibrium. This independence of the two motions is based on the fact that the motion after the impact is geometrical ; that is, it does not tend to increase or decrease the intensity of the impact, and it is only such because the bodies, being hard, etc. . . . "

Let U denote the velocity lost by a particle M during the impact and let V be its velocity after the impact.

By induction, starting from Torricelli's principle, Carnot states the law

(1) $$S M U V \cos (\widehat{V, U}) = 0:$$

Here indeed, Carnot makes appeal to continuous motions by starting from the axiom that " when the centre of gravity is lowest, the system is in equilibrium. " If p is the accelerating force, Carnot writes the condition for the equilibrium of a system, in continuous motion under the influence of the forces p, in the form

$$\mathrm{S}pMV \cos (\widehat{p,\ V}) = 0.$$

From this he deduces the law (1) by applying this principle to percussions.

Carnot verifies the law (1) for the particular impact of two hard bodies, using an analysis that is, this time, direct. He then extends the law to the impact of any number of hard bodies.

From these results, Carnot easily deduced the following theorem, with which his name is still associated.

" *In the impact of hard bodies, the sum of the living forces before the impact is always equal to the sum of the living forces after the impact together with the sum of the living forces that each of these bodies would have if it moved freely with only the velocity which it lost in the impact.* " [1]

Indeed, it was sufficient for him to write

$$\mathrm{S}MW^2 = \mathrm{S}MV^2 + \mathrm{S}MU^2 + 2\,\mathrm{S}MVU \cos (\widehat{V,\ U})$$

where W is the velocity before the impact and law (1) is applied.

Using d'Alembert's procedures throughout, Carnot treated problems of elastic impact as corollaries of problems of impact between " hard " bodies. The elasticity doubles the momentum lost without changing its direction. Thus, to Carnot, the conservation of living forces in the impact of perfectly elastic bodies is justified by his theorem on the impact of hard bodies.

From the general equation (1) Carnot also deduced the remarkable result that the sum of the living forces due to the velocities lost is a minimum in the impact of a system of hard bodies.

" *Among the motions to which a system of perfectly hard bodies is susceptible, when the bodies act on each other by a direct impact or by any mechanism without elasticity, so that there results a sudden change in the state of the system, the one that actually remains after the action is the geometrical motion which is such that the sum of the products of each of the masses by the square of the velocity that it loses is a minimum ; that is, less than the sum of the products of the masses and the square of the velocity that it would have lost if the system had acquired any other geometrical motion.* "

Carnot himself remarked that this result was directly connected with Maupertuis' application of the principle of least action to the impact of bodies.

---

[1] *Principes généraux de l'équilibre et du mouvement*, p. 145.

In this connection, Carnot emerges as an opponent of the doctrine of final causes. Indeed, he declares that his demonstration of this minimum law " is more general [than that of Maupertuis] because it includes bodies which have various degrees of elasticity. But it also demonstrates how insecure are those which are based on final causes, since it shows that the principle is not general, but restricted to systems of bodies which have the *same degree* of elasticity. "

Without carrying this analysis of Carnot's mechanics further, we shall indicate how he passed from the study of these problems of impact to problems in which continuous forces intervene.

" *When a system of hard bodies, free or acted upon by any mechanism without elasticity, and actuated by any moving forces, changes its motion by imperceptible degrees then if, it any instant of the motion, each one of the particles is called* m *; its velocity* V *; its motive force* P [1] *; the velocity that it would take if , ? actual motion were suddenly suppressed and replaced by another geometrical one,* u *; the element of time,* dt *; then there will obtain*

$$\mathrm{S}mud\,[V\cos{(\widehat{u,\,V})}] - \mathrm{S}mu Pdt\,\cos{(\widehat{u,\,P})} = 0.\text{ "}$$

This theorem is deduced from the general formula (1) by observing that

$$Pdt\,\cos{(\widehat{u,\,P})} - d\,[V\cos{(\widehat{u,\,V})}]$$

is the projection, on the direction of u, of the velocity lost by the mass m, due to the action of the other elements of the system.

Carnot also develops some very interesting considerations on the work of the internal forces in animal systems.

" An animal, like the inanimate bodies, is subject to the law of inertia. That is, the general system of parts which compose it cannot produce by itself any progressive motion in any direction. . . . In the whole system of the animal, the principle of the equality of the action and the reaction is applicable, as in inert matter. So that it is only by the friction of its feet on the ground that it can carry itself forward, thereby impressing on the earth on which it walks a quantity of motion equal and opposite to that which it assumes, but which is imperceptible to us.

" It therefore seems, as far as its physique is concerned, that the animal may be considered as an assembly of particles separated by springs which are more or less compressed and which, by this fact,

---

[1] Here it is necessary to read " accelerating force. "

store a certain quantity of living forces ; and that these springs, by extending, may be considered to convert this latent living force into real living force. . . .

" When a similar agency imparts living force to its own mass, although the quantity of motion which results in any direction may be zero, the living force is not zero. And if this agency is applied to a machine, its acquired living force will be, by means of this machine, transmitted to the resisting forces without loss—always with the reservation that there should be no impacts ; for what will be consumed will be wholly absorbed and will be precisely what we call the effect produced. " [1]

The general conclusion of Carnot's mechanics is the following one.

" For any system of bodies, animated by any motive forces, in which several external agents such as men or animals—either by themselves or by machines—are used to move the system in different ways, whatever may be the change produced in the system, the moment of activity consumed by the external powers in any time will always be equal to half the amount by which the sum of the living forces in the system of bodies to which they are applied will be increased during this time, less half the amount by which this same sum of living forces would be increased if each of the bodies had moved freely on the curve which it described—supposing that it had experienced the same motive force, at each point of this curve, as that which it actually experienced ; and provided always that the motion changes by imperceptible degrees, so that, if machines with springs are used, these springs are left in the same state of tension as at the beginning. "

Certainly Carnot's language did not approach the clarity of the great authors of the Century. But the foundation of his work is of an undisputed originality, at once physical and philosophical. In fact, Lazare Carnot was to inspire Laplace, Barré de Saint-Venant and probably Coriolis as well.

[1] *Principes généraux de l'équilibre et du mouvement*, p. 246.

CHAPTER ELEVEN

THE " MÉCANIQUE ANALYTIQUE " OF LAGRANGE

1. THE CONTENT AND PURPOSE OF LAGRANGE'S " MÉCANIQUE ANALYTIQUE. "

We now come to a piece of work which united and crowned all the efforts which were made in the XVIIIth Century to develop a rationally organised mechanics.

Coming from a Touraine family, Louis de Lagrange (1736-1813) started his career at Turin, where he had been born. After having come under the influence of Euler at the Academy of Berlin, he finally went to Paris in 1787 where, in particular, he inaugurated the teaching of analysis at the *École polytechnique*. Thus, by his descent and for an important part of his scientific career, Lagrange belonged to France.

The first edition of the *Mécanique analytique* appeared in 1788.[1] In it Lagrange accomplished the project, which had been conceived and partially executed by Euler, of a single treatise of rational science *(analytice exposita)* covering all branches of mechanics, statics and hydrostatics, dynamics and hydrodynamics.

Lagrange's reading covered everything. Apart from the works of his contemporaries, he had studied, with a remarkable objectivity, those of all the ancient and modern writers that were known in his time. This is witnessed by the historical references with which he enriched his treatise.

Lagrange eliminated the contradictions and the inarticulateness which abounded in the work of his predecessors. He adopted the concepts and the postulates of the great creators of the previous century (Galileo, Huyghens, Newton). He surpassed Euler and d'Alembert. And he became preoccupied with the organisation of mechanics, the

---

[1] The last edition to be published in LAGRANGE's lifetime appeared in 1811. In the present book we have made use of the edition of 1853-1855, which was amended by Joseph BERTRAND and used certain manuscripts which had not been published during LAGRANGE's life.

foundation of its principles, the perfection of its mathematical language and the isolation of a general analytical method for solving its problems. His clarity of mind, his mathematical insight, served him so well that he arrived at an almost perfect codification of mechanics in the classical field. In a detailed way, Lagrange made the following statement of his aims in an *Avertissement*.

" To reduce the theory of mechanics, and the art of solving the associated problems, to general formulae, whose simple development provides all the equations necessary for the solution of each problem.

" To unite, and present from one point of view, the different principles which have, so far, been found to assist in the solution of problems in mechanics ; by showing their mutual dependence and making a judgement of their validity and scope possible. "

As for the purely mathematical point of view which was Lagrange's principal interest, he made the following declaration.

" No diagrams will be found in this work. The methods that I explain in it require neither constructions nor geometrical or mechanical arguments, but only the algebraic operations inherent to a regular and uniform process. Those who love Analysis will, with joy, see mechanics become a new branch of it and will be grateful to me for thus having extended its field. "

2. LAGRANGE'S STATICS.

In the historical part of his work Lagrange makes special mention of Archimedes, Stevin, Galileo and Huyghens. In his view, the equilibrium of a straight and horizontal lever whose ends are loaded with equal weights and whose point of support is at the centre is " a truth that is evident on its own. " On the other hand, the principle of the *superposition of equilibria*, as fruitful as the principle of the superposition of figures in geometry, is essential for a treatment of the angular lever. This leads to the *principle of moments*, in which connection Lagrange cites Guido Ubaldo.

Lagrange refers to Stevin and to Galileo's mechanics in connection with the inclined plane. In the matter of the decomposition of a force into its components, he places Roberval before Stevin.

To Lagrange, Descartes' principle and that of Torricelli were put forward without proof by their authors.

Lagrange mentions Aristotle, Archimedes, Nicomedes and, among the moderns, Descartes, Wallis and Roberval, as having used the composition of motions. It was Galileo who had made first use of this concept in dynamics, in connection with the motion of projectiles.

But, with good reason, Lagrange attributes the composition of forces, in the proper sense of the term, to Newton, Varignon and Lamy. An immediate connection, which Varignon saw and demonstrated by the theory of moments, exists between the principle of the lever and that of the composition of forces.

Lagrange gives the following opinion on the justification of the rule of the parallelogram which had been given by Daniel Bernoulli. " By separating, in this way, the principle of the composition of forces from the principle of the composition of motion, the principal advantages of clarity and simplicity were lost, and the principle was reduced to being merely the result of geometrical constructions and analysis. "

Lagrange then comes to *the principle of virtual work*, which he states in the following way.

" Powers are in equilibrium when they are inversely proportional to their virtual velocities taken in their own directions."

Lagrange mentions Guido Ubaldo as having been concerned in the formation of this principle. He refers to the concept of *momento* as used in Galileo's statics, recalls the part played by Descartes and Torricelli and honours Jean Bernoulli for having been the first to formulate the principle in all its generality.

The justification of the principle of virtual work occupies a great deal of Lagrange's attention.[1]

" As for the nature of the principle of virtual velocities, it must be agreed that it is not sufficiently clear in itself to be formed into a first principle. But it can be regarded as the general expression of the laws of equilibrium, deduced from two principles [of the lever and of the composition of forces]. Further, in the demonstrations of this principle which have been given, it has always been made to depend on these by means which are more or less direct. But there is another general principle in statics which is independent of the principle of the lever and the principle of the composition of forces which, although it is customarily related to the others in mechanics, appears to be the natural foundation of the principle of virtual velocities —it can be called *the principle of pulleys*.

" If several pulleys are mounted together on a single frame this assembly is called a *polispaste* or pulley-block. The combination of two pulley-blocks—one fixed and the other moveable—which is wound with a single string, one end of which is permanently attached and the other, acted upon by a power, forms a machine in which the power is to the weight carried by the moveable pulley-block as unity is to the number of strands which converge on this pulley-block ; this,

---

[1] *Mécanique analytique*, Vol. I, p. 21.

if the strands are all supposed to be parallel and the friction and the stiffness of the strings is neglected.

 " By increasing the numbers of fixed and moveable pulley-blocks, and winding them all with the same string by means of various fixed and reversing pulleys, the same power, when it is applied to the moveable end, will be able to support as many weights as there are moveable pulley-blocks. Then, each weight will be to the power as the number of strands of the pulley-block supporting it is to unity.

 " For greater simplicity, make the last strand pass over a fixed pulley and let it support a weight instead of the power. We shall assume this weight to be unity. Also imagine that the different moveable pulley-blocks, instead of supporting weights, are attached to bodies—regarded as points—and arranged among each other so that they form any given system. In this way, by means of the string which is wound round all the pulley-blocks, the same weight will produce various powers, which act on the different parts of the system in the direction of the strings which converge on the pulley-blocks attached to these points. The powers will be to the weight as the number of strands is to unity. So that the powers themselves will be represented by the number of strands which come together and, by their tension, produce them.

 " Now it is clear that in order that the system drawn by these different powers may remain in equilibrium, *it is necessary that the weight should be unable to descent by any infinitely small displacement of the points of the system.* For since the weight always tends to descend, if there is any infinitely small displacement of the system which allows it to descend, it will necessarily do so and will produce this displacement of the system.

 " Denote the infinitely small distances which this displacement would make the different points of the system travel by α, β, γ, ... in the direction of the power which pulls them. Also denote the number of strands of the pulley-blocks applied at these points, to produce these powers, by P, Q, R, ... It can be seen that the distances α, β, γ, ... will also be those by which the moveable pulley-blocks approach the associated fixed pulley-blocks. Further, it can be seen that these movements will decrease the length of the string which is wound round all the pulley-blocks by the quantities $P\alpha$, $Q\beta$, $R\gamma$, ... So that, because of the fixed length of the string, the weight will descend throughout the distance

$$P\alpha + Q\beta + R\gamma + \ldots$$

" Therefore, in order that the powers represented by the numbers $P, Q, R, \ldots$ may be in equilibrium, it will be necessary that the equation

$$Pa + Q\beta + R\gamma + \ldots = 0$$

should obtain. This is the analytic expression of the general principle of virtual velocities. "

We remark here, with Jouguet,[1] that Lagrange's demonstration is based on physical facts—on certain simple properties of pulleys and strings. Lagrange also assumes the truth of the principle in a very particular case, which reduces to the hypothesis of Huyghens and Torricelli.

We owe to Lagrange the elegant method called that of multipliers. The object of this was to express, in a general way, the problems of statics by means of mathematical equations.[2]

Lagrange expressed the constraints of the system by equations of the type

$$L = 0 \qquad M = 0 \qquad N = 0 \ldots$$

where L, M, N are finite functions of the coordinates of the points of the system.

Differentiating these conditions, Lagrange writes

$$dL = 0 \qquad dM = 0 \qquad dN = 0 \ldots$$

(He does not exclude equations of constraint between differentials that are not " exact differences "—these are the constraints that are now called *non-holonomic*.)

Lagrange declares, " These equations should only be used to eliminate a similar number of differentials in the general formula of equilibrium, after which the coefficients of the remaining differentials all become equal to zero. It is not difficult to show, by the theory of the elimination of linear equations, that the same result will obtain if the various equations of condition

$$dL = 0, \quad dM = 0, \quad dN = 0, \ldots$$

are each multiplied by an indeterminate coefficient and simply added to the equation concerned ; if then, the sum of all the terms which are multiplied by the same differential are equated to zero, which will give as many particular equations as there are differentials ; and if, finally, the indeterminate coefficients by which the equations of con-

[1] *L. M.*, Vol. II, p. 179.
[2] *Mécanique analytique*, Vol. I, p. 69 *et seq.*

dition have been multiplied are eliminated from the last set of equation. "

Whence the rule stated by Lagrange for finding the conditions of equilibrium of any system—

" The sum of the *moments* [that is, apart from sign, the *virtual works*] of all the powers which are in equilibrium will be taken, and the differential functions which become zero because of the conditions of the problem will be added to it, after each of these functions has been multiplied by an indeterminate coefficient ; then the whole will be equated to zero. Thus will be obtained a differential equation which will be treated as an ordinary equation of *maximis et minimis*. From this will be deduced as many equations as there are variables. These equations, being then rid of the indeterminate coefficients by elimination, will provide all the conditions necessary for equilibrium.

" The differential equation concerned will therefore be of the form

$$Pdp + Qdq + Rdr + \ldots + \lambda dL + \mu dM + \nu dN + \ldots = 0$$

in which λ, μ, ν are the indeterminate quantities. In the sequel we shall call this the *general equation of the equilibrium*.

" Corresponding to each coordinate of each body of the system, such as x, this equation will give an equation of the form

$$P\frac{\partial p}{\partial x} + Q\frac{\partial q}{\partial x} + R\frac{\partial r}{\partial x} + \ldots + \lambda\frac{\partial L}{\partial x} + \mu\frac{\partial M}{\partial x} + \nu\frac{\partial N}{\partial x} + \ldots = 0.$$

Therefore the number of these equations will be equal to the number of all the coordinates of all the bodies. We shall call these the *particular equations of the equilibrium.* "

It only remains to eliminate the multipliers λ, μ, ν. Taking account of the equations of constraint, the problem of the determination of the coordinates of the different elements of the system is thus solved.

Lagrange did not confine himself to this abstract analysis, but gave it a physical interpretation. The terms λdL, μdM, νdN " must be regarded as representing the moments [of virtual works] of certain forces applied to a system. "

Thus dL is written in the form

$$dL\,(x', y', z', x'', y'', z'' \ldots) = dL' + dL'' + \ldots$$

In this equation (x', y', z'), (x'', y'', z''), etc. . . . represent the coordinates of each particle, and dL', dL'', etc. . . . only depend on (x', y', z'), (x'', y'', z''), etc. . . . respectively. Lagrange then verifies that the term λdL is equivalent to the effect of different forces

$$\lambda \sqrt{\left(\frac{\partial L}{\partial x'}\right)^2 + \left(\frac{\partial L}{\partial y'}\right)^2 + \left(\frac{\partial L}{\partial z'}\right)^2}$$

$$\lambda \sqrt{\left(\frac{\partial L}{\partial x''}\right)^2 + \left(\frac{\partial L}{\partial y''}\right)^2 + \left(\frac{\partial L}{\partial z''}\right)^2}.$$

applied, respectively, at the points (x', y', z'), (x'', y'', z''), etc. . . . and normal to the different surfaces defined by the equation $dL = 0$. In this equation the variation is first performed with respect to (x', y', z'), then with respect to (x'', y'', z''), etc. . . .

Lagrange concludes, " It follows from this that each equation of condition is equivalent to one or more forces applied to the system in given directions. So that the state of equilibrium of the system will be the same whether the consideration of forces is used, or whether the equations of condition themselves are used.

" Conversely, these forces must take the place of the equations of condition resulting from the nature of the given system, so that by making use of these equations it will be possible to regard the bodies as entirely free and without any restraint. And from this is seen the metaphysical reason why the introduction of the terms $\lambda dL + \mu dM + \ldots$ in the general equation of equilibrium ensures that this equation can then be treated as if all the bodies were entirely free. . . .

" Strictly speaking, the forces in equation take the place of the resistances that the bodies would suffer because of their mutual constraint or because of obstacles which, by the nature of the system, could oppose their motion ; or rather, these forces are merely the same forces as the resitances, which are equal and directly opposite to the pressures exerted by the bodies. As is seen, our method provides a means of determining these forces and resistances. . . . "

The considerable progress achieved by Lagrange in the analytical application of the principle of virtual work is very evident.

Lagrange does not become inordinately eloquent on the concept of force itself. He confines himself to saying, " By *force* or *power* is understood, in general, the cause which imparts, or tends to impart, motion to the bodies to which it is supposed to be applied ; further, it is by the quantity of motion imparted, or which may be imparted, that the force must be represented. In the state of equilibrium the force does not have actual effect ; it only provides a tendency to motion. But it can always be measured by the effect that it would produce if it were not arrested. " [1]

---

[1] *Mécanique analytique*, Vol. I, p. 1.

3. Lagrange and the history of dynamics.

In Lagrange's view, dynamics is " the science of accelerating or retarding forces and the varying motions which they must produce. This science we owe entirely to the moderns, and Galileo is the one who laid its first foundations. . . . Huyghens, who seems to have been destined to perfect and complete most of Galileo's discoveries, supplemented the theory of heavy bodies by the theories of the motion of pendulums and centrifugal forces, and thus prepared the way for the great discovery of universal gravitation. Mechanics became a new science in the hands of Newton, and his *Principia*, which appeared in 1687, was the occasion of this revolution. " [1] Thus, neglecting all the vicissitudes of Aristotelian mechanics and the few inspirations of the Schoolmen, Lagrange acknowledged a century of evolution in the subject that he was to codify.

Lagrange ascribes the two principles of the *force of inertia* (that is, inertia) and the *composition of motions* to Galileo. He analyses the method followed by Huyghens in his work on the centrifugal force in the following way.

" For the estimation of forces, it suffices to consider the motion produced in any time, finite or infinite, provided that the force may be regarded as constant during this time. Consequently, whatever the motion of the body and the law of acceleration may be, since, by the properties of the differential calculus the action of every accelerating force may be regarded as constant during an infinitely small time, it will always be possible to find the value of the force which acts on the body at each instant. This is done by comparing the velocity produced in this instant with the duration of the same instant ; or by comparing the distance which the body travels with the square of the duration of the same instant. It is not necessary, even, that the distance should be actually travelled by the body, it is sufficient that it may be supposed to have been travelled by a compound motion, since the effect of the force is the same in one case as in the other. " [2]

In a careful analysis of the use of mathematics, Lagrange remarks that " Newton made constant use of the geometric method as simplified by the consideration of the first and last ratios. " Euler's *Mechanica* (1736) is, to Lagrange, the first great work in which Analysis was applied to the science of motion. As for MacLaurin's *Treatise on Fluxions* (1742), this was the first work which systematically used

[1] *Mécanique analytique*, Vol. I, p. 207.
[2] *Ibid.*, p. 210.

the rectangular components of the force instead of their tangential and normal components.

Lagrange then comes to a principle which allows the determination *of the force on bodies in motion*, having regard to their mass and velocity. " This principle consists in that, in order to impart to a given mass a certain velocity in some direction, whether the mass be at rest or in motion, the necessary force is proportional to the product of the mass and the velocity and its direction is the same as that of the velocity. " [1]

Here Lagrange cites Descartes as having first realised the existence of this principle, but as having deduced from it incorrect rules about the impact of bodies. On the other hand, Wallis made successful use of the principle to discover the laws of the transfer of motion in the impact of hard or elastic bodies. And Lagrange continues, " Just as the product of the mass and the velocity represents the finite force of a body in motion, so the product of the mass and the accelerating force—which we have seen to be represented by the element of velocity divided by the element of time—will represent the elementary or nascent force. And this quantity, if it is considered as the measure of the effect that the body can exert because of the velocity which it has assumed, or which it tends to assume, constitutes what is called *pressure* ; but if it is regarded as a measure of the force or power necessary to impart this same velocity, it is then what is called *motive force.* "

In modern language, the finite force of a body in motion is represented by the product $m\vec{v}$, and the " elementary or nascent force " by $m \dfrac{d\vec{v}}{dt}$.

Lagrange does not openly take sides between Euler's thesis— based on the law $\vec{F}dt = md\vec{v}$ (where $\vec{F}$ is the *static* force)—and d'Alembert's thesis. This matter of principle interested him less than the formal organisation of dynamics, which was the primary object of his own treatise. Because of the work of his predecessors, the mechanics of a particle had no mystery for him. Primarily, he sought to provide statics, and then the dynamics of systems, with the general method that they still lacked.

In turn, Lagrange analyses the four principles of dynamics— the conservation of living forces ; the conservation of the motion of the centre of gravity ; the conservation of moments or the principle of areas ; and the principle of the least quantity of action.

Lagrange says, legitimately, that the first of these principles goes back to Huyghens " in a form a little different from that in which it

is presented now. " Jean Bernoulli, following Leibniz, fashioned it
into the principle of the *conservation of living forces*. Daniel Bernoulli,
after applying it to fluids, extended it (in the Memoires of Berlin for
1748) to a system of bodies attracting each other, or tending towards
fixed centres, according to any law which is a function of distance.

The second principle is due to Newton and was revived by d'Alembert.

The third principle, discovered by Euler,[1] Daniel Bernoulli,[2]
and d'Arcy,[3] is only the generalisation of a theorem of Newton
concerning several particles attracted by the same centre.

D'Arcy went further and sought to make the principle of areas
into a principle of the conservation of action. Lagrange protests,
" As if this vague and arbitrary nomenclature were the essence of the
laws of nature and could, by some secret property, elevate the simple
results of the known laws of mechanics into final causes. "[4]

The criticism which Lagrange directs against Maupertuis' principle
merits quotation.

" Finally I come to the fourth principle, which I call that of *least
action* by analogy with that which Maupertuis gave under the same
name, and which the writings of many illustrious authors have since
made so well-known. This principle, looked at analytically, consists
in that, in the motion of bodies which act upon each other, the sum
of the products of the masses with the velocities and with the distances
travelled is a minimum. The author deduced from it the laws of
the reflection and refraction of light, as well as those of the impact
of bodies.

" But these applications are too particular to be used for establishing the truth of a general principle. Besides, they have a somewhat
vague and arbitrary character, which can only render the conclusions
that might have been deduced from the true correctness of the principle
unsure. Further, it seems to me that it would be wrong to place
this principle, presented in this way, among those which we have
just given. But there is another way in which it may be regarded,
more general, more rigorous, and which itself merits the attention
of the geometers. Euler gave the first hint of this at the end of his
Traité des isopérimètres, printed at Lausanne in 1744. He demonstrated, in the trajectories described under the action of central forces,
that the integral of the velocity multiplied by the element of the curve

[1] *Opuscules*, Vol. I, 1746.
[2] *Mémoires de Berlin*, 1746.
[3] *Mémoires de l'Académie des Sciences*, 1747.
[4] *Mécanique analytique*, Vol. I, p. 228.

is always a maximum or a minimum.   By means of the conservation
of living forces I have extended this property, which Euler discovered
in the motion of isolated bodies and which seemed confined to these
bodies, to the motion of any system of bodies which interact in any
way.   From this has come a new general principle, that the sum of
the products of the masses with the integrals of the velocities, each
of which is multiplied by the element of distance travelled, is invariably
a maximum or a minimum.

" This is the principle which I now give, however improperly,
the name of least action.   I regard it not as a metaphysical principle,
but as a simple and general result of the laws of *mechanics*. " [1]

4. LAGRANGE'S EQUATIONS.

Lagrange was able to put the equations of dynamics into a very
general and valuable form which has now become classical.

For each element, of mass m, of a system, Lagrange defines " the
forces parallel to the axes of coordinates which are used, directly,
to move it, " to be

$$m\,\frac{d^2x}{dt^2} \qquad m\,\frac{d^2y}{dt^2} \qquad m\,\frac{d^2z}{dt^2}.$$

He regards each element of the system as acted upon by similar
forces, and concludes that the sum of the moments [2] of these forces
must always be equal to the sum of the given accelerating forces which
act on each element.   Thus he writes

$$\mathrm{S}m\left(\frac{d^2x}{dt^2}\,\delta x + \frac{d^2y}{dt^2}\,\delta y + \frac{d^2z}{dt^2}\,\delta z\right) + \mathrm{S}m\,(P\delta p + Q\delta q + R\delta r + \ldots) = 0$$

the given forces P, Q, R, ... being supposed to act on each element
along the lines p, q, r, ...

Lagrange transforms the first sum by using the identity

$$d^2x\delta x + d^2y\delta y + d^2z\delta z = d\,(dx\delta x + dy\delta y + dz\delta z) - \frac{1}{2}\delta\,(dx^2 + dy^2 + dz^2).$$

By a change of variables in which each differential dx, dy, dz, ...
is expressed as a linear function of the differentials $d\xi$, $d\psi$, $d\varphi$, ...,

---

[1] *Mécanique analytique*, Vol. I, pp. 229, 230.
[2] In the sense already encountered in LAGRANGE'S statics.

Lagrange establishes that if Φ is the transform of the quantity

$$\frac{1}{2}(dx^2 + dy^2 + dz^2)$$

then the following equation is identically true.

$$d^2x\delta x + d^2y\delta y + d^2z\delta z = \left(-\frac{\partial\Phi}{\partial\xi} + d\,\frac{\partial\Phi}{\partial d\xi}\right)\delta\xi + \left(-\frac{\partial\Phi}{\partial\psi} + d\,\frac{\partial\Phi}{\partial d\psi}\right)\delta\psi + \cdots$$

Lagrange confines himself to forces P, Q, R, ... for which the quantity

$$P\delta p + Q\delta q + R\delta r + \cdots$$

is integrable, which, he declares, " is probably true in nature. " This enables him to suppose that

$$\mathrm{S}m\,(P\delta q + Q\delta q + R\delta r + \ldots) = \delta\,\mathrm{S}m\Pi\,(\xi,\,\psi,\,\varphi\ldots).$$

The general equations of dynamics are then written in the form

$$\Xi\delta\xi + \Psi\delta\psi + \ldots = 0$$

by putting

$$\Xi = d\,\frac{\partial T}{\partial d\xi} - \frac{\partial T}{\partial\xi} + \frac{\partial V}{\partial\xi}$$
$$\cdots$$

with

$$T = \frac{1}{2}\,\mathrm{S}m\left(\frac{dx^2}{dt^2} + \frac{dy^2}{dt^2} + \frac{dz^2}{dt^2}\right) \quad\text{and}\quad V = \mathrm{S}m\Pi.$$

Having arrived at these results, Lagrange examines the particularly interesting circumstance in which the variables ξ, ψ, ... are exactly sufficient to characterise the motion of the system after all the equations of constraint have been eliminated.

" If, in the choice of the new variables ξ, ψ, ..., regard has been paid to the equations of condition provided by the nature of the proposed system, so that the variations are now completely independent of each other and that, consequently, their variations $\delta\xi$, $\delta\psi$, ..., remain absolutely indeterminate, then the particular equations

$$\Xi = 0 \qquad \Psi = 0 \ldots$$

will serve to determine the motion of the system, since these equations are equal in number to the variables ξ, ψ, ... on which the position of the system at each instant depends. " [1]

[1] *Mécanique analytique*, Vol. I, p. 291.

Lagrange connects this analysis with the method of multipliers which he introduced in statics. If the variables $\xi, \psi, \ldots$ are greater in number than the degrees of freedom of the system, they will be related by the equations

$$L = 0 \qquad M = 0 \qquad N = 0 \ldots$$

Then Lagrange's general formula becomes

$$\Xi\delta\xi + \Psi\delta\psi + \ldots + \lambda\delta L + \mu\delta M + \nu\delta N + \ldots = 0$$

whence the equations of motion

$$\begin{cases} \Xi + \lambda \dfrac{\partial L}{\partial \xi} + \mu \dfrac{\partial M}{\partial \xi} + \nu \dfrac{\partial N}{\partial \xi} + \ldots = 0 \\[2mm] \Psi + \lambda \dfrac{\partial L}{\partial \psi} + \mu \dfrac{\partial M}{\partial \psi} + \nu \dfrac{\partial N}{\partial \psi} + \ldots = 0 \\[2mm] \ldots \end{cases}$$

which must be associated with the equations of constraint.

The method of multipliers, which Lagrange himself only applied here to the systems of constraints which are now called *holonomic*, is easily extended to *non-holonomic* constraints—that is, to constraints which cannot be expressed finitely as functions of $\xi, \psi, \ldots$

5. The conservation of living forces as a corollary of Lagrange's equations.

Better than d'Alembert had been able to do, Lagrange established that the conservation of living forces is a consequence of the equations of dynamics, as long as the constraints are without friction and independent of time.

For this purpose, Lagrange considers the true motion of the system between the time t and the time $t + dt$; that is, he substitutes dx, dy, dz, $\ldots$ and dp, dq, dr, $\ldots$ for δx, δy, δz, $\ldots$ and δp, δq, δr, $\ldots$ in the general formula. This enables him to write

$$\mathrm{S}m \left(\frac{dxd^2x + dyd^2y + dzd^2z}{dt^2} + Pdp + Qdq + Rdr + \ldots \right) = 0$$

If the quantity

$$Pdp + Qdq + Rdr + \ldots$$

is integrable, then

$$\mathrm{S} \left(\frac{dx^2 + dy^2 + dz^2}{2dt^2} + \Pi \right) m = H.$$

" This equation includes the principle known by the name of the *conservation of living forces.* Indeed, since $dx^2 + dy^2 + dz^2$ is the square of the distance which the body travels in the time dt, then $\dfrac{dx^2 + dy^2 + dz^2}{dt^2}$ will be the square of the velocity and $m \dfrac{dx^2 + dy^2 + dz^2}{dt^2}$ will be its living force. Therefore

$$S \left(\frac{dx^2 + dy^2 + dz^2}{dt^2} \right) m$$

will be the living force of the whole system, and it is seen, by means of the equation concerned, that this living force is equal to the quantity $2H - 2\ S\Pi m$, which only depends on the accelerating forces which act on the bodies and not on their mutual constraints. So that the living force is always the same as that which the bodies would have acquired if they had moved freely, each along the line that it described, under the influence of the same powers. " [1]

Thus Lagrange discovers the same principle as that formulated by Huyghens to be a simple corollary of his general equations.

6. THE PRINCIPLE OF LEAST ACTION AS A COROLLARY OF LAGRANGE'S EQUATIONS.

Lagrange starts from the equation of living forces

$$S m \left(\frac{u^2}{2} + \Pi \right) = H \qquad \left(u^2 = \frac{dx^2 + dy^2 + dz^2}{dt^2} \right)$$

and differentiates it to obtain

$$S m \left(u\delta u + \delta \Pi \right) = 0$$

or

$$S m \left(P\delta p + Q\delta q + R\delta r + \ldots \right) = -\ S m u \delta u.$$

Substitution in the general formula leads to

$$S m \left(\frac{d^2 x}{dt^2} \delta x + \frac{d^2 y}{dt^2} \delta y + \frac{d^2 z}{dt^2} \delta z - u\delta u \right) = 0$$

or

$$S m \left[\frac{d\left(dx\delta x + dy\delta y + dz\delta z \right)}{dt^2} - u^2 \frac{\delta ds}{ds} - u\delta u \right] = 0$$

[1] *Mécanique analytique*, Vol. I, p. 268.

or again

$$S m \left[\frac{d \left(dx \delta x + dy \delta y + dz \delta z \right)}{dt} - \delta (uds) \right] = 0$$

and finally

$$\delta S m \int uds = S m \left(\frac{dx}{dt} \delta x + \frac{dy}{dt} \delta y + \frac{dz}{dt} \delta z \right).$$

If it is supposed that the variations δx, δy, δz are zero at the ends of the ranges of integration, then

$$\delta S \int muds = 0.$$

Lagrange concludes, " In the general motion of any system of bodies, actuated by mutual forces of attraction, or by attractions towards fixed centres which are proportional to any function of the distance, the curves described by the different bodies, and their velocities, are necessarily such that the sum of the products of each mass by the integral of the product of the velocity and the element of the curve is necessarily a maximum or a minimum ; provided that the first and last points of each curve are regarded as fixed, so that the velocities of the corresponding coordinates at those points are zero. " [1]

Maupertuis' principle is thus found to be valid, in a more general form than that which Euler gave it. Moreover, this principle expresses the *extremal* character of the living force between two known configurations of the system. This Lagrange establishes in the following way.

" Since $ds = udt$, the formula

$$S m \int uds$$

which is either a maximum or a minimum, can be put in the form

$$S m \int u^2 dt \quad \text{or} \quad \int dt S m u^2.$$

Here $S m u^2$ represents the living force of the whole system at any time. Thus the principle reduces to— the sum of the instantaneous living forces of all the bodies, from the moment that they start from given points to that when they arrive at other given points, is a maximum or a minimum. It could be called, with more justice, the *principle of the greatest or least living force*, and this way of regarding it would have the advantage of being general, since the living force of a system is always greatest or least in the equilibrium condition. " [2]

[1] *Mécanique analytique*, Vol. I, p. 276.
[2] *Ibid.*, p. 281.

7. On some problems treated in the "Mécanique analytique."

Mécanique analytique includes a study of a great number of problems which we are not able to treat in this book.

We note, however, that Lagrange initiated a general method of approximation in dynamical problems which was based on the variation of arbitrary constants; that he developed the theory of small motions; that he studied the stabilty of equilibrium, and stated that equilibrium is necessarily stable when the potential of the given forces is a minimum. This demonstration was to be perfected by Lejeune-Dirichlet.

Lagrange also studied in detail the motion of a heavy solid of revolution which was suspended from a point on its axis, and expressed the solution in terms of elliptic integrals.

8. Lagrange's hydrodynamics.

After describing the historical development of hydrodynamics, Lagrange made a very important contribution to the subject.

As a supplement to Euler's variables, Lagrange introduced the variable with which his name is still associated into the kinematics of continuous media. The *actual* coordinates of an element of the medium are considered as functions of the time and of the *initial* coordinates a, b, c, of the same element.

Lagrange established a fundamental theorem on the permanence of the irrotational property in fluid motion.

If the fluid is first supposed to be incompressible and homogeneous, and its density is taken equal to unity, Lagrange [1] also assumed that the accelerating forces X, Y, Z which act on the elements of the fluid are such that $X dx + Y dy + Z dz$ is an exact differential dV. Lagrange writes

$$dp - dV = \left(\frac{\partial u}{\partial t} + u \frac{\partial u}{\partial x} + v \frac{\partial u}{\partial y} + w \frac{\partial u}{\partial z}\right) dx$$
$$+ \left(\frac{\partial v}{\partial t} + u \frac{\partial v}{\partial x} + v \frac{\partial v}{\partial y} + w \frac{\partial v}{\partial z}\right) dy$$
$$+ \left(\frac{\partial w}{\partial t} + u \frac{\partial w}{\partial x} + v \frac{\partial w}{\partial y} + w \frac{\partial w}{\partial z}\right) dz.$$

The right-hand side of this equation, like the left hand side, must be an exact differential. Now the right-hand side can be written as

$$d\left(\frac{u^2 + v^2 + w^2}{2}\right) + \frac{\partial u}{\partial t} dx + \frac{\partial v}{\partial t} dy + \frac{\partial w}{\partial t} dz + \left(\frac{\partial u}{\partial y} - \frac{\partial v}{\partial x}\right) (v dx - u dy)$$
$$+ \left(\frac{\partial v}{\partial z} - \frac{\partial w}{\partial y}\right) (w dy - v dz) + \left(\frac{\partial w}{\partial x} - \frac{\partial u}{\partial z}\right) (u dz - w dx).$$

[1] *Mécanique analytique*, Vol. II, p. 268.

Lagrange remarks that this quantity will be an exact differential whenever $udx + vdy + wdz$ is such, " but as this is only a special supposition, it is necessary to inquire in what cases it can and must be appropriate. "

Lagrange then verifies that when u, v, w are expanded as functions of the time, in the form

$$u = u' + u''t + u'''t^2 + u^{IV}t^3 + \ldots \quad v = \ldots \quad w = \ldots,$$

it is necessary that, whenever $u'dx + v'dy + w'dz$ is an exact differential, that $u''dx + v''dy + w''dz$, $u'''dx + v'''dy + w'''dz$, etc. . . . should also be exact differentials.

He concludes, " From this it follows that if the quantity $udx + vdy + wdz$ is a total exact differential when $t = 0$, it must also be a total exact differential when t has any other value. Therefore, in general, since the origin of t is arbitrary, and since t can either be taken positive or negative, it follows that if the quantity $udx + vdy + wdz$ is a total exact differential at any time, it must be such at all other times.

" Accordingly, if there is a single instant at which it is not a total exact differential, it can never become such throughout the motion. For if it were a total exact differential at any instant, it would also be such at the first. "

This theorem of Lagrange is a fine example of discovery achieved by a procedure which appears to be purely mathematical.

In sympathy with the spirit of the time, which readily assumed that nature conformed to simple laws, Lagrange declared " that it is possible to ask whether there are motions for which $udx + vdy + wdz$ is not a total exact differential. "

To answer this question, he shows that in the motion

$$u = gy \quad v = -gx \quad w = 0$$

the condition that $udx + vdy + wdz$ should be a complete differential is not satisfied, although it is possible to write

$$p = V - \frac{g^2}{2}(x^2 + y^2) + \text{funct. } t.$$

Now " it is clear that these values of u, v, w represent the motion of a fluid which rotates with a constant angular velocity equal to g about the fixed axis of coordinates z. And it is known that such a motion can always take place in a fluid. From this it can be concluded that in the oscillations of the sea due to the attraction of the Sun and the Moon, it cannot be supposed that the quantity $udx + vdy + wdz$ is integrable, since it is not so when the fluid is at rest with respect to the Earth and only has the same rotational motion as the Earth. "

Lagrange extended his theorem to compressible fluids by introducing an " elasticity " that was a function of density alone, so that $\dfrac{d\varepsilon}{\varrho}$ is the differential of some function $E(\varrho)$.

We also mention Lagrange's study of the motion of a fluid in an almost horizontal shallow canal. This motion is governed by an equation similar to the equation of the propagation of sound. The wave velocity turns out to be proportional to the square root of the depth of the fluid if the canal has a uniform breadth.

PART FOUR

SOME CHARACTERISTIC FEATURES
OF THE EVOLUTION
OF CLASSICAL MECHANICS AFTER LAGRANGE

FOREWORD

It would seem that the valuable function of history is that analysing the paths which scientific thought has travelled in the creation, in a limited field like that of classical mechanics, of a rationally organised structure.

The material for this study thus consist of the vicissitudes encountered on these paths, the interaction between currents of thought which were in principle divergent, or even opposed. It is difficult material, sometimes deceptive, but always revealing of the profound difficulties of research and—for this reason—instructive.

After Lagrange, after the efforts of the students of the XVIIIth Century, mechanics had attained this rational structure. It lasted until the impact of the needs of relativistic and quantum physics.

The intervening period was a didactic one, which we are reluctant to deal with for fear of duplicating the books that have, rightly, become classical.

That is why, in the following pages, we shall confine ourselves to some characteristic features of the evolution of classical mechanics after Lagrange. We shall be concerned with discussions of the principles themselves, and with certain isolated facts to which it seems natural to attach a historical significance because of their influence on the later development of mechanics.

CHAPTER ONE

LAPLACE'S MECHANICS (1799)

1. LAPLACE AND THE PRINCIPLES OF DYNAMICS.

We shall discuss here only that part of Laplace's work which is directly concerned with the principles of dynamics.

Laplace referred the motion of bodies to " an infinite space, at rest and penetrable to matter. " [1]

The concept of *force* evoked the following comment.

" The mechanism of that remarkable agency, force, by which a body is moved from one place to another, is and always will be unknown. It is only possible to determine the laws which govern its behaviour. A force acting on a particle will necessarily set it in motion, if there is nothing to prevent this. The direction of the force is that of the straight line which the particle is made to describe. " [2]

Laplace defined *inertia* as the tendency of matter to remain in its state of rest or motion. That the direction of motion was constant appeared obvious to him ; with respect to its uniformity, he pointed out that the law of inertia was the simplest conceivable law, and that it was justified by astronomical and terrestrial observations.

The Laplace endeavoured to prove that " force " is proportional to the velocity. He is concerned here, of course, with the force of a moving body.

Let v be the velocity of the Earth, common to all bodies on its surface, and f be the force which a particle M experiences because of this motion. The ration $\frac{v}{f}$ is an unknown function of f, say $\varphi(f)$. The form of this function has to be found by a method which has recourse to experimental observation.

Suppose that M is also acted upon by another force, f', which

---

[1] *Mécanique céleste*, Part I, Book I, p. 3 (1799).
[2] *Ibid.*, p. 4.

combines with f to form a resultant F according to the parallelogram rule. Under these conditions the particle will acquire a certain velocity U.

Laplace now argued that f' could be considered as " infinitely small " compared with f. " The greatest forces that we are able to impress on bodies on the surface of the earth are much smaller than those that it experiences because of the motion of the earth. "

Accordingly, the relation

$$\vec{F} = \vec{f} + \vec{f'}$$

yields

$$F = f + \frac{\vec{f} \cdot \vec{f'}}{f}$$

if f'^2 is neglected in comparison with f. It follows that

$$\varphi(F) = \varphi(f) + \frac{\vec{f} \cdot \vec{f'}}{f} \varphi'(f) \quad \text{with} \quad \varphi'(f) = \frac{d\varphi}{dt}.$$

The relative velocity of M with respect to the Earth, $\vec{U} - \vec{v}$, is equal to

$$\vec{F}\varphi(F) - \vec{f}\varphi(f)$$

which can easily be shown to lead to the equation

$$\vec{U} - \vec{v} = \vec{f'} \cdot \varphi(f) + \vec{f} \left[\frac{\vec{f} \cdot \vec{f'}}{f} \varphi'(f) \right].$$

From this it follows, in the general case in which the directions of $\vec{f}$ and $\vec{f'}$ are not the same, that this relative velocity must have a component perpendicular to that of the impressed force f' unless $\varphi'(f)$ vanishes. (It is assumed that the scalar product $\vec{f} \cdot \vec{f'}$ is not zero.)

At this point Laplace appeals to experiment.

" Thus, imagine that a sphere at rest on a smooth horizontal plane is hit by the base of a right cylinder, moving along the direction of its axis which is supposed to be horizontal. The apparent relative motion of the sphere should not be parallel to this axis, for all positions of the axis with respect to the horizon. Here is a simple means of finding out by experiment whether $\varphi'(f)$ has an appreciable value on the Earth. But the most accurate experiments do not demonstrate any deviation of the apparent motion of the sphere from the direction of the impressed force. From which it follows that on the Earth $\varphi'(f)$ is very nearly equal to zero. Its value, however inappreciable it might be, would make itself apparent in the length of the oscillations of a pendulum, which would vary according to the position of the plane of its motion with respect to the direction of the Earth's motion. The most accurate observations do not reveal any such difference. We must then conclude that $\varphi'(f)$ is inappreciable, and can be supposed to be zero, on the Earth.

" If the equation $\varphi'(f) = 0$ was obtained whatever the force f might be, $\varphi(f)$ would be constant and the velocity would be proportional to the force. It would also be proportional to the force if the function $\varphi(f)$ were only composed of a single term, since otherwise $\varphi'(f)$ would never be zero. Therefore, if the velocity were not proportional to the force, it would be necessary to suppose that in nature the function of the velocity which represents the force consists of several terms, which is unlikely. It would also be necessary to suppose that the velocity of the Earth is exactly that which the equation $\varphi'(f) = 0$ requires, which is against all probability. Moreover, the velocity of the Earth varies in the different seasons of the year—it is about a thirtieth greater in winter than in summer. This variation is still more considerable if, as everything seems to show, the solar system is in motion through space—for whether this progressive motion combines with that of the Earth or whether it is opposed to it, during the course of the year there must result large variations in the absolute motion of the Earth. This would modify the equation concerned and thus the relation of the impressed force to the absolute velocity, if this equation and this velocity were not independent of the motion of the Earth. However, observation has not revealed any appreciable variation. " [1]

Laplace concludes, " Here then are two laws of motion ; namely, the law of inertia and that according to which the force is proportional to the velocity, and these are provided by observation. They are the simplest and most natural that could be imagined and undoubtedly they derive from the nature of matter itself. But since this nature is unknown they are merely, for us, observed facts ; moreover, the only ones which mechanics borrows from observation. " [2]

Laplace next gave his attention to " forces which appear to act in a continuous manner, like gravity. " Here, like Carnot, Laplace considers that gravity acts in successive impulses at infinitely small intervals of time. " We suppose that the interval of time which separates two actions of some force is equal to the element of time dt. It is clear [that the instantaneous action of the force] must be supposed to be proportional to the intensity and to the element of time in which

[1] *Mécanique céleste*, Part I, Book I, p. 17.
[2] *Ibid.*, p. 18.

it is supposed to act. Thus, representing the intensity by P, it must be supposed, at the beginning of each instant dt, that the particle is actuated by a force Pdt and that it is moved uniformly during that instant." [1]

2. The general mechanics compatible with an arbitrary relation between the " force " and the velocity.

In the last § we have seen Laplace emphasise the communication of motion without seeking to elucidate the original causes of the motion. He thus belongs to the tradition of d'Alembert and Carnot. In particular, like Carnot, he asserted the experimental character of the laws of mechanics. His analysis—this time entirely original—of the notion of the force of a body in motion led him, by an innate propensity to purely mathematical generalisation, to an extension of dynamics which encompassed the ideas that the physicists were to use, a century later, in special relativity.

This extension was the subject of Chapter VI of the first part of Book I of the *Mécanique céleste*. The chapter is called " *On the laws of motion of a system of bodies associated with all possible mathematical relationships between the force and the velocity.* "

Laplace remarks (as we have mentioned in the preceding paragraph) that there is an infinite number of self-consistent ways of expressing the " force " in terms of the velocity. This infinity corresponds to all possible forms of the relation between the force and the velocity.

$$F = \varphi(v)$$

The general equation of the dynamics of systems is

$$\sum m \left\{ \delta x \left\{ d\left(\frac{dx}{dt}\right) - Pdt \right\} + \delta y \left\{ d\left(\frac{dy}{dt}\right) - Qdt \right\} + \delta z \left\{ d\left(\frac{dz}{dt}\right) - Rdt \right\} \right\} = 0$$

and is valid when the " force " is proportional to the velocity. In order to obtain the generalisation which is sought, it is sufficient to assume that the body of mass m is actuated by a " force " whose components parallel to the axes are

$$\varphi(v)\frac{dx}{ds} \qquad \varphi(v)\frac{dy}{ds} \qquad \varphi(v)\frac{dz}{ds}.$$

In the instant following this force becomes

$$\varphi(v)\frac{dx}{ds} + d\left(\varphi(v)\cdot\frac{dx}{ds}\right) \quad \text{etc.}$$

[1] *Mécanique céleste*, Part I, Book I, p. 19.

or
$$\varphi(v)\,\frac{dx}{ds} + d\left(\frac{\varphi(v)}{v}\cdot\frac{dx}{dt}\right) \quad \text{etc.}$$

Then the general equation of the dynamics of systems takes the form

$$\sum m \left\{ \delta x \left\{ d\left(\frac{\varphi(v)}{v}\,\frac{dx}{dt}\right) - Pdt \right\} + \delta y \left\{ d\left(\frac{\varphi(v)}{v}\,\frac{dy}{dt}\right) - Qdt \right\} \right.$$
$$\left. + \delta z \left\{ d\left(\frac{\varphi(v)}{v}\,\frac{dz}{dt}\right) - Rdt \right\} \right\} = 0.$$

Here $\dfrac{dx}{dt}$, $\dfrac{dy}{dt}$ and $\dfrac{dz}{dt}$ appear as products with the function $\dfrac{\varphi(v)}{v}$, which reduces to unity " for that natural law according to which the force is proportional to the velocity. "

Laplace remarked that this difference makes the solution of the problems of mechanics very difficult.

But the principle of the conservation of living forces, the principle of areas and the principles of the centre of gravity and of least action, can be extended to this case.

The extension of the principle of living forces is obtained by substituting dx, dy and dz for δx, δy and δz in the general equation. Thus, if U is the function of the forces,

$$\sum \int mv \cdot dv \cdot \varphi'(v) = U + h \quad \text{with} \quad \varphi'(v) = \frac{d\varphi(v)}{dv}.$$

" The principle of the conservation of living forces therefore obtains for all the possible mathematical relationships between the force and the velocity, provided that the *living force* of a body is understood as the product of its mass and twice the integral of the product of its velocity and the differential of the function of velocity which represents the force. "

In the same way, Laplace extended the theorem of quantities of motion to an isolated system. He generalised the principle of areas to the form

$$\sum m\left(\frac{xdy - ydx}{dt}\right) \cdot \frac{\varphi(v)}{v} = \text{constant.}$$

Finally, he wrote the generalised principle of least action as

$$\delta\left\{ \sum \int m\varphi(v)\,ds \right\} = 0.$$

Thus, as early as 1799, Laplace was able to formulate the general mechanics of which the dynamics of special relativity, in a given reference system, is only a particular case. However, there is a slight difference of meaning between Laplace's purely mathematical conception and that of the modern physicists. To Laplace, the mass m of a particle remained constant and it was the momentum which was no longer proportional to the velocity. In the physical theory of relativity, on the other hand, the mass M becomes a function of the velocity while the momentum remains in the form Mv.

In order to pass from one of these systems to the other, it is sufficient to put

$$M = m \frac{\varphi(v)}{v}.$$

3. LAPLACE AND THE SIGNIFICANCE OF THE LAW OF UNIVERSAL GRAVITATION.

In his *Exposition du Système du Monde* Laplace recalls that when Newton formulated the principle of universal gravitation, Descartes had precisely managed to " substitute the intelligible ideas of motion, impulse and centrifugal force for the occult quantities of the Aristotelians. "[1] His system of vortices met with the approval of the philosophers, " who rejected the obscure and meaningless doctrines of the Schoolmen, and who believed that they saw those occult features, that French philosophy had so legitimately banned, reborn in the attractions. "[2]

Laplace opined that Newton would have deserved this reproach if he had been content to attribute the elliptic motion of the planets and comets, the inequalities of the motion of the Moon, of the terrestrial degrees of latitude and gravity, to the universal attraction without showing the connection between this principle and these phenomena. But the geometers, rectifying and generalising Newton's demonstrations, had been able to verify the perfect agreement between the observations and the results of the analysis.

Laplace regarded " *this analytical connection of particular facts with a general fact* " as a properly constituted *theory*. And he flattered himself with having obtained one in the deduction on the effects of capillarity from a short range interaction between molecules ; a *true theory*, one which expresses the rigorous agreement of the calculation and the phenomena.

Here we see portrayed the *dogma of universal attraction*. However —and this is essential—it is deprived of the *a priori* character of the

[1] P. 377.
[2] P. 378.

assertion of a quality in the sense that the Schoolmen used. Although it must certainly have passed through his mind, Laplace did not use the term dogma in this connection. For his attitude, supported at many points by experiment, was not necessarily of a dogmatic character.

Laplace assumed, moreover, that the following question could be asked—" Is the principle of universal gravity a primordial law of nature, or is it merely the general effect of an unknown cause ? " [1]

Laplace also asked whether the propagation of the attraction was instantaneous. An attempt to explain the secular acceleration of the Moon's motion had led him to assume that, if the velocity of propagation was finite, it must be seven million times greater than that of light. . . . Thus he declared for an instantaneous propagation.

Further, he wrote, " Doubtless the simplicity of the laws of nature must not necessarily be judged by the ease with which we appreciate them. But since those that seem most simple to us agree perfectly with all the phenomena, we are well justified in regarding them as being rigorous. "

We see that Laplace's attitude was a moderate one, and that, to him, the certainty of a natural law depended on a kind of passage to the limit in the mathematical sense of the term.

[1] P. 384.

CHAPTER TWO

FOURIER AND THE PRINCIPLE OF VIRTUAL WORKS (1798)

We owe to Fourier a demonstration of the principle of virtual works which is based on the equilibrium of the lever, and which has been used as the basis of the presentations of the principle that are now classical.[1]

Fourier borrowed the notion of *virtual velocity* from Jean Bernoulli.[2] On the other hand, he called the *moment* of a force the product (change of sign) of this force with the virtual velocity of the point to which it is applied.

If a point is in equilibrium under the action of n forces, Fourier first verifies that the total moment of these forces is zero for an arbitrary displacement of the point.

He then seeks the total moment of two equal and opposed forces " applied at the ends of a straight inflexible line " and acting in the direction of this line.

" If, at first, the two points at which these forces act are regarded as entirely free, and if each of the points is taken as the fixed centre of the force which acts on the other, it will be easy to see that, since the distance apart of the points is a function of their coordinates, the virtual velocity of the first will be equal to differential of the distance when the variation is made with respect to the coordinates of this point alone. It will be the same for the second point. So that the total moment, which is proportional to the sum of the virtual velocities, will also be proportional to the sum of the partial differentials which represent these velocities—that is, proportional to the complete differential of the distance between the two points.

Thus the total moment of the two forces is zero if the distance between the two points is constant.

If the two forces are repulsive, the total moment is negative when

[1] *Mémoire sur la Statique, contenant la démonstration du principe des vitesses virtuelles et la théorie des moments, Journal de l'École polytechnique*, 5th cahier, 1798.

[2] See above, p. 232.

the distance between the two points increases and positive when this distance decreases. Converse results are obtained when the forces are attractive.

Fourier then considers " two inflexible (and perfectly smooth) surfaces which resist each other " and studies the total moment of their mutual reactions in any disturbance of the system. He considers two points on the common normal to the surfaces at the point of contact and such that one lies inside each surface. These two points cannot be closer together than they are in the equilibrium position. So that either their distance apart increases or it does not change when the system is disturbed.

" The first distance is the smallest of all those which occur when the position of the two surfaces is varied in such a way that they remain in contact with each other. Since the law of continuity is obeyed, it is necessary that the differential should be zero. " Since the total moment of the reactions is proportional to the variation of the distance of the two points at which they act, it therefore remains zero, as long as the surfaces remain in contact, whatever the displacement may be.

In order to generalise these results Fourier observes that the moments combine and decompose like forces (if a solid body is concerned).

If a solid body is considered to be in equilibrium under the action of n forces it is established that the total moment of the n forces is necessarily zero. The converse is also true.

Fourier then imagines a system of bodies to be connected by inextensible threads and acted upon by any forces which are such that there would be equilibrium independently of any external resistance. The forces which act on each body cancel each other out. Apart from the forces directly applied to the body, these forces comprise the tension of the threads between points of this body and points of neighbouring bodies.

" That is why, in considering simultaneously all the forces which act on all the bodies, their total moment can be said to be zero for all conceivable disturbances—even for those which the presence of the threads does not allow. It is now necessary to select, from among these disturbances, those which satisfy the equations of condition ; and, for these particular disturbances, to discover the value of the total moment of those forces which are due to the tensions alone. "

This value is zero. For each of the threads is acted upon by two equal and opposed forces, and the distance between the extremities is constant. From this it follows that the total moment of the applied forces alone is also zero.

If the distance between the ends of the threads does not remain constant, it can only become smaller. Since the forces of tension tend to decrease this distance, the total moment of the forces of tension is negative. Therefore, the sum of the moments of the applied forces alone can only be positive for disturbances of this kind.

Fourier next considers an " undefined assembly of hard bodies " whose shapes and dimensions are arbitrary and which are supported upon each other. Each body is in equilibrium under the action of the forces which are applied to it and the resistances of neighbouring bodies. If two neighbouring bodies always remain in contact during the disturbance of the system, albeit at different points, the moment of their reactions is zero. It is *negative* if the bodies happen to separate.

" In considering the combination of all the forces which act on all the bodies, it is certain that some of the moments must be zero for all the disturbances which can be imagined—even for those which may be prevented by the mutual impenetrability of the solids. Now, for displacements compatible with the latter condition, the moment of all the forces of pressure is either zero or negative. Therefore, for all the possible disturbances, the sum of the moments of the applied forces alone is either zero or positive [1]—it is zero when the equations which express the condition that contact must take place are satisfied, and positive whenever two bodies which touch each other, or act upon each other, become entirely separated. There is no possible disturbance for which the sum of the moments can be negative."

Fourier treats incompressible fluids by considering that their different points are subject to an interaction which opposes every variation of the distances between the points.

He then proceeds to the logical reduction of the theorem of virtual work to the principle of the lever. For this purpose he replaces the system by " a simpler body which can nevertheless be disturbed in the same way. "

Let $p, q, r, s, \ldots$ be the points of the given system to which the forces $P, Q, R, S, \ldots$ are applied. The displacement which gives the points $p, q, r, s, \ldots$ the initial virtual velocities $dp, dq, dr, ds, \ldots$ in the directions of the lines $p', q', r', s', \ldots$ is considered. The body substituted for the system will also pass through the points $p, q, r, s, \ldots$ Similarly, its elements will have virtual velocities $dp, dq, dr, ds, \ldots$ along $p', q', r', s', \ldots$

Fourier draws a plane perpendicular to the line p', and passing through the point p ; also a plane perpendicular to q' through the point

---

[1] Since FOURIER's *moment* is, apart from a change of sign, the modern *virtual work*, this conclusion is the one which is usually expressed in the form, " The virtual work of *given* forces is zero or negative for every displacement compatible with the constraints. "

q. These two planes intersect in the straight line *d.* A perpendicular, *h,* is dropped from *p* to *d.* At the intersection of *h* and *d,* and in the plane perpendicular to *q'* passing through *q,* the line *h'* is drawn perpendicular to *d.* From *q* the line *h"* is drawn perpendicular to *h'.* The straight lines *h* and *h'* are considered as the arms on an angular lever with axis *d.* The straight line *h"* is considered as a straight lever with axis δ (in the plane perpendicular to *q'* drawn through *q*).

If *p* is displaced along *p'* (by *dp*) the end of the arm *h'* is correspondingly displaced, and the axis δ can evidently be chosen in such a way that the displacement of *q* (required by the straight lever) is exactly equal to *dq.* An assembly on analogous levers can be imagined between the point *q* and the point *r,* between the point *r* and the point *s,* . . . so that the system of levers thus constructed is susceptible of the same displacements as those attributed to the original system, *and of this displacement alone.*

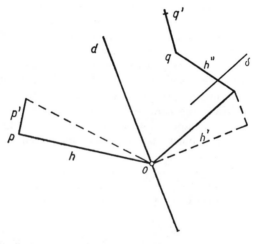

Fig. 107

Suppose that the forces *P, Q, R, S,* . . . have a total moment which is zero for the displacement *dp, dq, dr, ds,* . . . Because of the principle of the lever and the principle of the composition of forces, the forces *P, Q, R, S,* . . . will necessarily produce equilibrium in the system of levers constructed in the way that has been described.

We shall show, by *reductio ad absurdum,* that these same forces will leave the original system in equilibrium. Indeed, if the points *p, q, r, s,* . . . assume the velocities *dp, dq, dr, ds,* and if it is assumed that the point p of the system of levers is connected with the point p of the given system, the assembly of levers will be carried along in the displacement of the given system, and the points q and q, r and r, . . . of the two systems will not separate.

Therefore it can be supposed that there are connected not only the points p and p, but also the pairs of points q and q, r and r, s and s, . . . in the two systems. Accordingly the forces P, Q, R, S, . . . will produce the motion of the two systems connected at the points p, q, r, s, . . . Now the same forces cancel each other out when applied to the system of levers alone. The reunion of the two systems could not perturb this equilibrium. Whence it is impossible that the forces P, Q, R, S, . . . should produce the movement of the given system. This is true for any other displacement for which the total moment of the forces is zero. " And from this can be deduced the following particular conclusion, which includes the principle of virtual velocities. If, of all the possible displacements, there is none which corresponds to a zero moment, there must be equilibrium. "

Moreover it suffices that the *sum of the moments should not be negative*. Indeed, " it is easily proved, by the theory of the lever alone, that these forces applied [to the levers alone] cannot produce a displacement for which the total moment is positive. And since it is supposed that the presence of obstacles makes all other displacements impossible, it is necessary that when the forces act on the levers, they maintain them in equilibrium. This will still be true if the first system is applied to the second. Therefore these forces cannot separately produce the displacement in question in the first system. For this displacement would also accur if the second system were applied to the first, and we have just seen that it is then impossible.

" Conversely, if some powers maintain any material system in equilibrium, there can be no displacement of the system possible for which the sum of the moments can be negative. This is proved in the following way. If it is assumed that the system can move into such a position that the moment of the forces is negative, it must be concluded that equilibrium does not exist. For the equilibrium would not cease to exist if this displacement became the only possible one. It is easy to represent this last effect by imagining assemblies of levers, similar to those which have been described above, between all the points p, q, r, s, . . . of the system, and capable of the virtual velocities which correspond to the displacement concerned. It is unnecessary to show that the equilibrium will not be disturbed by the addition of these levers. Now it is impossible that there should not be motion, because the forces will find themselves applied to an assembly of levers which could not fail to be displaced if the sum of the moments of the forces were negative, just as it follows from the theory of the lever. Therefore it is necessary that the sum of the moments of the forces should never be negative. "

Finally Fourier introduced the distinction between *bilateral* and *unilateral* constraints.

" Whenever the displacemeñts of which the body is capable are determined by the equations of condition which they must satisfy, the total moment of the forces cannot be positive when the forces are in equilibrium. For if this moment were positive, the moment corresponding to the contrary displacement would be negative. Now as this latter displacement is equally possible, since it satisfies the equations of condition, the forces could not cancel each other out. . . . That is why it is necessary, in this case, that the sum of the moments of the forces must be zero in order that there should be equilibrium. This is the true meaning of the principle of virtual velocities. But if the displacements are not prohibited by the equations of condition —which often happens—the equilibrium can subsist without the moment of the forces being zero, provided that it is not negative. "

In connection with this demonstration, Jouguet has remarked that Fourier thus established the principle of the equivalence of constraints, which may be stated in the following way.

" Let there be a system of points acted upon by forces F and bound by constraints (L). Replace the constraints (L) by the constraints (L') which preserve the same elementary mobility as the constraints (L). In order that the system should be in equilibrium, it is sufficient and it is necessary that the forces F should be in equilibrium under the constraints (L'). " [1]

Fourier's analysis breaks down if the constraints (L) and (L') introduce resistances to motion, even if (L) and (L') assure the same kinematic mobility of the system. " It is not only the kinematic mobility which must be preserved but also, as it were, the dynamical mobility. " [2]

[1] *L. M.*, Vol. II, p. 171.
[2] *Ibid.*, p. 172.

CHAPTER THREE

THE PRINCIPLE OF LEAST CONSTRAINT (1829)

The principle of least constraint was stated by Gauss in a paper called *Über ein neues Grundgesetz der Mechanik* in Volume IV of the *Journal de Crelle* (1829).[1]

He wrote, " It is known that the principle of virtual velocities makes the whole of statics a matter of analysis, and that d'Alembert's principle, in its turn, reduces dynamics to statics. In the nature of things, there can exist no new principle in the science of equilibrium and motion which is not included in the two preceding principles, or which cannot be deduced from them.

" However, such a principle may not be without value. It is always interesting and instructive to regard the laws of nature from a new and advantageous point of view, so as to solve this or that problem more simply, or to obtain a more precise presentation.

" The great geometer [Lagrange] who succeeded so brilliantly in constructing mechanics from the principle of virtual velocities, had no disdain to generalise and to develop the principle of least action in Maupertuis' sense, but rather, he used it to great advantage. "

To Gauss, the principle of virtual velocities was the prototype of the principles of mechanics. But this principle was not intuitive, and demanded a special treatment in order that it might be extended from statics to dynamics. That is why Gauss believed it useful to state, in the following form, a new principle.

" The motion of a system of particles connected together in any way, and whose motions are subject to arbitrary external restrictions, always takes place in the most complete agreement possible with the free motion *(in möglich grösster Übereinstimmung mit der freien Bewegung)* or under the weakest possible constraint *(unter möglich kleinstem Zwange)*. The measure of the constraint applied to the

---

[1] *The complete Works of Gauss* (french edition), Vol. V, p. 25.

system at each elementary interval of time is the sum of the products of the mass of each particle with the square of its departure from the free motion. "

Let
$$m, m', m'' \ldots$$

be the masses of the points of the system and
$$a, a', a'' \ldots$$

be their positions at the time t. Let
$$b, b', b'' \ldots$$

be the positions which they would assume at the time $t + dt$ under the action of the forces which are applied to them, and because of their velocities at the time t, if it were supposed that they were completely free of all constraint. The actual positions, $c, c', c'', \ldots$ of the different points will be such that, while being compatible with the constraints, they minimise the sum
$$m(bc)^2 + m'(b'c')^2 + m''(b''c'')^2 + \ldots$$

The equilibrium is evidently a special case of the general law according to which
$$m(ab)^2 + m'(a'b')^2 + m''(a''b'')^2 + \ldots$$

is a minimum.

Gauss wrote, " This is how our principle can be deduced from principles already known. The force which is exerted on the particle m is evidently composed of two ; first, the force that, taking account of the velocity at the time t, brings the particle from a to c in the time dt ; secondly, the force which, in the same time, would bring the same element from c to b, if it were supposed to be free and to start from rest. [This is the same decomposition as that which d'Alembert used.] Similarly for all particles.

" By d'Alembert's principle the points m, m', m'' must be in equilibrium, because of the constraints of the system, under the sole action of the second forces acting along $cb, c'b', c''b'', \ldots$ According to the principle of virtual velocities this equilibrium requires that [the sum of the virtual works] should be zero for every virtual displacement which is compatible with the restraints. Or, more accurately, this sum should *never be positive*.

" Then let $\gamma, \gamma', \gamma'', \ldots$ be different positions of $c, c', c'', \ldots$ which are compatible with the constraints. Let $\theta, \theta', \theta'', \ldots$ be the angles that $c\gamma, c'\gamma', c''\gamma'', \ldots$ make with $cb, c'b', c''b'', \ldots$ Then
$$\sum mcb \cdot c\gamma \cdot \cos\theta$$

will be zero or negative.

" Since

$$\overline{\gamma b}^2 = \overline{cb}^2 + \overline{c\gamma}^2 - 2cb \cdot c\gamma \cdot \cos \theta$$

it is clear that

$$\sum m\overline{\gamma b}^2 = \sum m\overline{cb}^2 + \sum m\overline{c\gamma}^2 - 2 \sum mcb \cdot c\gamma \cdot \cos \theta.$$

Therefore

$$\sum m\overline{\gamma b}^2 - \sum m\overline{cb}^2$$

will always be positive.   Therefore, finally,

$$\sum m\overline{cb}^2$$

will always be a minimum.   Q. E. D. "

In conclusion, Gauss emphasised the fact that free motions, when they are incompatible with the constraints, are modified in Nature in the same way that experimental data are modified, by the method of least squares, so as to be compatible with a necessary relation between the measured quantities.

CHAPTER FOUR

RELATIVE MOTION
RETURN TO A PRINCIPLE OF CLAIRAUT
CORIOLIS' THEOREMS
FOUCAULT'S EXPERIMENTS

1. RETURN TO A PRINCIPLE OF CLAIRAUT (1742).

The first outline of a theory of relative motion appeared, as we have seen, in Huyghens' *De vi centrifuga*.[1]

Though he did not resolve all the difficulties of relative motion, Clairaut had the indisputable merit of generalising Huyghens' conceptions. This he did in a paper called *Sur quelques principes qui donnent la solution d'un grand nombre de problèmes* [2] which could not have escaped the attention of Coriolis.

Clairaut set out to find " what happens to any system of bodies, actuated by gravity or other accelerating forces, when this system is attached by some part to a plane and is carried with this plane in some curvilinear motion. " He introduced " the general principle for finding the motions of systems of bodies carried along by the planes on which they are placed " in the following way.

" Imagine that the rectangle *FGHI* is placed between two curves *AB* and *CD* and that, when the corner *G* is moved at will on the curve *AB*, the corner *I* follows the curve *CD*.

" Suppose now that one of the bodies *M* of the system given, accelerated by gravity or by any other forces, describes the curve *Mμ* because of the properties of the system.

" We seek the accelerating or retarding force that the motion of the plane *FGHI* gives the body *M*. We shall start by tracing the curve *PQ* that the point *M* would describe, during the motion of *FGHI*, if it were fixed in the plane *FGHI*. We shall then determine the velo-

---

[1] See above, Part II, p. 194.
[2] *Mémoires de l'Académie des Sciences*, 1742, p. 1.

city with which it would move along this line, which will only depend on the given velocity of G and the curves AB and CD. This done, we shall seek the accelerating forces which it is necessary to suppose distributed in the space AB, CD in order that the body M, left to itself with the velocity that it has at M on Mm, might travel the line PQ. Let MS, for example, represent what this force would be at M. I say that by producing MS and taking $MT = MS$, the straight line MT represents the force by which the motion of the plane $FGHI$ alters the velocity of M on $M\mu$.

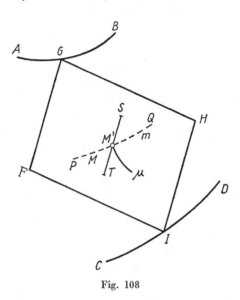

Fig. 108

" To prove this, I start by distinguishing the particle M from the fixed point of $FGHI$ which corresponds to it, and I call this fixed point M'. I then remark that if, at the instant that the body M has travelled along $M\mu$ and the body M' has travelled along $M'm$, the curves AB and CD were suddenly removed and the plane $FGHI$ were allowed to move uniformly with the velocity of M' along $M'm$, the system which is on the plane $FGHI$ would necessarily move in the same way as if this plane were fixed. I add to this remark that the reason why the motion along the arc Mm is altered in the curvilinear motion of the plane $FGHI$ is that, in order to produce the curvilinear motion, it is necessary to imagine that the body M' receives an impulse MS at the instant that it has travelled along $M'm$, and that the body M does not receive this impulse. For if the body M received this impulse, the motion of the system would be exactly the same as if the plane $FGHI$ were fixed. Given this, I say that it is the same whether M' receives an impulse and M does not, or whether M receives it in the opposite direction and M' does not.

" Therefore the plane $FGHI$ can be regarded as fixed and it can be supposed that the body experiences the action of the given forces as well as the action of the forces MT. "

In short, in this way Clairaut arrives at an estimate of the quantity $m\vec{\gamma}_r$ in the relative motion. This estimate is $\vec{F}_a - m\vec{\gamma}_e$, where $\vec{\gamma}_e$ is the dragging acceleration.

We know that this principle is incomplete. However, it led Clairaut to correct results when he confined himself to applying the principle of kinetic energy to relative motion. For it is known that Coriolis' complementary force of inertia does no work.

Incomplete though it may be, Clairaut's argument has a synthetic value and, for this reason, it is of some use to complete it. This, in fact was accomplished by Joseph Bertrand in 1848.[1]

We shall not follow Bertrand's analysis here, but shall present an alternative method by which the argument may be completed.

With respect to a fixed reference system in which the law $m\vec{\gamma}_a = \vec{F}_a$ is valid, the particle M describes the curve MM_1 between the times t and $t + dt$.

Under the same conditions its coincident point M' is connected to a moving reference system (S) which has any arbitrary continuous motion relative to an absolute reference system. It then describes the curve $M'M'_1$ between the times t and $t + dt$. If the particle M had retained the absolute velocity $\vec{v}_a$ which it had at the time t, it would have travelled to M_2 in the time dt, where $\overrightarrow{MM_2} = \vec{v}_a dt$. Similarly, if M' had retained its absolute velocity $\vec{v}_e$, it would have travelled

---

[1] *Journal de l'École polytechnique*, Vol. 32, p. 148, 1848.

In this connection, J. BERTRAND brings grist to the mill of history. " Too often, after having studied analytical mechanics, a man believes it is useless to seek to complete this study by the reading of the scattered works with which the predecessors of LAGRANGE enriched the academic collections of the XVIIIth Century. I believe that this tendency, unfortunately very common, is such as to destroy the progress of mechanics, and that it has already produced unfortunate results. The too common custom of deducing formulae has, to some extent, led to the loss of a proper respect for the truths of mechanics considered in themselves. "

In BERTRAND's opinion, " M. Coriolis, without knowing it, has done the same as the illustrious Clairaut. "

That is his opinion. It is true that CORIOLIS does not refer to CLAIRAUT. It is also true that he went further than him. But it seems unlikely to us that CLAIRAUT's paper could have been omitted from CORIOLIS' reading.

to M_2', where $\overrightarrow{M'M_2'} = \vec{v}_e dt$. Follow on let $\vec{v}_r$ be the relative velocity of M in the system (S) at the time t. In the triangle MM_2M_2', using the composition of velocities $(\vec{v}_a = \vec{v}_e + \vec{v}_r)$, it is seen that $\overrightarrow{M_2'M_2} = \vec{v}_r dt$.

Now give the reference system (S) and the particle M the *same absolute motion* defined in the following way— first, a translation $\overrightarrow{M_1'M_2'} = -\dfrac{1}{2}\vec{\gamma}_e dt^2$ which annuls the deviation $\overrightarrow{M_2'M_1'}$ of the point M'; secondly, the rotation $-\vec{\omega}dt$ at M_2', to annul the effect of the absolute rotation of the system (S), considered as a solid, between the times t and $t + dt$. Thus, taking account of the motion which it originally possessed, the system (S) will have experienced a rectilinear translation $\overrightarrow{M'M_2'}$ of velocity $\vec{v}_e$.

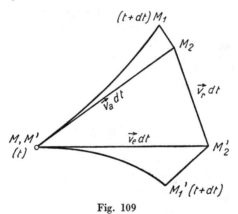

Fig. 109

Correspondingly, the particle M experiences the translation $-\dfrac{1}{2}\vec{\gamma}_e dt^2$ and, *to the third order*, a displacement

$$-\vec{\omega}dt \wedge \overrightarrow{M_2'M_2} = -(\vec{\omega} \wedge \vec{v}_r)dt^2.$$

These two displacements of the particle M in absolute space can be *fictitiously* imputed (as Clairaut realised) to forces $-m\vec{\gamma}_e$ and $-2m(\vec{\omega} \wedge \vec{v}_r)$ respectively.

Therefore, between t and $t + dt$, the fundamental law of can be written in the system (S), *corrected in its motion in this way* (that is, in rectilinear and uniform translation with velocity $\vec{v}_e$), in the form

$$m\vec{\gamma}_r = \vec{F}_a - (m\vec{\gamma}_e + 2m(\vec{\omega} \wedge \vec{v}_r)).$$

This completes and rectifies Clairaut's principle.

2. CORIOLIS' FIRST THEOREM.

Coriolis' name has remained associated with the law of the composition of accelerations. This law belongs to the domain of pure kinematics— that is the way it is taught at present, before its dynamical consequences are explored.

Historically, Coriolis was concerned with the theory of waterwheels when he embarked on his study of relative motion. This theory had already been studied by Jean Bernoulli, Euler, Borda, Navier and Ampère. To progress beyond the earlier work, it was necessary to study the following general problem.

" *To find the motion of any machine in which certain parts are moved in a given way.* "

Here we shall follow Coriolis' first paper, which was read to the *Académie des Sciences* on June 6th, 1831, and was printed in the *Journal de l'École polytechnique*.[1]

Coriolis considers two reference frames. One, $Ox_1y_1z_1$, is fixed —absolute. The other, $OXYZ$, is movable—relative. Let ξ, η, ζ be the absolute coordinates of the origin of the movable frame and (a, a', a''), (b, b', b''), (c, c', c'') be the direction cosines of the movable axes with the fixed axes.

The constraints which exist during the motion are supposed to be perfect and expressible in finite terms in the relative coordinates. Let $\vec{\mathcal{L}}$ be the force of constraint applied at a point of the system. Using the method of Lagrange multipliers, Coriolis writes the projections of the force on the movable axes as

$$\vec{\mathcal{L}}\begin{cases} \lambda\dfrac{\partial L}{\partial x} + \mu\dfrac{\partial M}{\partial x} + \cdots \\[2mm] \lambda\dfrac{\partial L}{\partial y} + \mu\dfrac{\partial M}{\partial y} + \cdots \\[2mm] \lambda\dfrac{\partial L}{\partial z} + \mu\dfrac{\partial M}{\partial z} + \cdots \end{cases}$$

Returning to the fixed axes, Coriolis calculates the total force —the given force and the force of constraint—acting on one point of the system. Briefly, this may be written

(A) $m\vec{\gamma}_a = \vec{P} + \vec{\mathcal{L}}$.

Coriolis sums the equations (A) after multiplying them by the *relative displacement ds_r*. Under these conditions the forces due to

---

[1] 21st cahier, 1832, p. 268.

the constraints vanish. But in order to perform a particular calculation, it is necessary to express the absolute accelerations $\vec{\gamma}_a$ as functions of the relative coordinates and velocities as well as the dragging motion.

For this purpose Coriolis distinguishes total differentials of the true motion, indicated by the symbol d, from the differentials obtained by varying only the quantities $(a, b, \ldots c'')$. This differentiation is indicated by the symbol d_e, and corresponds to a variation of the *orientation* of the movable system in which the quantities $x, y, z,$ ξ, η, ζ remain constant.

" If the points are not displaced relatively to the moving axes, they only have the *dragging motion* [1] corresponding to these axes, whose origin is supposed immovable and which only have a rotational motion about this origin. "

Thus, " by omitting to write the denominators dt^2 under the differentials," Coriolis writes the components of the absolute acceleration in the form

(B) $\begin{cases} d^2x_1 = ad^2x + bd^2y + cd^2z + 2dxda + 2dydb + 2dzdc + d_e^2x + d^2\xi \\ d^2y_1 = \ldots \\ d^2z_1 = \ldots \end{cases}$

Terms of the equation (A) such as

$$m\left(\frac{d_e^2x}{dt^2} + \frac{d^2\xi}{dt^2}\right), \ m\left(\frac{d_e^2y}{dt^2} + \frac{d^2\eta}{dt^2}\right), \ m\left(\frac{d_e^2z}{dt^2} + \frac{d^2\zeta}{dt^2}\right)$$

then appear to him as " *the components, with respect to the fixed axes, of the forces* F_e *which would produce the motion which each point would take if it remained in the same place with respect to the moving axes.* " We have quoted from the paper of Coriolis in order to illustrate the development of his thought, how he did not pause on the kinematic aspect of the problem, but went directly to the cause that would be able to produce the dragging motion (in the modern sense now).

Then Coriolis returns to the moving axes by first substituting the equations (B) in the equations (A), then by projecting on the moving axes. He obtains the relation

(C) $\begin{cases} md^2x + 2mdy\,(adb + a'db' + a''db'') + 2mdz\,(adc + a'dc' + a''dc'') \\ \qquad\qquad\qquad\qquad + X_e = X + \lambda\dfrac{\partial L}{\partial x} + \mu\dfrac{\partial M}{\partial x} + \ldots \\ md^2y + \ldots \\ md^2z + \ldots \end{cases}$

[1] This definition does not coincide with that which is now common.

By summing for all the points of the system, after multiplying by the relative displacements dx, dy, dz, he arrives at

$$\sum m \left(dx \frac{d^2x}{dt^2} + dy \frac{d^2y}{dt^2} + dz \frac{d^2z}{dt^2} \right) + \sum (X_e\, dx + Y_e\, dy + Z_e\, dz)$$
$$= \sum (X dx + Y dy + Z dz).$$

The cross-terms in dx, dy, dz, vanish in the summation because of the relations between the direction cosines.

If $\vec{V}_r$ is the relative velocity of any point of the system with respect to the moving axes, and $\vec{P}_e$ the force which is equal and opposite to $\vec{F}_e$, $\vec{P}$ the given force, Coriolis writes

$$\sum m \vec{V}_r \cdot d\vec{V}_r = \sum \vec{P} \cdot \vec{ds}_r + \sum \vec{P}_e \cdot \vec{ds}_r$$

or, by integration,

$$\sum m \left(\frac{V_r^2 - V_{0r}^2}{2} \right) = \sum \int (P + P_e)\, ds_r.$$

" *Thus the principle of living forces is still true for motion relative to moving axes, provided that there are added to the quantities of action* [that is, of work] $\int \mathrm{P} ds_r$ cos $(\widehat{\mathrm{P} \cdot \mathrm{ds}_r})$ *calculated from the given forces and the arcs* ds_r *described in the relative motion, other quantities of action which are due to the forces* P_e. *These forces are supposed equal and opposite to those which it would be necessary to apply to each moving point in order to make it take the motion that it would have had if it were in invariably connected to the moving axes.* "

This is *Coriolis' first theorem*, which essentially belongs to the dynamics of relative motion. Coriolis applied it to the " quantity of action " transmitted to the machine which carried the movable axes. We shall not follow him in this application, where he made simultaneous use of the theorem of kinetic energy in the absolute motion and the relative motion.

Coriolis remarked that when the question was that of the equilibrium of a fluid contained in a vessel turning about an axis, " it is immediately seen that it is necessary to introduce actions equal to the centrifugal forces. But it is not the same for the principle of living forces applied to the relative motion. It would be mistaken to regard this proposition as evident ; to proceed in this way for any other equation than that of the living forces would be to arrive at false results. "

In conclusion, Coriolis declared, " Such are the principal results of this paper. It seems that the principle from which they stem may find many applications in the theory of machines, provided that it is supplemented by a number of propositions from rational mechanics. "

3. CORIOLIS' SECOND THEOREM.

Coriolis expectation was more than fulfilled. For his first paper already contained the germ of the fundamental theorem with which his name is now associated. This vital point is his equation (C), which Coriolis did not himself analyse thoroughly, anxious as he was to calculate the quantity of action transmitted to his water-wheels.

In a second paper, *Sur les équations du mouvement relatif des systèmes de corps*,[1] Coriolis wrote—

" In this paper I give the following general proposition—that to establish an equation of the relative motion of a system of bodies or of any machine, it suffices to add to the existing forces two kinds of supplementary forces. The first are always those to which it is necessary to have regard for the equation of living forces ; that is, which are the forces opposed to those which are able to keep the particles constantly connected with movable planes. The second are directed perpendicularly to the relative velocities and to the axes of rotation of the movable planes ; they are equal to twice the product of the angular velocity of the movable planes and the relative quantity of motion on a plane perpendicular to this axis.

" The latter forces are most closely analogous to ordinary centrifugal forces. To display this analogy it suffices to remark that the centrifugal force is equal to the quantity of motion multiplied by the angular velocity of the tangent to the curve described ; that it is directed perpendicularly to the velocity and in the osculatory plane, this is, perpendicularly to the axis of rotation of the tangent. Thus in order to pass from ordinary centrifugal forces to the second forces which occur, multiplied by two, in the preceding statement, it is only necessary to replace the angular velocity of the tangent by that of the movable planes, and to substitute for the direction of the axis of rotation of this tangent, the direction of the axis of rotation of the same movable planes. In other words, it suffices to substitute everything which is related in magnitude and direction to the rotation of the tangent by what is related to the rotation of the movable planes, and to multiply the forces thus obtained by two.

" It is because of this analogy that I concluded that these forces must be named compound centrifugal forces. Indeed, they have some of the characteristics of the relative motion because of the quantity of motion and some of the characteristics of the motion of the movable planes through the use of their axes of rotation and angular velocity.

[1] *Journal de l'École polytechnique*, 24th cahier, 1835, p. 142.

" Therefore it will be said that, for an equation of relative motion which is not that of the living forces, it is necessary to introduce twice the compound centrifugal force.

" The theorem which I presented at the *Académie des Sciences* in 1831 consists of the *disappearance of the compound centrifugal* forces from the equation of the living forces. It now becomes a particular case of the more general statement on the introduction of these compound centrifugal forces. "

Coriolis' demonstration depends directly on the equation (C) already written above. Indeed, if p, q, r are the " three angular velocities of the movable planes about their axes, " (C) can be written in the form

$$(\text{C}) \begin{cases} m \frac{d^2x}{dt^2} = 2m \left(r \frac{dy}{dt} - q \frac{dz}{dt} \right) + X - X_e + \lambda \frac{\partial L}{\partial x} + \mu \frac{\partial M}{\partial x} + \dots \\ \dots \\ \dots \end{cases}$$

" The terms in p, q, r, dx, dy and dz in the above equation are twice the components, along the moving axes, of a force directed perpendicularly to the plane of the axis of rotation and the relative velocity. The magnitude of this force will be the product of the angular velocity $\sqrt{p^2 + q^2 + r^2}$ with the projection, on a plane perpendicular to the axis of rotation, of the quantity of motion due to the relative velocity of the particle. The sense in which this force will be carried, with respect to a motion which carries the axis of rotation towards the relative velocity, will be the same as that of the axis of rotation with respect to the velocity of rotation. "

" *The expressions for the forces which must be added to the given forces in order to obtain the expressions for the forces in the relative motions are— first, those which are opposed to the forces able to produce, for each particle, the motion which it would have if it were connected to movable planes ; secondly, twice the compound centrifugal forces.* "

This is valid for a particle of the system. Coriolis then considers the *virtual velocities*—the displacements δx, δy, δz in the relative motion—compatible with the relative constraints

$$L = 0 \qquad M = 0 \quad \text{etc.}$$

These relative constraints, supposed to be perfect, will disappear on combining the equations (C), giving

$$\sum m \left(\frac{d^2x}{dt^2} \, \delta x + \frac{d^2y}{dt^2} \, \delta y + \frac{d^2z}{dt^2} \, \delta z \right) + 2p \sum m \left(\frac{dy\delta z - dz\delta y}{dt} \right)$$

$$+ 2q \sum m \left(\frac{dz\delta x - dx\delta z}{dt} \right)$$

$$+ 2r \sum m \left(\frac{dx\delta y - dy\delta x}{dt} \right)$$

$$= \sum (X\delta x + Y\delta y + Z\delta z) - \sum (X_e\delta x + Y_e\delta y + Z_e\delta z).$$

If ds is the actual virtual displacement and δs the virtual relative displacement; if α, β and γ are the direction cosines of the instantaneous rotation $\omega(p, q, r)$ with respect to the moving axes and λ, μ, ν are the direction cosines of the normal to the plane $(ds, \delta s)$, the Coriolis' complementary term becomes

$$2\omega \sum m \, \frac{ds}{dt} \, \delta s \sin (ds, \delta s) (\alpha\lambda + \beta\mu + \gamma\nu).$$

Therefore, " *In order to obtain an equation of the relative motion it is necessary to add to the terms ordinarily existing for absolute motion —first, those which arise from the forces which are able to force the particles to remain connected to the movable planes; and, in addition, a term which is equal to twice the velocity of rotation multiplied by the algebraic sum of the projections, on a plane perpendicular to the axis of rotation of these planes, of all the areas of the parallelograms defined by the effective quantities of motion and the virtual velocities.* "

For the equation of the kinetic energy, each area is zero. For the virtual velocity coincides with the effective velocity (or rather, with the true displacement). Thus Coriolis' two theorems are linked together.

We have said enough to illustrate the development of Coriolis' thought. In fact, he complicated his task by isolating the law of the composition of the accelerations—singularly hidden—from the already difficult problem of the dynamics of systems. It is rather interesting to remark in passing that Coriolis composed two accelerations by summing

$$\frac{d_e^2x}{dt^2} + \frac{d^2\xi}{dt^2}, \quad \text{etc.}$$

Very fortunately, however, this procedure entailed no risk because it reduced to connecting together two terms of the unique dragging acceleration $\vec{\gamma}_e$ (in the modern sense).

We stress the fact that Coriolis did not deal, in fact, with kinematics. He argued exclusively from a dynamical point of view, using *forces*, and only endowed products such as $m\vec{\gamma}_e$ with a physical significance. His aim was to find an equation of the relative motion which might be independent of the constraints, supposed to be holonomic and perfect. It is for this reason that he first encountered the theorem of the kinetic energy, in which the compound centrifugal force vanished. Then he was able to give a more general equation in which the complementary term appeared. All this only makes his discovery more remarkable.

4. THE EXPERIMENTS OF FOUCAULT (1819-1868).

In the strict sense, mechanics which is referred to terrestrial axes should take account of Coriolis' compound centrifugal force. Nevertheless, we have already had occasion to remark [1] that the deviation of heavy bodies towards the East can be predicted by a very simple intuitive argument. Moreover, as early as 1833, Reich studied free fall in a mine-shaft at Freiberg (Saxony). The depth of the mine was 188 m., and he observed an *average* deviation of 28 millimetres in 106 separate observations.

In 1851 Foucault published a paper called *Démonstration physique du mouvement de rotation de la Terre au moyen du pendule.* [2]

This demonstration made no appeal to Coriolis' work—only after the event did occur a mathematical literature. Foucault, who had been a mediocre pupil at school, was a natural physicist and an incomparable experimenter. However, he started work as the scientific member of the staff of the *Journal des Débats.*

He set out to experiment on the direction of the plane of oscillation of a pendulum. If the observer is at first supposed at the pole (North or South) and the pendulum is reduced to a homogeneous spherical mass suspended from an *absolutely fixed point*, then if this point is exactly on the axis of rotation of the Earth, the plane of oscillation remains fixed in space. " The motion of the Earth, which forever rotates from west to east, will become appreciable in contrast with the fixity of the plane of oscillation, whose trace on the ground will seem to be actuated by a motion conforming to the apparent motion of the celestial sphere. And if the oscillations can continue for twenty-four hours, in this time the plane will execute a whole revolution about the vertical through the point of suspension. "

But Foucault also remarked that in reality it is necessary to " take support on moving earth ; the rigid pieces to which the thread of the

---

[1] See above, Part I, p. 63.
[2] *Comptes rendus de l'Académie des Sciences*, Vol. 32, p. 135 (February 3rd, 1851).

pendulum is attached cannot be isolated from the diurnal motion. It should be borne in mind that this motion, communicated to the thread and the mass of the pendulum, might alter the direction of the plane of oscillation. " Nevertheless, experiment shows that " provided that the thread is round and homogeneous, it can be made to turn rather rapidly on itself, in one sense or another, without appreciably affecting the plane of oscillation. So that, at the pole, the experiment must succeed in all its purity. " [1]

" But when our latitudes are approached, the phenomenon becomes complicated in a way that is rather difficult to appreciate. To the extent that the Equator is approached, the plane of the horizon has a more and more oblique direction with respect to the Earth. The vertical, instead of turning on itself as at the pole, describes a cone which is more and more obtuse.

" From this results a slowing down in the relative motion of the plane of oscillation. This becomes zero at the Equator and changes its sense in the other hemisphere. "

Without explicitly justifying the fact in his paper, Foucault assumed that the angular displacement of the plane of oscillation must be equal to the product of the angular motion of the Earth in the same time with the sine of the latitude. If the correspondence published in the collection of his works is studied in this connection, it is apparent that Foucault arrived at this relation semi-intuitively, before it had been obtained by calculations in mechanics.

At first Foucault worked on a relatively modest scale by suspending a sphere of 5 kg. from a steel wire two metres long. The point of support was a strong piece of casting fixed to the top of the roof of a cellar. He took the precautions of ridding the wire of torsion and ensuring that there was no torsional oscillation of the sphere. He " encircled the sphere with a loop of organic thread whose end is attached to a point fixed on the wall, and chosen so that the oscillation of the pendulum might be 15 to 20°. " He then burnt the organic thread. This is what he observed.

" The pendulum, subject to the force of gravity alone, sets off and provides a long sequence of oscillations whose plane is not slow to demonstrate an appreciable displacement. At the end of half an hour the displacement is such that it is immediately obvious. But it

----

[1] In another place FOUCAULT reassured himself, more objectively, that " whether or not the Earth, turning, draws along the point of attachment with the monument [where the experiment was performed], the thread experiences no torsion. This implies that the bob of the pendulum submits to this motion without dragging the plane of oscillation. "

is more interesting to follow the phenomenon closely, so as to be assured of the continuity of the effect. For this purpose a vertical point, consisting of a kind of style mounted on a support placed on the earth is fixed so that the appendicular projection of the pendulum, in its to and fro motion, grazes the fixed point when it comes to its extremity. In less than a minute, the exact coincidence of the two points ceases. The oscillating point is continuously displaced towards the observer's left, which indicates that the deviation of the plane of oscillation takes place in the same sense as the apparent motion of the celestial sphere. . . . In our latitudes the horizontal trace of the plane of oscillation does not complete a whole circuit in twenty-four hours. "

The liveliness and the accuracy of this account will be admired. As we have indicated, Foucault had started his work in a cellar. Thanks to Arago, who put at his disposal the meridian room at the *Observatoire* (Paris), he was later able to repeat his experiment with a pendulum 11 m. long. This provided a slower and more extensive oscillation. Finally, Foucault worked at the Panthéon (Paris) with a pendulum weighing 28 kg. suspended on a steel wire 67 m. long.

As Foucault remarked—and this is an example of his remarkable intuition—" *the pendulum has the advantage of accumulating the effects* [1] and carrying them from the field of theory into that of observation. "

At this point Foucault referred to a paper of Poisson,[2] in which the latter has studied the deviation of projectiles.

In the world of learning Foucault's experiment had the immediate success that it deserved. Notes accumulated in the *Comptes rendus* on the subject of the pendulum which had been revived in this way ; they included contributions from Binet, Sturm, Poncelet, Plana, Bravais, Quet, Dumas, etc. . . .

Nevertheless, however brilliant it may have been, Foucault's experiment remained rather mysterious to the general public, since it depended on the displacement of a plane of oscillation. Moreover, Foucault wished to give a still more tangible proof of the rotation of the Earth.

The gyroscope provided him with a means of doing this. He used a pendulum suspended by its centre of gravity and executing what is called in mechanics a motion *à la Poinsot.*

Foucault's gyroscope was a bronze fly-wheel mounted inside a metallic circle whose diameter contained a steel axis supporting the wheel. The gyroscope turned about one of its central axes of inertia, which remained fixed in space.

---

[1] Without this accumulation FOUCAULT would not have been able to detect a force that was only the 55,000th part of the weight of the *Panthéon* pendulum.

[2] *Comptes rendus de l'Académie des Sciences*, November 13th, 1837.

Foucault wrote[1] " The body can no longer participate in the diurnal motion which actuates our sphere.[2]   Indeed, although because of its short length, its axis appears to preserve its original direction relatively to terrestrial objects, the use of a microscope is sufficient to establish an apparent and continuous motion which follows the motion of the celestial sphere exactly. . . . As the original direction of this axis is disposed arbitrarily in all azimuths about the vertical, the observed deviations can be, at will, given all the values contained between that of the total deviation and that of this total deviation as reduced by the sine of the latitude. "

Foucault concludes, in a somewhat journalistic style that was probably natural to the reporter of the *Débats*—

" In one fell swoop, with a deviation in the desired direction, a new proof of the rotation of the Earth is obtained; this with an instrument reduced to small dimensions, easily transportable, and which mirrors the continuous motion of the Earth itself. . . . In your possession are pieces of material which are truly subject to the dragging of the diurnal motion. "

Thus Fourier achieved one of Poinsot's aims.

The compound centrifugal force in the sense of Coriolis, and Foucault's pendulum, are two essential achievements in mechanics ; the one has an origin which is purely mathematical, the other was the product of a physicist's brilliant intuition. Though they are united in the same rational exposition in the books that are now classical, they were born separately—it was not the reading of Coriolis that inspired Foucault's experiment.

---

[1] *Comptes rendus de l'Académie des Sciences*, Vol. 35, p. 421 (September 27th, 1852).

[2] More correctly, it is easy to give the gyroscope a very rapid proper rotation about its own axis, say $\vec{\omega}$, which is very large compared with the *absolute* rotation of the Earth, say $\vec{\varrho}$.

If $\vec{\Omega}$ is the *absolute* rotation of the gyroscope,

$$\vec{\Omega} = \vec{\omega} + \vec{\varrho}$$

and the axis remains directed *towards the fixed stars* $(\vec{\Omega} \sim \vec{\omega})$ as long as $\vec{\varrho}$ is negligible compared with $\vec{\omega}$.

CHAPTER FIVE

POISSON'S THEOREM (1809)

1. Poisson's theorem and brackets.

Poisson's theorem appeared among the investigations made immediately after the appearance of two papers by Lagrange. One of these papers appeared in 1808 and the other in 1809, and they were incorporated in the 1811 edition of the *Mécanique analytique*. Stimulated by the needs of the theory of perturbations in classical mechanics, they were concerned with the variation of arbitrary constants.

Here we shall follow a paper of Poisson which was read at the *Institut de France* on October 16th, 1809.[1]

Poisson starts from Lagrange's equations

$$\frac{d}{dt}\left(\frac{\partial T}{\partial q_i'}\right) - \frac{\partial T}{\partial q_i} + \frac{\partial V}{\partial q_i} = 0 \qquad (i = 1, 2, \ldots k).$$

Putting $R = T - V$, where V depends only on the q_i and not on the q_i', he obtains

$$\frac{\partial R}{\partial q_i'} = \frac{\partial T}{\partial q_i'} = u_i$$

whence

(1) $$\frac{du_i}{dt} = \frac{\partial R}{\partial q_i}.$$

" In this way the equations of motion are reduced to the simplest form that they can be given. "

[1] *Journal de l'École polytechnique*, cahier XV, 1809, p. 266.

While following the development of Poisson's analysis rigorously, we have taken the liberty of condensing its form by using the convention of the *summation of dummy suffixes*. This is commonly used in the absolute differential calculus and allows the direct consideration of a system of k degrees of freedom (rather than three, as Poisson did). Further, we have introduced the distinction between the symbols of partial and total derivatives.

The new variables u_i are functions of the q_i and the q_i'. Conversely, the q_i' can be regarded as functions of the q_i and the u_i.

Considering R as a function of the q_i and the u_i, Poisson denotes the partial derivatives of R when the independent variables are q_i and u_i by $\dfrac{\partial R}{\partial q_i}$; and the derivatives of R when the independent variables are q_i and q_i' by $\left(\dfrac{\partial R}{\partial q_i}\right)$. Thus equation (1) becomes

$$\frac{du_i}{dt} = \left(\frac{\partial R}{\partial q_i}\right).$$

However

(2)
$$\frac{\partial R}{\partial q_i} = \left(\frac{\partial R}{\partial q_i}\right) + \frac{\partial R}{\partial q_k'}\frac{\partial q_k'}{\partial q_i}$$

so that

(3)
$$\frac{du_i}{dt} = \frac{\partial R}{\partial q_i} - u_k\frac{\partial q_k'}{\partial q_i}.$$

The partial derivatives of R may be written

(4)
$$\frac{\partial R}{\partial u_i} = u_k\frac{\partial q_k'}{\partial u_i}$$
$$\frac{\partial^2 R}{\partial u_i \partial q_j} = u_k\frac{\partial^2 q_k'}{\partial u_i \partial q_j}.$$

Finally, by calculating $\dfrac{\partial^2 R}{\partial u_i \partial u_j}$ in two different ways, it is found that

$$\frac{\partial^2 R}{\partial u_i \partial u_j} = u_k\frac{\partial^2 q_k'}{\partial u_i \partial u_j} + \frac{\partial q_j'}{\partial u_i}$$
$$= u_k\frac{\partial^2 q_k'}{\partial u_i \partial u_j} + \frac{\partial q_i'}{\partial u_j}.$$

From this is obtained the relation

(5)
$$\frac{\partial q_j'}{\partial u_i} = \frac{\partial q_i'}{\partial u_j}$$

which will be used in the sequel.

Given this, Poisson considers a first integral of the equations of motion containing a single arbitrary constant a. This integral equation, if solved for a, would lead to

$$a = \text{funct } (q_1 \ldots q_k, u_1 \ldots u_k, t).$$

Hence

$$0 = \frac{\partial a}{\partial t} + \frac{\partial a}{\partial q_k} q_k' + \frac{\partial a}{\partial u_k} u_k'$$

$$= \frac{\partial a}{\partial t} + \frac{\partial a}{\partial q_k} q_k' + \frac{\partial a}{\partial u_k} \left(\frac{\partial R}{\partial q_k} - u_r \frac{\partial q_r'}{\partial q_k} \right).$$

Differentiating with respect to q_i,

(6) $$0 = \frac{d}{dt} \left(\frac{\partial a}{\partial q_i} \right) + \frac{\partial a}{\partial q_k} \frac{\partial q_k'}{\partial q_i} + \frac{\partial a}{\partial u_k} \left(\frac{\partial^2 R}{\partial q_k \partial q_i} - u_r \frac{\partial^2 q_r'}{\partial q_k \partial q_i} \right).$$

If another first integral of the equations of motion, containing an arbitrary constant b, had been considered, then

(7) $$0 = \frac{d}{dt} \left(\frac{\partial b}{\partial q_i} \right) + \frac{\partial b}{\partial q_k} \frac{\partial q_k'}{\partial q_i} + \frac{\partial b}{\partial u_k} \left(\frac{\partial^2 R}{\partial q_k \partial q_i} - u_r \frac{\partial^2 q_r'}{\partial q_k \partial q_i} \right).$$

By multiplying equation (6) by $\dfrac{\partial b}{\partial u_i}$ and equation (7) by $- \dfrac{\partial a}{\partial u_i}$, summing over the dummy suffixes and adding the two equations,

(8) $$0 = \frac{\partial b}{\partial u_i} \frac{d}{dt} \left(\frac{\partial a}{\partial q_i} \right) - \frac{\partial a}{\partial u_i} \frac{d}{dt} \left(\frac{\partial b}{\partial q_i} \right) + \frac{\partial q_k'}{\partial q_i} \left(\frac{\partial a}{\partial q_k} \frac{\partial b}{\partial u_i} - \frac{\partial b}{\partial q_k} \frac{\partial a}{\partial u_i} \right).$$

Differentiating the equation from which (6) was obtained with respect to u_i, instead of q_i, there is obtained

(6′) $$0 = \frac{d}{dt} \left(\frac{\partial a}{\partial u_i} \right) + \frac{\partial a}{\partial q_k} \frac{\partial q_k'}{\partial u_i} + \frac{\partial a}{\partial u_k} \left(\frac{\partial^2 R}{\partial q_k \partial u_i} - u_r \frac{\partial^2 q_r'}{\partial q_k \partial q_i} \right) - \frac{\partial a}{\partial u_k} \frac{\partial q_i'}{\partial q_k}.$$

The third term vanishes because of (4). There remains

(6″) $$0 = \frac{d}{dt} \left(\frac{\partial a}{\partial u_i} \right) + \frac{\partial a}{\partial q_k} \frac{\partial q_k'}{\partial u_i} - \frac{\partial a}{\partial u_k} \frac{\partial q_i'}{\partial q_k}.$$

Also, for the other first integral,

(7″) $$0 = \frac{d}{dt} \left(\frac{\partial b}{\partial u_i} \right) + \frac{\partial b}{\partial q_k} \frac{\partial q_k'}{\partial u_i} - \frac{\partial b}{\partial u_k} \frac{\partial q_i'}{\partial q_k}.$$

Multiplying (6″) by $- \dfrac{\partial b}{\partial q_i}$ and (7″) by $\dfrac{\partial a}{\partial q_i}$, summing over the dummy suffixes and adding the two equations,

(8′) $$0 = \frac{d}{dt} \left(\frac{\partial b}{\partial u_i} \right) \cdot \frac{\partial a}{\partial q_i} - \frac{\partial b}{\partial q_i} \frac{d}{dt} \left(\frac{\partial a}{\partial u_i} \right)$$
$$+ \frac{\partial q_k'}{\partial u_i} \left(\frac{\partial b}{\partial q_k} \frac{\partial a}{\partial q_i} - \frac{\partial b}{\partial q_i} \frac{\partial a}{\partial q_k} \right) - \frac{\partial q_i'}{\partial q_k} \left(\frac{\partial b}{\partial u_k} \frac{\partial a}{\partial q_i} - \frac{\partial b}{\partial q_i} \frac{\partial a}{\partial u_k} \right).$$

The third term of this equation is zero because of the relation (5). If the suffixes i and k are interchanged in the fourth term, by adding (8) to (8') it follows that

$$\frac{\partial b}{\partial u_i} \frac{d}{dt}\left(\frac{\partial a}{\partial q_i}\right) - \frac{\partial a}{\partial u_i} \frac{d}{dt}\left(\frac{\partial b}{\partial q_i}\right) + \frac{\partial a}{\partial q_i} \frac{d}{dt}\left(\frac{\partial b}{\partial u_i}\right) - \frac{\partial b}{\partial q_i} \frac{d}{dt}\left(\frac{\partial a}{\partial u_i}\right) = 0$$

or

$$\frac{d}{dt}\left[\frac{\partial b}{\partial u_i} \frac{\partial a}{\partial q_i} - \frac{\partial a}{\partial u_i} \frac{\partial b}{\partial q_i}\right] = 0$$

or finally

$$\frac{\partial b}{\partial u_i} \frac{\partial a}{\partial q_i} - \frac{\partial b}{\partial q_i} \frac{\partial a}{\partial u_i} = \text{constant}$$

or

$$(b, a) = \text{constant}$$

where (b, a) is an expression which has become known as a " *Poisson bracket.* "

It is evident that

$$(b, a) = -(a, b) \text{ and that } (a, a) = 0.$$

Poisson concludes, " The analysis that we have just performed therefore leads us to this remarkable result—that if the values of the arbitrary constants on the integrals of the equations of motion of a system of bodies are expressed as functions of the independent variables (q_i) and the quantities (u_i), the combination of the partial differentials of these functions that is represented by (a, b) will always be a constant quantity. "

This proposition, which has become classical, evidently exhibits considerable aesthetic value. Its practical content is more limited. Indeed, Poisson's theorem seems to indicate that it is sufficient to know two first integrals of the equations of motion in order to be able to deduce a third from them ; by combining this with one of the first two, a fourth would be obtained, etc. . . . But if the bracket (a, b) is identically constant, or if it is a function of the integrals already known, this process contains nothing new.

2. The Lagrange-Poisson square brackets.

Poisson supposes that, to the right hand side of the Lagrange equations

$$(1) \qquad \frac{d}{dt}\left(\frac{\partial T}{\partial q_i'}\right) - \frac{\partial T}{\partial q_i} + \frac{\partial V}{\partial q_i} = 0$$

is added a term depending on the function of the perturbing forces, Ω. This yields

$$(2) \qquad \frac{d}{dt}\left(\frac{\partial T}{\partial q_i'}\right) - \frac{\partial T}{\partial q_i} + \frac{\partial V}{\partial q_i} = \frac{\partial \Omega}{\partial q_i} \qquad (i = 1, 2 \ldots k).$$

Since the variables u_i are always defined by $\dfrac{\partial T}{\partial q_i}$, Lagrange's equations take the form

$$(3) \qquad \frac{du_i}{dt} = \left(\frac{\partial R}{\partial q_i}\right) + \frac{\partial \Omega}{\partial q_i}.$$

The expression $\left(\dfrac{\partial R}{\partial q_i}\right)$ must always be interpreted as a derivative in which the q_i and the q_i' are chosen as the independent variables. This distinction does not apply to derivatives of Ω, which is supposed to be independent of the q_i'.

If the equations (1) are integrated completely, so that the solution contains $2k$ arbitrary constants a_s, it is desired to satisfy the equations (2) by varying the arbitrary constants. Since the number of these is twice the number of the equations (2), the $2k$ quantities a_s can be restricted by any k conditions that may be chosen.

Poisson supposed, " as in the theory of the Planets " that the differentials of the variables q_i kept the same form independently of whether the a_s were constants or not.

He goes on to express this condition by the k relations

$$(4) \qquad \delta q_i = \frac{\partial q_i}{\partial a_s}\, da_s = 0 \qquad \begin{array}{l}(i = 1, 2 \ldots k) \\ (s \text{ from } 1 \text{ to } 2k).\end{array}$$

On the other hand

$$du_i = \left(\frac{\partial R}{\partial q_i}\right) dt + \frac{\partial \Omega}{\partial q_i}\, dt.$$

But when the a_s are constants, the first two terms are equal because of (1). Accordingly, if only the a_s vary

$$(5) \qquad \delta u_i = \frac{\partial \Omega}{\partial q_i}\, dt.$$

The $2k$ equations (4) and (5), which are linear and of the first order in the quantities da_s, determine the differentials of the arbitrary constants.

Let $\qquad a_r = \text{funct } (t, q_1, \ldots q_k, u_1, \ldots u_k)$

be a first integral of the equations (1). Then put

$$da_r = \frac{\partial a_r}{\partial u_j}\, \frac{\partial \Omega}{\partial q_j}\, dt.$$

Now
$$\frac{\partial\Omega}{\partial q_j} = \frac{\partial\Omega}{\partial a_s} \cdot \frac{\partial a_s}{\partial q_j}.$$

Therefore

(6)
$$da_r = \frac{\partial a_r}{\partial u_j} \frac{\partial a_s}{\partial q_j} \frac{\partial\Omega}{\partial a_s} dt \qquad \binom{j \text{ from } 1 \text{ to } k}{s \text{ from } 1 \text{ to } 2k}.$$

But in
$$\Omega(q_j) = \Omega[a_s(q_j, u_j)]$$

the derivatives $\dfrac{\partial\Omega}{\partial u_j}$ are necessarily zero. Therefore

(7)
$$0 = \frac{\partial\Omega}{\partial u_j} = \frac{\partial\Omega}{\partial a_s} \frac{\partial a_s}{\partial u_j}$$

or, by summing the equations (7) after multiplying them by $\dfrac{\partial a_r}{\partial q_j}$,

(8)
$$0 = \frac{\partial\Omega}{\partial a_s} \frac{\partial a_s}{\partial u_j} \frac{\partial a_r}{\partial q_j}.$$

By subtracting (8) from (6)
$$da_r = \frac{\partial\Omega}{\partial a_s} dt \left(\frac{\partial a_r}{\partial u_j} \frac{\partial a_s}{\partial q_j} - \frac{\partial a_s}{\partial u_j} \frac{\partial a_r}{\partial q_j}\right)$$

and, using the definition of the Poisson brackets,
$$da_r = \frac{\partial\Omega}{\partial a_s} dt \,(a_r, a_s) \qquad \binom{r = 1, 2 \ldots 2k}{s \text{ from } 1 \text{ to } 2k}.$$

The brackets (a_r, a_s) are functions only of the arbitrary constants $a_1, \ldots, a_{2k}$. " It follows that, in the equations of mechanics, the first differentials of the arbitrary constants can be expressed by means of the partial differences of the function Ω, taken with respect to these quantities and multiplied by functions of these same quantities, which do not contain the time explicitly. This is the beautiful theorem that Mr. Lagrange and Mr. Laplace first discovered in connection with the differences of elliptic elements, and which Mr. Lagrange then extended to a system of any bodies subject to forces directed towards fixed or movable centres and whose intensities are functions of the distances of the bodies from these centres. "

Lagrange had arrived at the formulae
$$\frac{\partial\Omega}{\partial a_r} dt = [a_r, a_s]\, da_s$$

in which the square-bracket expression had the value
$$[a_r, a_s] = \frac{\partial q_j}{\partial a_r} \frac{\partial u_j}{\partial a_s} - \frac{\partial q_j}{\partial a_s} \frac{\partial u_j}{\partial a_r}.$$

CHAPTER SIX

ANALYTICAL DYNAMICS
IN THE SENSE OF HAMILTON AND JACOBI

1. HAMILTON OPTICS. ITS DOUBLE INTERPRETATION IN TERMS OF EMISSION AND WAVE PROPAGATION.

Hamilton's ideas on dynamics cannot be divorced from his ideas on optics. For this reason it is essential that we should, for a few moments, concern ourselves with the latter.

We shall follow the edition of Hamilton's works that has been published by the Royal Academy of Ireland. Apart from the papers which have been known and classical for some time, this edition, very fortunately, contains extracts from the numerous note-books which Hamilton kept and which had not been published before. No doubt the author, considering them minor works, had not wished to make them public—but they throw the work of this inspired Irishman into a new and very interesting light.[1]

In the first place we shall cite an article which appeared in the *Dublin University Review*[2] for October, 1833, called *On a general method of expressing the paths of light, and of the planets, by the coefficients of a characteristic function.*

At the time that Hamilton started his investigations in optics, neither the theory of waves nor the emission theory were generally accepted. Hamilton's geometric optics, which was essentially a new method of *formalising* the collection of results that had already been obtained, was capable of being interpreted in terms of wave propagation (in Huyghens' sense) and corpuscles (in the sense of the dynamical principle of least action). This was the essential merit of his theory,

---

[1] *The Mathematical Papers of Sir William Rowan Hamilton.* Vol. I, *Geometrical Optics* edited for the Royal Irish Academy by A. W. CONWAY and J. L. SYNGE (1931); Vol. II, *Dynamics* edited for the R. I. A. by A. W. CONWAY and J. McCONNEL (1940). Cambridge University Press.
[2] Pp. 795-826.

which can only seem more meritorious to our modern age in which a similar dualism has been established in theoretical physics.

A great admirer of Lagrange, Hamilton declared, in the article referred to above, that he was " struck by the imperfection of deductive mathematical optics. " He wished to give to optics, on the plane of formal theory, the same " beauty, power and harmony " with which Lagrange had been able to endow mechanics.

I repeat that it was certainly the formalism which concerned him. " Whether we adopt the Newtonian or the Huyghenian, or any other physical theory, for the explanation of the laws that regulate the lines of luminous or visual communication, we may regard these laws themselves, and the properties and relations of these linear paths of light, as an important separate study, and as constituting a separate science, called often *mathematical optics.* " [1]

Hamilton recalled the development which we have already studied, Fermat, Maupertuis, Euler, Lagrange. " But although the law of least action has thus attained a rank among the highest theorems of physics, yet its pretensions to a cosmological necessity, on the ground of economy in the universe, are now generally rejected. And the rejection appears just, for this, among other reasons, that the quantity pretended to be economised is in fact often lavishly expended. " This, for instance, is what is shown in the commonplace case of reflexion on a spherical mirror, where obviously if one of the rays issuing from a point is minimal, the other corresponds in fact to a maximal. We can therefore speak reasonably only of a *stationary* property of the action (or an extremal one, as understood in the calculus of variations).

" We cannot, therefore, suppose the economy of this quantity to have been designed in the divine idea of the universe : though a simplicity of some high kind may be believed to be included in that idea ".

Such are the rational motives which led Hamilton, at the same time as he retained the consecrated term *action*, to speak, in optics as in dynamics, of *stationary* or *varying* action, according to whether the extremities of the rays or trajectories are fixed by hypothesis or not.

I shall pass over the remarkable statement of the principles of the calculus of variations contained in the paper we are analysing, and come

---

[1] Incidentally HAMILTON did not hesitate to state his doctrine of scientific philosophy. Thus he distinguished a stage in which the *facts* are raised to *laws* by *induction* and *analysis*, and another in which the laws are used to obtain *consequences* by *deduction* and *synthesis*. This was formulated in the following remarkable passage.

" We must gather and group appearances, until the scientific imagination discerns their hidden laws, and unity arises from variety ; and then from unity we must rededuce variety, and force the discovered law to utter its revelations of the future. "

Better than a fine formula, this thesis is the expression of the method of work that HAMILTON always followed.

to the exact statement of the Hamiltonian principle of stationary action in optics :

" *The optical quantity called action, for any luminous path having* i *points of sudden bending by reflexion or refraction, and having therefore* i + 1 *separate branches, is the sum of* i + 1 *separate integrals,*

$$\text{Action} = V = \sum \int dV_i = V_1 + V_2 + \ldots + V_{i+1}$$

of which each is determined by an equation of the form

$$V_i = \int dV_i = \int v_i \sqrt{dx_i^2 + dy_i^2 + dz_i^2}$$

the coefficient v_i *of the element of the path, in the* i[th] *medium, depending, in the most general case, on the optical properties of that medium, and on the position, direction and colour of the element, according to rules discovered by experience.* (For example, if the *i[th]* medium is an ordinary medium, v_i is its refractive index.) *This quantity* V *is stationary in the propagation of the light.* "

The *law of varying action* is a generalisation of the stationary law in which the ends of an optical *(luminous)* path are allowed to vary. The conditions at the limits thus make necessary the intervention of finite difference equations of the type $\Delta V = \lambda u = 0$ on each surface $u = 0$ of reflection or refraction. In passing, Hamilton indicated that the remarkable permanence of what he called the *components of normal slowness* (inversely proportional to those of the velocity of wave propagation in Huyghens' sense) had been suggested to him by the observation that the characteristic function V is such that the wave surfaces satisfy the equation

$$V = \text{Constant}.$$

The components of *normal slowness* are nothing else than the partial derivatives $\dfrac{\partial V}{\partial x}, \dfrac{\partial V}{\partial y}, \dfrac{\partial V}{\partial z}.$ Thus is rediscovered the theorem—then disputed—of Huyghens according to which the rays of every homogeneous system, starting from a single point or normal to a surface, remain normal to a family of surfaces after they have been subjected to any number of reflections or refractions.

We learn a little more about Hamilton's procedures in optics by following the *Third Supplement to an Essay on the Theory of Systems of Rays.*

First, calling the initial and final coordinates of a ray (x', y', z') and (x, y, z), Hamilton writes

(A) $\delta V = \dfrac{\partial v}{\partial \alpha}\,\delta x + \dfrac{\partial v}{\partial \beta}\,\delta y + \dfrac{\partial v}{\partial \gamma}\,\delta z - \dfrac{\partial v'}{\partial \alpha'}\,\delta x' - \dfrac{\partial v'}{\partial \beta'}\,\delta y' - \dfrac{\partial v'}{\partial \gamma'}\,\delta z'$

(where α, β, γ and α', β', γ' are the direction cosines of the ray at its end and beginning respectively) or

$$\frac{\partial V}{\partial x} = \frac{\partial v}{\partial \alpha}, \dots, -\frac{\partial V}{\partial z'} = \frac{\partial v'}{\partial \gamma'}.$$

On the other hand, the conditions for an extremum require that

(B) $\quad \dfrac{\partial v}{\partial x} ds = d \dfrac{\partial v}{\partial \alpha}, \dots \quad \dfrac{\partial v}{\partial z} ds = d \dfrac{\partial v}{\partial \alpha} \quad \left(ds = \sqrt{dx^2 + dy^2 + dz^2} \right).$

The function v is homogeneous in α, β and γ and also depends on the frequency of the light. Hamilton expresses the latter fact by the introduction of a *chromatic index* χ.

Thus he arrives at the following two equations in the first partial derivatives

(C) $\quad \begin{cases} \Omega \left(\dfrac{\partial V}{\partial x}, \dfrac{\partial V}{\partial y}, x, y, z, \chi \right) = 0 \\[2mm] \Omega' \left(\dfrac{-\partial V}{\partial x'}, \dfrac{-\partial V}{\partial y'}, \dfrac{-\partial V}{\partial z'}, x', y', z', \chi \right) = 0. \end{cases}$

The similarity of the form of these equations with that of the equations of dynamics is evident—V corresponds to the action integral (in the Euler-Lagrange sense); the equation (C) corresponds to the equation of kinetic energy and χ to a certain function of the total energy. Moreover, Hamilton's optical equations can be easily written in the canonical form that he himself gave to the equations of dynamics. It is sufficient to denote the *components of normal slowness* (or the partial derivatives of V) by σ, τ and ν to write

$$\frac{dx}{dV} = \frac{\partial \Omega}{\partial \sigma} \dots \qquad \frac{d\sigma}{dV} = -\frac{\partial \Omega}{\partial x} \text{ etc.}$$

As we have already indicated, Hamilton interpreted the action V, in the language of the wave theory, as the time necessary for a wave of frequency χ which starts from the point (x', y', z') to travel to the point (x, y, z).

If the wave velocity *(ondulatory velocity)* of propagation along the corresponding radius is called u, the relation

(D) $$u = \frac{1}{v}$$

or, more generally,

$$u = \frac{1}{vf(\chi)}$$

allows V to be written as

$$V = \int_{x', y', z'}^{x, y, z} \frac{ds}{u\,(\alpha, \beta, \gamma, x, y, z, \chi)}.$$

Since the rays are identical in the two theories, to the extrema

$$\delta V = \delta \int v ds = 0$$

of the emission theory there corresponds the extremal

$$\delta \int \frac{ds}{u} = 0$$

which is *Fermat's principle*.

There is, here, the germ of the transcription which Schrödinger was to turn to good use in dynamics, in generalising equation (*D*) by the introduction of a *group velocity* (in Rayleigh's sense) identified with *v*.

2. The dynamical law of varying action in Hamilton's sense.

Historically, Hamilton's first work in dynamics is contained in a manuscript dated 1833 and called *The Problem of Three Bodies by my Characteristic Function.*[1] He treated the problem of the Sun, Jupiter and Saturn and introduced, from the beginning, the characteristic function.

$$V = \int_0^t 2T dt.$$

(The living force accumulated from the origin of time to the time *t*.) Hamilton showed that this function must satisfy two equations in the partial derivatives of the first order. He then compared an approximate solution of this problem with that obtained by Laplace, studied the perturbations, determined the characteristic function of elliptical motion and established the equation

$$\frac{\partial V}{\partial h} = t \qquad (h, \text{ constant of living forces}).$$

He then proved that the two equations connecting the partial derivatives of V have a common solution, and directed his attention to the determination of this solution by successive approximations.

Therefore this paper already contained essential results. We shall not, however, further discuss it, for Hamilton undertook the codification of these investigations in two fundamental papers which were published in 1834. We propose to analyse these.

[1] Note-book 29.

In his statement of the intentions of his *First Essay on a General Method in Dynamics* [1] Hamilton recalled that the determination of the motion of a system of free particles, subject only to their mutual attraction or repulsion, depended on the integration of a system of $3(n-1)$ ordinary differential equations of the second order or, by a transformation due to Lagrange, on a system of $6(n-1)$ ordinary differential equations of the first order.

Hamilton reduced this problem to the " search and differentiation of a single function " which satisfied two equations of the first order in the partial derivatives.

From this transference of the difficulties, even if it is thought that no practical advantage results, " an intellectual pleasure may result from the reduction of the most complex and, probably, of all researches respecting the forces and motions of bodies, to the study of one characteristic function, the unfolding of one central relation.... " And Hamilton adds, " this dynamical principle is only another form of that idea which has already been applied to optics in the *Theory of systems of rays*.... "

Starting from the classical equation

$$(1) \qquad \sum m \, (x''\delta x + y''\delta y + z''\delta z) = \delta U \quad \text{with} \quad U = \sum mm'f(r),$$

in which U is the function of forces, Hamilton denotes the living force of the system by $2T = \sum m \, (x'^2 + y'^2 + z'^2)$, and writes the law of living forces in the form

$$T = U + H.$$

The quantity H—which it has become customary to call the Hamiltonian of the system—is independent of the time in a *given* motion of the system. But when the initial conditions are varied, H varies correspondingly according to the equation

$$\delta T = \delta U + \delta H.$$

On multiplying by dt, integrating from 0 to t, using equation (1) and the equation that defines the kinetic energy, Hamilton obtains

$$\int_0^t \sum m \, (dx\delta x' + dy\delta y' + dz\delta z') = \int_0^t \sum m \, (dx'\delta x + dy'\delta y + dz'\delta z) + \int_0^t \delta H dt.$$

Then, by means of the calculus of variations

$$(A) \qquad \delta V = \sum m \, (x'\delta x + y'\delta y + z'\delta z) - \sum m \, (a'\delta a + b'\delta b + c'\delta c) + t\delta H$$

---

[1] *Phil. Trans. Roy. Soc.* (1834), II, p. 247.

where (x, y, z) and (a, b, c) are the final and initial coordinates of the points of the system, and V is the function

(B) $$V = \int_0^t \sum m \left(x' dx + y' dy + z' dz \right) = \int_0^t 2T dt.$$

By considering the *characteristic function* V as a function of the initial and final coordinates and of the quantity H, and by starting from the fundamental equation (A), the following system of equations is obtained.

(C) $$\begin{cases} \dfrac{\partial V}{\partial x_1} = m_1 x_1' \ldots & \dfrac{\partial V}{\partial x_n} = m_n x_n' \\[2mm] \dfrac{\partial V}{\partial y_1} = m_1 y_1' \ldots & \dfrac{\partial V}{\partial y_n} = m_n y_n' \\[2mm] \dfrac{\partial V}{\partial z_1} = m_1 z_1' \ldots & \dfrac{\partial V}{\partial z_n} = m_n z_n' \end{cases}$$

(D) $$\begin{cases} \dfrac{\partial V}{\partial a_1} = - m_1 a_1' \ldots & \dfrac{\partial V}{\partial a_n} = - m_n a_n' \\[2mm] \dfrac{\partial V}{\partial b_1} = - m_1 b_1' \ldots & \dfrac{\partial V}{\partial b_n} = - m_n b_n' \\[2mm] \dfrac{\partial V}{\partial c_1} = - m_1 c_1' \ldots & \dfrac{\partial V}{\partial c_n} = - m_n c_n' \end{cases}$$

(E) $$\frac{\partial V}{\partial H} = t.$$

The elimination of H allows $3n$ relations to be obtained between the $3n$ variable coordinates, the time and the $6n$ initial conditions. Thus the problem is reduced to the determination and the differentiation of a single function V.

The fundamental equation (A) expresses the *law of varying action*.

Hamilton remarked that Lagrange has already obtained an extremal law, called that of *least action*, which was equivalent to (A) ; but that Lagrange only used it to form the ordinary equations of the second order, which can be obtained by other means. Further, Lagrange, Laplace and Poisson had, for this reason, considered the principle of least action as being of only small practical importance in dynamics.

The equation of living forces, combined with the systems (C) and (D), takes the form

(F) $$\frac{1}{2} \left\{ \sum \frac{1}{m} \left\{ \left(\frac{\partial V}{\partial x} \right)^2 + \left(\frac{\partial V}{\partial y} \right)^2 + \left(\frac{\partial V}{\partial z} \right)^2 \right\} \right\} = U + H$$

(final coordinates)

(G) $$\frac{1}{2}\left\{\sum\frac{1}{m}\left\{\left(\frac{\partial V}{\partial a}\right)^2+\left(\frac{\partial V}{\partial b}\right)^2+\left(\frac{\partial V}{\partial c}\right)^2\right\}\right\} = U_0 + H$$

(initial coordinates).

These two equations must be satisfied identically by V.

Hamilton verifies that, by differentiating equations (C) and taking account of (F), there are retrieved the ordinary equations of motion

$$m_1 x_1'' = \frac{\partial U}{\partial x_1} \text{ etc.} \dots$$

If the rectangular coordinates are replaced by the Lagrange variables defined in the following way,

$$\begin{cases} \eta_1, \eta_2, \dots & \text{instead of the final coordinates} \\ e_1, e_2, \dots & \text{instead of the initial coordinates} \end{cases}$$

the fundamental equation (A) takes the form

$$\delta V = \sum\frac{\partial T}{\partial \eta'}\,\delta\eta - \sum\frac{\partial T}{\partial e'}\,\delta e + t\delta H$$

whence

$$\frac{\partial V}{\partial \eta_i} = \frac{\partial T}{\partial \eta_i'},\dots\frac{\partial V}{\partial e_i} = -\frac{\partial T}{\partial e_i'}.$$

From this it is easy to deduce Lagrange's equations

$$\frac{d}{dt}\left(\frac{\partial T}{\partial \eta_i'}\right) - \frac{\partial T}{\partial \eta_i} = \frac{\partial U}{\partial \eta_i}.$$

Hamilton then generalised equation (A) to the case of motion described by $3n + k$ Lagrange variables which are connected by k finite equations. Then he determined the function V for binary systems with $U = m_1 m_2 f(r)$ and turned his attention to the extension of this analysis to systems of more than two particles by decomposing V into a principal part and a perturbing part.

At the very end of his first *Essay* Hamilton introduced the function S that he was later to call the *principal function*. This was defined in terms of the characteristic function by the equation

$$V = tH + S.$$

This function,

$$S = \int_0^t (T + U)\,dt\,,$$

leads, by the calculs of variations, to

$$\delta S = \sum m\,(x'\delta x + y'\delta y + z'\delta z - a'\delta a - b'\delta b - c'\delta c) - H\delta t$$

which constitutes another form of the law of varying action.

From this definition follows the system of equations

$$\frac{\partial S}{\partial x_1} = m_1 x_1' \text{ etc.} \dots \frac{\partial S}{\partial a_1} = - m_1 a_1' \text{ etc.} \dots$$

as well as the equations in the first order partial differentials, which S must satisfy identically,

$$\frac{\partial S}{\partial t} + \sum \frac{1}{2m} \left\{ \left(\frac{\partial S}{\partial x}\right)^2 + \left(\frac{\partial S}{\partial y}\right)^2 + \left(\frac{\partial S}{\partial z}\right)^2 \right\} = U$$

(final coordinates)

$$\frac{\partial S}{\partial t} + \sum \frac{1}{2m} \left\{ \left(\frac{\partial S}{\partial a}\right)^2 + \left(\frac{\partial S}{\partial b}\right)^2 + \left(\frac{\partial S}{\partial c}\right)^2 \right\} = U_0$$

(initial coordinates).

In the *Second Essay on a General Method in Dynamics*,[1] Hamilton systematically developed the use of the *principal function* S and also established the *canonical equations* of motion.

Using Lagrange coordinates, he first wrote the living force equation in the form

$$2T = \sum \eta' \frac{\partial T}{\partial \eta'}.$$

This supposes that the Hamiltonian H does not depend on the time. From this the variation of the living force is obtained in the form

$$\delta T = \sum \left(\eta' \delta \frac{\partial T}{\partial \eta'} - \frac{\partial T}{\partial \eta} \delta \eta \right).$$

By considering T as a function of the quantities

$$\bar{\omega}_i = \frac{\partial T}{\partial \eta_i'}$$

together with the variables η_i, the equations of Lagrange may be written in the form

$$\frac{d\bar{\omega}_i}{dt} = \frac{\partial}{\partial \eta_i} (U - T).$$

Alternatively, by the introduction of the Hamiltonian H which is considered as a function of $(\bar{\omega}_i, \eta_i)$, the equations of dynamics may be written in their canonical form

[1] *Phil. Trans. Roy. Soc.*, 1835, Part I, p. 95.

$$\begin{cases} \dfrac{d\overline{\omega}_i}{dt} = -\dfrac{\partial H}{\partial \eta_i} \\[2mm] \dfrac{d\eta_i}{dt} = -\dfrac{\partial H}{\partial \overline{\omega}_i} \end{cases}$$

This is a system of $6n$ equations of the first order, if n particles are concerned, and the initial conditions are (e_i, p_i).

These equations are integrated by a knowledge of the principal function written in the new form

$$S = \int_0^t \left(\sum \overline{\omega} \frac{\partial H}{\partial \overline{\omega}} - H \right) dt = \int_0^t S' dt$$

whose variation reduces to the especially simple form

$$\delta S = \sum (\overline{\omega} \delta \eta - p \delta e)$$

in which p and e are the initial values of $\overline{\omega}$ and η. From this variation, the system

$$\overline{\omega}_i = \frac{\partial S}{\partial \eta_i} \qquad p_i = \frac{\partial S}{\partial e_i}$$

follows.

The variation of the integral S enjoys the double property of giving the equations of motion (in the form of the Lagrange equations) when the limits are fixed, and of giving the integrals of the canonical equations when the limits are variable.

Hamilton then turned his attention to the approximate evaluation of the function S. This was done by separating it into a principal part and a perturbing part, according to $S = S_1 + S_2$.

3. The significance of the Hamiltonian dynamics.

In a paper published by the *British Association* in 1834 and called *On the application to dynamics of a general mathematical method previously applied to optics* Hamilton explained the essential characteristics of his dynamical work. We quote the following passage as being particularly characteristic—it is written in the impersonal form that was sometimes demanded by custom.

" Professor Hamilton's solution of this long celebrated problem contains, indeed, one unknown function, namely, the *principal function S*, to the search and the study of which he has reduced mathematical dynamics. This function must not be confounded with that so beautifully conceived by Lagrange for the more simple and elegant expression of the known differential equations. Lagrange's function *states*,

Mr. Hamilton's function would *solve* the problem. . . . To assist in pursuing this new track and in discovering the form of this new function, Mr. Hamilton remarks that it must satisfy the following partial differential equation of the first order and second degree, . . . which may be transformed

$$S = S_1 + \int_0^t \left(U - \frac{\partial S_1}{\partial t} - \sum \frac{1}{2m} \left\{ \left(\frac{\partial S_1}{\partial x}\right)^2 + \left(\frac{\partial S_1}{\partial y}\right)^2 + \left(\frac{\partial S_1}{\partial z}\right)^2 \right\} \right) dt$$
$$+ \int_0^t \sum \frac{1}{2m} \left\{ \left(\frac{\partial S}{\partial x} - \frac{\partial S_1}{\partial x}\right)^2 + \left(\frac{\partial S}{\partial y} - \frac{\partial S_1}{\partial y}\right)^2 + \left(\frac{\partial S}{\partial z} - \frac{\partial S_1}{\partial z}\right)^2 \right\} dt$$

S_1 being any arbitrary function of the quantities t, m, x, y, z, a, b, c, supposed only to vanish (like S) at the origin of time. If this arbitrary function S_1 be so chosen as to be an approximate value of the sought function S (and it is always easy so to choose it), then the two definite integrals in the formula above are small, but the second is in general much smaller than the first; it may, therefore, be neglected in passing to a second approximation, and in calculating the first definite integral the following approximate equations may be used,

$$\frac{\partial S_1}{\partial a} = -m_1 a' \qquad \frac{\partial S_1}{\partial b} = -m_1 b' \qquad \frac{\partial S_1}{\partial c} = -m_1 c'.$$

" In this manner a first approximation may be successively and indefinitely corrected. . . . "

Beginning in 1836, Hamilton developed a method of successive approximations for the calculation of S. This he called the *calculus of principal relations*.

In short, jealous of the formal perfection which Lagrange had been able to give to dynamics, and which optics lacked, Hamilton undertook the rationalisation of geometrical optics. He did this by developing a formal theory which was free of all metaphysics and which, moreover, succeeded in accounting for all the experimental facts. Nevertheless, it did not decide the debate between the protagonists of the corpuscular and the wave hypotheses.

Then, returning to dynamics, Hamilton presented the law of *varying action* in a form very like that which he had discovered in optics. Thus he reduced the general problem of dynamics (for conservative systems) to the solution of two simultaneous equations in partial derivatives, or to the determination of a single function satisfying these two equations.

Hamilton's principal function " would solve " the problems of dynamics. But, as a general rule, it could only be hoped to find this function by successive approximations. It is for this reason that Hamilton devoted so much effort to that task.

Hamilton's guiding idea is continuous from his optical work to his work in dynamics— in this fact lies his greatness and his power. Here was a synthesis that Louis de Broglie was to rediscover and turn to his own account; a synthesis that was, it appears, to be Schrödinger's direct inspiration. Thus the furrow that Hamilton ploughed was to bear fruit outside the domain of classical mechanics. This power of extension of a formalism which, in the classical field, was equivalent to d'Alembert's principle deserves emphasis— never have the formal tools of analytical mechanics been more used by physicists than since the time when the classical structure was shaken by the intervention of quanta.

4. JACOBI'S CRITICISM.

In Volume XVII of the *Journal de Crelle*, Jacobi showed that every *complete* integral—in Lagrange's sense—of Hamilton's equation in the partial derivatives, is sufficient to determine the trajectories and the law of motion even when the Hamiltonian depends explicitly on the time. Jacobi simplified and also extended Hamilton's theory in a form that has become classical.

Jacobi declared " Hamilton has thrown his fine discovery into a false light " *(in ein falsches Licht)* and " limited and artificially complicated the question." Indeed, he did not, at first, rigorously prove that the function V (or S) *simultaneously* satisfied the two equations in the partial derivatives that he wrote. . . . The introduction of the initial and final coordinates led to the complication of the problem of integration.

Finally, it is easy to extend Hamilton's analysis to include forces which depend on the time.

We shall follow here the well-known *Vorlesungen über Dynamik* which Jacobi gave between 1842 and 1843 at Koenigsberg.[1]

First, Jacobi showed that the second equation in the partial derivatives which Hamilton obtained (that which contains the initial conditions) is quite superfluous *(vollkommen überflüssig)*.

He wrote the other Hamiltonian equation (involving the actual coordinates) in the form

$$(1) \qquad \frac{\partial V}{\partial t} + \Psi = 0.[2]$$

In this way Jacobi arrived at the following theorem.

---

[1] These lectures were edited by CLEBSCH and published in 1866 (Berlin, Reimer).
[2] This is concerned with the *principal function*, HAMILTON's S.

" Let there be a system of n ordinary differential equations between the $n + 1$ variables t, x_1, x_2, ... x_n ; let x_1^0, x_2^0, ... x_n^0 be the initial values of the variables x_1, x_2, ... x_n at the time. Finally, let

$$(A) \qquad \begin{cases} x_1 = f_1 (t, \tau, x_1^0, x_2^0, \ldots x_n^0) \\ \ldots \ldots \\ x_n = f_n (t, \tau, x_1^0, x_2^0, \ldots x_n^0) \end{cases}$$

be a solution of this system. By interchanging the variables t, x_1, x_2, ... x_n and τ, x_1^0, x_2^0, ... x_n^0 a second solution is obtained in the form

$$(B) \qquad \begin{cases} x_1^0 = f_1 (\tau, t, x_1, x_2, \ldots x_n) \\ x_n^0 = f_n (\tau, t, x_1, x_2, \ldots x_n) \end{cases}$$

which dispenses with all processes of elimination. " [1]

What happens to the function V in this interchange ? Since the equations of the dynamical problem depend on the *supposition* that equation (1) is *integrable* in the form

$$\begin{cases} q_1 = \chi_1 (t, \alpha_1, \alpha_2, \ldots \alpha_{2n}) \qquad p_1 = \overline{\omega}_1 (t, \alpha_1, \alpha_2, \ldots \alpha_{2n}) \\ \vdots \\ q_n = \chi_n (t, \alpha_1, \alpha_2, \ldots \alpha_{2n}) \qquad p_n = \overline{\omega}_n (t, \alpha_1, \alpha_2, \ldots \alpha_{2n}) \end{cases}$$

then, at the initial instant, it is possible to write

$$q_i^0 = \chi_i (\tau, \alpha_1, \alpha_2, \ldots \alpha_{2n}) \qquad p_i^0 = \overline{\omega}_1 (\tau, \alpha_1, \alpha_2, \ldots \alpha_{2n})$$

where i runs from 1 to n. Now

$$V = \int_\tau^t \varphi \, dt$$

where φ is a function of the variables t, q_1, q_2, ... q_n, p_1, p_2, ... p_n.
Therefore

$$(2) \qquad V = \Phi (t, \alpha_1, \alpha_2, \ldots \alpha_{2n}) - \Phi (\tau, \alpha_1, \alpha_2, \ldots \alpha_{2n}).$$

The function V will be a *complete integral* of the equation (1) when the $2n$ constants α are eliminated by means of the $2n$ equations which express the q_i and q_i^0 in terms of them. But these $2n$ equations are reduced to n by the exchange of the variables (t, q_i) and (τ, q_i^0). Therefore each of the constants α, considered as a function of the q_i and q_i^0, must remain invariant in this exchange. Therefore, by (2), the exchange transforms V into $- V$.

To return to the only case treated by Hamilton, suppose that φ does not contain the time explicitly. This is tantamount to assuming

---

[1] For the demonstration, see *Vorlesungen über Dynamik*, p. 153.

that the function of the forces, U, does not depend on the time nor the Hamiltonian $H = \psi = T - U$.

In the canonical system corresponding to the given problem,

$$\frac{dt}{1} = \frac{dq_1}{\dfrac{\partial \psi}{\partial p_1}} = \ldots = \frac{dq_n}{\dfrac{\partial \psi}{\partial p_n}} = \frac{dp_1}{-\dfrac{\partial \psi}{\partial q_1}} = \ldots = \frac{dp_n}{-\dfrac{\partial \psi}{\partial q_n}}$$

t is immediately eliminated. Then, *after integration*, all the variables are expressed as functions of t by

$$dt = \frac{dq_1}{\dfrac{\partial \psi}{\partial p_1}} \quad \text{or} \quad t - \tau = \int_{q_1^0}^{q_1} \frac{dq_1}{\dfrac{\partial \psi}{\partial p_1}}.$$

Finally, all the variables of the problem (q_i) are expressed as functions of the variable

$$\theta = t - \tau.$$

In the exchange of $t, q_1, q_2, \ldots q_n$ for $\tau, q_1^0, q_2^0, \ldots q_n^0$, V transforms into $-V$, θ into $-\theta$; $\dfrac{\partial V}{\partial \theta}$ does not change. Moreover,

$$\frac{\partial V}{\partial \theta} = \frac{\partial V}{\partial t} = -\frac{\partial V}{\partial \tau}.$$

Therefore the equation

$$\frac{\partial V}{\partial t} + \psi = \frac{\partial V}{\partial \theta} + \psi = 0$$

transforms into

$$\frac{\partial V}{\partial \theta} + \psi_0 = -\frac{\partial V}{\partial \tau} + \psi_0 = 0.$$

Consequently, Hamilton's second equation is deduced from the first by the interchange of $t, q_1, q_2, \ldots q_n$ and $\tau, q_1^0, q_2^0, \ldots q_n^0$, which establishes its superfluous character. Q. E. D.

5. JACOBI'S FUNDAMENTAL THEOREM.

" *Let*

(1) $$\frac{\partial V}{\partial t} + \psi = 0$$

be an arbitrary equation in the partial derivatives of the first order which does not contain V *explicitly, so that* ψ *is any function of the variables*

$t,\ q_1,\ q_2,\ \cdots q_n,\ p_1,\ p_2,\ \cdots p_n \left(where\ p_i = \dfrac{\partial V}{\partial q_i} \right).$ *If a complete integral* V *of equation* (1) *is known—that is, if an integral which, apart from the arbitrary constant which can be incorporated in* V, *depends on the* n *arbitrary constants* $\alpha_1,\ \alpha_2,\ \cdots \alpha_n$ — *and if the relations*

(2) $$\frac{\partial V}{\partial \alpha_1} = \beta_1 \qquad \frac{\partial V}{\partial \alpha_2} = \beta_2 \quad \cdots \quad \frac{\partial V}{\partial \alpha_n} = \beta_n$$

are assumed, where the β *are new arbitrary constants, then these equations, together with the equalities*

(3) $$\frac{\partial V}{\partial q_1} = p_1 \qquad \frac{\partial V}{\partial q_2} = p_2 \quad \cdots \quad \frac{\partial V}{\partial q_n} = q_n$$

solve the following system of differential equations.

(4) $$\frac{dq_i}{dt} = \frac{\partial \psi}{\partial q_i} \qquad \frac{dp_i}{dt} = -\frac{\partial \psi}{\partial q_i} \qquad (i = 1, 2, \ldots n).\ "\ [1]$$

To prove this theorem, Jacobi shows that when a complete integral which is supposed to be known is substituted for V in equation (1), the first term becomes a function of t, $q_1, q_2, \ldots q_n$, $\alpha_1, \alpha_2, \ldots \alpha_n$, which vanishes identically. Consequently its partial derivatives vanish.

In order to retrieve the first half of the equations (4) by starting from the equations (2), Jacobi differentiates the latter with respect to the time and obtains

(5) $$0 = \frac{\partial^2 V}{\partial \alpha_i \partial t} + \frac{\partial^2 V}{\partial \alpha_i \partial q_1} \frac{dq_1}{dt} + \ldots + \frac{\partial^2 V}{\partial \alpha_i \partial q_n} \frac{dq_n}{dt} \qquad (i = 1, 2, \ldots n).$$

If this system, which is linear in the quantities $\dfrac{dq_i}{dt}$, is supposed to be solved, the solution must be identified with the derivatives $\dfrac{\partial \psi}{\partial p_i}$ in order to retrieve the first of the equations (4). For this purpose, Jacobi differentiates equation (1) with respect to the α_i (which are, evidently, only contained in the p_i) and obtains

(6) $$0 = \frac{\partial^2 V}{\partial t \partial \alpha_i} + \frac{\partial \psi}{\partial p_1} \frac{\partial p_1}{\partial \alpha_i} + \ldots + \frac{\partial \psi}{\partial p_n} \frac{\partial p_n}{\partial \alpha_i} \qquad (i = 1, 2, \ldots n)$$

or again

(7) $$0 = \frac{\partial^2 V}{\partial t \partial \alpha_i} + \frac{\partial \psi}{\partial p_1} \frac{\partial^2 V}{\partial q_1 \partial \alpha_i} + \ldots + \frac{\partial \psi}{\partial p_n} \frac{\partial^2 V}{\partial q_n \partial \alpha_i} \qquad (i = 1, 2, \ldots n).$$

[1] *Vorlesungen über Dynamik*, p. 157.

The identity of the two systems (5) and (7), and consequently of the derivatives $\dfrac{dq_i}{dt}$ and $\dfrac{\partial \psi}{\partial p_i}$ is established.

In order to establish the second half of the set of equations (4), Jacobi appeals to equations (3),

$$\frac{\partial V}{\partial q_i} = p_i \, ,$$

and differentiates them with respect to the time to obtain

$$\frac{dp_i}{dt} = \frac{\partial^2 V}{\partial q_i \partial t} + \frac{\partial^2 V}{\partial q_i \partial q_1} \frac{dq_1}{dt} + \cdots + \frac{\partial^2 V}{\partial q_i \partial q_n} \frac{dq_n}{dt}$$

or

$$\frac{dp_i}{dt} = \frac{\partial^2 V}{\partial q_i \partial t} + \frac{\partial p_1}{\partial q_i} \frac{dq_1}{dt} + \cdots + \frac{\partial p_n}{\partial q_i} \frac{dq_n}{dt} \qquad (i = 1, 2, \ldots n).$$

Or again, by using the equalities already obtained,

$$\frac{dq_i}{dt} = \frac{\partial \psi}{\partial p_i}$$

(8) $$\frac{dp_i}{dt} = \frac{\partial^2 V}{\partial q_i \partial t} + \frac{\partial p_1}{\partial q_i} \frac{\partial \psi}{\partial p_1} + \cdots + \frac{\partial p_n}{\partial q_i} \frac{\partial \psi}{\partial p_n} \qquad (i = 1, 2, \ldots n).$$

Now, by differentiating the equation (1) with respect to q_i the system of equations

(9) $$0 = \frac{\partial^2 V}{\partial t \partial q_i} + \frac{\partial \psi}{\partial p_1} \frac{\partial p_1}{\partial q_i} + \cdots + \frac{\partial \psi}{\partial p_n} \frac{\partial p_n}{\partial q_i} + \frac{\partial \psi}{\partial q_i} \qquad (i = 1, 2, \ldots n)$$

is obtained.

Comparing equations (8) and (9), it is seen that

$$\frac{dp_i}{dt} = - \frac{\partial \psi}{\partial q_i}$$

which are the second half of the set of equations (4).

But, strictly speaking, the system which is linear in the derivatives $\dfrac{dq_i}{dt}$ should have one and only one solution. Therefore the determinant of the coefficients of $\dfrac{dq_i}{dt}$ in the system (5),

$$R = \frac{D\left(\dfrac{\partial V}{\partial \alpha_1}, \, \ldots \dfrac{\partial V}{\partial \alpha_n}\right)}{D\left(q_1, q_2, \ldots q_n\right)} = \frac{D\left(\dfrac{\partial V}{\partial q_1}, \, \ldots \dfrac{\partial V}{\partial q_n}\right)}{D\left(\alpha_1, \ldots \alpha_n\right)}$$

must be different from zero. Otherwise, there would exist a relation

$$F\left(\frac{\partial V}{\partial q_1}, \ldots, \frac{\partial V}{\partial q_n}, q_1, \ldots q_n, t\right) = 0$$

not containing the α_i. But this is not possible if V is truly a *complete integral* depending effectively on n arbitrary constants. The theorem is therefore established.

6. The canonical equations and Jacobi's multiplier.

Jacobi wrote the equations of motion in the canonical form

$$\frac{dq_i}{dt} = \frac{\partial T}{\partial p_i} \qquad \frac{dp_i}{dt} = -\frac{\partial T}{\partial q_i} + Q_i$$

with

$$Q_i = \sum_k \left(X_k \frac{\partial x_k}{\partial q_i} + Y_k \frac{\partial y_k}{\partial q_i} + Z_k \frac{\partial z_k}{\partial q_i}\right).$$

These equations are more general than those of Hamilton, for they do not necessarily presume the existence of a function of the forces. These equations can also be put in the form

$$\frac{dt}{1} = \frac{dq_1}{\dfrac{\partial T}{\partial p_1}} = \ldots = \frac{dq_n}{\dfrac{\partial T}{\partial p_n}} = \frac{dp_1}{-\dfrac{\partial T}{\partial q_1} + Q_1} = \ldots = \frac{dp_n}{-\dfrac{\partial T}{\partial q_n} + Q_n}.$$

Jacobi's multiplier M (a generalisation of the *integrating factor* in Euler's sense) is defined by the relation

$$0 = \frac{d \log M}{dt} + \sum_k \frac{\partial \left(\dfrac{\partial T}{\partial p_k}\right)}{\partial p_k} + \sum_k \frac{\partial \left(-\dfrac{\partial T}{\partial q_k} + Q_k\right)}{\partial p_k}.$$

If the forces only depend on the coordinates, and not on the velocities, then

$$\frac{\partial Q_k}{\partial p_k} = 0,$$

identically.
Then

$$-\frac{d \log M}{dt} = \sum_k \left(\frac{\partial^2 T}{\partial p_k \partial q_k} - \frac{\partial^2 T}{\partial q_k \partial p_k}\right) = 0.$$

Therefore

$$M = \text{const.}$$

Evidently M can be chosen equal to unity, so that the multiplier has the same value as if the system were completely free.[1]

With the reservation that the forces do not depend on the velocities, but without it being necessary that there should exist a function of forces, unity is therefore, in Jacobi's sense, the multiplier of the canonical equations of the general problem of dynamics.

7. Geometrisation of the principle of least action.

Euler, Lagrange and Hamilton gave the principle of least action a precise mathematical significance in the field of the dynamics of a particle and of systems. The task of *geometrising* the principle remained, and was carried out by Jacobi.

The action integral is

$$\int \sum m_i v_i ds_i = \int \sum m_i \frac{ds_i^2}{dt}.$$

The energy equation is

$$\frac{1}{2} \sum m_i v_i^2 = U + h$$

and may be written in the form

$$\sum m_i \left(\frac{ds_i}{dt}\right)^2 = 2(U + h).$$

Whence, by the elimination of the time

$$\frac{1}{dt} = \sqrt{\frac{2(U + h)}{\sum m_i ds_i^2}}$$

Jacobi obtained the expression

$$\int \sum m_i v_i ds_i = \int \sqrt{2(U + h) \sum m_i ds_i^2}$$

for the action integral.

Whence the expression of the principle of least action becomes

$$\delta \int \sqrt{2(U + h) \sum m_i ds_i^2} = 0 \,.[2]$$

Jacobi declared that " in this true form " it is difficult " to assign a metaphysical cause to this principle. " Strictly speaking, Hamilton and, before him, Lagrange had rid the principle of least

---

[1] *Vorlesungen über Dynamik*, p. 141.
[2] *Ibid.*, p. 44.

action of all metaphysics. But they had not geometrised it. This geometrisation, which is obtained by considering trajectories which correspond to the same *total energy*, and which explains the part played by this principle in many physical theories, was to be the concern, after Jacobi, of Liouville (1856), Lipschitz (1871), Thomson and Tait (1879), Levi-Civita (1896) and Darboux. The last-named devoted two chapters of his *Leçons sur la théorie générale des surfaces* to this topic.

CHAPTER SEVEN

NAVIER'S EQUATIONS

1. THE MOLECULAR HYPOTHESIS.

In this chapter we shall follow a paper by Navier (1785-1836) which was read to the *Académie des Sciences* on March 18th, 1822.[1] In view of " the considerable or total disagreements that appear, in certain cases, between the natural effects and the results of known theories "—that is, the disagreements between experiment and the theory of perfect fluids—Navier believed it necessary to apply different ideas, notably those of " certain *molecular actions* which principally appear in the phenomena of motion. "

Navier confined his analysis to incompressible fluids.

To him, a fluid was " an assembly of molecules placed at a very small distance from each other, and able to alter their positions with respect to each other almost freely. " The mechanism is then the following one. " A pressure is exerted on the surface of the fluid and penetrates into the interior of the body. It tends to bring neighbouring parts together, and these resist this action by means of the *repulsive forces* which exist between neighbouring molecules. If the fluid is at rest, each molecule is in equilibrium because of the repulsive forces and the external forces, such as gravity, which are applied to it. . . .

" If the fluid is in motion, which in general presumes that the neighbouring molecules approach or separate from each other, it seems to us natural to assume that the repulsive forces . . . are modified by this circumstance. Indeed, we imagine that, in a state in which the fluid is at rest, the neighbouring molecules are placed at distances from each other which are determined by the mutual annihilation of the forces of repulsion and compression. It is this which has determined the size of the volume occupied by the body, due to the temperature and the external pressure to which it is subjected. Now all

---

[1] *Mémoires de l'Académie royale des Sciences de l'Institut de France*, 1823, p. 389.

the phenomena indicate that the actions exerted from molecule to molecule, in the interior of bodies, vary with the separation of the molecules. ... A liquid resists an effort which tends to take neighbouring parts away from each other much less than a solid does, but experiment has shown that the resistance to separation is not zero. In accordance with these considerations, we shall assume that, in a fluid in motion, two molecules which approach each other will repel each other, and two molecules which are separated from each other repel each other less strongly, than they would have done if their actual resistance had not varied. We shall take it to be a principle, in the following investigations, that by the effect of the motion of a fluid, *the repulsive actions of the molecules are increased or diminished by an amount proportional to the velocity with which the molecules approach, or separate from, one another.*"

2. Equilibrium of Fluids.

Navier considers two molecules, M and M', which are infinitely close together and have coordinates (x, y, z) and $(x + \alpha, y + \beta, z + \gamma)$ respectively. The repulsive force between M and M' depends on the distance $r = \sqrt{\alpha^2 + \beta^2 + \gamma^2}$. It is of the form $f(r)$ and decreases very rapidly as r increases. Each molecule M is thus influenced by all the molecules M' which surround it. Moreover, it is subjected to a force of components P, Q, R which has the dimensions of a weight per unit volume.

If an elementary displacement is imparted to the fluid so that the components of the displacement of M are δx, δy and δz, M' will be displaced by

$$\begin{cases} \delta x + \dfrac{\partial \delta x}{\partial x} \alpha + \dfrac{\partial \delta x}{\partial y} \beta + \dfrac{\partial \delta x}{\partial z} \gamma = \delta x + \delta \alpha \\[2mm] \delta y + \dfrac{\partial \delta y}{\partial x} \alpha + \dfrac{\partial \delta y}{\partial y} \beta + \dfrac{\partial \delta y}{\partial z} \gamma = \delta y + \delta \beta \\[2mm] \delta z + \dfrac{\partial \delta z}{\partial x} \alpha + \dfrac{\partial \delta z}{\partial y} \beta + \dfrac{\partial \delta z}{\partial z} \gamma = \delta z + \delta \gamma. \end{cases}$$

Whence, by an easy calculation,

$$\delta r = \frac{1}{r} \left[\frac{\partial}{\partial x} (\delta x) \cdot \alpha^2 + \frac{\partial}{\partial y} (\delta x) \alpha\beta + \frac{\partial}{\partial z} (\delta x) \alpha\gamma + \frac{\partial}{\partial x} (\delta y) \alpha\beta + \right.$$
$$\left. \frac{\partial}{\partial y} (\delta y) \beta^2 + \frac{\partial}{\partial z} (\delta y) \beta\gamma + \frac{\partial}{\partial x} (\delta z) \alpha\gamma + \frac{\partial}{\partial y} (\delta z) \beta\gamma + \frac{\partial}{\partial z} (\delta z) \gamma^2 \right].$$

The product $f(r)\,\delta r$ represents the *moment of the mutual actions* of M and M'. It is necessary to integrate this moment over all the fluid. We shall omit the details of the calculation whereby Navier obtains, for the *mutual actions of M and of all the milecules M' of the fluid*, the moment

$$\frac{4\pi}{3}\int_0^\infty r^3 f(r)\,dr \cdot \left[\frac{\partial}{\partial x}(\delta x) + \frac{\partial}{\partial y}(\delta y) + \frac{\partial}{\partial z}(\delta z)\right].$$

By summing *for all the molecules M of the fluid*, and dividing by two in order to count the moment of the mutual actions of the different molecules once only, Navier obtains the equations of equilibrium

$$0 = \int\int\int dx\,dy\,dz \left[p\left(\frac{\partial}{\partial x}(\delta x) + \frac{\partial}{\partial y}(\delta y) + \frac{\partial}{\partial z}(\delta z)\right) + P\delta x + Q\delta y + R\delta z\right]$$

where

$$p = \frac{2\pi}{3}\int_0^\infty r^3 f(r)_i^!dr$$

is a function of x, y and z.

An integration by parts gives the equilibrium equations

$$\frac{\partial p}{\partial x} = P \qquad \frac{\partial p}{\partial y} = Q \qquad \frac{\partial q}{\partial z} = R.$$

In this way Navier rediscovers Clairaut's equations, making them " depend more directly on the physical notions that can be formed on the nature of these bodies. "

3. The molecular forces in the motion of a fluid.

Therefore the introduction of molecular forces in Navier's sense does not result in any modification of the general equations of equilibrium of fluids. It is not the same for the equations of motion.

Indeed, in order to proceed from Clairaut's hydrostatic equations to Euler's hydrodynamical equations

$$P - \frac{\partial p}{\partial x} = \varrho\left(\frac{\partial u}{\partial t} + u\frac{\partial u}{\partial x} + v\frac{\partial u}{\partial y} + w\frac{\partial u}{\partial z}\right), \text{ etc.}\dots$$

it is necessary that the *repulsive forces*, in the sense that Navier used them, should not be modified. But according to the principle expressed in § 1, " it is necessary to assume the existence of new molecular forces which are produced by the state of motion. "

Always considering two neighbouring molecules $M(x, y, z)$ and M' $(x + \alpha, y + \beta, z + \gamma)$, Navier writes that the *relative velocity* of M' with respect to M, projected on MM', has a value

$$V = \frac{\alpha}{r}\left(\frac{\partial u}{\partial x}\alpha + \frac{\partial u}{\partial y}\beta + \frac{\partial u}{\partial z}\gamma\right) + \frac{\beta}{r}\left(\frac{\partial v}{\partial x}\alpha + \frac{\partial v}{\partial y}\beta + \frac{\partial v}{\partial z}\gamma\right) + \frac{\gamma}{r}\left(\frac{\partial w}{\partial x}\alpha + \frac{\partial w}{\partial y}\beta + \frac{\partial w}{\partial z}\gamma\right).$$

He represents the force which exists between the molecules M and M' by $f(r) V$.

At this point Navier supposes that the fluid in motion is given " an impulsion by the effect of which the velocities u, v, w have varied by an amount δu, δv, δw respectively. "

Under these conditions the alteration of V is

$$\delta V = \frac{\alpha}{r}\left[\delta\left(\frac{\partial u}{\partial x}\right)\cdot\alpha + \delta\left(\frac{\partial u}{\partial y}\right)\beta + \delta\left(\frac{\partial u}{\partial z}\right)\cdot\gamma\right] + \cdots$$

and the quantity

$$f(r) V \cdot \delta V$$

represents the total " moment " of the force acting on M because of M', and of the opposite force acting on M' because of M.

Now

$$f(r) V \cdot \delta V = \frac{f(r)}{r^2}\left[\alpha\left(\frac{\partial u}{\partial x}\alpha + \frac{\partial u}{\partial y}\beta + \frac{\partial u}{\partial z}\gamma\right) + \cdots\right]$$
$$\left[\alpha\left(\delta\left(\frac{\partial u}{\partial x}\right)\alpha + \delta\left(\frac{\partial u}{\partial y}\right)\beta + \delta\left(\frac{\partial u}{\partial z}\right)\gamma\right) + \cdots\right].$$

At this point it is necessary to integrate this " moment " for all the fluid.

By a calculation whose details we shall omit, Navier obtains, for the *sum of the moments of the mutual actions of M and all the molecules M' of the fluid*, the expression

$$\frac{8\pi}{30}\int_0^\infty dr \cdot r^4 f(r)\left[3\frac{\partial u}{\partial x}\delta\left(\frac{\partial u}{\partial x}\right) + \frac{\partial u}{\partial y}\delta\left(\frac{\partial u}{\partial y}\right) + \frac{\partial u}{\partial z}\delta\left(\frac{\partial u}{\partial z}\right) + \frac{\partial v}{\partial y}\delta\left(\frac{\partial u}{\partial x}\right) + \right.$$
$$\frac{\partial v}{\partial x}\delta\left(\frac{\partial u}{\partial y}\right) + \frac{\partial w}{\partial z}\delta\left(\frac{\partial u}{\partial x}\right) + \frac{\partial w}{\partial x}\delta\left(\frac{\partial u}{\partial z}\right) + \frac{\partial u}{\partial x}\delta\left(\frac{\partial v}{\partial y}\right) + \frac{\partial u}{\partial y}\delta\left(\frac{\partial v}{\partial x}\right) +$$
$$\frac{\partial v}{\partial x}\delta\left(\frac{\partial v}{\partial x}\right) + 3\frac{\partial v}{\partial y}\delta\left(\frac{\partial v}{\partial y}\right) + \frac{\partial v}{\partial z}\delta\left(\frac{\partial v}{\partial z}\right) + \frac{\partial w}{\partial y}\delta\left(\frac{\partial v}{\partial z}\right) + \frac{\partial w}{\partial z}\delta\left(\frac{\partial v}{\partial y}\right) +$$
$$\frac{\partial u}{\partial x}\delta\left(\frac{\partial w}{\partial x}\right) + \frac{\partial u}{\partial z}\delta\left(\frac{\partial w}{\partial z}\right) + \frac{\partial v}{\partial y}\delta\left(\frac{\partial w}{\partial z}\right) + \frac{\partial v}{\partial z}\delta\left(\frac{\partial w}{\partial y}\right) + \frac{\partial w}{\partial x}\delta\left(\frac{\partial w}{\partial x}\right) +$$
$$\left.\frac{\partial w}{\partial y}\delta\left(\frac{\partial w}{\partial y}\right) + 3\frac{\partial w}{\partial z}\delta\left(\frac{\partial w}{\partial z}\right)\right].$$

By summing *for all the molecules M of the fluid* and dividing by two in order to count the mutual actions of different molecules once only, Navier obtains the equation

$$0 = \iiint dxdydz \left\{ \left[P - \frac{\partial p}{\partial x} - \varrho \left(\frac{\partial u}{\partial t} + u\frac{\partial u}{\partial x} + v\frac{\partial u}{\partial y} + w\frac{\partial u}{\partial z} \right) \right] \delta u + \right.$$
$$\left[Q - \frac{\partial p}{\partial y} - \varrho \left(\frac{\partial v}{\partial t} + u\frac{\partial v}{\partial x} + v\frac{\partial v}{\partial y} + w\frac{\partial v}{\partial z} \right) \right] \delta v +$$
$$\left. \left[R - \frac{\partial p}{\partial z} - \varrho \left(\frac{\partial w}{\partial t} + u\frac{\partial w}{\partial x} + v\frac{\partial w}{\partial y} + w\frac{\partial w}{\partial z} \right) \right] \delta w \right\}$$
$$- \iiint \varepsilon \, dxdydz \left[3\frac{\partial u}{\partial x}\delta\left(\frac{\partial u}{\partial x}\right) + \frac{\partial u}{\partial y}\delta\left(\frac{\partial u}{\partial y}\right) + \ldots + 3\frac{\partial w}{\partial z}\delta\left(\frac{\partial w}{\partial z}\right) \right]$$

with

$$\varepsilon = \frac{4\pi}{30}\int_0^\infty dr \cdot r^4 f(r).$$

An integration by parts transforms the second term of the equation into

$$+ \varepsilon \iiint dx\,dy\,dz \left[\left(3\frac{\partial^2 u}{\partial x^2} + \frac{\partial^2 u}{\partial y^2} + \frac{\partial^2 u}{\partial z^2} + 2\frac{\partial^2 v}{\partial x\partial y} + 2\frac{\partial^2 w}{\partial x\partial z} \right) \delta u + \right.$$
$$\left(2\frac{\partial^2 u}{\partial x\partial y} + \frac{\partial^2 v}{\partial x^2} + 3\frac{\partial^2 v}{\partial y^2} + \frac{\partial^2 v}{\partial z^2} + 2\frac{\partial^2 w}{\partial y\partial z} \right) \delta v +$$
$$\left. \left(2\frac{\partial^2 u}{\partial x\partial z} + 2\frac{\partial^2 v}{\partial y\partial z} + \frac{\partial^2 w}{\partial x^2} + \frac{\partial^2 w}{\partial y^2} + 3\frac{\partial^2 w}{\partial z^2} \right) \delta w \right].$$

In this expression the terms obtained from the limits are omitted, and only these which occur in the *indefinite equations* of motion are included.

Taking account of the condition of incompressibility

$$\frac{\partial u}{\partial x} + \frac{\partial v}{\partial y} + \frac{\partial w}{\partial z} = 0$$

and of the conditions which are obtained by differentiating this one, Navier obtains the *indefinite equations* with which his name is associated. These are

$$\begin{cases} P - \frac{\partial p}{\partial x} = \varrho\left(\frac{\partial u}{\partial t} + u\frac{\partial u}{\partial x} + v\frac{\partial u}{\partial y} + w\frac{\partial u}{\partial z} \right) - \varepsilon\left(\frac{\partial^2 u}{\partial x^2} + \frac{\partial^2 u}{\partial y^2} + \frac{\partial^2 u}{\partial z^2} \right) \\ \ldots \\ \ldots \end{cases}$$

To be accurate, Navier also gave his attention to the conditions at the limits and studied at length " the actions exerted between the molecules of the fluid and those of the solid walls. " In this way he succeeded in introducing, as well as the coefficient ε, a coefficient E which was directly related to the *sliding of the layer of liquid on the boundary wall*.

This hypothesis of sliding has now been given up. It has been so since the work of Stokes and the experiments of Poiseuille (1846) on the flow of viscous liquids in capillary tubes. Navier used it, for his part, to show that his theory was in agreement with the empirical formulae of Girard, and even added " It would be misleading to think that the preceding formulae might be accepted if E were supposed zero. " [1]

4. Remark on the origin of the general equations of elasticity.

Without pretending to write the history of elasticity, which goes outside the compass of this book, we think it of some value to give some indication of the origin of the *general equations* of elasticity. We have seen that Navier's work had already established a connection between these and the equations of fluid motion.

In fact, if the XIXth Century had not had to discover the theory of the elasticity of bodies in one or two dimensions, it had to create the general equations of elastic media in three dimensions which have become classical.

Navier was concerned with elasticity even before he formulated his equations of hydrodynamics. His paper on the *Lois de l'équilibre et du mouvement des corps solides élastiques* was read to the *Académie des Sciences* on May 14th, 1821. It was therefore about a year earlier than his work on the equilibrium and the motion of fluids.

Navier's starting point was, here again, a *molecular* hypothesis, which he formulated very precisely.

" A solid body is regarded as an assembly of material molecules placed at extremely small distances. These molecules exert on each other two opposed forces—namely, a proper force of attraction and a force of repulsion due, in principle, to heat. Between a molecule M and any other of the neighbouring molecules, say M', there exists an action, P, which is the difference between these two forces. In the natural state of all bodies the actions P are zero, or nullify each other, since the molecule is at rest. When the shape of the body has been altered, the action P assumes a different value Π and there is equilibrium

[1] Navier, *loc. cit.*, p. 435.

between all the forces Π and the forces which were applied to the body to produce the change of shape. The forces Π can be imagined to be divided into two parts, π and π', by supposing that the first part, π, is such that there would be equilibrium between all the forces P in the natural state of the body if they existed alone. Since therefore the forces π nullify each other, it will be necessary that equilibrium exists between the resisting forces π' and the forces applied to the body. Given this, we shall take as a principle that these latter forces, π',—which are produced between any two molecules M and M' by the alteration of the shape of the body and which must, alone, be in equilibrium with the forces applied to the body—are respectively proportional to the amount by which the alteration (supposed very small) of the body's shape has altered the distance MM' between the two molecules. The force π' is an attraction if the distance MM' has increased, and a repulsion if the distance MM' has decreased. Moreover, we regard the molecular actions concerned as only existing between molecules which are close together and as having values which decrease very rapidly, according to an unknown law, for molecules which are further and further away from each other. "

Navier carried out an analysis analogous to that whose basic method we have described in § 3 of the present Chapter. Calling ε the quantity

$$\varepsilon = \frac{4\pi}{30} \int_0^\infty dr \cdot r^4 f(r)$$

where r denotes the distance of the molecule $M(xyz)$ from an arbitrary molecule M', he arrived at the general equations

$$\begin{cases} -X = \varepsilon \left(3\frac{\partial^2 x}{\partial a^2} + \frac{\partial^2 x}{\partial b^2} + \frac{\partial^2 x}{\partial c^2} + 2\frac{\partial^2 y}{\partial a \partial b} + 2\frac{\partial^2 z}{\partial a \partial c} \right) \\ -Y = \ldots \\ -Z = \ldots \end{cases}$$

In these equations X, Y, Z are the components of the force applied to the molecule M, whose initial coordinates are a, b, c and whose coordinates after the displacement are $a + x$, $b + y$, $c + z$.

Navier had to sustain an extremely lively criticism from Poisson, and his quarrel with the latter was conducted in the columns of the *Annales de Chimie et de Physique* for 1828 and 1829. Poisson showed a determined obstinacy in not recognising Navier's priority—however incontestable it may have been—in this matter [1] and, in the circumstances, does not seem to have played the better part.

---

[1] In order to illustrate the tone of this polemic, we shall quote from NAVIER.

" It seems to us that justice demands that the geometers who now concern themselves with these developments and who use neither new principles nor different differ-

As editor of the *Annales*, Arago had to settle this dispute. He absolved Navier from the accusation, which Poisson had made, of having heedlessly omitted to define the *natural state* of an elastic body. But he emphasised that Navier's text—the passage that we have quoted was the one concerned—contained some obscurity. It might be thought, as Poisson had done, that only the resultant of the actions P was zero in the natural state. Navier, at the end of the quarrel, was forced to clarify his ideas in the following way.

" 1. In the natural state of a body the attractions and repulsions between any two molecules nullify each other....

" 2. In the deformed state, when the original distance, r, between two molecules has become r', between these two molecules there is an action proportional to $r' - r$ which, moreover, decreases very rapidly to the extent that r is greater. So that this action is expressed by $(r' - r)f(r)$, where the function $f(r)$ has the property of becoming very small as soon as r itself assumes a value which is not small. In this way there is no difficulty in visualising the natural state of a body. It is seen that its present shape exists and maintains itself as a system in a stable equilibrium state, where the forces which nullify each other cease to do so if the positions of its parts are changed in relation to each other. " [1]

Poisson thought to improve Navier's solution by introducing an explicit form of the function $f(r)$, of the form

$$f(r) = ab^{-\left(\frac{r}{nd}\right)^m}$$

and by replacing Navier's integration in the coordinate r by a sum of discrete terms. But Lamé justly remarked that this improvement was illusory. For if Poisson eschewed the integration in r, he carried out an integration for the other two polar coordinates in space. Now Lamé saw a contradiction in the assumption of the *continuity of matter* (whatever coordinate might be concerned) together with a *molecular hypothesis*, to which Poisson was as firmly attached as Navier.

ential equations, will not cast our work into oblivion. Usually some appreciation has been accorded to our efforts, which have resulted in the establishment of the principles and the analytic forms by means of which a particular class of phenomena was made amenable to calculation for the first time. " (*Annales de Chimie et de Physique*, Vol. 38, 1828, p. 304.)

Or again, " M. Poisson's equations, having come seven years too late, might be said to be of the same form as the equations that had appeared first.... In order to rob me of the merit of having given the differential equations concerned, it would be necessary to show that my principles are contradictory in themselves or with the natural facts. It is not sufficient to say that the same equations have been found in another way and to claim, without proof, that this way is better than mine. "

[1] *Annales de Chimie et de Physique*, Vol. 40, 1829, p. 103.

Evidently it was desirable, as Cauchy and Lamé had done, to refrain from explicitly formulating this or that molecular hypothesis as the basis for the general equations of elasticity. It was preferable to " approach the problems connected with molecular mechanics by leaving indeterminate the reciprocal influence between the different kinds of matter, without the direct intervention of attractions or repulsions that follow certain hypothetical laws. If the problems are formulated as equations in this way, the nature of the influence concerned, the forces which express it and their exact laws, will be deduced as consequences. In this way will be reproduced the development of theoretical Astronomy, in which the universal attraction, rather than being put forward as the starting-point, is presented as a necessary consequence of the laws of motion. " [1]

In this passage we recognise Lamé's classical formulation of the equations of the theory of elasticity. Navier's molecular hypothesis did not endure, and his elasticity and hydrodynamics did not hold the interest of other students of the subject. Nonetheless, this hypothesis was the keystone of Navier's analysis, which is, certainly, a milestone in the history of general mechanics.

[1] LAMÉ, *Leçons sur la théorie mathématique de l' Élasticité des corps solides*, 1852, p. 37.

CHAPTER EIGHT

CAUCHY
AND THE FINITE DEFORMATION
OF CONTINUOUS MEDIA

We have already remarked in the preceding part of this book, in connection with the hydrodynamics of perfect fluids, that Euler, and later Lagrange, had obtained essential results in the kinematics of continuous media.

Cauchy had the merit of elucidating the problem of the finite deformation of a continuous medium, and of calling attention to the " *mean rotation* " which accompanied such a deformation. In an infinitely small deformation this mean rotation merges with what is now called the " *vortex.* "

As the kinematics of continuous media is the language of the mechanics of fluids and of elasticity, Cauchy's work is of considerable importance.

Cauchy returned to this kinematics on several occasions during twenty or more years from 1822. At first he concerned himself, like his predecessors, with the dilatations and the condensations (contractions), and only later did he discover the mean rotation.

We shall not occupy ourselves with the earlier papers, but follow a paper called *Sur les dilatations, les condensations et les rotations produites par un changement de forme dans un système de points matériels.*[1]

In an " original state "—that is, before deformation—let x, y, z be the coordinates of a " molecule " μ. Let $x + \varDelta x$, $y + \varDelta y$, $z + \varDelta z$ be those of a second molecule m. Let r be the radius vector drawn from μ to m, and a, b, c the direction cosines of this radius. Cauchy writes

$$(1) \qquad r^2 = \varDelta x^2 + \varDelta y^2 + \varDelta z^2$$

$$(2) \qquad a = \frac{\varDelta x}{r} \qquad b = \frac{\varDelta y}{r} \qquad c = \frac{\varDelta z}{r}$$

$$(3) \qquad a^2 + b^2 + c^2 = 1.$$

---

[1] *Œuvres complètes de Cauchy*, Vol. XII, 2nd series, p. 343.

In a " second state "—that is, after finite deformation and displacement—the coordinates of μ become $x + \xi$, $y + \eta$, $z + \zeta$ and those of m become $x + \Delta x + \xi + \Delta \xi$, etc. . . . The radius vector in this state becomes $r + \varrho$ and has for components $\Delta x + \Delta \xi$, etc. . . . and its direction cosines α, β, γ. Whence the three equations

(4) $$(r + \varrho)^2 = (\Delta x + \Delta \xi)^2 + (\Delta y + \Delta \eta)^2 + (\Delta z + \Delta \zeta)^2$$

(5) $$\alpha = \frac{\Delta x + \Delta \xi}{r + \varrho} \qquad \beta = \frac{\Delta y + \Delta \eta}{r + \varrho} \qquad \gamma = \frac{\Delta z + \Delta \zeta}{r + \varrho}$$

(6) $$\alpha^2 + \beta^2 + \gamma^2 = 1.$$

Cauchy calls the quantity

(7) $$\varepsilon = \frac{\varrho}{r}$$

the " *linear dilatation* " or " *condensation*, " according to whether it is positive or negative.

Fom this definition and equations (1) to (6) it follows that

(8) $$(1 + \varepsilon)^2 = \left(a + \frac{\Delta \xi}{r} \right)^2 + \left(b + \frac{\Delta \eta}{r} \right)^2 + \left(c + \frac{\Delta \zeta}{r} \right)^2$$

(9) $$\alpha = \frac{1}{1 + \varepsilon} \left(a + \frac{\Delta \xi}{r} \right) \text{ etc.} \ldots$$

In the deformation from the original to the second state the molecule μ goes from O to O'. If $\overline{OA}$ is the axis carrying the radius r, and $\overline{O'A'}$ that carrying the radius $r + \varrho$, the angle δ between $\overline{O'A'}$ and $\overline{OA}$ can be introduced, together with its projections φ, χ, ψ on the coordinate planes. These are the three angles contained between the projections of $r + \varrho$ and those of r. Now

$$\cos \delta = a\alpha + b\beta + c\gamma$$

or

(10) $$\sin^2 \delta = 1 - \cos^2 \delta = (a^2 + b^2 + c^2)(\alpha^2 + \beta^2 + \gamma^2) - (a\alpha + b\beta + c\gamma)^2 = (b\gamma - c\beta)^2 + (c\alpha - a\gamma)^2 + (a\beta - b\alpha)^2.$$

On the other hand

(11) $$tg\varphi = \frac{\dfrac{\gamma}{\beta} - \dfrac{c}{b}}{1 + \dfrac{c\gamma}{b\beta}} = \frac{b\gamma - c\beta}{b\beta + c\gamma} \qquad tg\chi = \frac{c\alpha - a\gamma}{c\gamma + a\alpha} \qquad tg\psi = \frac{a\beta - b\alpha}{a\alpha + b\beta}.$$

Moreover, because of (9)

(12) $$b\gamma - c\beta = \frac{b\Delta\zeta - c\Delta\eta}{(1+\varepsilon)r} \quad \text{etc.} \ldots$$

Make m tend towards μ in such a way that r, $\Delta\xi$, $\Delta\eta$, $\Delta\zeta$ tend to zero. That is, consider two infinitely close molecules which are both subject to the same finite deformation of the medium. The quantities ξ, η, ζ may be written as functions f of x, y, z

$$\Delta f = f(x + \Delta x, y + \Delta y, z + \Delta z) - f(x, y, z)$$

and consequently

(13) $$\lim \frac{\Delta f}{r} = \frac{f(x + ar, y + br, z + cr) - f(x, y, z)}{r} \quad \text{for} \quad r \to 0$$
$$= D_r f = aD_x f + bD_y f + cD_z f.$$

Therefore we shall put

(14) $$\begin{cases} \lim \dfrac{\Delta\xi}{r} = aD_x\xi + bD_y\xi + cD_z\xi \\[2mm] \lim \dfrac{\Delta\eta}{r} = aD_x\eta + bD_y\eta + cD_z\eta \\[2mm] \lim \dfrac{\Delta\zeta}{r} = aD_x\zeta + bD_y\zeta + cD_z\zeta. \end{cases}$$

Returning to equations (8), (9), (12), it follows that

(15) $$(1 + \varepsilon)^2 = [a(1 + D_x\xi) + bD_y\xi + cD_z\xi]^2 +$$
$$[aD_x\eta + b(1 + D_y\eta) + cD_z\eta]^2 + [aD_x\zeta + bD_y\zeta + c(1 + D_z\zeta)]^2$$

(16) $$\begin{cases} \alpha = \dfrac{1}{1+\varepsilon} [a(1 + D_x\xi) + bD_y\xi + cD_z\xi] \\[2mm] \beta = \dfrac{1}{1+\varepsilon} [aD_x\eta + b(1 + D_y\eta) + cD_z\eta] \\[2mm] \gamma = \dfrac{1}{1+\varepsilon} [aD_x\zeta + bD_y\zeta + c(1 + D_z\zeta)] \end{cases}$$

(17) $$b\gamma - c\beta = \frac{1}{1+\varepsilon} (aD_x + bD_y + cD_z)(b\zeta - c\eta) \quad \text{etc.} \ldots$$

from which δ, φ, χ, ψ may be obtained by the use of (10) and (11).

To clarify this, Cauchy considers the particular case where the axis $\overline{OA}$ of the radius vector r is parallel to the plane of yz in the original state. That is

$$a = 0 \quad b = \cos\tau \quad c = \sin\tau.$$

In these circumstances, by equation (11)

$$(18) \qquad tg\varphi = \frac{(bD_y + cD_z)\,(b\zeta - c\eta)}{1 + (bD_y + cD_z)\,(b\eta + c\zeta)}$$

or

$$(19) \qquad tg\varphi = \frac{(\cos \tau D_y + \sin \tau D_z)\,(b\zeta - c\eta)}{1 + (\cos \tau D_y + \sin \tau D_z)\,(b\eta + c\zeta)}.$$

The rotation φ of $\overline{OA}$ about Ox depends on τ. Its mean value is

$$(20) \qquad \left|\varphi\right| = \frac{1}{2\pi}\int_0^{2\pi} \varphi d\tau$$

and represents the *mean rotation* in Cauchy's sense.

The angle φ is the difference $\tan^{-1}\left(\dfrac{\gamma}{\beta}\right) - \tan^{-1}\left(\dfrac{c}{b}\right)$. Here $\dfrac{c}{b} = \tan \tau$.
Further, by (16),

$$(21) \qquad \frac{\gamma}{\beta} = \frac{aD_x\zeta + bD_y\zeta + c\,(1 + D_z\zeta)}{aD_x\eta + b\,(1 + D_y\eta) + cD_z\eta)} = \frac{\cos \tau D_y\zeta + \sin \tau\,(1 + D_z\zeta)}{\cos \tau\,(1 + D_y\eta) + \sin \tau D_z\eta}.$$

We shall therefore put

$$(22) \qquad tg\,(\varphi + \tau) = \frac{D_y\zeta + (1 + D_z\zeta)\,tg\tau}{1 + D_y\eta + D_z\eta tg\tau}$$

and obtain $\tan (\chi + \tau)$ and $\tan (\psi + \tau)$ by a cyclic permutation. From these expressions the *mean rotations* φ, χ, ψ, can be obtained.

We shall not discuss the geometrical interpretations which Cauchy gave of the expressions $(1 + \epsilon)^{-1}$, $(1 + \epsilon)$, $[(1 + \epsilon) \sin \delta]^{-1}$, $\tan \varphi$, $\tan \chi$ and $\tan \psi$, but shall return to the simple case of *infinitely small* deformations.

From equation (15)

$$\varepsilon = (aD_x + bD_y + cD_z)\,(a\xi + b\eta + c\zeta).$$

On the other hand, by approximating $\tan \varphi$ by φ, to the second order

$$\begin{aligned}
\varphi &= (\cos \tau D_y + \sin \tau D_z)\,(\zeta \cos \tau - \eta \sin \tau)\\
&= \cos^2 \tau D_y\zeta - \sin^2 \tau D_z\eta - \sin \tau \cos \tau(D_y\eta - D_z\zeta)\\
&= \frac{1}{2}\,(D_y\zeta - D_z\eta) + \frac{1}{2}\,(D_y\zeta + D_z\eta) \cos 2\tau - \frac{1}{2}\,(D_y\eta - D_z\zeta) \sin 2\tau.
\end{aligned}$$

If it sought to calculate the *mean rotation* $\left|\varphi\right|$,[7] it is sufficient to observe that

$$\int_0^{2\pi}\cos 2\tau d\tau = \int_0^{2\pi}\sin 2\tau d\tau = 0.$$

Therefore

$$\left| \varphi \right| = \frac{1}{2} (D_y \zeta - D_z \eta)$$

$$\left| \chi \right| = \frac{1}{2} (D_z \xi - D_x \zeta)$$

$$\left| \psi \right| = \frac{1}{2} (D_x \eta - D_y \xi).$$

Cauchy declared, " These are the mean rotations which will be executed by the system of points about the three semi-axes drawn through μ parallel to the three semi-axes of positive coordinates. "

CHAPTER NINE

HUGONIOT
AND THE PROPAGATION OF MOTION
IN CONTINUOUS MEDIA

1. Nature of the problem.

As an example of a profoundly original investigation, and in the spirit of the introduction to this Part,[1] we have chosen Hugoniot's studies on the propagation of motion in continuous media.[2]

These studies are a continuation of the work of d'Alembert, Lagrange, Laplace, Poisson, Cauchy, Riemann, Poncelet, Philips, Saint Venant, etc.... More recently, the mathematical apparatus of Hugoniot's work has been codified and perfected, and other investigations of a physical nature have followed. Nevertheless, these studies mark an epoch in the history of the mechanics of continuous media.

Hugoniot's paper starts with a mathematical introduction on the characteristics (in Monge's sense) of equations containing partial derivatives. He then comes to the *theory of the propagation of motion* in a continuous medium, and starts with the consideration of very simple examples.

Consider an elastic and homogeneous rod whose motion is governed by the equation

$$(1) \qquad \frac{\partial^2 u}{\partial t^2} = a^2 \frac{\partial^2 u}{\partial x^2}.$$

The rod is initially a rest and an uniformly accelerated motion is imposed at its end $x = \lambda$. For all $t \geq 0$, the conditions at the end are therefore

$$(2) \qquad u = 0 \quad \text{for} \quad x = 0 \quad \text{and} \quad u = \frac{\alpha t^2}{2} \quad \text{for} \quad x = \lambda.$$

---

[1] See above, p. 353.

[2] *Mémoire sur la propagation du mouvement dans les corps et spécialement dans les gaz parfaits, Journal de l' École polytechnique*, cahier 57, 1887, p. 3 and cahier 58, 1889, p. 1. This paper, published posthumously, had been deposited by the author with the Secretary of the *Institut* on October 26th, 1885.

The behaviour of the rod is governed by the solution $u = 0$ at the part near the origin of x, and by the solution

$$(3) \qquad u = \frac{\alpha}{2a^2}(x + at - \lambda)^2$$

which clearly satisfies the equation of motion and the condition imposed at $x = \lambda$, in the upper part of the rod. This solution advances towards the origin with the velocity a. It joins the other solution $u = 0$ at the point

$$x = \lambda - at$$

and thus continually extends itself at the expense of the solution $u = 0$, which represents the fact that the lower part of the rod is at rest. When the solution (3) arrives at the origin—at the time $\frac{\lambda}{a}$ —the condition (2) intervenes. Then there originates a new solution

$$(4) \qquad u = \frac{2\alpha x}{a^2}(at - \lambda)$$

which is completely determined by (3) and the condition $u = 0$ (for $x = 0$).

Once more the rod is divided into two parts. That nearer the origin has the motion (4) and the other retains the motion (3). These two parts meet at the point $x = at - \lambda$. The solution (4) propagates itself with velocity a and reaches the end $x = \lambda$ at the time $t = \frac{2\lambda}{a}$.

A new integral originates at this end, namely

$$(5) \qquad u = \frac{\alpha}{2a^2}(x + at - \lambda)^2 + \frac{2\alpha x}{a^2}(at - \lambda) + \frac{2\alpha\lambda}{a^2}(2\lambda - x - at)$$

which is determined by (4) and the condition $u = \frac{\alpha t^2}{2}$ for $x = \lambda$. This processes continues.

Hugoniot concludes, " It is seen how the solutions can be born at the ends of the body. When the motion of the prism is represented by a determinate solution and the condition imposed at the end is not compatible with the motion defined by this solution, another solution originates at the end and develops at the expense of the first, propagating itself with a velocity which (for elastic prisms) is always equal to a. The new solution, it is clear, can only depend on the original solution and the condition imposed at the end. "

But solutions can originate in the interior of the body. In order to be convinced of this, it is sufficient to consider the same rod, initially at rest, and to impose the following conditions at the ends—

$u = \dfrac{1}{2}\beta t^2$ for $x = 0$ and $u = \dfrac{1}{2}\alpha t^2$ for $x = \lambda$, whenever $t \geqq 0$.

Between $t = 0$ and $t = \dfrac{\lambda}{2a}$, the motion of the rod is represented by three solutions

$$u = \frac{\beta}{2a^2}(at - x)^2, \quad u = 0 \quad \text{and} \quad u = \frac{\alpha}{2a^2}(at + x - \lambda)^2.$$

Between $t = \dfrac{\lambda}{2a}$ and $t = \dfrac{\lambda}{a}$, these three solutions become

$$u = \frac{\beta}{2a^2}(at - x)^2, \quad u = \frac{\alpha}{2a^2}(at + x - \lambda)^2 + \frac{\beta}{2a^2}(at - x)^2$$

$$\text{and} \quad u = \frac{\alpha}{2a^2}(at + x - \lambda)^2.$$

It is seen that only the central solution is modified.

Between $t = \dfrac{\lambda}{a}$ and $t = \dfrac{3\lambda}{2a}$, the central solution remains and the two solutions at the ends are modified, etc. . . .

Thus at the time $t = \dfrac{\lambda}{2a}$, at the central point of the rod the initial rest gives way to a motion represented by a solution which is the sum of two primitive solutions. (This is a particular case.) The new solution originates because the two extreme solutions cannot coexist. It propagates itself, on one side and on the other of the central point, with the velocity a and to the detriment of the other two solutions. It arrives at the ends at the time $\dfrac{\lambda}{a}$. Then is born a solution at each end, then a new solution at the centre of the rod, etc. . . .

2. COMPATIBILITY OF TWO SOLUTIONS. VELOCITY OF PROPAGATION OF ONE SOLUTION IN ANOTHER. HUGONIOT'S THEOREM.

Hugoniot made clear what should be understood by *propagation of one solution in another*. Let $\xi(t)$ be the abcissa of the point which separates the two solutions. If the motion continues to be represented by the same two solutions at the time $t + dt$, these will be said to be *compatible*. The point ξ is displaced by $d\xi$ in a determinate direction, and $\dfrac{d\xi}{dt}$ will be, by definition, the velocity of propagation of one of these solutions in the other.

If the two solutions are *incompatible*, there will be born a new solution which is necessarily *compatible with the first two*.

How is this condition of compatibility to be presented ?

The motion is supposed to be governed by the linear equation of the second order,

$$L \frac{\partial^2 u}{\partial t^2} + 2K \frac{\partial^2 u}{\partial x \partial t} + H \frac{\partial^2 u}{\partial x^2} + M = 0.$$

Here L, K, H, M are functions of x, t, u, $\frac{\partial u}{\partial t}$, $\frac{\partial u}{\partial x}$. Let there be two solutions u_1 and u_2 which represent the motion of the two parts of the body at the time t, and which meet at the point $\xi(t)$.

The *compatibility* $\left(\text{absence of discontinuity in } u, \frac{\partial u}{\partial x} \text{ and } \frac{\partial u}{\partial t}\right)$ is therefore expressed by

$$\begin{cases} u_1 - u_2 = 0 \\ \dfrac{\partial}{\partial x} (u_1 - u_2) = 0 \\ \dfrac{\partial}{\partial t} (u_1 - u_2) = 0 \end{cases}$$

for all values of t and at the point $\xi(t)$. The second and third of these equations may be differentiated by regarding ξ as a function of t defined by $u_1 - u_2 = 0$, which gives

$$\begin{cases} \dfrac{\partial^2}{\partial x \partial t} (u_1 - u_2) + \dfrac{\partial^2}{\partial x^2} (u_1 - u_2) \dfrac{d\xi}{dt} = 0 \\ \dfrac{\partial^2}{\partial t^2} (u_1 - u_2) + \dfrac{\partial^2}{\partial x \partial t} (u_1 - u_2) \dfrac{d\xi}{dt} = 0. \end{cases}$$

On the other hand, by the equation of motion

$$L \frac{\partial^2}{\partial t^2} (u_1 - u_2) + 2K \frac{\partial^2}{\partial x \partial t} (u_1 - u_2) + H \frac{\partial^2}{\partial x^2} (u_1 - u_2) = 0.$$

Whence it is concluded that if the three second derivatives are not always equal to each other at the point ξ (in which case the solutions u_1 and u_2 would not be distinct), then

(C) $$L \left(\frac{d\xi}{dt}\right)^2 - 2K \frac{d\xi}{dt} + H = 0.$$

Knowing u_1, the corresponding values of L, K, H are taken. Then the preceding equation gives $\xi(t)$. From this the solution u_2 is obtained, using the condition

$$u_1 - u_2 = 0$$

in which x is replaced by $\xi(t)$.

The equation (C) is that of the characteristics of the equation of motion. Therefore the two solutions have a common characteristic.

The velocity of propagation, $\dfrac{d\xi}{dt}$, is given by the condition (C).

In general it has two values at each point and at all times and, in general, it depends on the solution considered.

If $u = 0$ is a solution of the equation of motion, the velocity of propagation compatible with the body at rest is known *a priori*. More generally, if a solution, u, of the equation of motion is known, the velocity of propagation of the solutions compatible with u is known, and these solutions represent all the motions which can propagate themselves in the first without the introduction of discontinuity.

Thus a *sequence of problems* had to be treated in a determined order, step by step. This sequence of problems reduces to the following one—

" To determine the ensemble of integral surfaces which correspond to a given solution along a given characteristic. "

Hugoniot observes, " This problem is basically equivalent to that of the integration of the equation in the partial derivatives which represents the motion, which does not appear to have been even partially solved. But it must be remarked that it is possible to find particular solutions, each one of which represents a case of propagation in another and which, accordingly, constitutes another step in the study of the natural phenomena. "

In the simple case of the motion of an elastic rod, whose equation of motion is

$$\frac{\partial^2 u}{\partial t^2} = a^2 \frac{\partial^2 u}{\partial x^2}$$

the general solution is known to be

$$u = \varphi(x + at) + \psi(x - at).$$

The condition of compatibility is easily interpreted with the help of the arbitrary functions φ and ψ.

Two compatible solutions, u_1 and u_2, have a common characteristic. Accordingly one of the quantities $x + at$ and $x - at$ remains constant. For example suppose that $x - at = K$ represents the equation of the line of intersection. This can be expressed by

$$\varphi_1(x + at) - \psi_1(K) = \varphi_2(x + at) - \psi_2(K)$$

or $\qquad \varphi_1(x + at) - \varphi_2(x + at) = -\psi_2(K) + \psi_1(K).$

Therefore the difference $\varphi_1 - \varphi_2$ reduces to a constant, which can be supposed to be zero (with a suitable adjustment of ψ_2 or ψ_1 if this is necessary). Thus when two solutions fit together along a characteristic $x - at = K$, the two corresponding functions φ are equal. An analogous conclusion follows for the functions ψ of two solutions which fit together along a characteristic $x + at = K'$.

When a solution is compatible with two others, it borrows the function φ from one and the function ψ from the other and is completely known apart from a constant which can be determined from the condition that the values of u should be the same at the common points.

As an application of this, Hugoniot solved the following problem, which had been attempted by Poncelet, in finite terms. To find the motion of an elastic rod which is originally in equilibrium under the action of its own weight, and from which a weight, regarded as an unchanging solid, is suspended.

Hugoniot then turned his attention to the velocity of propagation of motion in fluids.

The equation of motion of a perfect gas, in adiabatic motion in a cylindrical container, may be written as

$$\varrho_0 \frac{\partial^2 u}{\partial t^2} = \gamma p_0(x) \left(1 + \frac{\partial u}{\partial x}\right)^{-(\gamma+1)} \frac{\partial^2 u}{\partial x^2}$$

if the external forces are neglected.

The velocity of propagation follows from this—

$$\left(\frac{d\xi}{dt}\right)^2 = \frac{\gamma p_0}{\varrho_0} \left(1 + \frac{\partial u}{\partial x}\right)^{-(\gamma+1)}.$$

If the gas is at rest, $u = 0$ is a solution for which $\dfrac{\partial u}{\partial x} = 0$. Therefore, in a gas at rest,

$$\left(\frac{d\xi}{dt}\right)^2 = \frac{\gamma p_0}{\varrho_0}.$$

Hugoniot wrote, " An expression which has been known since Laplace, but which has never been rigorously demonstrated. "

In order to treat a gas in motion, Hugoniot starts by modifying the definition of the velocity of propagation. In the time dt the point ξ which is common to both solutions moves from the interval whose *initial* abcissa is ξ to the interval whose initial abcissa is $\xi + d\xi$. But at the time t these intervals occupy the positions $\xi + u$ and

$\xi + d\xi + u + \dfrac{\partial u}{\partial x} \, d\xi$. Therefore the *distance actually travelled in the fluid* by the common point is $d\xi \left(1 + \dfrac{\partial u}{\partial x}\right)$.

Given this, Hugoniot states the following theorem, with which his name is still associated.

When one motion propagates itself in another, the velocity of propagation in gases is always equal to $\sqrt{\dfrac{\gamma p}{\varrho}}$, *where* p *is the pressure and* ϱ *the density at the point which separates the two motions* (with the reservation that there should not be discontinuities at the point common to both solutions).

An immediate transformation, which makes apparent the velocity of propagation corresponding to the *actual state* of the gas, gives

$$\left(1 + \frac{\partial u}{\partial x}\right)^2 \left(\frac{d\xi}{dt}\right)^2 = \frac{\gamma p_0}{\varrho_0} \left(1 + \frac{\partial u}{\partial x}\right)^{-\gamma+1}.$$

But

$$p = p_0 \left(1 + \frac{\partial u}{\partial x}\right)^{-\gamma} \text{ and } \varrho = \varrho_0 \left(1 + \frac{\partial u}{\partial x}\right)^{-1}.$$

Therefore

$$\left[\frac{d\xi}{dt}\left(1 + \frac{\partial u}{\partial x}\right)\right]^2 = \frac{\gamma p}{\varrho}$$

whence follows the theorem.

The same law is applicable to the motion of a gas in spherical layers. It can also be generalised to the adiabatic motion of a homogeneous fluid—if no external forces, friction or viscosity are present—whose law of motion is

$$\varrho_0 \frac{\partial^2 u}{\partial t^2} = \varphi \left(\frac{\partial u}{\partial x}\right) \frac{\partial^2 u}{\partial t^2}.$$

We cannot deal with the numerous problems which Hugoniot studied, and we shall confine ourselves to the principle of his studies of discontinuities.

3. Discontinuities in the propagation of motion.

Hugoniot distinguished three kinds of discontinuities which can occur in the propagation of motion.

1) Those which pre-exist in the initial state.

2) Those which arise from the conditions at the limits (for example, from external impact).

3) Those which can arise in the course of the motion itself, apart from any discontinuity of the first two kinds.

In this connection Hugoniot developed the following interesting observations.

" It may be asked whether the discontinuities encountered in the theory of the propagation are merely an analytic fiction, or whether they correspond to physical reality. This is a question which it is rather difficult to answer in the present state of science.

" If only the discontinuities at the ends were concerned, the answer would not be uncertain. The discontinuous analytic form that is given to these conditions does not correspond to the facts. Thus, in the example of the impact of a body on the end of an elastic rod, the action of the body on this end starts before contact is established, because of the action which the material particles at the end exert at a distance. This action at first negligeable, increases very rapidly when the distance between the body and the rod becomes extremely small, and the end of the rod only attains the velocity V [assumed by the consideration of instantaneous impact] after having passed through all the intermediate velocities. For a very rapid variation of this velocity—whose law is unknown—the hypothesis of a sudden variation is substituted for simplification.

" It is much more difficult to decide, if the conditions at the end and in the initial state are continuous, when a sudden appearance of a discontinuity of the motion is seen. At first sight it even seems that the discontinuity must be, in reality, produced. Nevertheless, it must be observed that while the partial differential equations which are used in the study of the motion are based on hypotheses which are very close to the truth, if they were replaced by the true laws they would lead to very different equations. For example, in order to establish the equations of motion of a gas it has been supposed that the conductivity of the fluid was accurately equal to zero. Now it is well known that while this conductivity is very small, it nevertheless has a measurable value. Therefore the equation in the partial derivatives which has been used is only an approximation.

" However it may be, discontinuities force themselves upon us in the problems of mathematical physics. Whether they are regarded as having a real existence or are accepted as being merely an approximation, it is necessary to investigate the influence which they have on the phenomena of the propagation of motion. "

Hugoniot considered two bodies A and A′ which are actuated by motions in layers which are parallel to Ox and bound to remain in contact along a plane B. The motion of each of these bodies is

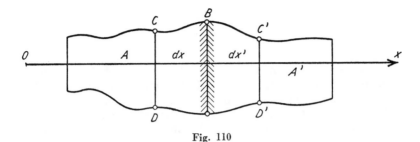

Fig. 110

represented by a solution of a certain equation in the partial derivatives of the second order. At the time t considered the velocity and the pressure are not the same in the two layers in contact. For A, let v be the velocity, $z = \dfrac{\partial u}{\partial x}$ the dilatation, p the pressure of the layer in contact with A', and let v', z' and p' be the analogous quantities for A'.

The condition of the two layers dx and dx' will be modified by mutual reaction. At the end of the time dt the layer dx of A will have taken the velocity v_1, the dilatation z_1 and the pressure p_1; the analogous quantities for the layer dx' in A' will be v_1', z_1' and p_1'. Thus the problem contains eight unknowns— v_1, v_1', z_1, z_1', p_1, p_1', $\dfrac{dx}{dt}$ and $\dfrac{dx'}{dt}$.

With the hypothesis that A and A' must always remain in contact, *differences of velocity and pressure cannot exist across B.*

Therefore

(1) $$v_1 = v_1'$$

(2) $$p_1 = p_1'.$$

Applying the principle of the conservation of momentum to each layer dx and dx',

(3) $$\varrho_0(v_1 - v)\,\frac{dx}{dt} = p - p_1$$

(4) $$\varrho_0'(v_1' - v')\,\frac{dx'}{dt} = p_1' - p'.$$

The increase of the length of the layer dx during the time dt is $(v_1 - v)dt$. Its original length $dx(1 + z)$ becomes $dx(1 + z_1)$. Therefore

(5) $$(z_1 - z)\,\frac{dx}{dt} = v_1 - v$$

and similarly

(6)
$$(z_1' - z') \frac{dx'}{dt} = v' - v_1'.$$

Thus there are six equations which are independent of the nature of the bodies A and A'. The problem is determinate if there is, in each medium, a relation between p and z—that is, $p = \varphi(z)$, and $p' = \psi(z')$.

In particular, if the two bodies A and A' are *identical* then $\varphi = \psi$, $z_1 = z_1'$ because of (2), and $\varrho_0 = \varrho_0'$. *Moreover, suppose that one of the velocities of propagation* $\left(\dfrac{dx'}{dt}\right)$ *vanishes.*

Using the preceding relations, an elementary calculation gives

(C)
$$\varrho_0(v' - v)^2 = [\varphi(z) - \varphi(z')] \, (z' - z)$$

and, incidentally, $v_1 = v_1'$ and $z_1 = z'$.

Then two cases are to be distinguished.

1) The condition (C) is not satisfied. The two layers dx and dx', which have assumed equal velocities and pressures, are moved in the same way. The body is divided into three parts— one, to the left of CD, is represented by a certain solution of the partial differential equation which, moreover, is known in advance. It is the same for the part to the right of $C'D'$. Between these two is an infinitely small part, of length $dx + dx'$, in which the velocity is v and the dilatation z. The original discontinuity at B has given rise to two other discontinuities.

2) The condition (C) is satisfied. One of the velocities $\left(\text{here } \dfrac{dx'}{dt}\right)$ is zero. The intermediate portion $CDC'D'$ disappears. The original discontinuity has only altered its position and the motion of the body is still represented by two unique solutions which allow a discontinuity in the velocities and dilatations to exist at the common point. This point is displaced by a distance dx—one of the solutions has propagated itself in the other with velocity $\dfrac{dx}{dt}$, without giving rise to any new phenomenon. Hugoniot would say that the two solutions are *compatible*. The section at which the solutions meet is defined by the identity of the displacements—by

$$u = u'.$$

The condition (C) can also be written as

$$\varrho_0 \left(\frac{dx}{dt}\right)^2 = \frac{\varphi(z) - \varphi(z')}{z' - z} = \frac{p - p'}{z - z'}.$$

The curve on which the two solutions fit together is no longer a characteristic—it depends on both solutions.

Nevertheless, if the coefficients of the derivatives of the second order in the partial differential equation do not depend on z or v, or $\dfrac{du}{dt}$—that is, if $p = A(x, t, u)$, it is discovered that

$$\varrho_0 \left(\frac{dx}{dt}\right)^2 = - A.$$

Thus the velocity of propagation is the same for both solutions. Therefore the velocity of propagation is the same whether or not there may be a discontinuity. Therefore the integral surfaces fit together along a characteristic.

We return to the case of condition (C) not being satisfied. There is a solution which governs the motion of the layer $dx + dx'$. On the other hand

$$\varrho_0 \left(\frac{dx}{dt}\right)^2 = \frac{\varphi(z) - \varphi(z_1)}{z_1 - z}$$

$$\varrho_0 \left(\frac{dx'}{dt}\right)^2 = \frac{\varphi(z_1) - \varphi(z')}{z' - z_1}.$$

Therefore this solution is *compatible* with both the original solutions and propagates itself in these without giving rise to new phenomena.

Hugoniot then turns to the more general case of a relation between the pressure and the dilatation of a layer which depends on the transformations which this has undergone. The conditions of compatibility are then of number two—

If two incompatible motions A and B are present, a new one develops in each of the layers in contact. The two new motions, A' and B', propagate themselves in opposite directions at the expense of the original motions. The layer common to the motions A' and B' remains the same. The velocities of the two motions are equal there, as are the pressures, but this does not apply to the dilatations.

We cannot follow Hugoniot in these developments, or in the several problems to which he applied his theory. We believe that we have said enough to illustrate the nature and the originality of this paper, in which, starting from a given motion, the author studies the accidents that can happen to it by iterating a process of propagation, or of creation of different motion. However remarkable this analysis may be, it is always subordinate to the physical sense which guided its author's thought.

CHAPTER TEN

HELMHOLTZ AND THE ENERGETIC THESIS
DISCUSSION OF THE NEWTONIAN PRINCIPLES
(SAINT-VENANT, REECH, KIRCHHOFF, MACH, HERTZ,
POINCARÉ, PAINLEVÉ, DUHEM)

1. HELMHOLTZ AND THE ENERGETIC THESIS.

Before coming to the discussion of the newtonian principles which distinguished the second half of the XIXth Century and the beginning of this Century (before relativity) it seems valuable to devote a little time to the so-called *energetic thesis*. This certainly lies outside the scope of rational mechanics, founded as it was on the contemporary thermodynamics, electricity and magnetism. But it also aimed—not without illusion—at replacing the classical exposition of the concepts and axioms of dynamics. In this connection Poincaré made a very apt criticism, to which we shall return. For the moment, we confine ourselves to summarising the essential content of the energetic thesis in the field of mechanics alone.

In 1847 Helmholtz published at Berlin a paper *On the Conservation of Force*.[1] From the philosophical, not to say from the metaphysical, point of view Helmholtz assigned to the theoretical sciences the task of inquiring into the " constant causes " of phenomena. Even if nature provides evidence of free will, science must seek the regularities susceptible of being reduced to the effects of fixed causes.

In some ways Helmholtz appears as a Cartesian. Having endowed matter with extension, and with quantity or mass, he was of the opinion that matter can only recognise changes of position in space—that is, movement. But matter shows *activity* in movement. Therefore the energetic thesis is concerned with the properties of " *active matter*. "

Helmholtz confined himself to the study of *conservative* systems, whose elements are only subject to their interactions (functions of

---

[1] We have used the French translation by PÉRARD, Masson, 1869.

their distances apart alone) or to forces emanating from fixed centres (and functions only of their distances from these centres). The kinetic energy of such a system does not depend on the positions of the elements in space (*relative* positions of different elements and distances of the elements from the fixed centres).

Calling φ the intensity of a force acting in the direction of a vector r, and reckoning it positive when attractive and negative when repulsive, Helmholtz called the integral

$$\int_r^R \varphi dr$$

the *quantity of tensions* action between the limits r and R. He wrote, " The increase of the kinetic energy of a particle in its motion under the influence of a central force is equal to the quantity of tensions which corresponds to the relative variation from the centre of action. "

He called the quantity of disposable tension the *potential energy* after Rankine ; the kinetic energy became the *actual energy*.

Thus, for a particle subject to a central force (a function of the distance) emanating from a fixed centre, the *sum of these two energies remains constant.*

This result extends to " all cases of the motion of free particles under the influence of their attractive and repulsive forces, whose intensities only depend on the distances, " in the sense that " the decrease of the potential energy is always equal to the increase of the kinetic energy " and conversely.

Helmholtz recalled that the principle of the conservation of kinetic energy had been applied to all motions due to the influence of universal gravitation, and in the transmission of motion by means of incompressible bodies (solids or liquids) when friction and impact with inelastic bodies were not involved.

He congratulated himself with having thus put mechanics " into the form of an almost popular rule— in all motion transmitted and modified by mechanical powers, there is a loss of force proportional to the velocity gained. " [1]

Helmholtz's conservation principle also extends to solid bodies and perfectly elastic liquids. The principle has been used for a considerable time in application to solid bodies. In fluids (liquid or gaseous) the propagation of motion is accomplished in the form of waves of velocity

$$u = a \cos \left\{ \frac{2\pi}{\lambda} (x - at) \right\}.$$

[1] *Loc. cit.*, p. 82. The intention is clear, but no purpose is served by emphasising the extent to which the terminology is not accurate.

In a wave motion the kinetic energy of a particle Δm is $\frac{1}{2} \cdot \Delta m \cdot a^2$ and therefore, is proportional to the intensity a^2.

If two waves of intensities a^2 and b^2 do not interfere, they communicate the intensity $a^2 + b^2$ to all the points they encounter. If they interfere, the maxima $(a + b)^2$ and the minima $(a - b)^2$ are greater or less than $a^2 + b^2$ by the same quantity $2ab$. The *kinetic energy is not destroyed, but distributed differently.*

Kinetic energy is only lost in the absorption of waves. But then there is friction, or impact with inelastic bodies, and release of heat.

Helmholtz then went on to the equivalence of " force " and heat, invoking Joule, Henry, Davy, Clapeyron, etc. . . ; then to the mechanical equivalent of electrical, magnetic and electro-magnetic phenomena. But it is as well to remember his intention—" To reduce *all natural phenomena* to invariable forces, attractive or repulsive, whose intensity only depends on the distance from centres. "

There is some illusion in the pursuit of such an ideal and this, to Helmholtz himself, had, at least in part, the character of a wish. If this wish could have been fulfilled, he would have arrived at " the necessary form of the conception of nature, which can be called *objective truth.* "

2. BARRÉ DE SAINT-VENANT.

We return to our task of outlining the discussion of the newtonian principles which appears in the systematic or critical works of several authors.

Barré de Saint-Venant's attitude was related to that of Lazare Carnot. Indeed forces were to him " kinds of agencies of an occult or metaphysical nature " which did not appear either in the data or the results of a problem of terrestrial or celestial mechanics.[1]

Observation shows that in order that a body should assume an acceleration (and therefore a velocity) it is necessary that other bodies change their situation with respect to it. It is the same if the body is already actuated with some velocity. The acceleration arising

---

[1] *Principes de Mécanique fondés sur la Cinématique* (1851).
More clearly still, in an article on DU BUAT published at Lille in 1866, SAINT-VENANT wrote, " It is very possible that forces—these problematical beings or rather substantive adjectives, which are neither matter nor spirit but blind and unthinking beings that must nevertheless be endowed with the wonderful faculty of estimating distances and of exactly proportioning their intensities accordingly—may be more and more expelled and separated from the mathematical sciences. They would take the place of *laws*—not only geometrical but also physical, etc. . . . "

from such a change is *independent* of the velocity *possessed*. In the impact of two identical bodies the velocity gained by one is always equal to the velocity supposed gained by the other. If the two bodies differ in matter or volume, the velocities gained in the impact are unequal but in a constant ratio.

Barré de Saint-Venant expressed these experimental facts in the following general law.

" Bodies move as systems of points which have, at each instant, in space, accelerations whose geometrical components—which are directed along the lines which join the points, and are variable with the lengths of these lines but not with the velocities of the points—are always equal and opposed for the two points between which each line is a measure of the distance. "

Barré de Saint-Venant then introduced the concepts of mass and force based on this general law.

" *Masses* are those numbers which are proportional to the numbers of the elementary points that it is necessary to suppose in the bodies, comparing one with another, to explain their various motions by means of the statement of the general law.

" Attractive or repulsive *forces* of bodies, considered two by two, are the lines proportional to the resultants of the reciprocal accelerations of their elementary points according to the same law. Generally it is supposed, for simplicity, that the constant and arbitrary relationship of the forces with the resultants is the same as the constant relationship of the masses with the number of points. "

Then Barré de Saint-Venant arrives at the following definitions.

" *Masses.* — The mass of a body is the ratio of two numbers expressing how many times this body and another body, chosen arbitrarily and always the same, contain parts which, being separated and colliding with each other two by two, communicate opposite equal velocities to each other.

" *Forces.* — The force, or the attractive or repulsive attraction, of a body on another is a line whose length is the product of the mass of the second body and the mean acceleration of its points towards those of the first ; its direction is that of the acceleration. "

In order that these definitions may be appreciated, it must be emphasised, as Jouguet has done,[1] that Barré de Saint-Venant was a " convinced atomist. " From his " entirely practical " point of view, he explicitly refused to discuss whether the masses were in any way related to the quantities of matter of different heterogeneous bodies, and whether the forces were in any way related to the *efficient causes* of motion.

[1] *L. M.*, Vol. II, p. 81.

3. REECH AND THE " SCHOOL OF THE THREAD. "

In contrast, Reech adopted Euler's point of view. That is, he took the concept of force as his starting-point. We shall follow here his *Cours de Mécanique d'après la nature généralement flexible et élastique des corps* (1852).

To Reech, the word *force* did not denote a cause of motion, but " that effect of any cause that is called pressure or traction and which we appreciate with such a high degree of clarity in a stretched thread, supposed to be deprived of its material quality or mass. "

Reech pictured a particle attached to a thread. By causing an elongation of the thread, force is produced and the motion of the particle is modified.

The particle opposes a force of inertia (in Euler's sense) to the thread.

The force of the thread is directly measurable by the elongation of the thread. The force of inertia of the particle, mf (where m is its mass), is deduced from

$$(1) \qquad \vec{F} = m\vec{f}.$$

As Huyghens did in *De vi Centrifuga*, Reech supposes that the thread is suddenly cut. The acceleration of the particle suffers a discontinuity

$$(2) \qquad \vec{f} = \vec{\gamma} - \vec{\gamma}'$$

and the experiment yields the value of $\vec{\gamma}'$.[1]

Reech intended to verify the identity of $\vec{f}$ in the two formulae (1) and (2). It appeared evident to him that $\vec{F}$ is proportional to the mass. As for the identity of the direction of $\vec{F}$ and $\vec{f}$, he substantiated this by the fact that the lateral motion of the particle in a direction perpendicular to the thread could not change. Briefly

$$\vec{F} = m\vec{f}\,\psi\,(f\ldots)$$

where ψ can depend, *a priori*, on the time, position and velocity of the point.

Reech excluded any dependence on the position of the particle— " the properties of matter must be thought of as being the same everywhere. " Observation of a plumb-line and the arbitrary character of the system of reference exclude the time and the velocity of the particle. Therefore

$$\vec{F} = m\vec{f}\,\psi\,(f).$$

---

[1] For the sake of brevity, we have taken liberties with REECH's notation.

The principle of the independence of the partial effects of several simultaneous forces makes it impossible that ψ should be anything else than a constant, which can be taken equal to unity by a suitable choice of the mass.

Finally the law of motion

$$\vec{F} = m\left(\vec{\gamma} - \vec{\gamma}'\right)$$

is obtained.

Very properly, Reech makes use here of " the physical and experimental sense of things. " This law " will not be one of the elementary truths that may be assumed at the beginning of the science, in the guise of an axiom. Neither will it be a purely abstract, or purely experimental truth. In it will be found much of one and much of the other, and the subsequent applications of the science which will depend on it will prove its perfect legitimacy. "

If the force F vanishes, according to Reech's law it follows that

$$\vec{\gamma} = \vec{\gamma}'$$

for the motion of a particle which is completely free.

With axes fixed on the earth, experiment shows that

$$\vec{\gamma}' = \vec{g}.$$

Accordingly, trajectories are parabolic.

At this point Reech observed that he had no need to invoke " that well-known law of inertia concerning the state of uniform and rectilinear motion of a particle which is not acted upon by any force. "

This statement follows from the fact that Reech went back to the source of things by only applying the word " force " to the traction of a thread, and by accepting every trajectory that experiment could assign to an apparently free particle as a natural state of motion.

Thus, in terrestrial mechanics, this trajectory would be a parabola. " If it were only the action of gravity that could show itself, impalpably, to our senses, there is no doubt that we would not think of modifying this point of view in any way. For we could regard the parabolic state of the motion of bodies at the surface of the Earth as being the natural state of all kinds of matter in the absence of obstacles, and from this point of view there would be nothing more to seek. "

But experiment shows that there are also causes which are electrical, magnetic, etc. . . . and " which set bodies in motion quite as impalpably and mysteriously as gravity does. "

To include these, Reech was led to decompose the acceleration $\vec{\gamma}'$ in the form

$$\vec{\gamma}' = \vec{\gamma}'_0 + \vec{\gamma}''_0$$

and to write the law of motion

$$\vec{F} = m\,(\vec{\gamma} - \vec{\gamma_0} - \vec{\gamma_0''})$$

or, alternatively,

$$\vec{F} + m\vec{\gamma_0'} = m\,(\vec{\gamma'} - \vec{\gamma_0''}).$$

The left-hand side of the last equation is identified as the *total force*, comprising the *effective force* F and the " mysteriously acting " force $m\vec{\gamma_0'}$. In order that this may not be a " vain complication, " Reech chooses, *by convention*, the rectilinear and uniform motion in the case of a free particle—a particle which is not subject to any *total* force. This reduces to the annullment of $\vec{\gamma_0''}$ and yields the law

$$\vec{F} + m\vec{\gamma_0'} = m\vec{\gamma} = \vec{F_0} = \text{total force}.$$

To what extent does Reech's convention permit independence of the notion of absolute motion (or, what is more accurate here, of the notion of privileged frames of reference *in space*) ?

Andrade has remarked [1]—and this is, moreover, a classical proposition in kinematics—that as long as the velocities do not suffer any discontinuity, the difference $\vec{\gamma} - \vec{\gamma'}$ is independent of the chosen reference frame, while both $\vec{\gamma}$ and $\vec{\gamma'}$, taken separately, do depend upon it.

Therefore Reech's fundamental law

$$\vec{F} = m\,(\vec{\gamma} - \vec{\gamma'})$$

is independent of the chosen reference system. More accurately, this law is the same for two reference systems in continuous (and arbitrary) motion relative to each other.

Andrade goes further and seeks the effect of a modification of the time on Reech's law. Let j and γ be the initial accelerations with respect to two clocks, a and α, which record times t and θ respectively. Let j' and γ' be the final accelerations with respect to the same clocks. If u denotes the velocity with respect to the clock α, by simple differentiation

$$\vec{\gamma} = \vec{j}\left(\frac{dt}{d\theta}\right)^2 + \vec{u}\,\frac{d^2t}{d\theta^2}$$

$$\vec{\gamma'} = \vec{j'}\left(\frac{dt}{d\theta}\right)^2 + \vec{u}\,\frac{d^2t}{d\theta^2}.$$

Whence

$$\vec{\gamma} - \vec{\gamma'} = (\vec{j} - \vec{j'})\left(\frac{dt}{d\theta}\right)^2$$

[1] *Leçons de Mécanique physique*, 1898.

Andrade concludes, " If, for example, experiment shows that a known static force $\vec{F}$ produces variations $\gamma' - \gamma$ of the acceleration in the natural motion of a single particle, at different experimental epochs θ ; and if it were true that $\vec{F} = H(\theta)\,(\gamma' - \gamma)$—where H is a positive coefficient which depends on θ [alone]—a time t could always be defined such that

$$F = A(j' - j). \qquad (A = \text{const})$$

Then it would be sufficient to define the progress of the clock a, with respect to the clock α, by the relation

$$t = \int \frac{d\theta \sqrt{A}}{\sqrt{H}}$$

and t could be called the " absolute time. "

But, as Jouguet has remarked,[1] the existence of privileged reference frames is not eliminated from mechanics in this way. It reappears with the principle of the equality of action and reaction. It is necessary to suppose that there exists a reference frame in which all the particles exert upon each other forces which are equal in pairs.

4. KIRCHHOFF AND THE LOGISTIC STRUCTURE OF MECHANICS.

Kirchhoff made an attempt to develop a *logistic* structure of mechanics.[2]

The question is that of constructing mechanics with the notions of space, time and matter, and using, if necessary, concepts like force and mass which are derived from these three.

The motion of a particle may be described with the help of its coordinates, as functions of the time.

It can also be described with the help of the components of its velocity, as functions of the time, at every instant by starting from a knowledge of its coordinates at a given instant.

The motion of a particle can also be determined with the help of the components of its acceleration at every instant, starting from a knowledge of the position and velocity of the particle at a given instant.

Evidently this procedure could be carried out with the help of derivatives of the third, or a higher, order. " But experiment shows that natural motions are such that simplicity would not be gained,

[1] *L. M.*, Vol. II, p. 151.
[2] *Mechanik* (Teubner, Leipzig), 1876.

but that, on the contrary, it would be lost, by this introduction. This is the result of the fact that in all natural motion, as experiment has shown, the second derivatives of the coordinates are functions of the coordinates which do not contain the initial values of the coordinates and components of velocity. "

Kirchhoff writes the equations of motion

$$\frac{d^2x}{dt^2} = X \qquad \frac{d^2y}{dt^2} = Y \qquad \frac{d^2z}{dt^2} = Z$$

and considers the right-hand side of each of these equations as being one of the components of an *accelerating force* acting on a particle of unit mass. A particle is said to be subject to a system of forces if its motion is effected in accordance with the system

$$\frac{d^2x}{dt^2} = X_1 + X_2 + \ldots \quad \frac{d^2y}{dt^2} = Y_1 + Y_2 + \ldots \quad \frac{d^2z}{dt^2} = Z_1 + Z_2 + \ldots$$

A system of forces acting on a point is always equivalent to a unique force, which is the resultant of the system

$$X = X_1 + X_2 + \ldots \qquad Y = Y_1 + Y_2 + \ldots \qquad Z = Z_1 + Z_2 + \ldots$$

When the system consists of only two forces, these equations are the analytical expression of the theorem of *the parallelogram of forces*.

At this point Kirchhoff makes the following important observation— all the forces $(X_1, Y_1, Z_1), (X_2, Y_2, Z_2) \ldots$ *except one* can be chosen arbitrarily, and the remaining non-arbitrary force is then chosen in such a way that the resultant is equal to the acceleration. Mechanics is unable to give a complete definition of the concept of force. However, " experiment shows that, in natural motions, there can always be found systems whose separate forces are given more easily than their resultant. " At this point Kirchhoff cites the conservative motion of n particles under the action of their mutual forces, in accordance with Newton's law.

Kirchhoff then turns to the expression of the bound motion of a particle. The constraint is written as

$$\varphi(x, y, z, t) = c$$

and the equation of motion becomes

$$\frac{d^2x}{dt^2} = X + \lambda \frac{\partial \varphi}{\partial x} \quad \frac{d^2y}{dt^2} = Y + \lambda \frac{\partial \varphi}{\partial y} \quad \frac{d^2z}{dt^2} = Z + \lambda \frac{\partial \varphi}{\partial z}.$$

The convenience of this choice justifies itself by the fact that the equations preserve their form in any system of rectangular axes if the given force (X, Y, Z) does not depend on the coordinate system.

Kirchhoff then generalised this result to motion taking place with friction and to a system of particles subject to constraints expressible in finite terms.

Whatever the worth of his synthesis may have been, Kirchhoff's purely logical exposition of mechanics only crowned a structure that was already complete.

5. MACH.

Although the first edition of his *Mechanics* was only published in 1883, Mach explicitly announced that he owed nothing to Kirchhoff. And as his personality was a very independent one, he must be readily believed.

Mach also announced that his book was " a work of critical explanation animated by an *anti-metaphysical* spirit. "

He was categorical on the experimental character of the axioms of mechanics.

" The principles of mechanics cannot be in any way considered as mathematically demonstrated truths ; rather, they must be considered as propositions which not only assume but also *demand* the continual control of experiment. " [1]

Mach made a close criticism of the principle of equality of action and reaction, as well as the concept of mass.

He refused to appeal to the notion of " quantity of matter, " just as much as to the atomistic hypothesis.

Making use of the principle of symmetry, Mach assumed that two identical bodies communicate to each other equal and opposite accelerations, directed along the straight line which joins them. Given this, he called " *bodies of the same mass, two bodies which communicate to each other equal and directly opposite accelerations when they interact.* " [2]

If the bodies A and B communicate to each other accelerations $- \varphi$ and $+ \varphi'$ when they interact, it will be said that the body B has $- \dfrac{\varphi}{\varphi'}$ as much mass as A. Alternatively, choosing the mass of the body A as unity, it will be said that a body is of mass m if this body, when it acts on the body A, communicates to the latter an acceleration equal to m times that which it itself receives by the action of A.

[1] *M.*, p. 230.
[2] *Ibid.*, p. 211.

Mach adds, " In this conception of mass there is no theory. The ' quantity of matter ' (in Newton's sense) is quite useless. It consists of nothing else than the fixation, the designation and the naming, of a fact."[1]

We draw attention, here, to the fact that even if Mach was one of the most lucid critics of mechanics, he did not succeed in banning all metaphysics from his thought, as his programme would have had him do. Science appeared to him, indeed, to be entirely inspired by a principle of economy which applied not to the ways of nature themselves, but to those of the mind. " All science sets out to replace and economise experiments by means of the reproduction and representation of the facts in the mind. This representation is more easily handled than experiment itself and can be substituted for it in most connection. "[2]

Even more explicitly, Mach wrote, " According to us all science has the mission of replacing experiment. Consequently it must remain partly in the domain of experiment and must partly go beyond this, always awaiting corroboration or denial from the latter. Where it is impossible to corroborate or deny, science has nothing to do. It always moves in the domain of *incomplete* experiment. . . . The agreement between theory and experiment can always be improved by the perfection of observational techniques.

" Isolated experiments, without the thoughts that accompany it would always remain unknown to us. The *most scientific thoughts* are those which remain valid in the *most extensive* domains and which supplement and enrich experiment *most*. In research we proceed by the principle of *continuity*, for only this can provide a useful and economical conception of experiment. "[3]

Mach felt himself able to assert that the recognition of this economic function, pervading the whole being of science, would expel " all mysticism " from the field of science. Here there is some self-deception. In the terminology of modern philosophy Mach, reducing science to a well-formed language, would be called a pan-mathematician.

6. Hertz.

Under the title of *Die Prinzipien der Mechanik in neuem Zusammenhang dargestellt*,[4] Hertz constructed a mechanics from the three fundamental concepts of time, space and mass.

[1] *M.*, p. 212.
[2] *Ibid.*, p. 449.
[3] *Ibid.*, p. 457.
[4] Leipzig, 1894.

Hertz declared that with the help of external objects—and with the object of predicting the future by means of the past—we construct " certain images or internal symbols " in such a way that the necessary logical consequences of these images might be in agreement with the reality they represent.

The requisite properties of these images are—the absence of logical contradiction *(Zulässigkeit)*, agreement with experiment *(Richtigkeit)* and finally, convenience *(Zweckmässigkeit)*.

According to Hertz, the commonly accepted presentation of mechanics proceeds from four concepts (space, mass, force and motion) which are connected by Newton's laws and d'Alembert's principle.

To Hertz, this sytem does not appear free from contradiction. The circular motion of a stone at the end of a cord requires a force according to Newton's second law of motion. Newton's third law requires a reaction equal to the force exerted by the hand on the stone (through the agency of the cord). This reaction is called the *force of inertia* or the *centrifugal force*. But does it differ from the inertia of the stone ? Is the latter not counted twice, first as mass and a second time as force ? If, on the other hand, force is a cause of motion, existing before motion, is it possible to speak of a force which may be a consequence of motion ? In fact, " the designation of the force of inertia as a force is improper. Like that of living force, its name is a historical legacy, and reasons of convenience excuse its use more than they justify it. But what becomes of the requirements of the third law, which demands a force exerted by the inanimate hand on the stone, and which is only satisfied by a real force, not by a simple name ? "

Hertz does not go as far as to deny the *Zulässigkeit* of the classical system. The logical imprecisions of that system " merely depend on arbitrary incidental features added by us to the essential content provided by nature. "

He does not criticise the *Richtigkeit* of the classical system with regard to contemporary experiment. But every reservation must be made for experiment to come—" What arises from experiment may be overturned by it. We must not allow ourselves to be misled by the fact that the experimental elements in our principles are hidden in and fused with the unchanging and logically necessary elements. "

Finally, as far as the *Zweckmässigkeit* of the classical field is concerned, Hertz remarks that mechanics does not embrace all the properties of natural motion. The classical field could be both too narrow and too wide ; just as the concept of force is too wide, since we know that all forces must be subject to the conservation of energy. On the other hand, mechanics introduces many parasitical representations.

" The forces which our mechanics uses to treat all the problems of physics often resemble wheels turning in the vacuum. Thus, in celestial mechanics, direct observation of the forces of gravitation is never carried out ; only the positions of the stars are susceptible to it. Forces only appear as auxiliary quantities in the process of relating future experiments to past experiment, and disappear in the sequel. "

Moreover, Hertz dismisses all expositions of an energetic tendency, in which the concept of energy is substituted for that of force. Only retaining the three fundamental representations constituted by space, time and mass, he approaches an acceptance of Kirchhoff's ideas. But an addition appears necessary to him to account for the diversity of experiments possible with the help of these fundamental concepts alone.

" If we seek to understand the motion of the bodies around us, and to reduce them to simple and clear rules, but only consider what is directly under our eyes, our research will, in general, prove abortive. . . . If we wish to obtain a picture of the world which is closed on itself, referred to laws, *we must conjecture other invisible beings behind the things we see, and seek the hidden actors behind the barriers of our senses.* "

To Hertz, the concepts of force and energy in the classical presentation would be idealisations of this kind.

" But another path is open to us. We can assume that some hidden thing acts, and nevertheless deny that this something belongs to a particular category. We are free to assume that *what is hidden is nothing else than motion and mass*, not differing from the visible masses and motions but merely having other relations with us and our usual method of perception. "

By this addition to the visible masses of the universe, of hidden masses subject to the same laws, Hertz declares that the universe becomes understandable and governed by laws. " What we are accustomed to denote by the names of force and energy then reduces to an action of mass and of motion. But it is not necessary that this should always be the action of a mass or a motion which is perceptible by our gross senses. "

It would seem that this reduction of the concepts of mechanics was a reflection of the cartesian doctrine in Hertz's thought ; on the other hand, it made appeal to hypothetical elements.

" The masses of the hidden bodies can only be defined by hypothesis. "

For an *isolated system* composed of any number of material elements, Hertz posed the fundamental law of mechanics, *a priori*, in the following form.

" *The system travels along a trajectory of least curvature with a constant velocity—that is, it travels along a trajectory whose curvature at a point is less than that of every tangential trajectory.*"

Curvature must be understood as the quantity

$$Sm(x''^2 + y''^2 + z''^2)$$

extended to all the points of the system. Thus Hertz's fundamental law may be written

$$\delta \{ Sm(x''^2 + y''^2 + z''^2) \} = 0$$

where the variation δ is taken without allowing the variation of the coordinates or the velocities. This law reduces rather easily to the known laws of mechanics, especially to Gauss's principle of least constraint.[1]

Hertz considered a *non-isolated* system to be part of an isolated system. Any observable system is part of an isolated system, of which another part can remain hidden. The forces which act on the observable system result from the masses and the motions of the hidden part through the agency of the constraints (in Lagrange's sense).

Gauss himself, in formulating the principle of least constraint, had emphasised the impossibility of discovering, in the classical field, a principle that was basically distinct from that of d'Alembert. While greeting Hertz's conceptions as an " essential progress, " Mach criticised him for having reduced the physical content of mechanics —even more than Lagrange had done—to the extent of making it, on the surface, barely perceptible. In short, Mach complained of the abstract character of Hertz's mechanics. It seems that it was the *arbitrary* character of the hidden masses that caused his reluctance to adopt it, in spite of the attractive reduction in the number of fundamental concepts that it achieved.

Moreover, Hertz, above all, cared for logical and formal perfection in this matter, and willingly acknowledged that the classical presentation had a greater practical value than his own system.

7. POINCARÉ— CRITICISM OF THE PRINCIPLES AND DISCUSSION OF THE NOTION OF ABSOLUTE MOTION.

In the pages of his rightly famous interpretative works, Henri Poincaré carried out a penetrating discussion of the principles of classical mechanics.

---

[1] *Cf.* JOUGUET, *L. M.*, Vol. II, p. 275.

In *Science et Hypothèse* Poincaré remarked that " the English teach mechanics as an experimental science " while " on the Continent, it is always presented more or less as a deductive and *a priori* science. " His intention was to distinguish, in this branch of physics, " what is experiment, what is mathematical argument, what is convention and what hypothesis. "

Mechanics contains serious difficulties, for although it can only conceive of relative motions, it locates them in an *absolute space*. *Absolute time*, on the other hand, is nothing but a convention— simultaneity at two different localities is not an intuitive notion. Finally, it would seem permissible to resort to a non-euclidean space to describe motions, for the euclidean geometry itself is only a kind of convention.

The *principle of inertia* does not impose itself *a priori* (otherwise the Greeks would not have misunderstood it). No more is it an experimental fact, for it is never possible to experiment on bodies which are removed from the action of all force.

It is therefore necessary to pose as a principle, verified by its consequences, the fact that the motions of all material molecules of the Universe depend on differential equations of the second order.

Force, considered as a cause of motion, belongs only to metaphysics. In order that the concept may become useable, it is necessary to define a measure. As it is impossible, in order to estimate the equality of two forces, to detach a force that is applied to a body and attach it to another, " as a locomotive is uncoupled and coupled to another train, " force appeals to the *principle of the equality of action and reaction* —which has the character of a *definition* and not that of an experimental law.

Given this, if two bodies A and B interact, the acceleration of A multiplied by the mass of A is equal to the *action* of B on A. Conversely, the product of the acceleration of B and the mass of B is equal to the *reaction* of A on B. Since the action is equal to the reaction by definition, the masses of A and B are inversely proportional to their accelerations. Since this is the definition of the ratio of two masses, it is the task of experiment to verify whether this ratio is constant.

This would only be possible if A and B were the only masses present, " and were removed from the action of the rest of the world. " Therefore it is necessary to be able to *decompose* the acceleration of A in order to recognise the component arising from the action of B, with the exclusion of those of $C, D, E. \ldots$ This becomes possible if the *hypothesis of central forces* is assumed—that is, if the mutual attraction of two masses m and m' is assumed to be given by the law

$$k \frac{mm'}{r^2}$$

(where r is the distance between the two masses).

But nothing ensures that one day experiment will not invalidate this law. And if it must be given up, we no longer have a means of recognising the respective actions of B, C, D, E ... on A. The rule for the measurement of mass becomes inapplicable.

Then we fall back on the *single law of rectilinear and uniform motion of the centre of gravity of an isolated system.* But this law itself is only rigorous if it is applied to the whole Universe. Since it is clearly impossible to observe the motion of the centre of gravity of the Universe, we end in a " confession of impotence. " Masses no longer appear except as " coefficients which it is convenient to introduce in the calculations. "

After having thus robbed us of all our illusions, Poincaré consoles us in the following way. Force is equal to the product of mass and acceleration *by definition.* Similarly, action is equal to reaction *by definition.* These principles are *unverifiable,* for there are no *perfectly* isolated systems in nature. But there do exist systems that are *approximately* isolated, and to these the newtonian principles apply approximately. " Thus it is explained how experiment can serve as the foundation of the principles of mechanics and yet will never be able to contradict them. "

Then Poincaré discusses Kirchhoff's thesis, " which [seems to have only] complied with the general tendency of mathematician's towards nominalism. " He asserts that the notion of force is a *primitive, irreducible and indefinable notion,* the direct intuition of which is the notion of effort.

At this point Poincaré attacks the " School of the thread, " or the mechanics of Reech and Andrade. He declares, " We have defined the force to which the thread is subject by the deformation experienced by the thread, which is very reasonable. Next we have assumed that if a body is attached to the thread, the effort which is transmitted to it by the thread is equal to the action which the body exerts on the thread. Thus we have made use of the principle of the equality of action and reaction, and have considered it not as an experimental truth, but as the very definition of force. "

According to Poincaré, this definition appears just as conventional as that of Kirchhoff and has the disadvantage of being much less general. For example, if it were assumed that the Earth were attached to the Sun by some invisible thread, it would clearly be impossible to measure the extension of this thread.

However, Poincaré concedes that the Reech-Andrade mechanics has the advantage of being more understandable. "The reflections that it suggests to us show how the human mind has risen from a naive anthropomorphism to the present conceptions of science."

From all this discussion, Poincaré concludes that the law of acceleration and the rule of the parallelogram of forces are only conventions. But these conventions are not arbitrary. They are the products of imperfect experiments which, nevertheless, suffice to justify them. Further, the experimental origin of these conventions must not be forgotten.

In the following chapter of *Science et Hypothèse* Poincaré goes on to discuss relative and absolute motion. He understands the *principle of relative motion* to be the conservation of the laws of mechanics, whether they are referred to fixed axes or to axes required to move uniformly and rectilinearly. This principle is corroborated by the most everyday experiment.

It follows that the acceleration must not depend on the absolute velocity. Or, if it is prefered, the accelerations of the different bodies of an isolated system only depend on the differences of their velocities and the differences of their coordinates.

Thus the principle of relative motion appears as a kind of generalisation of the principle of inertia— it allows of the same discussion and therefore does not have the status of an *a priori* decision or that of a direct result of experiment.

Why is this principle only true when the motion of the movable axes is rectilinear and uniform ? In particular, why is it no longer applicable when the motion of the axes reduces to a rotation, even a uniform one ?

"If the sky were always covered with clouds, if we had no means of observing the stars, we would nonetheless conclude that the earth rotates. We would be informed by its flattening, or better still, by Foucault's pendulum experiment.

"And yet, in this case, would it be meaningful to say that the earth rotates? If absolute space does not exist, is it possible to rotate without rotating with respect to something, and on the other hand, how could we accept Newton's conclusion and believe in absolute space ? . . .

"Return to our picture. Thick clouds hide the stars from men, who cannot observe them and are unaware of their existence. How will men know that the earth rotates ? Even more than our ancestors, they will regard the ground that supports them as fixed and immovable; they will wait much longer for the coming of a Copernicus. But finally this Copernicus would arrive. How could he come ? "

Poincaré explains that, in mechanics, the centrifugal force and the compound centrifugal force could be attributed a real existence without contradicting the principle of generalised inertia.

But if an effectively isolated system were ever achieved, the centre of gravity would have a curvilinear trajectory. If centrifugal force were then invoked and were attributed to the mutual actions of bodies, to the extent that the isolation of the system was perfected, this force would seem to increase indefinitely with the distance, not to vanish.

Clearly an *ether* exerting a repulsive action on bodies could be imagined. But the laws of motion would never be symmetrical—they would distinguish between left and right.

From complexity upon complexity there would spring " something which would not be more extraordinary than Ptolemy's spheres of glass " until the Copernicus came to say—" It is much more simple to assume that the Earth rotates. " More simple ; because the laws of mechanics would then be expressed in a much more convenient language.

This does not confer objective existence on absolute space. Experimentally speaking, the statement " the earth rotates " has no meaning. Or, more accurately, it only has the following meaning— " It is more convenient to suppose that the earth rotates. "

8. POINCARÉ AND THE ENERGETIC THESIS.

In Chapter VII of *Science et Hypothèse* Poincaré discussed the energetic thesis. We have outlined Helmholtz's conception of this above.

In Poincaré's opinion, the energetic thesis offered two distinct advantages.

1) It is less incomplete, in that it excludes certain motions which (according to Helmholtz) do not occur in nature, and which would be incompatible with the classical theory.

2) It dispenses with the hypothesis of atoms (in the classical field, which is the only one that concerns us here) which is almost unavoidable in the customary presentation.

But, in compensation, it is hardly easier to define the two energies, potential and kinetic, than it is to define force and mass in the usual presentation.

In conservative systems U (potential energy), which only depends on positions, and T (kinetic energy), which only depends on velocities, can easily be distinguished. But if the forces depend on the velocities, the distinction between U and T becomes artificial.

Moreover, in the conservation of energy it is necessary to take account of all forms of energy. Together with T and U, it is necessary to introduce a factor Q which represents the molecular *internal energy* present in thermal, chemical and electrical forms, and, in general, to write

$$\varphi(T + U + Q) = \text{const.}$$

Now the electrostatic energy depends on the positions, and the electrodynamic energy on the positions and the velocities.

" We do not have any way of sorting out the terms which must become T, U and Q and of separating the three parts of the energy. "

In short, we are left with this statement of the energy principle—there is *something that remains constant.* In this form it is outside the reach of experiment and is a kind of tautology. It is clear that if the world is governed by laws there will be quantities that will remain constant. Like Newton's principles, and for an analogous reason, the principle of energy, founded on experiment could no more be invalidated by it. "

On the whole, in passing from the classical system to the energetic system some progress is achieved. But this progress is not sufficient.

In passing, Poincaré stresses that the very statement of the principle of least action has some quality which offends the mind.

" In order to move from one point to another a material molecule which is removed from the action of all force, but bound to move on a surface, will take the geodesic line—that is the shortest path.

" This molecule appears to know the point to which it is wished to bring it, appears to know the time it will take to reach there by following this or that path, and then appears to choose the most suitable path. In a sense, this statement holds the molecule up to us as a living and free being. "

Poincaré desired a less offensive statement " in which final causes would not seem to substitute themselves for efficient causes. "

This criticism also applies to Maupertuis' thesis of least action. But, before Maupertuis, Fermat had been subjected to a similar reproach from the Cartesians. He had replied, we recall, that he did not attempt to penetrate the hidden and obscure ways of Nature, and that he only offered her " a small geometrical assistance, " supposing that she had need of it. Lagrange and Hamilton had taken care to eliminate the metaphysical content of Maupertuis' thesis. As a corollary of the laws of dynamics, the principle of least action is not in itself of a finalist essence.

9. Painlevé and the principle of causality in mechanics.

We owe to Painlevé a profound discussion of the principles of mechanics, not only in the classical field, but also in that of relativity. In the next Part of this book we shall have occasion to return to Painlevé's conceptions. For the moment we shall confine ourselves to showing the application of the *principle of causality* that he made to classical dynamics.[1]

To Painlevé, the principle of causality and the notion of absolute motion has led to the discovery of all the axioms of mechanics. If it is now possible to separate these axioms from all " metaphysical motives, " it was necessary for the creators of mechanics to assume, *a priori*, certain properties of absolute motion.

Absolute motion distinguishes itself from all *apparent* motion by the fact that it alone corresponds to the principle of causality. How did the Schools understand this ?

A) The *Copernicans* (Kepler, [2] Galileo and Newton) assumed

1) that the absolute motion of the particles which compose the universe is the same after the time t and after the time t_1, if, at both these times, the particles occupy the same absolute positions and have the same absolute velocities;

2) that if at two times t and t_1 the same conditions, apart from a transport through space, are realised, the absolute motion of the particles which compose the universe is the same after the time t and after the time t_1, apart from the same transport through space;

3) that the absolute motion of a particle is not modified if the positions and the velocities of particles *infinitely removed* from the particle considered are, alone, modified.

It follows that a particle which is infinitely separated from all others is actuated by a rectilinear absolute motion. Moreover, it is postulated that this rectilinear motion is uniform. From these principles the Copernicans deduced that the Earth cannot be absolutely fixed and that it turns about an axis whose direction is fixed.

B) The *Scholastics* (those who adhere to aristotelian dynamics) assumed—explicitly or not—the same principles, but with the essential difference that *position alone* suffices to determine the subsequent motion of a moving body.

Every particle which is infinitely removed from all other elements

---

[1] *Les Axiomes de la Mécanique et le principe de causalité, Bulletin de la*¬*Société française de philosophie*, Vol. V, 1905. *Les Axiomes de la Mécanique*, Gauthier Villars, 1922.

[2] A reservation must be made with regard to Kepler, for, in dynamics, he remained faithful to the scholastic discipline.

of the universe must remain at absolute rest. If it is placed near the other elements it immediately acquires a certain velocity.

Of all the frames of reference that may be adopted, there is one with respect to which the motions of the Universe palpably enjoy the properties of absolute motion in the copernican sense (Galileo, Newton). This frame is actuated by a certain motion of rectilinear and uniform motion with respect to axes passing very nearly through the Sun, and having fixed directions with respect to the stars.

In Painlevé's opinion, the experimental confirmations (motion of planets, flattening of the Earth, variation of g with latitude, Foucault's pendulum, etc. . . .) tend to " confirm the *objective value* of the notion of absolute motion in the copernican sense. "

In order to appreciate Painlevé's thought it is necessary to repeat " the common form of the principle of causality which we apply, more or less unconsciously, all the time. "

" *If, at two times, the same conditions are realised, only transported in space and time, the same phenomena will reproduce themselves, only transported in space and time.* "

Painlevé contrasts this form with that usually assumed by the philosophers—

" *If the conditions of a phenomenon are determined at a given instant and in a given position, the phenomenon is determined.* "

The first principle implies the second, but the converse is not true. The second principle does not require—as the first does—that space and time cannot be *efficient causes*.

Painlevé himself recognises that he is translating the ideas of Copernicus and his successors into modern language. However, he gives them a form which the Copernicans were never able to give them—for Copernicus and Galileo did not know of the differential calculus.

Painlevé states the axioms of mechanics, " in a purely positive form, " in the following way.[1]

" *Axiom of inertia.* — A material element infinitely removed from all others describes a straight line with a constant velocity.

" *Axiom of action and reaction.* — [Painlevé states this axiom in terms of acceleration. If M and M_1 are the only elements present, the acceleration γ of M (caused by M_1) and the acceleration γ_1 of M_1 (caused by M) are *directly opposed*. The accelerations γ and γ_1 are determined in magnitude and direction when, at the time t, the distance r between the two points M and M_1 and their relative velocity

---

[1] *Les Axiomes de la Mécanique*, p. 65 *et sec.*

are known. The ratio $\frac{\gamma}{\gamma_1}$, called the *relative mass* of M_1 with respect to M, only depends on the position and velocity of M *relative* to M_1.

For three arbitrary elements M_1, M_2, and M_3, the relative mass of M_2 with respect to M_1 is the ratio of the relative mass of M_2 with respect to M_3 and the relative mass of M_1 with respect to M_3.]

" *Axiom of the independence of the effects of material elements.* — The acceleration of a material element M in the presence of a medium formed of the elements M_1, M_2, ... M_n is the geometrical sum of the accelerations produced by M_1, M_2, ... M_n respectively. "

If there exists a reference frame in which these axioms are satisfied, there will exist an *infinity of others*.

Indeed, all the axioms depend exclusively on the *accelerations* (including the axiom of inertia, which expresses the absence of acceleration for an isolated material element) and the accelerations are not altered by a rectilinear and uniform translation of the reference frame.

The *absolute mass* (or more simply, the mass) is defined by means of the relative mass by choosing, once and for all, an invariable element M_0.

The *absolute force*, which is here a derived notion, is by definition the product of the mass and the absolute acceleration.

We return to the significance of the principle of causality in mechanics. To many physicists the word *causality* has the restricted meaning of determination of future events. On the contrary, Painlevé insists on the necessity of not confusing causality and the hypothesis of determinism. To him the *principle* of causality in mechanics resides in the possibility of a certain transference of motion in space and time ; neither space nor time can be an efficient cause.

The term " causality " has also acquired the right to be cited in mathematics, thanks to Mr. Bouligand[1]— a demonstration is said to be *causal* if it succeeds in escaping all parasitical hypotheses of the kind that necessarily involve the use of certain algorithms. Moreover, to every group there corresponds a *domain of causality*, the hypotheses invariant under the transformations of the group entailing conclusions which are invariant under the same conditions.

The *domain of causality* of classical mechanics is determined by what is now called the *galilean group* (the group of transformations which conserve the absolute reference systems) while, for example, the *Lorentz group* defines the domain of causality of special relativity.

---

[1] *La causalité des théories mathématiques*, Hermann, Paris, 1934.

10. DUHEM AND THE EVOLUTION OF MECHANICS.

We shall conclude this review of the dicussion of the newtonian principles by extracting some of Duhem's interpretations of the various theories of mechanics from his book *Évolution de la Mécanique*.[1]

Descartes tried to ban the scholastic notion of quality from the science and to construct a universal mathematics in which quantity was the dominant entity. But according to Duhem, cartesian matter is incapable of motion and Cartesian motion is insufficient to be fashioned into a true mechanics. The relativity of all motion would be an illusion— the form of mechanics changes with the reference system.

Newtonian mechanics encountered the criticism of the Cartesians and the Atomists, who had no wish to hear talk of quality and, therefore, of attraction. But if Leibniz considered attraction as an " incorporeal and inexplicable property, " he saw the necessity of the notion of force—irreducible to extension and motion—to which he unfortunately gave the essentially metaphysical character of " substantial form. " This was to be harmful to the notion of force in the mind of many a physicist until the modern epoch.

Duhem contrasts the *analytical mechanics* of Lagrange with the *physical mechanics* of Poisson. While the first assumes both actions at a distance and forces of constraint, the second is based exclusively on *molecular actions* between *free* points. Following Poisson, de Saint-Venant remained a convinced atomist— to him, pressure was not a force of constraint in the sense of Lagrange, but the *mean repulsion* of fluid molecules. For himself, Duhem is a Lagrangian, to the extent that he does not consider mechanics as a science which is incomplete in itself.

First, after Lamé, he refers to the contradiction in Poisson's work of treating a continuous body by the use of integrals and not discrete summations. Then he remarks on the difficulty, already seen by Boscovitch, of obtaining equilibrium in assemblies of attractive interaction without the fusion of the constitutive particles. To remove this difficulty, Poisson and Navier assumed the eclipse of the mutual actions in the *natural state* of a body. Further, the demarcation between an isotropic elastic solid and an incompressible fluid is not apparent in this theory. Was not Poisson obliged to fall back on *secondary actions* —which restrict the mobility of the molecules—in order to rediscover Lagrange's constraints in an indirect way ?

Lagrange's mechanics must assume, apart from shape and motion, the notion of mass and the notion of force. But it only studies " rever-

[1] Published by Joanin, Paris, 1903.

sible" motions. It must be supplemented by passive resistances involving an *essentially negative* work.

Duhem is severe in his criticism of Kirchhoff's system. " The source of its rigour is also the source of its sterility, for it only writes identities. " It lacks the fecundity of intuition.

Finally Duhem comes to Hertz's endeavour, in comparison with W. Thomson's adynamic and gyrostatic ether, and reduces it to the following principle— "At each instant the forces of inertia applied to an independent system are such that, in every virtual displacement of the system, they do no work. "

To Duhem, this theory lacked complete application to the solution of concrete problems, and a determination of the hidden masses and motions that took the place of all force. Hertz's mechanics is less a doctrine than the programme of a doctrine. " This programme itself reduces, in the last analysis, to this statement— All forces introduced into the equations of dynamics can be regarded as the forces of constraint due to certain hypothetical bodies, or as the forces of constraint due to certain supposed motions. "

Without leaving the confines of this work we could not follow Duhem in the chapters in which he seeks to define a general mechanics based on thermodynamics, and which comprises both " a reaction against the atomistic and cartesian ideas " and " a very unexpected return to those which have contributed to this general mechanics to the most profound principles of the aristotelian doctrines ".

Duhem judges the classical theses in relation to this general mechanics, assessing the chance that each of them might lead to the solution of physical problems. Besides, with his accustomed candour, he adds, " I do not pride myself on impartiality. "

Duhem, writing in 1903, concludes with a remarkable prophecy—

" All that it is possible to say is that there is no reason in logic that allows the theories that have so far been outlined to be regarded as the only possible theories. In particular, the study of the various radiations, while presenting the experimenters with discovery upon discovery, has revealed to them such strange effects, so difficult to submit to the laws of our thermodynamics, that it would not be surprising to see a new branch of the science arise out of this study. "

11. CONCLUSION OF THIS CHAPTER.

The controversies on the value of the newtonian principles and concepts which arose during the second half of the XIXth Century and the beginning of the present Century are of considerable interest. They show clearly that the classical structure which appeared to have been completed by the work of Lagrange, could not be regarded as perfect. They proclaimed the necessity of a revision in the light of new experimental data.

This revision has taken place in the XXth Century, in the form of the modern physical theories of mechanics whose principles we are about to study.   To be accurate, most of the students of classical mechanics had not seen the necessity of this revision ; for this, it would be necessary to use reasons taken from optics, electromagnetism and the laws of radiation.   Moreover, certain of these put up a determined opposition to this revision—they were deeply convinced of the value of mechanics in their own traditional field.   But the revision was made possible by their own dissensions, which had shown the fallacy of attaching to the axioms of mechanics any other significance than that of contingent truths.

PART FIVE

THE PRINCIPLES
OF THE MODERN PHYSICAL THEORIES
OF MECHANICS

FOREWORD

In the preceding parts of this book we have treated the history of classical mechanics. We are about to attempt to carry this study into the field of the modern physical theories of mechanics.

There is no lack of objection to such an attempt. In the first place the shifting-sands of recent history may be feared. That is, it may be said that the freshly acquired facts are too unstable, and that there is no perspective from which to assess them.

Then it may be held that these new theories are confined to borrowing from classical dynamics certain elements of its symbolism, in order to construct models for an exclusively physical purpose, and that they have the ephemeral and brittle character of theories of this kind. Classical mechanics, firmly based on its principles, would have nothing to do with these variations.

The first objection is easy to rebutt. First, Michelson's experiment, and the first attempts to account for it, already lie a half-century behind us. The wave of incomprehension and pseudo-paradoxes stirred up by the special theory of relativity is now almost gone. And if wave and quantum mechanics is still developing technique, the associated axioms have already acquired a certain stable character. It is therefore not too soon to consign to history one of the most rapid and profound of the motions which shook the classical structure.

As for the second objection, it suffices to recall that classical mechanics often profited from close contact with physical theories. To cite only three names, Huyghens, Newton and Hamilton, who were physicists as well as students of mechanics, did not erect an artificial frontier between these two sciences.

It is true that in certain fields, like the dynamics of fluids or the dynamics of gases, the addition of physical or thermodynamical elements was indispensable and did not disturb the newtonian structure, while relativity and quanta cannot be accepted without a profound revision of the classical postulates. But we know, from history itself, that classical science was not born with that codified character that is now given it in its didactic presentation. It has known many vicissitudes. Its axioms are neither obvious nor of logical necessity.   It was experiment, and experiment alone, that enabled the classical science to escape from the scholastic doctrines.   Now, by its nature, experiment is always subject to revision.

If then it is agreed to dismiss these preliminary questions, the task of the historian and the critic can be undertaken.   So as not to make this dull, we shall suppose that the material of classical mechanics is known, even when this has not been referred to in the earlier pages of this book.   Modern theoretical physics makes constant and systematic use of this material.   Nor shall we attempt to validate the absolute differential calculus and the algebra of matrices.   These eliminations are justified by the fact, on which Painlevé has remarked, that it is misleading to tackle the modern mechanics without a serious foundation of classical culture and without a certain mathematical equipment.

Again, we shall not insist on the technical developments, sometimes considerable, which are often demanded in the treatment of concrete problems, even when these seem, at first sight, the simplest that could be posed.

We shall confine ourselves, then, to an attempt to follow the evolution of the modern physical theories of mechanics in the field of essential principles.   For this purpose it will suffice to analyse and compare certain fundamental texts, taken from the original papers themselves. As for the criticism, we shall most often find it in the writings of those of a classical tendency, like Painlevé, who were concerned with these questions, and in the attempts at interpretation that are due to the creators of the new mechanics themselves.

Conceived in this way, our task will not be a useless repetition of the didactic books, or of those whose object is to popularise and which, for practical reasons, often depart from the historical accuracy to which it is our only ambition to remain faithful.

CHAPTER ONE

SPECIAL RELATIVITY

A. *PRESENTATION*

1. IMMEDIATE ANTECEDENTS OF THE SPECIAL THEORY OF RELATIVITY.

The special theory of relativity was born of the difficulties of the optics of bodies in motion. In Fresnel's conception, waves of light were carried by the ether, an immaterial medium distributed through the vacuum and passing through material bodies. To Maxwell, optics lay within the compass of electromagnetic phenomena— a light wave is characterised by the vibration of two vectors $\vec{E}$ and $\vec{H}$ representing the electric and magnetic fields. The ether continues to exist in Maxwell's theory in the sense that the velocity of light with respect to the ether keeps a constant meaning. But it is no longer necessary to regard the ether as anything else than a reference system to which an absolute motion may be referred.

The state of motion of a system, in contrast with the state of absolute rest with respect to the ether, must involve optical effects accessible to experiment. In this way the theory accounts satisfactorily for effects of the *first order* ; that is, effects whose results depend on the ratio $\dfrac{v}{c}$, where v is a velocity of translation with respect to the ether, and c is the velocity of light in the vacuum. Such are, for example, the phenomenon of aberration (Bradley, Airy) ; the Doppler-Fizeau effect ; Fizeau's experiment on the transmission of light in a moving fluid ; the experiments of Wilson, Wien, Rowland, Roentgen and Eichenwald, and Sagnac and Harress.

But the theory was unable to deal with effects of the *second order*, that is, of the order of $\dfrac{v^2}{c^2}$, as soon as there were experiments accurate enough to allow these effects to be tackled. The special theory of relativity arose from this impasse.

2. MICHELSON'S EXPERIMENT AND LORENTZ'S HYPOTHESIS OF CONTRACTION.

Consider a system of reference S in which the ether is at absolute rest. Let S' be a system of reference actuated by a rectilinear and uniform translation, of velocity $\vec{v}$, with respect to S.

Consider two points A and B which are connected to the system S' and separated by a distance $AB = l$. Suppose that the direction of AB makes an angle θ with the direction of $v\,(0 < \theta < \pi)$. Let c be the velocity of light in the vacuum.

According to the wave theory, the velocity of propagation of light between A and B will be

$$t = \frac{l}{c^2 - v^2} \left(v \, \cos \, \theta + \sqrt{c^2 - v^2 \sin^2 \theta}\right).$$

Clearly the time taken for the propagation of light from B to A is deduced from this by the substitution of $\pi - \theta$ for θ.

Therefore the time taken for the propagation of light from A to B and then *back from B to A* will be

$$T = \frac{2l}{c^2 - v^2}\sqrt{c^2 - v^2 \sin^2 \theta} = \frac{2l}{c}\left[1 + \frac{v^2}{c^2}\left(1 - \frac{\sin^2 \theta}{2}\right) + \dots\right].$$

This time therefore differs from $\dfrac{2l}{c}$ by a quantity of the order of $\dfrac{v^2}{c^2}$.

This second order effect had been announced by Maxwell. It is much too small to lend itself to direct measurement— if, for example, the velocity $\vec{v}$ is that of the instantaneous translation of the Earth in its orbit, the effect is of the order of 10^{-8}.

But the effect is perfectly measurable by interference methods. The experiment was performed by Michelson in 1881 and repeated, with the collaboration of Morley, in 1887.[1] Two rays were made to interfere after having, separately, travelled *forward and backward* paths in the directions $\theta = \dfrac{\pi}{2}$ and $\theta = 0$ (and π). The difference of the times of propagation for these two paths was

$$\frac{l}{c} \cdot \frac{v^2}{c^2}.$$

---

[1] MICHELSON, *American Journal of Science*, Vol. 22, 1881, p. 120 and MICHELSON and MORLEY, *American Journal of Science*, Vol. 34, 1887, p. 333.

The result of the Michelson-Morley experiment was entirely negative, without the accidental errors of measurement having been implicated.

To assume, as Michelson did, that this result was due to a dragging of the ether would be to give up the explanation of the phenomenon of astronomical aberration provided by the wave theory.

Lorentz then put forward the following hypothesis.

" Assume that the arm of Michelson's apparatus lying in the direction of the Earth's motion contracts by an amount $l\left(\dfrac{v^2}{2c^2}\right)$ and that, at the same time, the effect of the translation conforms with Fresnel's theory. The result of Michelson's experiment is then completely explained. "

The field of hypotheses of this kind is, moreover, very wide. " For example, it can be supposed that the dimensions of a solid change in the ratio of 1 to $1 + \delta$ in the direction of motion, while the dimensions in the direction perpendicular to the direction of motion change in the ratio 1 to $1 + \varepsilon$. It must then be that

$$\varepsilon - \delta = \frac{v^2}{2c^2}$$

which allows the choice of, for example, one of the following solutions

$$\varepsilon = 0 \qquad \delta = -\frac{v^2}{2c^2}$$

$$\varepsilon = \frac{v^2}{2c^2} \qquad \delta = 0$$

$$\varepsilon = \frac{v^2}{4c^2} \qquad \delta = \frac{-v^2}{4c^2}. \text{ "}$$

Lorentz himself agrees that this hypothesis of contraction is most paradoxical at first sight. On the practical plane, he first remarks that this contraction is very small (6.5 cm. for the diameter of the Earth, or 1/200th of a micron per metre). Theoretically, he suggested that molecular actions, like electromagnetic forces, can be transmitted by the ether. If this is true, a translation will probably modify the reciprocal actions of two molecules or atoms in the same way as the separation or bringing-together of two charged particles. Since, in the last analysis, the shape and the dimensions of a solid body are conditioned by the intensity of molecular actions, the possibility of an alteration of the dimensions due to a translation is in no way excluded.

Lorentz then refers to a law of electrostatic interaction which he has discovered.[1] By extending this law to two systems of molecules S_1 and S_2, the second of which is at rest with respect to the ether and the first moving with velocity $\vec{v}$ in the direction of the axis of x, the following result is obtained. In both systems the components of the forces in the directions of the axis of x are the same, while the components in the directions of y and z are, in the moving system, multiplied by the coefficient $\sqrt{1 - \dfrac{v^2}{c^2}}$. Now it is clear that the forces must be in equilibrium in S_1 just as they are in S_2. The translation would therefore have the effect of reducing the dimensions of the molecules in the direction of the axis of x in the ratio of l to $\sqrt{1 - \dfrac{v^2}{c^2}}$. This leads to the choice of the solution

$$\varepsilon = 0 \qquad \delta = -\frac{v^2}{2c^2}$$

from among the hypotheses to which Michelson's experiment leads. To Lorentz himself we owe the information that Professor Fitzgerald, in his lectures, had already formulated the hypothesis on an analogous contraction. This same hypothesis had been considered by Lodge in 1893 in a study of aberration (*Phil. Trans. Roy. Soc.*, Vol. 184, p. 727).

3. The Lorentz transformation.

We now take up the study of a fundamental paper of Lorentz called *Electromagnetic phenomena in a system moving with any velocity less than that of light* (Proc. Acad Sc., Amsterdam, Vol. 6, 1904, p. 809). Lorentz starts from the equations of the theory of electrons

(E)
$$\begin{cases} \operatorname{div}\ \vec{E} = \varrho \quad \operatorname{div}\ \vec{H} = 0 \\ \operatorname{curl}\ \vec{H} = \dfrac{1}{c}\left(\dfrac{\partial \vec{E}}{\partial t} + \varrho \vec{u}_1\right) \\ \operatorname{curl}\ \vec{E} = -\dfrac{1}{c}\dfrac{\partial \vec{H}}{\partial t}. \end{cases}$$

These equations refer to a reference system $S(0_1 x_1 y_1 z_1)$ which is fixed with respect to the ether. Consider a reference system $S'(Oxyz)$ moving with respect to S with a uniform translation of velocity $\vec{v}$ directed along the axis of x—

$$x = x_1 - vt \qquad y = y_1 \qquad z = z_1.$$

---

[1] We shall encounter this law on p. 469.

The velocity $\vec{u}_1$ of an electron then transforms into $\vec{u} + \vec{v}$ and the operator $\dfrac{\partial}{\partial t}$ into $\dfrac{\partial}{\partial t} - v\dfrac{\partial}{\partial x}$. This transformation of variables leads to the equations

$$
\left\{
\begin{aligned}
&\quad \text{div } \vec{E} = \varrho \quad \text{div } \vec{H} = 0 \\
&\frac{\partial H_z}{\partial y} - \frac{\partial H_y}{\partial z} = \frac{1}{c}\left(\frac{\partial}{\partial t} - v\frac{\partial}{\partial x}\right)E_x + \frac{\varrho}{c}\,(v + u_x) \\
&\frac{\partial H_x}{\partial z} - \frac{\partial H_z}{\partial x} = \frac{1}{c}\left(\frac{\partial}{\partial t} - v\frac{\partial}{\partial x}\right)E_y + \frac{\varrho}{c}\,u_y \\
&\frac{\partial H_y}{\partial x} - \frac{\partial H_x}{\partial y} = \frac{1}{c}\left(\frac{\partial}{\partial t} - v\frac{\partial}{\partial x}\right)E_z + \frac{\varrho}{c}\,u_z \\
&\quad \text{curl } \vec{E} = -\frac{1}{c}\left(\frac{\partial}{\partial t} - v\frac{\partial}{\partial x}\right)\vec{H}.
\end{aligned}
\right.
$$

For the variables (x, y, z, t), which, we note in passing, would correspond to an ordinary galilean transformation in classical mechanics, Lorentz substitutes the variables

(T)
$$
\left\{
\begin{aligned}
x' &= \beta l x = \beta l\,(x_1 - vt) \\
y' &= l y = l y_1 \\
z' &= l z = l z_1 \\
t' &= \frac{l}{\beta}\,t - \beta l\,\frac{vx}{c^2} = l\beta\left(t - \frac{v}{c^2}x_1\right).
\end{aligned}
\right.
\qquad \text{with } \beta^2 = \cfrac{1}{1 - \cfrac{v^2}{c^2}}
$$

Here l is a function of v such that $l(0) = 1$, and which differs from unity by a quantity of the order of $\dfrac{v^2}{c^2}$.

The variable t' is what Lorentz calls the *local time (Ortszeit)*. It only coincides with t for $v = 0$.

With the intention of rediscovering the form of the equations (E) with the variables $(x'\ y'\ z'\ t')$ Lorentz defines the quantities $\vec{E'}$, $\vec{H'}$, $\vec{u}'$ by means of their components along the axes $Oxyz$, namely

(C)
$$
\left\{
\begin{array}{llll}
E' \ldots & \dfrac{1}{l^2}\,E_x & \dfrac{\beta}{l^2}\left(E_y - \dfrac{v}{c}\,H_z\right) & \dfrac{\beta}{l^2}\left(E_z + \dfrac{v}{c}\,H_y\right) \\[2ex]
H' \ldots & \dfrac{1}{l^2}\,H_x & \dfrac{\beta}{l^2}\left(H_y + \dfrac{v}{c}\,E_z\right) & \dfrac{\beta}{l^2}\left(H_z - \dfrac{v}{c}\,E_y\right) \\[2ex]
u' \ldots & \beta^2 u_x & \beta u_y & \beta u_z
\end{array}
\right.
$$

and the new density ϱ' by

$$
\varrho' = \frac{1}{\beta l^3}\,\varrho.
$$

Thus Lorentz succeeds in putting the equations (E) into the form

$$(\text{E}')\quad\begin{cases} \operatorname{div}' \vec{E}' = \left(1 - \dfrac{vu_x'}{c^2}\right)\varrho' \quad \operatorname{div}' \vec{H}' = 0 \\[2mm] \operatorname{curl}' \vec{H}' = \dfrac{1}{c}\left(\dfrac{\partial \vec{E}'}{\partial t'} + \varrho'\vec{u}'\right) \\[2mm] \operatorname{curl}' \vec{E}' = -\dfrac{1}{c}\dfrac{\partial \vec{H}'}{\partial t'} \end{cases}$$

where the symbols *div'* and *curl'* denote that the corresponding operations are effected with the help of the variables (x', y', z', t').

In passing we remark that the transformation (T) had been encountered by Voigt in 1887,[1] in a paper on the Doppler-Fizeau effect, from which it arose in the conservation of the wave equation

$$\Box \psi = \frac{\partial^2 \psi}{\partial x^2} + \frac{\partial^2 \psi}{\partial y^2} + \frac{\partial^2 \psi}{\partial z^2} - \frac{1}{c^2}\frac{\partial^2 \psi}{\partial t^2} = 0.$$

Historically this transformation, which is associated with the name of Lorentz, should be called the *Voigt-Lorentz transformation*. This all the more so as the conservation of the wave equation is one of the most remarkable of properties and is, moreover, directly related to the invariance of the quantity $x^2 + y^2 + z^2 - c^2 t^2$.

Lorentz himself acknowledged that the equations (E') only imperfectly achieved the end that he had set himself, and which Einstein was to accomplish—in fact there remains, in the equations (E'), a term $\dfrac{vu_x'}{c^2}$ which Lorentz did not succeed in removing.

The equations (E') entail the consequence that $\vec{E}$ and $\vec{H}$ separately depend on a scalar potential φ' and a vector potential $\vec{A}'$ which satisfy the equations

$$\begin{cases} \Box\,\varphi' = -\varrho' \\[2mm] \Box\,\vec{A}' = \dfrac{1}{c^2}\varrho'\vec{u}' \end{cases}$$

in such a way that

$$\begin{cases} \vec{E}' = -\dfrac{1}{c}\dfrac{\partial \vec{A}'}{\partial t'} - \overrightarrow{\operatorname{grad}}\,\varphi' + \dfrac{v}{c}\overrightarrow{\operatorname{grad}}\,A_x' \\[2mm] \vec{H}' = \operatorname{curl}\vec{A}'. \end{cases}$$

Without exhausting the consequences of these equations, we confine ourselves here to a consideration of an electrostatic system $\sum$ whose

[1] *Göttinger Nachrichten*, 1887, p. 41.

only motion is a simple translation of velocity $\vec{v}$. Then we can write

$$\vec{u}' = 0 \quad \vec{A}' = 0 \quad \vec{H}' = 0 \quad \vec{E}' = - \operatorname{grad} \varphi' \quad \Delta_2' \varphi' = - \varrho'$$

The force $\vec{F}$ exerted by unit charge on the element of volume of an electron has, as a general rule, the value

$$\vec{F} = \vec{E} + \frac{1}{c}(\vec{u_1} \wedge \vec{H}) \quad \text{where} \quad \vec{u_1} = \vec{u} + \vec{v}.$$

When they are expressed in terms of the accented quantities (E', H', u'), the components of this force along the axes of $Oxyz$ reduce, for the particular system $\sum$, to

$$l^2 E_x', \quad \frac{l^2}{\beta} E_y', \quad \frac{l^2}{\beta} E_z'.$$

Lorentz then associates with the system $\sum$ a system $\sum'$ which *is at rest in the space $x'\, y'\, z'$*, and whose dimensions are obtained by multiplying the corresponding dimensions of $\sum$ in the directions parallel to Ox, Oy, Oz by

$$\beta l, \quad l, \quad l,$$

respectively.

The deformation $\sum \to \sum'$ can be referred to as $(\beta l, l, l)$. Moreover, a density $\varrho' = \dfrac{\varrho}{\beta l^3}$ can be attributed to $\sum'$ so that the charges of corresponding elements of volume of the electrons of $\sum$ and $\sum'$ remain equal. The forces $F(\sum)$ applied to the electrons of $\sum$ are then deduced from the force $\vec{E}'$ applied to the electrons of $\sum'$ by means of the law

(F) $$F(\sum) = \left(l^2, \frac{l^2}{\beta}, \frac{l^2}{\beta}\right) F(\sum').$$

Lorentz then introduces the hypothesis " *that the electrons, which have the shape of a sphere in the state of rest, suffer, under the effect of the translation, the deformation* $\left(\dfrac{1}{\beta l}, \dfrac{1}{l}, \dfrac{1}{l}\right)$, *each element of volume conserving its charge.* "

The deformation $\sum \to \sum'$ or $(\beta l, l, l)$ then restores the spherical shape of the electrons. The relationships of the forces are determined by the law (F).

Lorentz postulates that these relationships *also govern the forces between the elementary particles of every heavy body*. These forces must be in equilibrium with each other, from which it follows that a solid body suffers the deformation $\left(\dfrac{1}{\beta l}, \dfrac{1}{l}, \dfrac{1}{l}\right)$ under the effect of the application of a translation of velocity $\vec{v}$.

Applied to the accelerations $\vec{\gamma}(\textstyle\sum)$ and $\vec{\gamma}(\textstyle\sum')$ the transformation (T) leads to the law

$$\vec{\gamma}(\textstyle\sum) = \left(\dfrac{l}{\beta^3}, \dfrac{l}{\beta^2}, \dfrac{l}{\beta^2}\right)\vec{\gamma}(\textstyle\sum').$$

Comparing this with the law (F), the following relation for the masses is obtained.

(M) $$m(\textstyle\sum) = (l\beta^3, l\beta, l\beta)\, m(\textstyle\sum').$$

Lorentz considers the electromagnetic momentum

$$\vec{G} = \frac{1}{c}\int (\vec{E} \wedge \vec{H})\, d\tau$$

where $d\tau$ is the element of volume of the space x, y, z. For an electrostatic system, and by the use of the inverted forms of e equations (C), this quantity reduces to

$$G_x = \frac{\beta^2 l^4 v}{c^2}\int (E_y'^2 + E_z'^2)d\tau = \frac{\beta l v}{c^2}\int (E_y'^2 + E_z'^2)\, d\tau'$$

where $d\tau'$ is the element of volume of the space x', y', z'.

If, in particular, $\sum$ reduces to a *single electron* whose charge e is supposed to be uniformly distributed (at rest) over a sphere of radius R, then

$$\int (E_y'^2 + E_z'^2)\, d\tau' = \frac{2}{3}\int E'^2 d\tau' = \frac{e^2}{6\pi}\int_R^\infty \frac{dr}{r^2} = \frac{e^2}{6\pi R}.$$

Therefore

$$G_x = \frac{e^2}{6\pi R c^2}\,\beta l v.$$

Now, $\vec{G}$ has the direction of $\vec{v}$ because of the symmetry of the system. Therefore

$$\vec{G} = \frac{e^2}{6\pi R c^2}\,\beta l \vec{v}.$$

This formula is only valid for a uniform translation of velocity $\vec{v}$. Lorentz assumes that it is still applicable to all times when the accelerations are sufficiently small. All oscillation in the motions entails the existence of a force

$$\vec{F} = \frac{d\vec{G}}{dt}.$$

Now, if it observed that

$$\frac{d}{dt}(\beta l\vec{v}) = \frac{d}{dt}(\beta l)\vec{v} + \beta l\frac{d\vec{v}}{dt}$$

it is reasonable to assign to the electron a *longitudinal mass* (in the direction of the motion)

$$m_l = \frac{d(\beta l v)}{dv} \cdot \frac{e^2}{6\pi Rc^2}$$

and a *transverse mass* (in a direction perpendicular to that of the motion)

$$m_t = \beta l \cdot \frac{e^2}{6\pi Rc^2}.$$

Both these masses are of an electromagnetic nature. Lorentz assumes that the electron does not possess any other " real " or " material " mass. In the state of rest

$$m_l = m_t = \frac{e^2}{6\pi Rc^2}.$$

In short, in this particular case

$$m(\textstyle\sum) = \left(\frac{d(\beta l v)}{dv},\ \beta l,\ \beta l\right) m(\textstyle\sum').$$

Comparing this with the law (M) already obtained for every system $\sum$, namely

(M) $$m(\textstyle\sum) = (l\beta^3, \beta l, \beta l)$$

there is obtained

$$\frac{d(\beta l v)}{dv} = \beta^3 l.$$

Now, by the definition of

$$\frac{d(\beta v)}{dv} = \beta^3$$

identically. Therefore

$$\frac{dl}{dv} = 0.$$

and, since $l(0) = 1$, $l \equiv 1$.

Lorentz is therefore led to assume *that the influence of a translation of velocity $\vec{v}$ on the dimensions of a single electron* (and, by extension, of a heavy body) *reduces to a relative contraction β parallel to the direction of motion.*

We have already seen that this influence explains the result of Michelson's experiment. It also explains the negative result of the experiment attempted by Trouton and Noble in 1903,[1] intended to demonstrate the existence of the couple, of the order of $\frac{v^2}{c^2}$, which would tend to orientate the plates of a charged condenser in the direction of the Earth's motion.

Finally the law of the variation of the mass of the electron introduced by Lorentz is in agreement with the experiments, carried out by Kaufmann,[2] on the deflection of β—rays in an electric or magnetic field, although the accuracy of these experiments does not allow the rejection of the values predicted by Abraham.[3]   More accurate experiments which confirmed Lorentz's were made much later, by Guye and Lavanchy.[4]

4. INTRODUCTION TO EINSTEIN'S ELECTRODYNAMICS.

In the statement of his intentions which appears in his paper *Zur Elektrodynamik bewegter Körper*,[5] Einstein indicated the asymetries which result from the application of Maxwell's theory to bodies in motion.   Thus the mutual actions of a magnet and a current do not depend exclusively on their relative motion, but differ according to whether the magnet is at rest and the body in motion or whether the converse situation obtains.

These difficulties, together with the acknowledged impossibility of demonstrating the instantaneous translation of the Earth with respect to the *medium* in which light waves are propagated, lead to the belief that, just as much in electrodynamics as in ordinary mechanics, no directly observable phenomenon can be connected with the notion of absolute rest.

Einstein therefore intends to construct a new electrodynamics of bodies in motion which might be free of contradiction, as simple as possible, and compatible with the laws formulated by Maxwell for bodies at rest.   In such a theory the consideration of an *ether*—the medium or the support of the vibrations of light—becomes superfluous *(überflüssig)*, for no special property arises to characterise a reference system which would be at absolute rest with respect to such a medium.

Einstein's paper, which we are about to analyse, contains a first kinematic part (definition of simultaneity, relativity of lengths and times, transformations of the coordinates of space and time, composition

[1] *Proc. Roy. Soc.*, Vol. 72, 1903, p. 165.
[2] *Phys. Zeitschr.*, Vol. 4, 1902, p. 55.
[3] *Ann. der Physik*, Vol. 10, 1903, p. 105.
[4] *Arch. de Genève*, Vol. 41, 1916, p. 286.
[5] *Ann. der Physik*, Vol. 17, 1905, p. 891.

of velocities) ; then a part devoted to electrodynamics (the conservation of Maxwell's equations) ; and finally, a part devoted to the dynamics of a slowly accelerated electron.

5. DEFINITION OF SIMULTANEITY.

Einstein founds his kinematics on the following definition of simultaneity.

Consider a system of reference, to be called the "*fixed system*," in which the laws of newtonian mechanics are valid.

If there is a clock at the point A in space, an observer at A can observe the time at which an event occurs in the immediate neighbourhood of A. Further, if there is an identical clock at the point B, an observer at B can observe the time at which an event occurs in his own immediate neighbourhood.

We thus provide for a " *local time of A* " and a " *local time of B*," but without a special convention we cannot compare the observations at A and B.

We assume *by definition* that the " time " taken for light to go from A to B is equal to the " time " taken for light to go from B to A. If a light-ray starts from A at the instant t_A (local time of A) and arrives at B at the instant t_B (local time of B), and is there reflected to return to A at the instant t_A' (local time of A), the clocks A and B will be said to be *synchronous* if

$$t_B - t_A = t_A' - t_B.$$

In addition, we suppose that

if the clock B is synchronous with the clock A, the converse is true ;

if the clock A is synchronous with the clocks B and C, then B and C are also synchronous.

The preceding " experiment of thought " involves a definition of *simultaneity* and a definition of time. The instant at which an event occurs is that which is measured, in the immediate neighbourhood of the point of space at which the event occurs, by a stationary clock which is synchronous with another stationary clock, chosen once and for all.

We regard it as established by the experiment that the velocity of light in the vacuum,

$$\frac{2AB}{t_A' - t_A} = c$$

is a universal constant.

The " time " thus defined relates to the " fixed system. "

6. Relativity of lengths and times.

Einstein starts from the following two principles.

a) *Principle of relativity.* — *The laws which govern the physical phenomena are the same in two systems of reference actuated, one with respect to the other, by a rectilinear and uniform translation.*

b) *Constancy of the velocity of light.* — *The* " time " *being defined as above, light is propagated in the* " fixed system " *with a velocity* c *which is independent of the motion of the source.*

Consider, in the fixed system, a measuring-rod which is at rest and whose length l is measured with a standard, also at rest.

When the same measuring-rod is moved with a rectilinear and uniform velocity $\vec{v}$, what does its length become ?

In this connection, the following two experiments can be conceived of.

a) The observer, carrying his standard, is displaced with the rod. The measurement is therefore carried out at relative rest in a moving system of reference, actuated by a rectilinear and uniform velocity $\vec{v}$ with respect to the fixed system.

b) The observer determines, *by means of synchronous clocks at rest in the fixed system,* the points of this system at which the ends A and B of the moving rod are found *at a given instant.* He then measures the distance between these two points by means of a standard which is at rest in this system.

It follows directly from the principle of relativity that the first measurement must give the result l.

As for the second measurement, which is concerned with the length r_{AB} of the moving rod as *observed from the fixed system,* we shall prove that it does not provide the result l.

On the contrary, classical mechanics implicitly assumes that the two measurements give the same result.

Now, imagine two clocks which are synchronous with the clocks of the fixed system and which are placed at the ends, A and B, of the moving rod. The readings of these two clocks correspond to the time in the fixed system in the positions that they occupy at a given instant. In addition, suppose that one observer accompanies each clock and applies the criterion of simultaneity to these. If r_{AB} is the length of the moving rod as measured in the fixed system, the two observers will find, because of the principle of the constancy of the velocity of light, that

$$t_B - t_A = \frac{r_{AB}}{c - v} \quad \text{and} \quad t_A' - t_B = \frac{r_{AB}}{c + v}. \,^1$$

---

[1] In these equations t_A, for example, is the reading of the moving clock when it occupies the position A. Therefore it is, by hypothesis, the time of the fixed system.

The two observers will conclude from this that the two clocks are not synchronous, while observers in the fixed system will pronounce them synchronous. Therefore *simultaneity* does not have an absolute meaning, and two events which are simultaneous in a given system are no longer so in a system in uniform translation with respect to the first.

7. Transformation of the coordinates of space and time.

Let there be a " fixed " system of reference $S(Oxyz)$ and a movable system of reference $S'(O'x'y'z')$ moving, with respect to the first, with a rectilinear and uniform velocity $\overline{v}$ directed along the axis Ox. In each system there is arranged a standard of length and, by graduation against synchronous clocks, a measure of time (t and t'). The latter is accomplished separately for S and S' by the procedure of the exchange of light signals.

An event in S is characterised by a system of values (x, y, z, t) and in S' by a system of values (x', y', z', t').

Einstein assumes that because of the homogeneity of space and time, the relations between the quantities $(x, y, z, t,)$ and (x', y', z', t') are *linear*. If $\xi = x - vt$, to a set of values (ξ, y, z) corresponds a point at relative rest in S'. We seek the relation which defines t' as a function of (ξ, y, z, t).

A light ray leaves the origin O' of the system S' at the time t'_0 and travels in the direction of the axis Oxx'. When it comes to the point $(\xi, 0, 0)$ at the time t'_1, this ray is reflected and returns to O' at the time t'_2. We must write

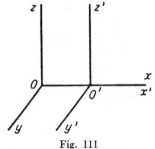

Fig. 111

$$\frac{1}{2}(t'_0 + t'_2) = t'_1.$$

Whence, in accordance with the principle of the constancy of light in the fixed system, this equation may be written

$$\frac{1}{2}\left[t'(0, 0, 0, t) + t'\left(0, 0, 0, t + \frac{\xi}{c-v} + \frac{\xi}{c+v}\right)\right] = t'\left(\xi, 0, 0, t + \frac{\xi}{c-v}\right).$$

If ξ is considered to be infinitely small

$$\frac{1}{2}\frac{\partial t'}{\partial t}\left(\frac{1}{c-v}+\frac{1}{c+v}\right)=\frac{\partial t'}{\partial \xi}+\frac{1}{c-v}\frac{\partial t'}{\partial t}.$$

Whence

$$\frac{\partial t'}{\partial \xi}+\frac{v}{c^2-v^2}\frac{\partial t'}{\partial t}=0.$$

Taking account of the fact that in a given direction $O'y'$ or $O'z'$ in S', perpendicular to the translation, light emitted from O' appears, when it is observed from the fixed system, to propagate itself with velocity $\sqrt{c^2-v^2}$, an analogous calculation can be performed and yields the result

$$\frac{\partial t'}{\partial y}=\frac{\partial t'}{\partial t}=0.$$

It then remains that

$$t'=a(v)\left[t-\frac{v\xi}{c^2-v^2}\right].$$

The principle of relativity combined with the principle of the constancy of the velocity of light in the fixed system, requires that the velocity of light *measured in the moving system* should be equal to c. Thus, for a ray travelling along the axis $O'x'$ and starting at time $t=t'=0$, it will be true that

$$x'=ct' \quad \text{or} \quad x'=ca(v)\left[t-\frac{v\xi}{c^2-v^2}\right].$$

But by observing this propagation in the fixed system, it is possible to write

$$\frac{\xi}{c-v}=t.$$

Therefore

$$x'=c^2a(v)\frac{\xi}{c^2-v^2}.$$

Similarly, considering a ray of light emitted from O' in the direction of $O'y'$, it turns out that

$$y'=ct'=ca(v)\left[t-v\frac{\xi}{c^2-v^2}\right] \quad \text{with} \quad \frac{y}{\sqrt{c^2-v^2}}=t$$

whence

$$y'=a(v)\frac{c}{\sqrt{c^2-v^2}}\,y \qquad \text{and} \qquad z'=a(v)\frac{c}{\sqrt{c^2-v^2}}\,z.$$

Collecting all these results, putting $\beta = \sqrt{1 - \dfrac{v^2}{c^2}}$ and replacing ξ by its value $x - vt$, it tums out that

$$(\text{T}) \quad \begin{cases} x' = \varphi(v)\beta(x - vt) \\ y' = \varphi(v)y \\ z' = \varphi(v)z \\ t' = \varphi(v)\beta\left(t - \dfrac{vx}{c^2}\right). \end{cases}$$

This transformation is identical with Lorentz's transformation (T). Further, Einstein verifies that a spherical light wave emitted at the time $t = t' = 0$ from the common origin of the systems S and S', will be observed in S as

$$(1) \qquad x^2 + y^2 + z^2 = c^2t^2.$$

Then, by the application of the transformation (T), it follows that

$$(2) \qquad x'^2 + y'^2 + z'^2 = c^2t'^2.$$

Thus an identical form is obtained in S'. He observes that the transformation (T) would be obtained more simply by starting from the condition that equation (1) should imply equation (2).

Einstein then considers another system S'' whose origin coincides with those of S' and S at the time $t = 0$, *and which may be moving with a rectilinear and uniform translation of velocity* $-\vec{\text{v}}$ *directed along* $O'x'$. A second application of the preceding analysis yields

$$\begin{cases} x'' = \varphi(-v)\beta(x' + vt') = \varphi(v)\varphi(-v)x \\ y'' = \varphi(-v)y' \qquad\quad = \varphi(v)\varphi(-v)y \\ z'' = \varphi(-v)z' \qquad\quad = \varphi(v)\varphi(-v)z \\ t'' = \varphi(-v)\beta\left(t' + \dfrac{vx'}{c^2}\right) = \varphi(v)\varphi(-v)t. \end{cases} \quad T(-v)T(v)$$

It is clear that this double transformation is equivalent to the identity transformation. Therefore $\varphi(v)\varphi(-v) \equiv 1$.

Einstein observes at this point if a segment of length l (measured in S') on the axis $O'y'$ is subjected to the transformation (T), the length of the same segment, measured in S, is found to be $\dfrac{l}{\varphi(v)}$. By reasons of symmetry, this result is independent of the sense of $\vec{v}$ and only depends on the magnitude of the velocity. Therefore

$$\frac{l}{\varphi(v)} = \frac{l}{\varphi(-v)} \quad \text{or finally} \quad \varphi(v) = \varphi(-v) \equiv 1.$$

The transformation of the coordinates of space and time therefore takes the reduced form

$$\begin{cases} x' = \beta\,(x - vt) \\ y' = y \\ z' = z \\ t' = \beta\Big(t - \dfrac{vx}{c^2}\Big). \end{cases}$$

8. Contraction of lengths and correlative dilation of times.

1) A sphere whose equation is

$$x'^2 + y'^2 + z'^2 = R^2$$

is at relative rest in the system S'. In the system S it appears as an ellipsoid of the equation

$$\beta^2 x^2 + y^2 + z^2 = R^2.$$

The dimensions of a sphere which is motionless in S'—that is, moving with velocity $\vec{v}$ with respect to S—are therefore unaltered in a direction perpendicular to $\vec{v}$ but seem, *viewed from S*, to be contracted in the ratio of 1 to $\sqrt{1 - \dfrac{v^2}{c^2}}$ in the direction of $\vec{v}$. [It follows that a velocity $v > c$ has no physical meaning.] *The converse is true*, when a sphere which is at rest in S is observed from S'.

2) A clock placed at O' and at relative rest in S' keeps the time t'. *Seen from S*, what time does this clock keep ? Here we have $x = vt$. Therefore $t' = \dfrac{t}{\beta}$ or $t' = t - \Big(1 - \dfrac{1}{\beta}\Big)\,t$. Therefore, seen from S, this clock runs slow by $\Big(1 - \dfrac{1}{\beta}\Big)$ seconds per second. *The reciprocal is true* when a clock, at rest in S, is observed from S'.

9. Composition of velocities.

Consider a particle moving in the direction of the axis $O'x'$ with a velocity $\vec{v}'$ relative to the system S', and starting from the origin O' at the time $t' = 0$. What is the velocity of the particle relative to the system S ?

Here we have $x' = v't'$. Therefore, returning to the variables (x, t) by means of the transformation (T),

(V)
$$x = \frac{v + v'}{1 + \dfrac{vv'}{c^2}}\, t.$$

The velocity c thus assumes the character of a *ceiling* which cannot be surpassed. Indeed, if two velocities v and v', which are both less than c, are compounded in this way the resultant velocity is less than c; if the velocity c is compounded with a velocity $v \leqq c$, the result is c.

10. Transformation of Maxwell's equations in the vacuum. Electrodynamic relativity.

Einstein starts from Maxwell's equations, written, in the " fixed system " S in the form

$$\begin{cases} \operatorname{curl} \vec{H} = \dfrac{1}{c}\, \dfrac{\partial \vec{E}}{\partial t} \\[2mm] \operatorname{curl} \vec{E} = -\dfrac{1}{c}\, \dfrac{\partial \vec{H}}{\partial t}. \end{cases}$$

The transformation (T), applied to the variables (x, y, z, t), implies that

$$\begin{cases} \dfrac{\partial}{\partial t} = \beta\, \dfrac{\partial}{\partial t'} - v\beta\, \dfrac{\partial}{\partial x'} \\[2mm] \dfrac{\partial}{\partial x} = -\dfrac{v\beta}{c^2}\, \dfrac{\partial}{\partial t'} + \beta\, \dfrac{\partial}{\partial x'}. \end{cases}$$

Then in the system S' Maxwell's equations take the form

$$\begin{cases} \operatorname{curl} \vec{H}' = \dfrac{1}{c}\, \dfrac{\partial \vec{E}'}{\partial t'} \\[2mm] \operatorname{curl} \vec{E}' = -\dfrac{1}{c}\, \dfrac{\partial \vec{H}'}{\partial t'} \end{cases}$$

if E'_x, H'_x, etc. . . . are given by

(C)
$$\begin{cases} E'_{x'} = E_x & H'_{x'} = H_x \\[2mm] E'_{y'} = \beta\left(E_y - \dfrac{v}{c}\, H_z\right) & H'_{y'} = \beta\left(H_y + \dfrac{v}{c}\, E_z\right) \\[2mm] E'_{z'} = \beta\left(E_z + \dfrac{v}{c}\, H_y\right) & H'_{z'} = \beta\left(H_z - \dfrac{v}{c}\, E_y\right). \end{cases}$$

That is, the fields transform among themselves in the manner advocated by Lorentz.

In Einstein's interpretation, *these formulae imply a complete symmetry between the systems S and S'—that is, an electrodynamic relativity.* The electrodynamic and magnetomotive forces become simple effects of the choice of reference system—a purely electric field in S becomes, in S', an electromagnetic field defined by the equations (C).

11. TRANSFORMATION OF MAXWELL'S EQUATIONS INCLUDING CONVECTION CURRENTS.

Einstein starts from Maxwell's equations including a convection current of velocity $\vec{u}$, of the form

$$\begin{cases} \operatorname{curl} \vec{H} = \dfrac{1}{c}\left(\dfrac{\partial \vec{E}}{\partial t} + \varrho\vec{u}\right) \\ \operatorname{curl} \vec{E} = -\dfrac{1}{c}\dfrac{\partial \vec{H}}{\partial t}. \end{cases}$$

These equations also embody the results of Lorentz's electron theory.

If these equations are transformed under the transformation (T), so that they may be appropriate for the system S', the form of the equations is preserved as

$$\begin{cases} \operatorname{curl} \vec{H}' = \dfrac{1}{c}\left(\dfrac{\partial \vec{E}'}{\partial t'} + \varrho'\vec{u}'\right) \\ \operatorname{curl} \vec{E}' = -\dfrac{1}{c}\dfrac{\partial \vec{H}'}{\partial t'} \end{cases}$$

if, at the same time, the transformation (C) is applied to the field and if it is supposed that

$$\varrho' = \operatorname{div} \vec{E}' = \beta\left(1 - \frac{vu_x}{c^2}\right)\varrho$$

and

$$u_{x'}' = \frac{u_x - v}{1 - \dfrac{u_x v}{c^2}} \qquad u_{y'}' = \frac{u_y}{\beta\left(1 - \dfrac{u_x v}{c^2}\right)} \qquad u_{z'}' = \frac{u_z}{\beta\left(1 - \dfrac{u_x v}{c^2}\right)}.$$

Thus Einstein arrives at a perfect symmetry between the systems S and S'. The formulae giving the components of $\vec{u}'$ along the axes of S' are obtained by composition (in the sense of § 9) of the velocities $\vec{u}$ and $-\vec{v}$, where the velocity $\vec{u}$ can have any orientation with respect to the axes $Oxyz$. It was due to this composition that Einstein was able to demonstrate the complete conservation of the equations, which had been lacking in Lorentz's paper.

12. Dynamics of the slowly accelerated electron.

Einstein considers an electron which is a relative rest in the system S' at the time t'. When the motion of the electron is accelerated, the system of reference which accompanies it (Lorentz's *proper* system) changes at each instant. At times close to $t' = 0$, at which the system S' is a proper system for the electron considered, Einstein postulates that the ordinary equations of classical mechanics are valid in S'; that is, that

$$\left\{ \begin{aligned} m_0 \, \frac{d^2 x'}{dt'^2} &= e \, E'_{x'} \\ m_0 \, \frac{d^2 y'}{dt'^2} &= e \, E'_{y'} \\ m_0 \, \frac{d^2 z'}{dt'^2} &= e \, E'_{z'} \end{aligned} \right.$$

where m_0 is the mass of the electron. These equations obtain as long as the velocity of the electron, measured in S', remains small. Now the coordinates of the fixed system S, with respect to which the electron moves with a velocity which is only a little different from $\vec{v}$, are introduced. It turns out that

$$\left\{ \begin{aligned} m_0 \beta^3 \, \frac{d^2 x}{dt^2} &= e \, E'_{x'} \\ m_0 \beta^2 \, \frac{d^2 y}{dt^2} &= e \, E'_{y'} \\ m_0 \beta^2 \, \frac{d^2 z}{dt^2} &= e \, E'_{z'}. \end{aligned} \right.$$

For, as Lorentz remarked, the accelerations transform according to

$$\vec{\gamma}(S) = (\beta^3, \beta^2, \beta^2) \vec{\gamma}(S').$$

This relation is deduced from the transformation (T). Further, applying the transformation (C), the charge remaining constant, it turns out that

$$\left\{ \begin{aligned} m_0 \beta^3 \frac{d^2 x}{dt^2} &= e E_x \\ m_0 \beta \frac{d^2 y}{dt^2} &= e \left(E_y - \frac{v}{c} H_z \right) \\ m_0 \beta \frac{d^2 y}{dt^2} &= e \left(E_z + \frac{v}{c} H_y \right). \end{aligned} \right.$$

These equations lead to the suggestion that the electron be assigned a *longitudinal* mass $m_0\beta^3$ and a *transverse* mass $m_0\beta$.[1]

Einstein observes that this variation of the mass with the velocity extends to heavy particles.

" For, by the addition of an *arbitrarily small* electric charge, a particle endowed with mass can be compared to an electron (in the sense that we attach to the term). "

Finally, Einstein calculates the kinetic energy of an electron, initially at rest at the origin O, which is displaced along the axis Ox under the action of an electrostatic force eE_x. The energy taken from the electrostatic field is therefore $\int eE_x dx$. Since the electron is " slowly accelerated, " no energy is lost by radiation and, consequently, the energy of the field is entirely transformed into kinetic energy T of the electron. Under these conditions

$$T = \int_o^x eE_x dx = \int_o^v m_0\beta^3 v\,dv = m_0 c^2 \left[\frac{1}{\sqrt{1 - \dfrac{v^2}{c^2}}} - 1 \right] = m_0 c^2(\beta - 1).$$

13. SPACE-TIME IN THE SENSE OF MINKOWSKI.

Minkowski[2] uses the term *world* to denote the space-time continuum in the four dimensions x, y, z, t. A *world-point* represents an event occurring at the time t and the point x, y, z. For each particle the coordinates x, y, z are functions of the time t, whence the existence of a *world-line*.

Minkowski introduces the hypersurface

$$c^2 t^2 - x^2 - y^2 - z^2 = 1$$

which consists of two sheets separated by $t = 0$. Consider the sheet situated in the region $t > 0$ and seek the linear homogeneous transformation of x, y, z, t into x', y', z', t' which leaves this sheet invariant.

---

[1] To be accurate, in his paper of 1905 EINSTEIN suggests that the components of the force in the fixed system be defined as $\left[eE_x, e\beta\left(E_y - \dfrac{v}{c}H_z\right), e\beta\left(E_z + \dfrac{v}{c}H_y\right)\right]$. This leads to the mass $m_0\,(\beta^3, \beta^2, \beta^2)$, while LORENTZ's suggestion is $m_0\,(\beta^3, \beta, \beta)$. At the suggestion of PLANCK, it is the second of these that is now accepted, and which arises naturally by writing $\vec{F} = \dfrac{d}{dt}(\vec{mv})$ instead of $\vec{F} = m\vec{\gamma}$.

[2] MINKOWSKI's paper was first given at a conference at Cologne on the 21st September, 1908, and then reproduced in a volume, including papers by LORENTZ and EINSTEIN, called *Das Relativitätsprinzip* (Leipzig-Berlin, Teubner ; 4th ed., 1922).

Suppose that the coordinates y and z are unaffected in the transformation and, further, put $u = ct$. Then the intersection of the sheet of the hypersurface with the plane (x, u) reduces to a branch of the rectangular hyperbola

$$u^2 - x^2 = 1 \quad \text{with} \quad u > 0.$$

Let OA' be a vector through the origin which intersects the hyperbola and let $A'B'$ be the tangent limited to the asymptote on the side which corresponds to $x > 0$.

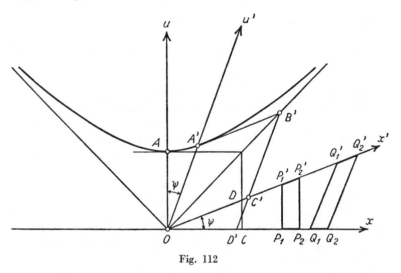

Fig. 112

Construct the parallelogram $A'B'C'O$, whose side $B'C'$ meets Ox at D'.

If the hyperbola is considered as being referred to the axes OC', OA', and if OC' is supposed equal to unity, then the equation is still

$$u'^2 - x'^2 = 1 \quad \text{with} \quad u' > 0.$$

Thus the transformation $(x, y, z, t) \rightarrow (x', y', z', t')$ fulfills the requirements imposed. Putting $\tan \psi = \dfrac{v}{c}$, where the angle $\widehat{uOu'}$ (or the angle $\widehat{x'Ox}$) is equal to ψ, then it can be easily verified that this transformation is identical with a Lorentz transformation of parameter v.

These transformations form a group G_c, the suffix c referring to the velocity of light in the vacuum. " If it is not wished to use the word ' space ' and the word ' vacuum, ' c can be defined as being the ratio of the electromagnetic and electrostatic units of quantity of electricity. " Minkowski then states the following principle.

The expression of the laws of physics is not modified by a transformation of the group G_c.

The world-line of a particle at rest in the system (x, u) is a straight line parallel to the axis of u. At a point on the world-line on any particular the tangent is parallel to a radius vector like OA'. If OA' is taken as the axis of u', the particle will appear at rest in the system (x', u'). In order that this may be possible, it is necessary that

$$c^2 dt^2 - dx^2 - dy^2 - dz^2 > 0 \quad \text{or} \quad v < c.$$

" The essential reason for taking the group G_c into consideration lies in the fact, recognised by Voigt, that this group leaves invariant the equation of the propagation of light waves in free space. On the other hand, the concept of solid body only has meaning in classical mechanics (group G_∞). The simultaneous consideration of the group G_c (for optics) and the group G_∞ (for mechanics) would define a *privileged direction* for time and, under these conditions, optical experiments performed with rigid instruments would necessarily be able to demonstrate the influence of the orientation with respect to the terrestrial translation. In particular, Michelson's experiment would have a positive result. . . . Lorentz's hypothesis seems ' *äusserst phantastisch,* ' for the contraction is not caused by the ether but is a pure gift from the gods. . . . This hypothesis is equivalent to our new conception of space and time, which makes the contraction much more comprehensible. "

In fact the ends P_1 and P_2 of a segment which is parallel to Ox and at rest in $S(x, u)$ describe world-lines which are parallel to Ou'. Suppose that the length of the segment $P_1 P_2$, *measured in S*, is equal to $l(P_1 P_2 = l \cdot OC)$. Similarly, suppose that the length of the segment $Q_1' Q_2'$, *measured in S'*, is equal to $l(Q_1' Q_2' = l \cdot OC)$. From the figure it is seen that $Q_1 Q_2 = l \cdot OD'$. But an elementary calculation yields $OD' = OC \sqrt{1 - \dfrac{v^2}{c^2}}$, where $\dfrac{v}{c} = \tan \psi$. Therefore

$$\frac{P_1 P_2}{Q_1 Q_2} = \beta = \frac{1}{\sqrt{1 - \dfrac{v^2}{c^2}}}.$$

The length $Q_1 Q_2$, or the measure in S of a moving segment, of length l, which is at rest in S', is therefore subject to the Lorentz contraction. The reciprocal proposition is also true, for

$$\frac{Q_1'Q_2'}{P_1'P_2'} = \frac{l \cdot OC'}{l \cdot OD} = \frac{OC}{OD'} = \beta.$$

Minkowski remarks, " Neither Lorentz nor Einstein grappled with the notion of space. Undoubtedly this was because the reduced transformation in which the plane (x', t') covers the plane (x, t) can be interpreted by keeping the same direction in space for the axis of x. To meddle with the concept of space in the face of such an agreement may be regarded as a distortion of the mathematical attitude. Yet this step is indispensable for an appreciation of the true significance of the group G_c. "

Minkowski considers the term "*postulate of relativity*," used to state an invariance with respect to the group G_c, to be very inadequate *(sehr matt)*. The meaning of this postulate is that it is only possible to contemplate a space-time world whose space and time sections allow of a certain amount of freedom. Accordingly Minkowski suggests the term *postulate of the absolute world* or, briefly, *world-postulate (Weltpostulat)*.

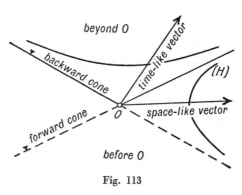

Fig. 113

Let O be the origin of space time. Then, by means of the equation

$$u^2 - x^2 - y^2 - z^2 = 0,$$

Minkowski defines a *backward cone* (on one side of O), in which $t < 0$, and a *forward cone* (on the other side of O), in which $t > 0$. The first region consists of world-points that " send light to O " while the second consists of world points that " receive light from O ".

The sheet of the hypersurface

(N) $$u^2 - x^2 - y^2 - z^2 = 1 \quad (u > 0)$$

which has been considered above, belongs to the second region. A branch of a *hyperbola*, as (H) for example, whose centre is O is contained between the two cones. It may be the world-line of a particle whose velocity tends to c as t tends to $\pm \infty$.

A *vector* such as OP, which is drawn from the origin to the point (x, y, z, t), has four components x, y, z, t.

A *time-like* vector is said to be a vector drawn from the origin towards the sheet (N). A *space-like* vector is a vector drawn from O towards the surface

$$(-N) \qquad\qquad x^2 + y^2 + z^2 - u^2 = 1.$$

A world-point in the *intermediate region* contained between the backward and forward cones can be represented as an event which is earlier than, simultaneous with, or later than O by a suitable choice of axes.

Two vectors $\overline{V}$ and $\overline{V}_1$ are said to be orthogonal if

$$uu_1 - xx_1 - yy_1 - zz_1 = 0.$$

For a space-like vector the units are defined by the distance from O to the sheet (—N) ; for a time-like vector, by the distance from O to the sheet (N). The quantity $\int \delta\tau$, where $d\tau$ is defined by

$$cd\tau = \sqrt{du^2 - dx^2 - dy^2 - dz^2}$$

is called the *proper time* of the world-line (or of the particle describing this world-line).

Let the presence of a dot above a letter, as in $\dot{u}$ for example, denote the operation of differentiation with respect to the proper time.

If $\vec{P}$ is the vector from O to P, the velocity $\dfrac{d\vec{P}}{dt}$, of components $(\dot{x}, \dot{y}, \dot{z}, \dot{u})$, is a time-like vector which is tangent to the world line. For

$$\dot{u}^2 - \dot{x}^2 - \dot{y}^2 - \dot{z}^2 = c^2.$$

On the contrary the acceleration $\dfrac{d^2\vec{P}}{dt^2}$, of components $(\ddot{x}, \ddot{y}, \ddot{z}, \ddot{u})$, is a space-like vector orthogonal to the velocity. For

$$\dot{u}\ddot{u} - \dot{x}\ddot{x} - \dot{y}\ddot{y} - \dot{z}\ddot{z} = 0.$$

The momentum vector *(Impulsvektor)* has components $(m_0\dot{x}, m_0\dot{y}, m_0\dot{z}, m_0\dot{u})$, where m_0 is the *rest mass*. Similarly, the components of the force are identified with those of the vector $(m_0\ddot{x}, m_0\ddot{y}, m_0\ddot{z}, m_0\ddot{u})$.

The kinetic energy is identified as the product of c and the u-component of the momentum *(Impulsvektor)*, or

$$m_0 c^2 \frac{dt}{d\tau} = \frac{m_0 c^2}{\sqrt{1 - \dfrac{v^2}{c^2}}}$$

in the case of the reduced transformation.

B. *ANALYSIS AND INTERPRETATION*

$2a.$[1] On Michelson's experiment.

Apart from the form of its presentation the calculation of § 2 is found in all the books on relativity. It must be remembered that the calculation supposes that the wave theory is accepted or, more accurately, that it is assumed that the velocity of the propagation of light in the sidereal vacuum is independent of the motion of the source. If, on the contrary, the emission theory is taken as the starting-point, it is immediately concluded that the experiment must give a negative result.[2]

We have already remarked that the hypothesis of a dragging of the ether—which would explain the result of Michelson's experiment—would make it impossible to account satisfactorily for the phenomenon of aberration in terms of the wave theory. As a curiosity we draw attention to the interpretation of the phenomenon of aberration that was suggested by P. Lenard.[3] This author introduces two ethers, namely—

the basic ether *(Uräther)*, filling all space and at absolute rest as a whole. This medium is totally independent of matter and is that in which electromagnetic waves are propagated with velocity c;

the ether *(Äther)*, " pertaining to all matter yet distinct from it, " contiguous with every material atom and whose distribution accords with that of matter. In the neighbourhood of the terrestrial sphere for example, the ether as a whole is at relative rest with respect to this and accompanies it in its motion.

---

[1] Throughout the present Part we shall use the *same numeral* to denote a paragraph in the presentation and the corresponding paragraph of analysis or interpretation, by allowing this numeral to be followed by letters $a, b, c. \ldots$

[2] *Cf.* Painlevé, *Les Axiomes de la Mécanique* (Gauthier-Villars), 1922, pp. 81 to 96.

[3] Lenard, *Über Äther und Uräther* (Hirzel), Leipzig, 1921.

In this scheme the aberration is explained in classical terms. The light from the stars comes to us through the *Uräther* ; the ether associated with the earth does not disturb this propagation any more than the air, or any other fluid, that the light may have to pass through before it arrives at the observing telescope. Lenard explains Michelson's experiment without appeal to a Lorentz contraction— for this experiment is performed entirely in the ether associated with the earth, and which accompanies the earth in its instantaneous translation. . . .

2*b*.

Following up an idea of Poincaré,[1] J. Le Roux remarked that Michelson's experiment, depending not on a simple light path but on a forward and backward path, only substantiated the *waves of interference*. The *progressive waves* can be ellipsoids of revolution in which the moving source is at one of the foci. In other words, " the isotropy of the wave of interference does not imply the isotropy of the progressive wave. " [2]

J. Le Roux's analysis, based on a systematic study of waves emitted by a moving source, is mathematically incontestable. But it would lead to much more complicated physical hypotheses than Einstein's theory and, here, the author gives us no assistance. " The ellipsoidal form of the progressive wave no longer allows it to be supposed that the propagated field may be isotropic in the restricted domain in which experiments are carried out. Therefore, in this domain, the Earth may have an influence on the field and, consequently, on the propagation of light. As for the form and nature of this influence, the choice of hypotheses is very wide. We shall formulate none. " [3]

3*a*. DYNAMICS OF THE ELECTRON IN POINCARÉ'S SENSE.

Poincaré [4] devoted a paper which appeared in 1906 to the study of Lorentz's calculations. As Einstein had done, he corrected the law of the transformation of the density of electric charge. Moreover, he simplified Lorentz's analysis by remarking that the function $l(v)$ which figures in the transformation of the coordinates of space and time and which is, *a priori*, arbitrary, must necessarily reduce to unity when the condition that these transformations form a group is imposed.

[1] POINCARÉ, *Science et méthode*, p. 99. *La Mécanique nouvelle* (Conference held at Lille in 1909), p. IX.

[2] LE ROUX, *Relativité restreinte et systèmes ondulatoires*, *Journal de Math.*, Vol. I, 1922, p. 205.

[3] *Ibid.*, p. 231.

[4] *Rendiconti di Palermo*, Vol. 21, 1906, p. 129.

Poincaré declared that Lorentz's hypothesis is " the only one compatible with the impossibility of making absolute motion evident. " We must therefore " fall back on the theory of Lorentz, but if we wish to retain it and avoid intolerable contradictions, a *special force* must be introduced which will explain both the contraction and the constancy of two of the axes. I have found that this force can be assimilated to a constant external pressure acting on the deformable and compressible electron and whose work is proportional to the variations in volume of this electron. "

So it was only grudgingly that Poincaré adopted Lorentz's theory and he considered it was necessary to complete it by introducing a supplementary potential proportional to the volume of the electron. He again expressed his scepticism and his hope of seeing the theory transformed by the following formula : " How do we constitute our measurements ? By applying the objects considered as invariable solids on one another, it will be answered at first. But that is not true if we admit the Lorentzian contraction. In this theory, two equal lengths are, by definition, two lengths that light takes the same time to traverse. It would suffice perhaps to abandon this definition for the theory of Lorentz to be as completely upset as was Ptolemy's system by the intervention of Copernicus. "

3b. FROM LORENTZ TO EINSTEIN.

Lorentz's theory already embodies the essential results of special relativity, namely the transformation (T) of the coordinates of space and time, the law (C) of the transformation of electrical and magnetic fields, the law (M) of the variation of the mass with the velocity. To these results, Einstein's first paper was to add only the law (V) of the composition of the velocities and the formula relating the mass to the energy.

But, in the passage of Lorentz's theory to that of Einstein, what is essential is the *novelty of the Einsteinian point of view*. Lorentz's theory is not relativistic, in this sense that a system at absolute rest in relation to the ether continues to enjoy special properties there. The Lorentzian contraction, which possesses material significance, results from the application itself of an instantaneous translation of velocity $\vec{v}$. The same applies to the laws (C) and (M).

On the contrary, there is perfect symmetry between two systems of reference, like S and S', in Einstein's theory. From this follows a complete relativity of mechanical and electromagnetic phenomena. For the mechanical phenomena, whose relativity in the classical field had been recognised for a considerable time, Minkowski made clear that it is only necessary to substitute the group G_c for the group G_∞, and to carry out the correlative modification of the kinematics (that is, to introduce the law (V)). In these circumstances the Lorentz contraction is no more than a simple consequence of the choice of reference system. For electromagnetic phenomena the relativity, connected with the group G_c, is entirely novel and removes all the asymmetries of the classical theories, making both the electromagnetic force and the " magnetomotrice " force simple consequences of the choice of reference system.

4a. THE ETHER MADE SUPERFLUOUS.

In this presentation it has been already necessary to speak of the ether on several occasions. It might be said that the physics of the XIXth Century was completely bathed in it. That the ether may have ceased, after Fresnel and Maxwell, to play the part of a medium and have taken the more abstract one of a system of reference suited to the definition of absolute rest, does not imply that it ceased to function as the substratum of thought in physics.

The reasons which led Einstein to declare the ether superfluous are well founded in logic ; but such a declaration made on the threshold of his theory could only alienate him from the physicists imbued with the classical representation. The example of Lenard which we have quoted above (§2a) is typical— rather than accept the disappearance of the ether, Lenard prefers to create a new one. And with two ethers he explains the phenomena. Rather than give up a myth, he preferred to multiply it.

5a. DIFFICULTIES OF EINSTEIN'S NOTION OF SIMULTANEITY.

The definition of simultaneity which Einstein placed at the foundation of his theory has the merit of demonstrating the universal part played by the exchange of signals of velocity between different observers. But nevertheless the definition remains a matter of some delicacy. As it appears in its popular presentations, it has given rise to numerous misunderstandings, or has been an obstacle to the understanding of the theory by many people. We shall not go as far as to complain, as certain critics do, that Einstein made use of magical clocks and enchanted measuring-rods in order to return, in the last analysis, to Lorentz's transformation (T). But there is no doubt that these ideal clocks and rods can only be used in idealised experiments. Further, we remark that when, in general, the clocks of the system S' keep the time t', Einstein does not hesitate to arrange for the system S' to be accompanied by clocks which, at each point in space, keep the local time t of the system S ; this in order to demonstrate the fact that the simultaneity he introduces only has a relative meaning.

6a. FIELD OF VALIDITY OF THE PRINCIPLE OF SPECIAL RELATIVITY— " GALILEAN " SYSTEMS OF REFERENCE.

Much more important, from the theoretical point of view, is the exact definition of the field of validity of the principle of special relativity. Several authors, basing themselves on the statement of the principle as it is formulated by Einstein (§ 6, p. 474), have written that *any* two systems S and S' which are in uniform and rectilinear translation with respect to each other are equivalent with regard to the principle. Numerous popularisers have lightly taken this step. Certainly Einstein, before stating his principle, is explicit that the " laws of newtonian mechanics must be valid " in the system S which is considered as " fixed " ; but this essential point of view needs to be most clear.

P. Frank seems to have been the first author to concern himself with this question.[1] Einstein himself returned to it in an interpretative work [2] which we are going to follow.

A system of reference in which the principle of inertia is valid —where the motion of an *isolated* particle is rectilinear and uniform— is called " galilean. " [3] The principle of relativity in classical mechanics shows that if S is a galilean system of reference, so also is any system of reference S' which is actuated by a rectilinear and uniform translation with respect to S. But this relativity of classical mechanics does not extend to the class of optical and electrodynamic phenomena.

Clearly there can be no question of giving up the classical form of relativity mechanics, which, with high accuracy, accounts for all the astronomical motions. If, on the other hand, this relativity breaks down in electromagnetism, it is more or less necessary to assume that there exists a privileged system S_0 in which the laws of nature are the most simple. For this reason this system could be said to

---

[1] P. FRANK, *Die Stellung des Relativitätsprinzips im System der Mechanik und Electrodynamik*, Wiener Sitzungsber., Vol. 118, 1909, p. 337.

[2] EINSTEIN, *Über die spezielle und die allgemeine Relativitätstheorie* (Gemeinverständlich), Vieweg, Braunschweig, 1916.

[3] This term, to which too precise a historical significance must not be attached, has now become part of the every day language of higher teaching. The meaning which EINSTEIN attached to it is moreover, perfectly definite.

be *at absolute rest*, all other systems being *in motion*. In these circumstances the direction of motion would assume a particular significance. But it has always been impossible to demonstrate such a defect of isotropy of the physical space in the neighbourhood of the Earth, in spite of the care lavished on the relevant experiments. This failure is a strong argument in favour of the adoption of a new principle of relativity.

Einstein concludes, " We assume that there exists a galilean reference system *S* in which the principle of inertia is valid [or where the velocity of light is the same in all directions, and is independent of the motion of the source]. Referred to *S*, the laws of nature must be as simple as possible.[1] But systems of reference like *S'*, which are actuated by a rectilinear and uniform translation with respect to *S*, must be equivalent to *S* as far as the expression of the laws of nature is concerned. All these systems are regarded as ' galilean ' systems, and the principle of relativity in its restricted sense *(spezielles Relativitätsprinzip)* only applies to such systems. "

Painlevé [2] has a penetrating discussion of the same question in a remark on the propagation of light. " The theory of relativity rests on the same fundamental postulate as classical mechanics and Fresnel's optics ; namely

" *The Kepler-Fresnel Postulate.* — It is possible to define, once and for all and for the whole universe, a measure of time, a measure of length and a frame of reference such that

" 1) *The motion of every particle which is very distant from all others is rectilinear and uniform* (Principle of inertia).

" 2) *Far from all matter the propagation of light is rectilinear and uniform and has the same velocity in all directions* (Fresnel's Principle).

" According to the classical doctrine this frame of reference will be the one adopted by the group of observers on a star *A*, which is very distant from all others and without rotation with respect to the fixed stars, *if the absolute velocity of this star is zero.* But the relativists add the following essential complement—

" *Postulate of relativity.* — If the Kepler-Fresnel postulate is true *for the observers of the star* A *(choosing this star as reference body), it is also true for the observers of a star* B [also very distant from all others and without rotation with respect to the fixed stars, but possibly moving rectilinearly and uniformly with respect to *A*] *choosing this star as reference body.* "

---

[1] We shall return to this matter later.
[2] Painlevé, *Les Axiomes de la Mécanique*, Gauthier-Villars, 1922, p. 98.

After explaining the transformation (T) of the group G_c which makes it possible to go from a relativistic system of reference S to another relativistic system of reference S', Painlevé concludes—

" Such is the principle of special relativity. It is seen that this theory, contrary to what is affirmed or implied by a very large number of expositions, supposes the existence of privileged axes. In the special theory of relativity as in the classical theory, there exists at least a measure of time and a measure of length, and a frame of reference such that Kepler's principle and Fresnel's principle are true in the sidereal vacuum. But in the classical theory this frame of reference is unique ; in the theory of relativity there exists an infinite number of such frames whose axes of reference (carrying their observers) are actuated by a rectilinear and uniform translation with respect to each other, and the formulae for the passage from one system to the other are [Lorentz's formulae (T)]. *These privileged frames coincide with those that classical mechanics describes as absolute axes.* Alternatively, in the present state of our measurements, these axes are the *axes of Copernicus* (axes of reference fixed with respect to the distant stars and whose origin is at the centre of gravity of the solar system) or the *axes of Galileo*, which are actuated by a rectilinear and uniform translation with respect to the first. "

The field of validity of the special principle of relativity being thus defined, what significance can be attached to it? Einstein, speaking of the " *heuristic value* " of the principle of relativity, made the following declaration.

" Every law of nature must be so formulated that it is expressed in the same formal terms when, by means of a Lorentz transformation, the variables (x, y, z, t) of a system S are replaced by the variables (x', y', z', t') of a system S'. In short, the laws of nature are covariant under Lorentz transformations.

" *This is the precise mathematical condition that the theory of relativity imposes on a natural law* (einem Naturgesetz vorschreibt). It can therefore be used heuristically in the investigation of the laws of nature. If there were found a general law of nature which did not agree with the above condition, at least one of the foundations of the theory would be contradicted. " [1]

Even the form of such a declaration would have offended a theorist of mechanics like Ernest Mach. It is better to confine oneself to establishing that the novelty introduced by Lorentz and Einstein in classical mechanics permits the successful inclusion of mechanics, optics and

---

[1] *Über die spezielle und die allgemeine Relativitätstheorie*, p. 29.

electromagnetism in one single synthesis agreeing with the experimental facts than to regard it as a necessary criterion for the selection of every natural law.

7a. ON DIFFERENT MATHEMATICAL WAYS OF OBTAINING THE LORENTZ TRANSFORMATION.

In the presentation we have seen that Lorentz introduces the transformation (T) in an attempt to preserve the form of the equations of his theory of electrons. On the other hand, Einstein obtains the same transformation by starting from his definition of simultaneity.

1) It is also possible to return to the original idea of Voigt, as von Laue has done, and to seek the transformations which are linear and homogeneous, in x, y, z, t, and which leave invariant the equation

$$\Box \, \varphi = 0$$

where $\Box$ is the operator

$$\frac{\partial^2}{\partial x^2} + \frac{\partial^2}{\partial y^2} + \frac{\partial^2}{\partial z^2} - \frac{1}{c^2} \frac{\partial^2}{\partial t^2}.$$

2) The invariance of the expression $x^2 + y^2 + z^2 - c^2 t^2$ may be taken as the starting-point, and the question treated analytically.

3) We recall here that, in a geometrical way, Minkowski takes the invariance of the expression $x^2 - u^2 (u = ct)$ as the basis of his interpretation.

4) Further, putting $l = ict$, the question can be reduced to the invariance of $x^2 + y^2 + z^2 + l^2$; that is, to the consideration of imaginary rotations. We shall confine ourselves to the reduced transformation in which y and z remain unchanged. Putting $\alpha = \dfrac{v}{c}$ and $\tan \varphi = i\alpha$, it follows that $\cos \varphi = (1 - \alpha^2)^{-\frac{1}{2}}$ and $\sin \varphi = i\alpha (1 - \alpha^2)^{-\frac{1}{2}}$.

The Lorentz transformation then reduces to the imaginary rotation

$$\begin{cases} x' = x \cos \varphi + l \sin \varphi \\ l' = - x \sin \varphi + l \cos \varphi \end{cases}$$

the composition of such rotations yields directly the law of addition of velocities parallel to the axis of x. For example, if $\varphi = \varphi_1 + \varphi_2$,

$$tg\varphi = tg(\varphi_1 + \varphi_2) = \frac{i\alpha_1 + i\alpha_2}{1 + \alpha_1\alpha_2} \quad \text{or} \quad v = \frac{v_1 + v_2}{1 + \dfrac{v_1 v_2}{c^2}}.$$

5) As J. Le Roux has remarked (in the paper cited on page 487 above), the Lorentz transformation is a transformation in homogeneous coordinates and which leaves invariant the fixed sphere of unit radius, $\sum$, whose equation is $u^2 - x^2 - y^2 - z^2 = 0$. The centre O of this sphere corresponds to a frame of reference. If P is a point inside $\sum$ and the vector $\overrightarrow{OP} = \vec{v}$, the passage from the frame of reference S to the frame of reference defined by P will be characterised by the argument

$$v = \tanh \theta \quad \text{or} \quad \theta = \frac{1}{2} \log \frac{1 + v}{1 - v} = \frac{1}{2} \log \ (M, \ M', \ O, \ P)$$

where M and M' are the two points at which the line OP cuts the sphere $\sum$.[1] More generally, the argument which characterises the passage from the frame of reference $S_1(P_1)$ to the frame of reference $S_2(P_2)$ will be

$$\theta_{12} = \frac{1}{2} \log \ (M, \ M', \ P_1, \ P_2)$$

where M and M' are the two points at which the line $P_1 P_2$ cuts the sphere $\sum$. This argument is therefore directly connected with the *Cayley distance* and $\sum$ plays the part of the absolute.

8a. PSEUDO-PARADOXES IN SPECIAL RELATIVITY.

Our subject does not include the numerous paradoxes which it has been found possible to infer from the principle of special relativity. Fortunately these paradoxes are, in general, outside the field of validity of the principles—a field which is, as we have seen in § 6a, very narrowly determined. Such is, for example, the paradox described as that of the interstellar traveller. Supposed being launched in a bullet with a speed inferior by a twenty thousandth part to the speed c, he comes back to the Earth growing older only by one year in going and one year in returning, while the inhabitants of the Earth are older by hundred years. Some authors, like Eddington, have clearly seen the error of such a paradox and believed to elude it, saying : " *if the traveller finds the mean of returning to Earth.* " In fact, the traveller is bound in going to a galilean frame of reference S' (when the normal speed has been attained) and equally in his return journey to another galilean frame of reference S''. But what to say about the start, the curl and the landing ? Neither of these three phases of the motion are submitted to the principle of special relativity. This paradox is so a *pseudoparadox* : it is not in the field of validity of the Lorentz-Einstein theory.

[1] The bracket refers to the anharmonic ratio in this formula.

If, apart of this simple reason, one wishes to be absolutely in peace about this subject, one can assume, with Mr. Chazy, that during these three phases the traveller became older by 98 years in the whole.

Such is the paradox of the revolving disc, though one may perhaps oppose that Einstein discussed it. We shall come back to this paradox in general relativity, showing that Einstein argues about it only to conclude *negatively* on the immediate physical meaning of coordinates of time and space in this respect.

The only a priori paradoxical effect which lies really in the field of the principle is that of contraction of lengths and correlative dilatation of times.

But this effect is *reciprocal*, as shown in the following equations—

$$\begin{cases} x' = \beta\,(x - vt) \\ t' = \beta\!\left(t - \dfrac{vx}{c^2}\right) \end{cases} \qquad \begin{cases} x = \beta\,(x' + vt') \\ t = \beta\!\left(t' + \dfrac{vx'}{c^2}\right) \end{cases}$$

$$1)\quad x = \frac{x'}{\beta} \quad \text{for} \quad 't = 0 \qquad\qquad 1')\quad x' = \frac{x}{\beta} \quad \text{for} \quad t' = 0$$

$$2)\quad t = \beta t' \quad \text{for} \quad x' = 0 \qquad\qquad 2')\quad t' = \beta t \quad \text{for} \quad x = 0$$

That is, in detail,

1) A length in S', materialised by a measuring-rod $O'x'$ of length x', which is observed from S at the time $t = 0$ appears, there, to be contracted in the ratio $\dfrac{1}{\beta}$.

1') A length in S, materialised by a measuring-rod Ox of length x, which is observed from S' at the time $t' = 0$ appears, there, to be contracted in the ratio $\dfrac{1}{\beta}$.

2) A clock fixed in S', at $x' = 0$, which keeps the time t' appears, from S, to keep the time t which is dilated with respect to t' in the ratio β.

2') A clock fixed in S, at $x = 0$, which keeps the time t appears, from S', to keep the time t' which is dilated with respect to t in the ratio β

It seems natural to take the view that, because of this reciprocity between S and S', the double paradox *disappears by symmetry*.[1] On

---

[1] The opinion which we express is diametrically opposed to that of EDDINGTON, who regards a pseudo-paradox like that of the traveller as an illustration of the theory and who discovers, on the contrary, difficulties in the reciprocity of S and S' in the matter of the LORENTZ contraction. " It is the reciprocity of these happenings—each one of the bodies in motion fancies that it is the other which suffers a contraction—which is so difficult to understand." Or " where the Lilliputians would have seemed pygmies to Gulliver and Gulliver have seemed a pygmy to the Lilliputians. "

the contrary, the paradoxes of which we have spoken above are characterised by obvious asymmetries which are outside the scope of special relativity, whose very principle is that of the indistinguishability of two galilean frames of reference with respect to any mechanical or electrodynamical phenomenon.

9a. COMPOSITION OF VELOCITIES AND FIZEAU'S EXPERIMENT.

By means of a theory involving both matter and the ether which permeates it, Fresnel arrived at the formula

$$w = \frac{c}{n} + v\left(1 - \frac{1}{n^2}\right).$$

Here w is the velocity, with respect to the surface of the earth, of the propagation of light in a fluid current. The index of refraction of the fluid is n and its velocity (relative to the surface of the earth) is v. In the classical interpretation the phenomenon occurs as if the ether was partially carried along by the current. According to Einstein's formula, the composition of the velocities $\frac{c}{n}$ and v leads directly to the result

$$w_1 = \frac{\dfrac{c}{n} + v}{1 + \dfrac{cv}{nc^2}} \infty \frac{c}{n} + v\left(1 - \frac{1}{n^2}\right)$$

in perfect agreement with Fizeau's experiment.

12a. RETURN TO THE DYNAMICS OF VARIABLE MASS IN PAINLEVÉ'S SENSE.

As early as 1890, in his lectures at Lille, Painlevé suggested a generalisation of the dynamics of a particle which, in particular, included the dynamics of special relativity in a given system of reference. Painlevé generalises the principle of the equality of action and reaction. Starting from the classical kinematic equations

$$\frac{dv}{dt} = \gamma_t \qquad\qquad \frac{v^2}{\varrho} = \gamma_n$$

he writes the equations of motion of a particle P in the form

$$m_0 f(v) \frac{dv}{dt} = \Phi \cos\theta \qquad \frac{m_0 v^2}{\varrho} \psi(v) = \Phi \sin\theta.$$

Let there be two particles P and P_1 which are isolated from all others. The force acting on P, say $\vec{\Phi}$, acts in the direction PP_1 and makes an angle θ with $\vec{v}$. This force can depend on the coordinates of the points P and P_1 as well as their velocities $\vec{v}$ and $\vec{v}_1$. In accordance with the principle of the equality of action and reaction, P_1 is subject to the force $-\vec{\Phi}$. Each of the particles P and P_1 has a *longitudinal* mass and a *transverse* mass which are, separately, arbitrary.

The mechanics constructed in this way by Painlevé, thanks to a purely mathematical generalisation and " when there was still no question in physics of a longitudinal or transverse mass, which coincides with ordinary mechanics for $f(v) \equiv \psi(v) \equiv 1$, is compatible with the copernican axiom of causality, with its corollary, the axiom of symmetry, and finally, with the axiom of the composition of forces. " [1]

In the interest of the physical applications it is necessary to modify slightly the equations which Painlevé obtains. First, we shall write the equation of motion of a particle in the form

$$\frac{d}{dt}(m\vec{v}) = \vec{\Phi}$$

where m is an arbitrary function of v—that is, we consider the vectorial derivative of the generalised momentum and equate it to the force. Projecting on the tangent, then on the principal normal to the trajectory, and having regard to the Serret-Frenet formulae, it turns out that

$$\begin{cases} \dfrac{d}{dt}(mv) = m\dfrac{dv}{dt} + v\dfrac{dm}{dt} = \Phi \cos\theta \\[2mm] \dfrac{mv^2}{\varrho} = \Phi \sin\theta. \end{cases}$$

Therefore the function $m(v)$ is identified as the transverse mass. It follows that the transverse mass and the longitudinal mass are necessarily connected by

$$m_l = m_t\left(1 + \frac{v}{m_t}\frac{dm_t}{dv}\right).$$

Another innovation of Painlevé's mechanics must be considered if a *retarded potential* is used— in such a case the principle of the equality of action and reaction is no longer valid.

Finally the form of the dynamics of variable mass can be modified by considering the mass as a function of the quantity T, which immediately generalises the classical kinetic energy. Thus we write

[1] We know (see above, Part IV, Chapter one) that LAPLACE had already contemplated such a generalisation as early as 1799.

$$T = mv^2 - \mathcal{L} + \text{const} \qquad (T = 0 \quad \text{for} \quad v = 0)$$

with

$$\frac{d\mathcal{L}}{dv} = mv.$$

The function $\mathcal{L}$ (which can be called the generalised Lagrange function) is that which occurs, in the dynamics of variable mass, in the Lagrange equations

$$\frac{d}{dt}\left(\frac{\partial \mathcal{L}}{\partial q_k'}\right) - \frac{\partial \mathcal{L}}{\partial q_k} = \frac{\partial U}{\partial q_k}$$

which preserve their usual form. The function $\mathcal{L}$ also occurs in the statement of Hamilton's principle

$$\delta \int_{t_1}^{t_2}(\mathcal{L} + U)dt = 0.$$

On the other hand, the kinetic energy T occurs in the statement of the energy principle

$$T = U + W$$

where W is the total energy. *The necessary and sufficient condition that* $\mathbf{T} \equiv \mathcal{L}$, *or, in words, that the same function should appear in Hamilton's principle and in the energy principle, is that the mass should be constant.*

Thanks to the fact that T has been considered, $m(T)$ or $m(U + W)$ can be substituted in the expression for the mass, $m = m_0\varphi(v)$. Thus the generalisation of Jacobi's equation is obtained directly—in the space $ds^2 = g_{ik}dx_i dx_k$ Jacobi's equation for the particle of mass $m(T)$ is

$$\frac{1}{2} g^{ik}\frac{\partial S}{\partial x_i}\frac{\partial S}{\partial x_k} = \int m(U + W)d(U + W) = J(U + W).$$

Similarly, the principle of least action is written

$$\delta \int \sqrt{2J(U + W)g_{ik}dx_i dx_k} = 0.$$

We shall not concern ourselves with the objection (which could be called the objection of the mathematician) which consists in saying that it is mathematically useless to speak of variable mass when it is always possible to return to ordinary mechanics ($m \equiv m_0$) by modifying the expression for the force and by making the latter a suitable function of the velocity. This is perfectly accurate. But, from the physical point of view, it is not a matter of indifference whether it is the law of force or the law of mass which is made more complicated.

All this shows that the dynamics of variable mass had been conceived —and may be developed—apart from the special theory of relativity. The dynamics of special relativity in a given system of reference is contained in the more general class of the various systems of dynamics of variable mass ; but this dynamics imposes itself *physically*, by means of the kinematics of special relativity, by the transformation (C) of the electromagnetic field and by the validity of ordinary mechanics in the *proper system* of the particle at time t. Moreover, referring to the expression for the kinetic energy T in the general case in which the mass in an arbitrary function of the velocity, it is seen that the relativistic law $m = m_0 \left(1 - \dfrac{v^2}{c^2}\right)^{-\frac{1}{2}}$ is the *most simple* that could be considered, for it is linear in T. Indeed, in the relativistic case

$$m = m_0 \left(1 - \frac{v^2}{c^2}\right)^{-\frac{1}{2}} \qquad \mathcal{L} = -m_0 c^2 \left(1 - \frac{v^2}{c^2}\right)^{\frac{1}{2}}$$

$$T = mv^2 - \mathcal{L} + \text{const} = m_0 c^2 \left(1 - \frac{v^2}{c^2}\right)^{-\frac{1}{2}} + \text{const}$$

As $T = 0$ for $v = 0$, it follows that

$$T = m_0 c^2 \left[\left(1 - \frac{v^2}{c^2}\right)^{-\frac{1}{2}} - 1\right] = (m - m_0)c^2$$

whence

$$m = m_0 + \frac{T}{c^2} = m_0 + \frac{U + W}{c^2}.$$

The simplicity of Lorentz's equations (M) is apparent from the last form $m = m(T)$.

This consists of the first term of the development of the mass in powers of the generalised kinetic energy.

Jacobi's equation in special relativity is then quadratic in $U + W$, since here

$$J(U + W) = m_0(U + W) + \frac{(U + W)^2}{2c^2}.$$

13a. On the meaning of space-time.

In the first words of the paper of Minkowski which we have analysed in § 13, he further develops the meaning that he attaches to space-time. " The considerations on space and time that I develop have their roots in experimental physics. That is their strength. Their tendency is radical— space in itself, and time in itself, must from this moment fall into the background ; only a kind of association of these two concepts keeps a proper individuality. " In short, Minkowski was of the opinion that space-time is *indissoluble*. Undoubtedly there is matter for discussion in this interpretation. Certainly the variables of space and time appear independent in the transformation which allows the passage from a galilean system of reference S to another galilean system of reference S'. But, as L. de Broglie has very legitimately remarked, each observer still " *cuts up his space and time* " in the relativistic continuum. In another form, Eddington writes that the universe does not have 4 dimensions, but rather, $3 + 1$ dimensions ; this referes to the form of the interval $c^2t^2 — x^2 — y^2 — z^2$. Finally, in 1916, Einstein declared that in special relativity, just as in classical mechanics, the coordinates of space *and* of time have a direct physical meaning.

Therefore it seems that Minkowski, carried away by a very natural enthusiasm for the remarkable geometrical synthesis that he had discovered, had to some extent gone beyond the relativistic doctrine, which does not in any way forbid that an observer should reason and calculate, as in daily life, in terms of space and time.

In an analogous matter, it seems very unlikely that Minkowski was misled by the " mystical aspect " of the equation

$$3. \ 10^5 \ km = \sqrt{-1} \ \text{second}$$

which he obtained by introducing the imaginary time $\sqrt{-1}t$. This result has led some popularisers to misleading statements like " over space-time rules the imaginary "—clearly this is an abuse of words.

Freed of all its superfluous commentary, it remains that Minkowski's presentation contains a remarkable geometrical synthesis of the mathematical aspects of special relativity.

CHAPTER TWO

GENERALISED RELATIVITY

A. *PRESENTATION*

1. STATEMENT OF THE PRINCIPLE OF GENERALISED RELATIVITY.

Einstein was led to the theory of generalised relativity by the analysis of some simple questions, notably the following one. Consider a galilean frame of reference S in which, by definition, the principle of inertia is valid, and a system S' which is actuated by a uniformly accelerated motion, of acceleration relative to S.

In S' every particle has an accelerated motion of acceleration $\vec{\gamma}_r = -\vec{\gamma}_e$.

Can an observer connected to S' conclude that he is placed in a *truly* accelerated system ? The reply to this question is negative, for the motion of every isolated particle in S' can be explained just as well, in the following way—the system of reference is not accelerated, but there is effective in the corresponding space a field of uniform gravitational acceleration $-\vec{\gamma}_e$.

Therefore a uniformly accelerated system such as S, where no field of gravitation is effective, and a non-accelerated system, where a field of uniform gravitation is effective, can be considered as *physically equivalent*. A simple change of system of reference can therefore " create " a field of gravitation, which suggests the search for a generalisation of the principle of special relativity intended to contain the theory of gravitation. In this generalisation it is also natural to see the principle of the constancy of the velocity of light in the vacuum undergo a modification. This is a problem which had already occupied Newton's attention.[1]

We shall no treat the first attempts that Einstein [2] made, but

---

[1] Indeed, in his *Opticks*, NEWTON wrote, " Might not bodies act at a distance on light ? Does not this action deviate the rays of light and is it not, other things being equal besides, all the more stronger as the distance is less ? "

[2] *Jahrbuch für Radioaktivität und Elektronik*, Vol. 4, 1907. *Annalen der Physik*, Vol. 35, 1911.

shall follow a fundamental paper *Die Grundlage der allgemeinen Relativitätstheorie*.[1]

In this paper Einstein puts forward the following postulate, which he calls the *principle of generalised relativity*.

" *The laws of nature must be such that they are valid in an arbitrary system of reference.* "

He first recalls the fact that in special relativity, just as in classical mechanics, the coordinates of space and time have a direct physical meaning. This circumstance does not apply to more general cases, as the following example shows.

" In a region free from all gravitational field consider a galilean system of reference $S(x, y, z, t)$. Let $S'(x', y', z', t')$ be a system of reference in uniform rotation relative to S.

" The origins O and O' of the two systems, as well as their axes of z, permanently coincide. We wish to show that, in the system S', the coordinates of space and time cannot have direct physical significance. By reasons of symmetry it is clear that a circle of centre O, situated in the plane xy of the system S, must also be considered as a circle in the system S'. Given this, imagine that the circumference and the diameter of the circle are measured with an elementary measuring-rod which is very small compared with the diameter, and that the quotient of the two measurements obtained is formed. If the experiment is made with a measuring rod *at rest relative to S* the result is π. With a rod at rest in S', the result will be a number greater than π. This fact is verified by observing the method of carrying out the latter experiment *from the system S*— the rod placed at the periphery experiences the Lorentz contraction while, on the diameter, it experiences no contraction. Euclidean geometry is therefore not valid in the system S'. Therefore the direct interpretation of the space coordinates, which presupposes the validity of euclidean geometry, breaks down in the system S'.

" It is also impossible to define a time which has a direct physical significance that could be measured by means of identical clocks at rest in S'. To verify this, imagine that two clocks have been placed at the origin O and at the periphery of the circle respectively, when these two clocks are observed *from the system S*. A known result of special relativity shows that the peripheral clock is slow compared with the central clock, for the first is in motion and the second not. An observer placed at O and who will be able to observe, using light for transmission, the two clocks at the same time, will therefore see

[1] *Annalen der Physik*, Vol. 49, 1916.

the peripheral clock become slow compared with the central clock. As he will not be able to resolve to make the velocity of light along the path considered depend explicitly on the time, he will interpret his experiment by saying that the first clock *actually* retards in comparison with the central clock. He will therefore be reduced to defining the time in such a way that the speed with which a clock beats time depends upon its position.

" We therefore arrive at the following result— In the theory of generalised relativity the magnitudes of space and time *cannot be defined is such a way that differences of coordinates are directly measurable by means of an unit measuring-rod, and so that differences of time are directly measurable by means of a standard clock.* "

Einstein remarks that every measurement is expressed by the coincidence of two localised events *(Punktereignissen)* in space and time. The introduction of a system of coordinates has no other purpose than the facilitation of the description of such coincidences. The multiplicity of the four space-time variables x_1, x_2, x_3, x_4 is arranged so that a set of values of these variables corresponds to each event. A coincidence is then characterised by the identity of two sets of values of the coordinates. If the arbitrary single-valued functions x_1', x_2', x_3', x_4' of the x_i are introduced instead of the variables x_1, x_2, x_3, x_4, a given coincidence can still be expressed by the identity of two sets of values of the x_i'. There is therefore no reason to choose one system of reference rather than the other. This leads to the statement of the principle of generalised relativity in the form of universal covariance—

" *The laws of nature must be expressed in such a way that they are equivalent for all systems of reference; that is, that they are covariant under any substitution of coordinates.* "

Einstein then proceeds by induction, not concerning himself with making the theory of generalised relativity dependent on the smallest number of indispensable axioms.

First he assumes *that the special theory of relativity must be valid for every infinitely small region of the four dimensional world, provided that a suitable system of reference is chosen.*

Then let ξ_1, ξ_2, ξ_3, be the space coordinates and ξ_4 the time coordinate in this system.

In a system of units in which the velocity c reduces to unity, Minkowski's interval

$$ds^2 = d\xi_4^2 - d\xi_1^2 - d\xi_2^2 - d\xi_3^2$$

is an intrinsic quantity which can be obtained by means of measurements of space and time. To the linear element ds, as well as to every event localised in the infinitely small domain considered, there correspond, in an arbitrarily chosen system of reference x_1, x_2, x_3, x_4 the differentials dx_i, and in this system.

$$d\xi_i = \alpha_{ik}dx_k \ ^1$$

or

$$ds^2 = g_{ik}dx_idx_k \qquad (i, k = 1, 2, 3, 4).$$

This is the expression of the *invariant* ds^2 in an arbitrary system of reference. This invariant is a quadratic form in the dx_i, whose coefficients g_{ik} characterise the gravitational field associated with the chosen system.

When it is possible to choose a system of reference in a *finite* region of space-time so that the g_{ik} have the values

(E)
$$\begin{vmatrix} -1, & 0, & 0, & 0 \\ 0, & -1, & 0, & 0 \\ 0, & 0, & -1, & 0 \\ 0, & 0, & 0, & +1 \end{vmatrix}$$

the conditions of the special theory of relativity are satisfied.

2. Remark on the mathematical tools of generalised relativity.

Einstein had the good fortune of being able to make use of a mathematical tool which was suited to the formulation of the theory of generalised relativity. This was the *absolute* differential calculus, which was developed by Ricci and Levi-Civita in a paper dated 1899 which appeared in the *Mathematische Annalen* (Vol. 54). The authors described the origin of this calculus in the following terms.

" The algorithm of the absolute differential calculus lies entirely in a remark due to Christoffel.[2] But the methods themselves and the advantages of the calculus have their justification and origins in the intimate relationship which connects them with the notion of variety in n dimensions, which we owe to Gauss and to Riemann. . . . A variety V_n remains invariant under every transformation of coordinates. The absolute differential calculus, by operating on covariant or contravariant forms to the ds^2 of V_n to derive others of the same kind, is itself, in the calculations and the results, independent of the choice of independent variables. Being in the nature of things essentially

---

[1] Throughout this presentation we adopt the usual convention of the *summation of the dummy suffixes.*

[2] *Über die Transformation der homogen Differentialausdrücke zweiten Grades,* Journal de Crelle, Vol. 70, 1869.

associated with V_n, it is the natural instrument of all investigations which have such a variety as their subject, or in which a positive definite quadratic form of the differentials of n variables, or their derivatives, appears as a characteristic feature. "

The scope of this book does not allow us to develop systematically the technique of the absolute differential calculus. The reader can refer to Einstein's paper itself, to Eddington's book *(Space, Time and Gravitation)* or to a text-book. We shall confine ourselves to an explanation of the essentials as we need them.

3. THE EQUATIONS OF MOTION OF A FREE PARTICLE IN A GRAVITATIONAL FIELD.

In special relativity a free particle moves uniformly along straight lines. In generalised relativity it is the same in every region in which a frame of reference S_0 can be chosen so that the g_{ik} have the particular values (E).

Observe this motion from an arbitrary system of reference S. In S_0 the trajectory is a geodesic. As the definition of such a line does not depend on the coordinates, the equations of this line will also define the motion in S.

Now from

$$ds^2 = g_{ik}dx_idx_k$$

a classical variation calculation depending on the extremal

$$\delta \int ds = 0$$

(where the limits are fixed) leads to the equations

$$(L) \qquad \frac{d^2x_i}{ds^2} = - \begin{Bmatrix} \mu\nu \\ i \end{Bmatrix} \frac{dx_\mu}{ds} \frac{dx_\nu}{ds} \qquad (i = 1, 2, 3, 4).$$

The symbol on the right-hand side of this equation has the meaning

$$\begin{Bmatrix} \mu\nu \\ i \end{Bmatrix} = g^{i\alpha} \begin{bmatrix} \mu\nu \\ \alpha \end{bmatrix}$$

where $g^{i\alpha}$ is the fundamental contravariant tensor associated with the form ds^2 [1] and where, moreover,

$$\begin{bmatrix} \mu\nu \\ \alpha \end{bmatrix} = \frac{1}{2}\left(\frac{\partial g_{\mu\alpha}}{\partial x_\nu} + \frac{\partial g_{\nu\alpha}}{\partial x_\mu} - \frac{\partial g_{\mu\nu}}{\partial x_\alpha} \right).$$

---

[1] Let g be the determinant of the $g_{\mu\nu}$. The fundamental contravariant tensor $g^{\mu\nu}$ is equal to the minor the element $g_{\mu\nu}$ of the determinant divided by g. Alternatively,

$$g^{\mu\alpha} g_{\alpha\nu} = \delta_\nu^\mu = \begin{cases} 0 \text{ if } \mu \neq \nu \\ 1 \text{ if } \mu = \nu. \end{cases}$$

Einstein then makes the *hypothesis that the equations* (L) *also represent the motion of a free particle in a gravitational field*, even when it is impossible to find a frame of reference S_0 in which the special theory of relativity is valid in a *finite* region ; " if the $\left\{ \begin{matrix} \mu\nu \\ i \end{matrix} \right\}$ vanish, the particle moves uniformly along a straight line. Therefore these quantities characterise the difference between the true motion and a uniform motion— they are the components of the gravitational field. "

4. EQUATIONS OF THE GRAVITATIONAL FIELD IN THE ABSENCE OF MATTER.

Before taking up this question, we make an indispensable mathematical digression.

The *covariant differential* of a covariant vector A_i is said to be the quantity

$$dA_i - \left\{ \begin{matrix} is \\ r \end{matrix} \right\} A_r dx_s$$

and, similarly, the covariant differential of a contravariant vector A^i is said to be the quantity

$$dA^i + \left\{ \begin{matrix} rs \\ r \end{matrix} \right\} A^r dx_s.$$

It is said that a vector A_i (or A^i) suffers a *parallel displacement* if its covariant derivative remains zero.

In particular, consider two *elementary displacements* dx^i and δx^i (contravariant vectors) from the point P. It is possible to consider the parallel displacement of the vector dx^i along δx^i and the parallel displacement of the vector δx^i along dx^i. Now

$$\delta(dx^i)^{\cdot} = - \left\{ \begin{matrix} rs \\ i \end{matrix} \right\} dx_r \delta x_s = - \left\{ \begin{matrix} sr \\ i \end{matrix} \right\} dx_s \delta x_r = - d(\delta x^i).$$

Therefore we shall define an *elementary parallelogram* such that

$$(d\delta - \delta d) x^i = 0.$$

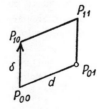

Fig. 114

Now consider an arbitrary covariant vector A_i and seek the difference ΔA_i between the components of this vector according to whether the point P_{11} has been reached by the operation δd or by the operation $d\delta$. Here

$$\delta d A_i = \delta \left\{ \begin{matrix} is \\ r \end{matrix} \right\} A_r dx_s + \left\{ \begin{matrix} is \\ r \end{matrix} \right\} \delta A_r dx_s + \left\{ \begin{matrix} is \\ r \end{matrix} \right\} A_r \delta dx_s$$

$$d\delta A_i = d \left\{ \begin{matrix} is \\ r \end{matrix} \right\} A_r \delta x_s + \left\{ \begin{matrix} is \\ r \end{matrix} \right\} d A_r \delta x_s + \left\{ \begin{matrix} is \\ r \end{matrix} \right\} A_r d\delta x_s.$$

Taking the difference, the last terms on the right-hand sides are eliminated and it remains that

$$\delta d A_i - d\delta A_i = \frac{\partial}{\partial x_j} \left\{ \begin{matrix} is \\ r \end{matrix} \right\} A_r dx_s \delta x_j + \left\{ \begin{matrix} is \\ r \end{matrix} \right\} \left\{ \begin{matrix} r\alpha \\ \beta \end{matrix} \right\} A_\beta \delta x_\alpha dx_s$$

$$- \frac{\partial}{\partial x_j} \left\{ \begin{matrix} is \\ r \end{matrix} \right\} A_r \delta x_s dx_j - \left\{ \begin{matrix} is \\ r \end{matrix} \right\} \left\{ \begin{matrix} r\alpha \\ \beta \end{matrix} \right\} A_\beta dx_\alpha \delta x_s$$

$$\Delta A_i = A_r \left[\frac{\partial}{\partial x_j} \left\{ \begin{matrix} is \\ r \end{matrix} \right\} - \frac{\partial}{\partial x_s} \left\{ \begin{matrix} ij \\ r \end{matrix} \right\} + \left\{ \begin{matrix} is \\ \alpha \end{matrix} \right\} \left\{ \begin{matrix} \alpha j \\ r \end{matrix} \right\} - \left\{ \begin{matrix} ij \\ \alpha \end{matrix} \right\} \left\{ \begin{matrix} \alpha s \\ r \end{matrix} \right\} \right] dx_s \delta x_j.$$

The quantity in square brackets is the mixed tensor of the 4th rank which is called the *Riemann-Christoffel tensor*; it is referred to by the symbol

$$B^r_{ijs}.$$

In an euclidean space this tensor is zero and, conversely, the vanishing of this tensor implies the euclidean character of the space.

The operation called that of *contraction* in the indices r and s yields the *contracted Riemann-Christoffel tensor*

$$B_{ij} = \frac{\partial}{\partial x_j} \left\{ \begin{matrix} is \\ s \end{matrix} \right\} - \frac{\partial}{\partial x_s} \left\{ \begin{matrix} ij \\ s \end{matrix} \right\} + \left\{ \begin{matrix} is \\ \alpha \end{matrix} \right\} \left\{ \begin{matrix} \alpha j \\ s \end{matrix} \right\} - \left\{ \begin{matrix} ij \\ \alpha \end{matrix} \right\} \left\{ \begin{matrix} \alpha s \\ s \end{matrix} \right\}. \quad {}^1$$

Given this, we return to Einstein's theory of gravitation.

Einstein distinguishes the gravitational field from the " matter," by which term he understands not only matter in the ordinary sense of the word, but also the electromagnetic field. To obtain the express-ion of the gravitational field in the absence of " matter, " he proceeds by an induction analogous to that which led him to the equations of motion of a free particle.

---

[1] It may be remarked here that the contracted three-index symbol $\left\{ \begin{matrix} is \\ s \end{matrix} \right\}$ reduces to $\frac{1}{2} g^{s\alpha} \frac{\partial g_{s\alpha}}{\partial x_i}$ or to $\frac{\partial \log \sqrt{-g}}{\partial x_i}$, where g is the determinant of the $g_{s\alpha}$. This simplifies the form of the tensor B_{ij} in a system of coordinates in which $\sqrt{-g} = 1$.

" In the particular case of special relativity the g_{ij} have constant values. If this circumstance is confirmed throughout a *finite* domain in a system of reference S_0, the components of the Riemann-Christoffel tensor will vanish in S_0 and, consequently, in every other system of reference.

" The equations of the gravitational field which are sought must therefore be verified in all the B'_{ijs} vanish. But this condition takes us too far. For it is clear that the gravitational field created by a particle, for example, cannot disappear for any choice of the coordinates. Therefore it occurs to one to substitute the nullity of the contracted tensor B_{ij} for that of the B'_{ijs}. This leads to 10 equations in the ten quantities g_{ij}. This choice offers a minimum of arbitrariness *(ein Minimum von Willkür)*. For, apart from B_{ij}, there exists no other tensor of the second rank, formed from the g_{ij} and their derivatives, which is linear and does not contain derivatives of higher order than the second, except the tensor $B_{ij} + \lambda g_{ij} \, (g^{\alpha\beta} \, B_{\alpha\beta})$. "

This hypothesis, together with the equations of the motion of a free particle, makes it possible for the newtonian law of attraction to be retrieved in the first approximation. The second approximation yields the displacement of the perihelion of Mercury, demonstrated by Le Verrier. To Einstein, these facts vouch for the physical value of the theory of generalised relativity.

5. GENERAL FORM OF THE EQUATIONS OF GRAVITATION.

In order to pass from the law of gravitation in the absence of " matter, "

$$(G_0) \qquad\qquad B_{ij} = 0$$

to the general case, Einstein introduces the energy tensor

$$T^{ij} = \varrho_0 \, \frac{dx_i}{ds} \, \frac{dx_j}{ds_{_i}}.$$

Here ϱ_0 is the proper density of the matter.

With this contravariant tensor can be associated the mixed tensor T^j_i, the covariant tensor T_{ij} and the invariant T.

By analogy with the conservation of energy in the classical field, it is natural to assign a zero *divergence* to the mixed tensor T^j_i; that is, to write

$$T^j_{ij} = 0.$$

On the other hand the energy tensor must belong to the category of tensors that can be deduced from the fundamental form of the

g_{ik}. Now, if the mixed tensor B_i^j and the invariant B are associated with the contracted Riemann-Christoffel tensor B_{ij}, it is verified that the expression

$$B_i^j - \frac{1}{2}\,\delta_i^j B$$

has a divergence equal to zero.

It is therefore natural to identify this expression (apart from a constant factor) with the mixed tensor T_i^j. This gives

$$B_i^j - \frac{1}{2}\,\delta_i^j B = -\,k T_i^j$$

or, alternatively,

(G) $$B_{ij} = -\,k\Big(T_{ij} - \frac{1}{2}\,g_{ij}\,T\Big).$$

This is the form which Einstein gives the equations of gravitation in the general case.

6. Reversion to Newton's theory in the first approximation.

To revert to the Newtonian mechanics in the first approximation, Einstein supposes

a) that the g_{ik} only differ from the euclidean values (E) by quantities of the first order, and then neglects quantities of the second and higher order;

b) that a suitable choice of coordinates allows the g_{ik} to be assigned the values (E) at infinity;

c) that the velocity of a particle is very small compared with the velocity of light in the vacuum, so that

$$\frac{dx_1}{ds} \qquad \frac{dx_2}{ds} \qquad \frac{dx_3}{ds}$$

are of the first order and that

$$\frac{dx_4}{ds}$$

is equal to unity (to the second order). Therefore the units are chosen so that $c = 1$.

Under these conditions the symbols

$$\left\{ \begin{matrix} \mu\nu \\ i \end{matrix} \right\}$$

are of the first order (at least) and the equations of a geodesic reduce to

$$\begin{cases} \dfrac{d^2x_i}{dt^2} = \begin{bmatrix} 4 & 4 \\ & i \end{bmatrix} \\ \dfrac{d^2x_4}{dt^2} = -\begin{bmatrix} 4 & 4 \\ & 4 \end{bmatrix} \end{cases} \qquad (i = 1, 2, 3);$$

d) finally is it supposed that the gravitational field is *quasi-static*, or that the particles which create the field only have velocities which are small compared with $c = 1$. Therefore the derivatives with respect to the time that appear in the square brackets are negligible in comparison with the derivatives with respect to the space coordinates. Then it follows that

$$\frac{d^2x_i}{dt^2} = -\frac{1}{2}\frac{\partial g_{44}}{\partial x_i} \qquad (i = 1, 2, 3).$$

These equations can be compared with the newtonian equations of the motion of a particle by making $\dfrac{g_{44}}{2}$ play the part of the potential.

Einstein then refers to the equations (G). To the chosen approximation the tensor T_{ij} reduces to its term

$$T_{44} = T = \varrho$$

and, by neglecting the partial derivatives in the time, it turns out that

$$\frac{\partial}{\partial x_1}\begin{bmatrix} ij \\ 1 \end{bmatrix} + \frac{\partial}{\partial x_2}\begin{bmatrix} ij \\ 2 \end{bmatrix} + \frac{\partial}{\partial x_3}\begin{bmatrix} ij \\ 3 \end{bmatrix} - \frac{\partial}{\partial x_4}\begin{bmatrix} ij \\ 4 \end{bmatrix}$$

$$= -\frac{1}{2}\left(\frac{\partial^2 g_{44}}{\partial x_1^2} + \frac{\partial^2 g_{44}}{\partial x_2^2} + \frac{\partial^2 g_{44}}{\partial x_3^2}\right) = -\frac{1}{2}\varDelta g_{44} = -\frac{k}{2}\varrho$$

or

$$\varDelta g_{44} = k\varrho.$$

The gravitational potential V deduced from this is

$$V = -\frac{k}{8\pi}\int\frac{\varrho\,d\tau}{r}.$$

Whence, by comparison with its classical expression and by taking account of the unit of time chosen,

$$V = -\frac{K}{c^2}\int\frac{\varrho\,d\tau}{r}$$

where K is the ordinary gravitational constant. This yields the value

$$k = \frac{8\pi K}{c^2}.$$

7. The conduct of measurements of space and time in a static
   gravitational field. Deviation of light rays. Displacement
   of the perihelion of the planets.

Merely the fact of reducing the expression of the gravitational
field to the single component g_{44} in the first approximation implies
that, if the condition $\sqrt{-g} = 1$ is respected, the other components
of the g_{ik} must be different from unity.

To define the field of a single mass m Einstein uses, at this point,
the solution

$$\begin{cases} g_{ik} = -\delta_{ik} - \alpha \dfrac{x_i x_k}{r^3} \text{ where } (i, k = 1, 2, 3) \text{ and } \delta_{ik} = \begin{cases} 0 & \text{if } i \neq k \\ 1 & \text{if } i = k \end{cases} \\ g_{i4} = 0 \\ g_{44} = 1 - \dfrac{\alpha}{r} \end{cases}$$

with $\qquad r = \sqrt{x_1^2 + x_2^2 + x_3^2} \qquad \alpha = \dfrac{km}{4\pi}.$

This takes account of the spherical symmetry about the origin and
satisfies the gravitational equations in the first approximation.

To explain the influence of the field of the mass m on the
measurements of space and time, it is suitable to compare the dx_i with
the measurements of length and time carried out in a *tangent euclidean
universe.*

For example, the measurement of an unit length along the axis of x_1

$$ds^2 = -1 \qquad dx_2 = dx_3 = dx_4 = 0 \qquad x_1 = r$$

implies the relation

$$-1 = g_{11}dx_1^2 = -\left(1 + \frac{\alpha}{r}\right) dx_1^2.$$

Whence

$$dx_1 = 1 - \frac{\alpha}{2r}.$$

Therefore the standard length placed *radially* appears contracted
in the system of coordinates (x_i).

On the contrary, if the coordinates of a standard length placed
tangentially are measured,

$$ds^2 = -1 \qquad dx_1 = dx_3 = dx_4 = 0 \qquad x_1 = r \qquad x_2 = x_3 = 0$$

and it is found that

$$-1 = g_{22}dx_2^2 = -dx_2^2.$$

The effect of the gravitational field on the standard placed tangentially is nothing.

Again, consider a clock at rest in a static field. Then

$$ds = 1 \qquad dx_1 = dx_2 = dx_3 = 0$$

$$dx_4 = \frac{1}{\sqrt{g_{44}}} = \frac{1}{\sqrt{1 + (g_{44} - 1)}} = 1 - \frac{g_{44} - 1}{2} = 1 + \frac{k}{8\pi} \int \frac{\varrho d\tau}{r}.$$

Therefore a clock runs slower when it is placed in a gravitational field. It follows from this that the spectral lines emmitted by the stars must appear to us to be displaced *towards the red*.

Finally, in a given direction in space for which the proportions

$$dx_1 : dx_2 : dx_3$$

are known, the equation

$$ds^2 = g_{ij} dx_i dx_j = 0 \qquad\qquad (i, j = 1, 2, 3, 4)$$

enables the velocity of light to be calculated

$$\gamma = \sqrt{\left(\frac{dx_1}{dx_4}\right)^2 + \left(\frac{dx_2}{dx_4}\right)^2 + \left(\frac{dx_3}{dx_4}\right)^2}$$

in the sense of euclidean geometry. For example, for a light ray parallel to the axis of x_2, it follows that

$$0 = g_{44} dx_4^2 + g_{22} dx_2^2 \quad \text{or} \quad \gamma = \sqrt{-\frac{g_{44}}{g_{22}}} = 1 - \frac{\alpha}{2r}\left(1 + \frac{x_2^2}{r^2}\right) \quad (c = 1).$$

At the distance $\varDelta$ from a mass m, the deviation of a light ray is thus given by

$$\frac{2\alpha}{\varDelta} = \frac{km}{2\pi\varDelta}.$$

In the neighbourhood of the Sun this deviation is about $1 \cdot 7''$, and in the neighbourhood of Jupiter, about $0 \cdot 02''$. Thus Newton's question on " gravitating light " is answered in the affirmative, the effect being double that which follows from the classical theory.

In order to calculate the displacement of the perihelion of the planets it is necessary, as Schwartzschild has done,[1] to take the calculation to a higher approximation. The quantity ds^2 was calculated for a *test particle* placed in the field of a single finite mass. The perihelion of the trajectory of the test particle experiences a displacement of

$$24\,\pi^3\,\frac{a^2}{T^2 c^2 (1 - e^2)}$$

[1] *Sitzungsberichte der Preussichen Akademie der Wissenschaften*, 1916, p. 189.

where a is the semi-major axis, e the eccentricity, T the period in seconds and c the velocity of light in the normal system of units.

For Mercury the effects amounts to $43''$ a century, which agrees with the result established by Le Verrier.

8. The spatially closed universe.

Einstein [1] starts by indicating the difficulties that follow from the classical theory. Consider a point O of the universe round which the gravitational field, taken as a whole, has spherical symmetry. Poisson's equation,

$$(1) \qquad \Delta\varphi = 4\pi K\varrho$$

shows that the mean density of matter must tend to zero more rapidly than $\dfrac{1}{r^2}$ if φ tends to a finite limit as r increases indefinitely. From this point of view Newton's universe appears finite even if its total mass can be infinitely great. Under these conditions the radiation emitted by the celestial bodies can, in part, escape beyond the limits of the newtonian universe. Is it possible that celestial bodies can entirely disappear in this way ? It is almost impossible to reply negatively to this question. Indeed, if φ tends to a finite limit as r increases indefinitely, a celestial body endowed with a finite force can move away to infinity, without hope of return, by overcoming the newtonian attractions.

It is impossible to escape this difficulty by giving the potential considerable value at infinity. For this would be in contradiction with observation and, in particular, with the data that are available on the velocity of the distant stars. Instead of Poisson's equation, Einstein proposes writing the equation

$$(2) \qquad \Delta\varphi - \lambda\varphi = 4\pi K\varrho$$

where λ is a universal constant. If ϱ_0 is the density of a UNIFORM distribution of mass, then

$$\varphi_0 = -\frac{4\pi K\varrho_0}{\lambda}$$

is a solution of (2) which corresponds to a true distribution of the fixed stars which would be homogeneous and of density ϱ_0. If, in fact, the distribution of matter contains local irregularities, a supple-

---

[1] *Kosmologische Betrachtungen zur allgemeinen Relativitätstheorie, Sitzungsberichte der Preussichen Akademie der Wissenschaften*, 1917.

mentary φ is added to the solution φ_0. In the neighbourhood of the discrete masses this can be more nearly approximated to by the newtonian law (1) as λ is small compared with $4\pi K\varrho$.

Returning to the theory of relativity, Einstein contemplates two possibilities for the conditions to be imposed on the law of gravitation at the limits—

a) either it is assumed that at infinity the g_{ik} reduce to their euclidean values for a suitable choice of coordinates;

b) or, as de Sitter suggests, the *a priori* imposition of any condition on the g_{ik} at infinity is not attempted.

Einstein declares that he is unable to reconcile himself to the attitude *b)*, which would have the character of a renunciation. On the other hand, hypothesis *a)* encounters many objections. In the first place, the fact that it resorts to a determined system of reference is contrary to the very spirit of the principle of relativity. Moreover, in a coherent theory of relativity inertia *(Trägheit)* in itself, "with respect to space," does not exist, but only an interaction between masses. If, for example, a mass is sufficiently separated from all the bodies of the universe, its *Trägheit* must vanish. The inertia of a particle of proper mass m depends explicitly on the g_{ik}[1]; but these differ so little from their values at infinity that, under these conditions, the inertia would be *influenced* but not *conditioned* by the matter situated at a finite distance. Finally, the same objections of statistical mechanics are encountered as in the newtonian case.

This is why, after all these detours, Einstein is led to postulate the existence of a *spatially closed universe (räumlich geschlossene Welt)* of such a kind that there is no need to suppose conditions at the limits. The notion of such a continuum is compatible both with the principle of generalised relativity and with the experimental fact of the smallness of the velocities of the stars. It requires a modification of the equations of gravitation.

It is supposed that there exists a system of reference in which matter might be considered to be permanently at rest. The contravariant energy tensor T^{ik} then reduces to a scalar

$$T^{44} = T = \varrho.$$

The mean density of distribution of matter, ϱ, is *a priori* a function of position. Suppose that this function reduces to a constant.

---

[1] Indeed, the momentum-energy vector of a particle of proper mass m has components

$$m \sqrt{-g}\, g_{i\alpha} \frac{dx_\alpha}{ds}.$$

The equations of motion of a free particle in a gravitational field are

$$\frac{d^2x_i}{ds^2} + \left\{ \begin{matrix} \mu v \\ i \end{matrix} \right\} \frac{dx_\mu}{ds} \frac{dx_\nu}{ds} = 0.$$

If the field is static it follows that, in the hypothesis of rest, g_{44} does not depend on the coordinates.

Therefore we can put

$$g_{44} = 1.$$

The *curvature of space* must be constant because of the supposed homogeneity of the distribution of masses. Moreover, as in any static field,

$$g_{4i} = 0 \qquad\qquad (i = 1, 2, 3).$$

The closed continuum that we seek is therefore a space which is spherical in x_1, x_2, x_3.

To define such a space we resort to an euclidean space of 4 dimensions

$$d\sigma^2 = d\xi_1^2 + d\xi_2^2 + d\xi_3^2 + d\xi_4^2$$

and, in that space, consider the hypersphere

$$\xi_1^2 + \xi_2^2 + \xi_3^2 + \xi_4^2 = R^2.$$

The points of this hypersphere form a spherical continuum of radius R. The euclidean space of four dimensions (ξ_i) is only used here as a means of defining the hypersphere. Eliminating ξ_4 and taking ξ_1, ξ_2, ξ_3 as independent variables, it turns out that

$$d\sigma^2 = \gamma_{ik} d\xi_i d\xi_k$$

with

$$\gamma_{ik} = \delta_{ik} + \frac{\xi_i \xi_k}{R^2 - \sigma_0^2}.$$

Here i, $k = 1, 2, 3$ and $\sigma_0^2 = \xi_1^2 + \xi_2^2 + \xi_3^2$.

Therefore the ds^2 of the universe, in the case of the spherical space, is defined by the following quantities g_{ik}—

$$\begin{cases} g_{44} = 1 \qquad g_{4i} = 0 \\ g_{ik} = -\left[\delta_{ik} + \dfrac{x_i x_k}{R^2 - (x_1^2 + x_2^2 + x_3^2)}\right] \end{cases} (i, k = 1, 2, 3).$$

The equations (G) of § 5 are not satisfied by these values of the g_{ik} and $T^{44} = \rho = $ const. For this reason, it is necessary to modify these equations in the same way as that suggested by Einstein for Poisson's equation, and to write

(G′)
$$B_{ij} - \lambda g_{ij} = - k \left(T_{ij} - \frac{1}{2} g_{ij} T \right).$$

As all the points of the continuum are equivalent, taking the hypotheses which have been made into account, it is sufficient to carry out the calculation for

$$x_1 = x_2 = x_3 = x_4 = 0.$$

At this point it is found that

$$B_{ij} = \frac{\partial}{\partial x_1} \begin{bmatrix} ij \\ 1 \end{bmatrix} + \frac{\partial}{\partial x_2} \begin{bmatrix} ij \\ 2 \end{bmatrix} + \frac{\partial}{\partial x_3} \begin{bmatrix} ij \\ 3 \end{bmatrix} + \frac{\partial^2 \log \sqrt{-g}}{\partial x_i \, \partial x_j}.$$

The equations (G) are satisfied for

$$-\frac{2}{R^2} + \lambda = -\frac{k\varrho}{2} \qquad -\lambda = -\frac{k\varrho}{2}.$$

Whence

$$\lambda = \frac{k\varrho}{2} = \frac{1}{R^2}.$$

Thus the constant λ determines the mean density ϱ and the radius of the spherical universe, R.

As the volume of this space has the value $2\pi^2 R^3$, the mass of the universe is also known to be

$$M = 4\pi^2 \frac{R}{k}.$$

9. GRAVITATION AND ELECTRICITY.

We owe to H. Weyl a remarkable mathematical synthesis of relativity, which is developed in his famous book *Raum, Zeit, Materie*. Here we shall confine ourselves to an analysis of the original article, published in 1918 in the *Berliner Sitzungsberichte*. In this, for the first time, the author lays the foundations of a unitary theory of gravitation and electricity.

Weyl draws his inspiration directly from the ideas of riemannian geometry, by taking a fundamental quadratic form

(1)
$$ds^2 = g_{ik} dx_i dx_k$$

the basis of the metric. But " in accordance with the spirit which animates the modern physics of contact actions, " he constructs a geometry of neighbourhoods *(Nahegeometrie)*. In this only the possibility of the transport of a vector from a point P to an infinitely close point P' arises, and it is not necessary to have regard to the integrability of the length of a vector in such a transport.

Parallel displacement in Levi-Civita's sense plays a fundamental part in this geometry. To a contravariant vector ξ^i at the point P is made to correspond, by parallel displacement to the infinitely close point P', a vector $\xi^i + d\xi^i$ whose components are linear in the ξ^r. That is

$$(2) \qquad d\xi^i = - d\gamma_r^i \xi^r.$$

The quantities $d\gamma_r^i$ are themselves linear in the dx_i, say

$$(3) \qquad d\gamma_r^i = \Gamma_{rs}^i dx_s.$$

The consideration of an elementary " parallelogram " (already considered in § 4) implies the condition of symmetry

$$(4) \qquad \Gamma_{rs}^i = \Gamma_{sr}^i.$$

In this geometry the scalar product

$$g_{ik}\xi^i\eta^k$$

of two contravariant vectors at P is only defined *apart from an arbitrary* (positive) *constant*. In the parallel displacement of each of the two vectors from P to P' the scalar product becomes

$$(5) \qquad (g_{ik} + dg_{ik})(\xi^i + d\xi^i)(\eta^k + d\eta^k) = g_{ik}\xi^i\eta^k(1 + d\varphi).$$

That is

$$(5') \qquad dg_{ik} - (d\gamma_{ik} + d\gamma_{ik}) = g_{ik}d\varphi$$

(where the γ_{ik} correspond to γ_k^r through the operation $g_{ir}\gamma_k^r$).

It follows that $d\varphi$ is a differential form

$$(6) \qquad d\varphi = \varphi_i dx_i$$

whence the form of the functions Γ is implied by the relation

$$(7) \qquad \Gamma_{i,\,kr} + \Gamma_{k,\,ir} = \frac{\partial g_{ik}}{\partial x_r} - g_{ik}\varphi_r.$$

To the consideration of the fundamental quadratic form (1) is thus added the consideration of the linear form (6). If the g_{ik} are multiplied by a function λ, which is arbitrary at each point, without changing the system of coordinates (x^i), the $d\gamma_k^i$ do not change, the $d\gamma_{ik}$ are multiplied by λ, and equation (5) then shows that

$$(8) \qquad d\varphi + \frac{d\lambda}{\lambda} = d\varphi + d\log\lambda.$$

That is, the differential form $d\varphi$ increases by a total differential. The antisymmetrical tensor

$$(9) \qquad F_{ik} = \frac{\partial \varphi_i}{\partial x_k} - \frac{\partial \varphi_k}{\partial x_i}$$

thus acquires an *absolute* significance—that is, its components are not affected by a change of *gauge* which transforms the g_{ik} into λg_{ik}. It is the same for the bilinear form

$$(10) \qquad F_{ik} dx_i \delta x_k = \frac{1}{2} F_{ik} \Delta x_{ik}$$

constructed from two arbitrary infinitesimal displacements at the point P, which define an element of surface of components

$$(11) \qquad \Delta x_{ik} = dx_i \delta x_k - dx_k \delta x_i.$$

In the particular case of g_{ik} which can be chosen in such a way that the φ_i vanish identically, a *unit length* defined arbitrarily at the point P can be transported by parallel displacement along a finite path and the result is independent of this path (integrability).

In these circumstances the quantities Γ_{rs}^{i} reduce to the Christoffel three-index symbols of ordinary riemannian geometry. The necessary and sufficient condition that this should occur is that the tensor F_{ik} vanishes identically.

Weyl then proposes *to interpret the φ_i as the four components of the electromagnetic potential*, the vanishing of the electromagnetic field leading back to Einstein's theory, in which only gravitation occurs.

In these circumstances the electromagnetic quantities, in a given system of reference, appear independent of an arbitrary change of gauge which, on the contrary, has an effect on the g_{ik}.

At this point Weyl introduces the notion of *weight* in application to a tensor. For example, let there be an invariant (with respect to every transformation of coordinates)

$$(12) \qquad a_{ik} dx_i \delta x_k.$$

If the g_{ik} are transformed to λg_{ik}, the a_{ik} become $\lambda^e a_{ik}$. The exponent e is called the weight of the tensor a_{ik}. Every *absolutely* invariant tensor (in the sense defined above for F_{ik}) necessarily has zero weight. We have already seen that the tensor F_{ik}, which satisfies the first system of Maxwell's equations

$$\frac{\partial F_{ik}}{\partial x_l} + \frac{\partial F_{kl}}{dx_i} + \frac{\partial F_{li}}{\partial x_k} = 0$$

is of this kind.

In Weyl's geometry it is possible—always using the notion of parallel displacement—to obtain the equation of a " geodesic, " which, it might be said in passing, loses all intuitive significance in this scheme. More generally, it is possible to develop the complete absolute differential calculus. For example, the covariant derivative of a tensor f_i (of rank 1 and weight 0),

$$\frac{\partial f_i}{\partial x_k} - \Gamma^r_{ik} f_r$$

is a tensor of rank 2 and weight 0.   This result is obtained by considering the infinitesimal parallel displacement of the invariant $f_i \xi^i$.

To construct the analogue of the Riemann-Christoffel tensor the following procedure is adopted.   At a point P_{00} the elementary " parallelogram " (there is a diagram on page 507) formed of the points P_{00}, P_{01}, P_{10}, P_{11} is considered.   The difference between the components of a vector ξ^i, according to whether the point P_{11} is reached from the point P_{00} by the path $d\delta$ or by the path δd, is

(13) $$\Delta \xi^i = \Delta R^i_j \xi^j.$$

The coefficient ΔR^i_j is independent of the vector ξ^i considered and depends linearly on the two elementary displacements d and δ. For

$$\Delta R^i_j = R^i_{jkl} dx_k \delta x_l = \frac{1}{2} R^i_{jkl} \Delta x_{kl}.$$

The tensor R^i_{jkl} is antisymmetric in k and l.   The associated tensor R_{ijkl} is of rank 4 and weight 1.   The tensor R can be decomposed into two terms in the following invariant manner—

(14) $$R^i_{jkl} = P^i_{jkl} - \frac{1}{2} \delta^i_j F_{kl}.$$

The first term is antisymmetric in i, j as well as in k, l.

The vanishing of the second term *expresses* the absence of electromagnetic field.   The vanishing of the first term *entails* the absence of gravitational field.   The simultaneous vanishing of both terms characterises euclidean space-time, containing neither gravitation nor electromagnetic field and in which parallel displacement alters neither the length nor the direction of a vector.

Consider again the invariant

$$L = \frac{1}{4} F_{ik} F^{ik}.$$

Since the tensor F_{ik} has weight zero, it is clear (by the definition of the g_{ik} itself) that the invariant L has weight -2. The element of volume

$$d\overline{\omega} = \sqrt{-g}\ dx$$

where g is the determinant of the g_{ik} and dx represents the product

$$dx_1 \cdot dx_2 \ldots dx_n,$$

n is the number of dimensions of the space, has the weight $\dfrac{n}{2}$.

Now it is well known that it must be possible to deduce Maxwell's equations from a principle of least action. In Weyl's geometry it must be possible to identify the electromagnetic action with the integral

$$\int L d\overline{\omega}.$$

It is clear that Weyl's geometry can be extended to a space of any number of dimensions. But if it is desired to attach an *absolute* significance to the integral above, it is necessary that its weight should be zero ; that is, that

$$-2 + \frac{n}{2} = 0 \qquad\qquad \text{where } n = 4.$$

In Weyl's conception the possibility of Maxwell's theory *is thus found to be connected with the fact that the universe has four dimensions.* We shall not pursue further the development of this conception in the purely electromagnetic domain.

B. *ANALYSIS AND INTERPRETATION*

1a to 5a. GENERAL OBSERVATIONS.

The observation already made in § 6a, page 493, on the " heuristic value " attributed to the special principle by Einstein is also applicable to the two statements which he gave of the generalised principle. Moreover we shall see in § 7c (page 531) that Painlevé formulated an analogous criticism on this same subject.

The example of the *rotating disc* has been much discussed and, it seems, is still discussed today. We do not believe it useful to describe these discussions for, in Einstein's original paper, this example only arose to end in a completely negative result— namely, the *impossibility* of directly expressing the intervals dx_i in physically measurable terms of space and time. This impossibility is a fundamental characteristic of the generalised theory, to which Painlevé does not seem to have attached all the importance that it deserves. It is this circumstance, together with the necessity of embodying the special theory of relativity, that makes it necessary, as soon as concrete measurements are concerned, to appeal to the tangent euclidean universe at a point in the space of the x_i.

We have seen that the logical field of application of the special principle is very narrowly defined but that, in compensation, all the variables are directly measurable. The generalised principle on the contrary is a very wide framework in which the difficulty is rather that of treating a concrete problem explicitly.

Although, in the preceding presentation, we have striven to make clear the successive postulates encountered by Einstein in the construction of his generalised theory, it may not be superfluous to summarise these postulates once more—

a) principle of covariance of the laws of physics under every substitution of the generalised coordinates x_i;

b) validity of special relativity in an infinitely small region about each point-event of the generalised theory;

c) representation of the motion of a free particle in a gravitational field by means of the equations of the geodesics of ds^2;

d) choice of the relation $B_{ij} = 0$ (or, more generally, $B_{ij} - \lambda g_{ij} = 0$) to express the law of gravitation;

e) the introduction of a tensor T_{ij}, satisfying the conservation law

$$T_{ij}^{j} = 0$$

to represent matter.

These postulates are, in the order in which they are stated, those which suggested themselves, in turn, to the author.

We recall that he, replying to the critics in advance, explicitly affirmed that he had not sought to found his theory on the smallest number of necessary axioms. But, for an understanding of the origins of the theory, the statement of these successive phases would seem more important than a rigorous axiomatic from the *a posteriori* point of view. It may be that Einstein, having proceeded in this way by successive inductions—some of which might appear, at least at first sight, rather daring—appreciated the necessity of retrieving the newtonian equations in the first approximation.

6a. On the geometrisation of classical mechanics.

It has often been said that Einstein " *geometrised* " the problem of gravitation by means of his generalised theory, just as it has been said that d'Alembert was able to reduce dynamics to statics. These statements are partially true, but the matter is too complicated for as brief a judgement to be tenable.

Without detracting in any way from the originality of the theory of generalised relativity, it must be recognised that this was not the first attempt at the geometrisation of dynamics. It was only—and this amply suffices to justify its at least theoretical interest—the first attempt at geometrisation *in space-time* which, moreover, assumed as the tangent universe that of special relativity characterised by the group G_c.

In fact the classical workers, in their own field, had already geometrised the general problem of dynamics by giving the principle of least action an appropriate form.

Inspired by the work of Jacobi, Thomson and Tait, Liouville and Lipschitz, Darboux devoted to this subject Chapters VI and VIII of the second Volume of his *Leçons sur la théorie générale des surfaces*. The geometrisation of the general problem of dynamics is obtained by considering, from among all the motions compatible with a given force-function U, those which correspond to the same value of the constant of the equation of kinetic energy h, or, if so desired, to the same *total energy*. This *energetic classification* has the virtue that the problem is susceptible to geometrisation, the investigation of the trajectories being reduced to the determination of the geodesics corresponding to a certain ds^2.

Let $2\,Tdt^2 = a_{ik}dq_idq_k$ (q_i, generalised variables of Lagrange) where T is the kinetic energy, be taken as the fundamental form. By introducing the momenta $p_i = a_{ij}q'_j$ the equation

$$2T = a^{ik}p_ip_k$$

is obtained. Jacobi's partial differential equation is then written as

$$a^{ik}\frac{\partial S}{\partial q_i}\frac{\partial S}{\partial q_k} = 2\,(U+h).$$

Let θ be a complete integral of this equation and $\theta_1,\,\theta_2,\,\ldots\,\theta_{n-1}$ the $(n-1)$ *distinct* integrals of the equation

$$a^{ik}\frac{\partial\theta}{\partial q_i}\frac{\partial F}{\partial q_k} = 0$$

which is linear in F.

Then, after Lipschitz (*Journal de Crelle*, Vol. 74, 1871),

$$2\,(U+h)\,a_{ik}dq_idq_k = d\theta^2 + f(d\theta_1,\,\ldots\,d\theta_{n-1})$$

whence it follows that, for the true motion ($d\theta_1 = \ldots d\theta_{n-1} = 0$),

$$\delta\int\sqrt{2\,(U+h)\,a_{ik}dq_idq_k} = 0.$$

This equation is an expression of the principle of least action. The determination of the trajectories is then reduced to the search for the geodesics of the ds^2 given by

$$ds^2 = 2\,(U + h)\,a_{ik}dq_idq_k.$$

We mention here that, in an analogous stream of thought, Ricci and Levi-Civita, in their paper on the absolute differential calculus that has been quoted above, applied this calculus to the study of first integrals of the equations of dynamics. They started from the fundamental form $a_{ik}dq_idq_k$ and wrote Lagrange's equations in the form

$$q_i'' = X^{(i)} - \left\{ \frac{rs}{i} \right\} q_r'\, q_s'.$$

7a. Expression and interpretation of Schwartzschild's " ds^2. "

We have already remarked that, in order to calculate the planetary orbits, Schwartzschild found it necessary to carry the solution to a higher approximation than that of Einstein's original solution.

Without reproducing all the calculation that leads to Schwartzschild's expression of " ds^2 ", we shall enumerate the hypotheses which, together with the law of gravitation in the absence of masses

$$B_{ij} = 0$$

determine this calculation—

1) The field is supposed *static*. That is, the variable t only occurs in the expression ds^2 as the square of the differential, as dt^2.

2) The mass generating the field is supposed spherical or of spherical symmetry, immovable and without rotation relative to the stars.

3) The ds^2 must reduce to the euclidean form—

on the one hand, at a great distance from the mass generating the field;

on the other hand, when the density of this mass tends to zero.

Given this ds^2, whose form is, *a priori*,

$$ds^2 = -\,A(r)dr^2 - B(r)[d\theta^2 + \sin^2\theta d\varphi^2] + C(r)dt^2,$$

assumes the form

$$ds^2 = \frac{-\,dr^2}{1 - \dfrac{2\mu}{r}} - r^2\,(d\theta^2 + \sin^2\theta d\varphi^2) + \left(1 - \frac{2\mu}{r}\right)dt^2$$

when the law of gravitation is taken into account. Here the unit of time is chosen in such a way that $c = 1$ and μ is a constant related to the mass generating the field.

The interpretation that must be assigned to this ds^2 is conveniently found by considering the measurements that can be carried out in a tangent euclidean universe. Here we shall follow M. Chazy's presentation.[1]

Space. — In Schwartzschild's space ($dt = 0$), θ and φ can retain their usual significance of euclidean polar coordinates. Moreover, for $dr = 0$ the ds^2 takes the form of the square of the linear element of a sphere, namely

$$r^2(d\theta^2 + \sin^2\theta d\varphi^2).$$

Therefore, at each point M of Schwartzschild's space, the variable r can be interpreted as the length of the circumferences which pass through M and have centre O, divided by 2π.

But the *distance* dD between two spheres of centre O in Schwartzschild's space, corresponding to the values r and $r + dr$, differs from the quantity dr.

In fact, along the radius $d\theta = d\varphi = 0$,

$$- ds^2 = \frac{dr^2}{1 - \dfrac{2\mu}{r}} = dD^2.$$

Therefore in the gravitational field considered there is contraction of a standard of length placed radially.

The section of Schwartzschild's space by a meridian plane ($d\varphi = 0$) has the same element,

$$\frac{dr^2}{1 - \dfrac{2\mu}{r}} + r^2\, d\theta^2$$

as a surface of revolution generated in euclidean space by the parabola

$$z^2 = 8\mu\, (x - 2\mu)$$

rotating about its directrix Oz.[2]

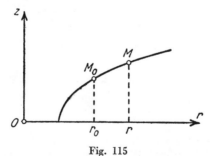

Fig. 115

[1] *La théorie de la relativité et la mécanique céleste* (Gauthier-Villars), 1928-1930.
[2] The same remark is found in WEYL's *Raum, Zeit, Materie.*

Therefore, outside the mass generating the field, the radial distance D may be regarded as the parabolic arc $\overset{\frown}{M_0 M}$ of projection $r - r_0$ (Fig. 115). As an indication of the order of magnitude, the parameter of the parabola in the case of the Sun is about 6 km.

Time. — The time t is said to be the *cosmic time* of the gravitational field considered. It corresponds to the *absolute time* of classical mechanics. But here, on the contrary to what is assumed in the classical domain, this time does not coincide with the *proper time* recorded by a clock at rest in the gravitational field. The interval ds of proper time at a given point of Schwartzschild's space is less than the corresponding interval of cosmic time, being related to the latter by

$$ds = \sqrt{1 - \frac{2\mu}{r}} \, dt.$$

Thus a clock appears to be retarded in a gravitational field. These interpretations are quite analogous to those we have already encountered in § 7, pp. 512, 513.[1]

7b. Generalised relativity in comparison with experiment.

Three groups of experiments have been made in order to test the generalised theory of relativity.

1) In the first place, it has been sought to demonstrate the *displacement towards the red* of spectral lines in a gravitational field. According to the theory, the frequency of radiation emitted by an atom on the surface of a celestial body must be slightly less than the frequency emitted by the same atom in empty space (or on the surface of a less massive body). It must therefore be possible to establish the displacement towards the red of lines emitted at the surface of stars in comparison with the lines emitted at the surface of the earth. This displacement should be given by the relation

$$\frac{\nu - \nu_0}{\nu_0} = - \frac{KM}{c^2 r}.$$

For solar light the expected displacement is very small (about two millionths of a wave-length). Grebe, Bachem, Evershed and Schwartzschild concluded that this displacement exists while Saint-

---

[1] Apart from SCHWARTZSCHILD's original paper already referred to on p. 513, see— DE SITTER, *Monthly Notices*, R. A. S., Vol. 77, 1916, p. 712 ; DROSTE, *Verslag, Amsterdam*, Vol. 25, 1916, p. 166 ; HILBERT, *Gœtt. Nachrichten*, 1917, p. 70 ; DARMOIS, *Annales de Physique*, 1924, p. 77. For the determination of ds^2 in the *interior* of a spherical fluid mass, we mention the names of SCHWARTZSCHILD, SCHRÖDINGER, BAUER, WEYL, BRILLOUIN, DE DONDER, HAAG, NUYENS, DARMOIS.

John, working with 17 cyanogen lines, measured a mean displacement that was accurately zero. In France Pérot obtained a positive result after taking a great number of precautions, including the elimination of the influence of pressure by utilising a terrestrial vacuum.

2) Next, it has been sought to measure the deviation of luminous rays in the vicinity of the Sun. The experiment was first made during the eclipse of 29 May 1919, by taking advantage of the transit of the Sun in the constellation of the Bull, across a part of the Hyades. A first expedition (Crommelin, Davidson) worked at Sobral (Brazil), while a second (Cottingham, Eddington) observed on Prince Island (Guinea). The two expeditions obtained the following results :

$$\text{Sobral} \quad : 1 \cdot 98'' \pm 0 \cdot 12''$$
$$\text{Prince I.} : 1 \cdot 61'' \pm 0 \cdot 30''$$

while Einstein's theory predicted $1 \cdot 87''$ and Newton's $0 \cdot 87''$. However, several astronomers discussed these experiments and gave contradictory interpretations of them. In the same way, the measurements made by Campbell and Trumpler during the eclipse of 21 September 1922 were not considered unanimously conclusive.

3) Lastly, if it is accepted, with a great many astronomers, that it is advisable, in the motion of a planet in relation to the Sun, to replace the inverse square law of the distances by the law deduced from Schwartzschild's ds^2, and to *add the correction thus obtained to the perturbations of the classical theory,* we arrive, according to Chazy, at the following discussion.

Le Verrier (1856) had estimated the secular advance of the perihelia of Mercury as $527''$, $38''$ of which were unexplained by the theory of perturbations. Newcomb (1895), taking into account the corrected values of the masses, calculated the secular advance of the perihelia of Mercury as $530 \cdot 46''$ and as $41 \cdot 24''$ the divergence not explained by perturbations. Moreover, Newcomb estimated the divergence between theory and experiment for the secular advance of the perihelia of Mars as $8 \cdot 03''$ and the similar divergence for the secular advance of the node of Venus as $10 \cdot 14''$.

On the other hand, the consideration of Schwartzschild's ds^2 introduces a secular advance of $42 \cdot 09''$ of the perihelia for Mercury, and of $1 \cdot 35''$ for Mars. The concordance is therefore perfect for Mercury ; for Mars it is of the same nature, but of the order of one-sixth only of the value given by Newcomb. Lastly, the theory of relativity does not imply any perceptible secular inequality of the nodes of the planetary orbits and leaves unexplained the third divergence noticed by Newcomb with the classical theory.

Such, briefly summarised, is the experimental balance-sheet of the generalised theory of relativity.

There still, in fact, controversial matters. However, it does not seem that these discussions could destroy the theoretical interest of the synthesis built up by Einstein.

7c. PAINLEVÉ'S CRITICISM— THE SEMI-EINSTEINIEN THEORY OF GRAVITATION.

Painlevé adopted a critical attitude towards the generalised theory of relativity.[1] It is of some value to devote a little attention to this, if only to clarify the divergences and the common points between this theory and the newtonian theory of gravitation.

Painlevé first recalls the fundamental principles of newtonian mechanics ; namely—

The postulate of *initial conditions*.

The axiom of *causality*, which is stated in the following way— If, in an *isolated* material system in which each element remains identical to itself, the initial conditions (positions and velocities) reproduce themselves apart from a transport in space and time, the motion of the system will be reproduced, apart from the same transport in space and time.

The corollaries of the preceding axiom—

The axiom of *symmetry*, expressing the persistence, during the motion, of all the symmetry of the initial conditions.

Kepler's principle, or the principle of *inertia*.

Painlevé then considers the problem of a *test body* P placed in the field of a mass S, which has spherical symmetry about a centre O. He refers the motion to axes $Oxyz$ whose directions relative to the distant stars are fixed. The coordinates r, θ, φ are the *ordinary* polar coordinates with centre O and the time t is that of an observer on S. Painlevé enumerates the following *converging postulates* of the two theories, classical and relativistic.

1) Principle of inertia.

2) Plane trajectory ; equations of motion of the form

$$(1) \qquad r'' = G\,(r, r', \theta') \qquad \theta'' = H\,(r, r', \theta')$$

not containing either θ or t explicitly.

---

[1] *La mécanique et la théorie de la relativité*, Comptes rendus de l'Académie des Sciences (Paris), Vol. 173, 1921, p. 677. *La gravitation dans la mécanique de Newton et dans celle d'Einstein*, Comptes rendus de l'Académie des Sciences (Paris), Vol. 173, 1921, p. 873.

3) Conservation of the equations of motion under the transformation of t into $-t$ *(reversibility)*.

4) The validity of the law of motion whatever the element P may be.

5) Propagation of light in the sidereal vacuum with a constant velocity for a given direction.

On the other hand, Painlevé presents the *diverging postulates* in the form of the following diptych.

Classical theory	*Einstein's theory*

1) If, without changing the time, any curvilinear coordinates u, v, w are introduced the trajectories of the motion are the geodesics of a

$$d\sigma^2 = (U + h) \, d\sigma_1^2.$$

Here $d\sigma_1^2$ is a quadratic form in du, dv, dw, and whose coefficients satisfy the invariant conditions (second order partial differential equations) which express the fact that $d\sigma_1^2$ is euclidean. Also U (u, v, w), outside the sphere S, satisfies Laplace's equation in curvilinear coordinates—a linear and homogeneous second order equation in which U does not appear explicitly.

The motions of P are defined by a ds^2 in four dimensions, whose coefficients must satisfy certain conditions invariant under any change of the space-time variables. These equations are of the second order in the partial derivatives, and allow the problem a certain degree of indeterminacy which experiment appears to corroborate.

2) In the wave theory the velocity of light far from all material bodies is only the same in all directions if S is absolutely fixed, and not in translational motion relative to the ether.

In the emission theory the velocity of light is the same in every direction if the source of light is absolutely fixed relative to the axes $Oxyz$.

The velocity of light, far from all material bodies, is the same in all directions and the geodesics for which ds^2 is zero define the trajectories and the motion of the light.

Painlevé next summarises the conclusions of all the postulates. In classical mechanics

$$\text{(A)} \qquad d\sigma^2 = \left(\frac{\mu_1}{r} + h\right)[dr^2 + r^2(d\theta^2 + \sin^2\theta d\varphi^2)]$$

and in relativity

$$\text{(B)} \qquad ds^2 = \left[1 - \frac{2\mu}{f(r)}\right]dt^2 - f^2(r)[d\theta^2 + \sin^2\theta d\varphi^2] - \frac{f'^2(r)\,dr^2}{1 - \dfrac{2\mu}{f(r)}}.$$

Here μ is a constant and $f(r)$ is an arbitrary function of r whose derivative $f'(r)$ is always positive and tends to unity when $\dfrac{\mu}{r}$ tends to zero. The units are so chosen that $c = 1$.

The expression (B) is more general than Schwartzschild's ds^2 because it includes the arbitrary function $f(r)$. In compensation, Painlevé assumes—and it is for this reason that his analysis may be called the *semi-Einsteinien theory of gravitation*—that *all three* variables r, θ, φ preserve their euclidean significance (and consequently their direct physical significance). This in spite of the fact, which we have seen that in the general theory of relativity as it is strictly interpreted, the abstract variable r differs from the euclidean distance and must be *interpreted* (cf. § 7a above) in the *tangent euclidean universe*. This interpretation gives a different result according to whether tangential or radial measurements are considered. Thus Painlevé dispenses with one of the essential characteristics of the generalised theory of relativity which must, it seems, be numbered among the postulates assumed by Einstein.

Painlevé then discusses the consequences of the expression of ds^2 in the light of astronomical observations. He puts

$$f(r) = r\left[1 + \varepsilon\frac{\mu}{r}\right]$$

where $\varepsilon \to 0$ with $\dfrac{\mu}{r}$. Thus he easily succeeds in accounting for the advance of the perihelion of Mercury. But, in another connection, he can particularise the function $f(r)$ in such a way that the contractions of standards of length in a gravitational field, or the deviation of light rays at the limb of the Sun, are no longer discovered, and thus diverge from Einstein's theory.

It would seem rather difficult to follow Painlevé in all his conclusions. The fact of preserving the euclidean significance of the variable r

appears as a hypothesis which is *a priori* admissible but which does not impose itself logically. It would seem better to assume that Painlevé, rejecting certain consequences of Einstein's theory, wished to construct a dynamics which might be intermediate between that theory and newtonian mechanics, and yet compatible with the data of the orbit of Mercury. Indeed the hope " that of the relativistic doctrines there will remain a body of formulae which, without contradicting it, will be founded in the classical science ; but that there will not remain those philosophico-scientific principles or conclusions that have been, according to individual opinion, the scandal or the miracle of relativity. "

Without going as far as to share this opinion, certain of Painlevé observations merit thought. For example, " [Einstein's followers state] the *principle of invariance*— All positive consequences of science can be given a form which is invariant under an arbitrary transformation of the four variables which define space-time.

" This principle cannot be contested if it is clarified— for my part I shall state it in the following way.

" *It is possible to deduce from the laws of Nature consequences that are invariant under any transformation of the space-time frame of reference and which define these laws, apart from such a transformation.*

" But precisely because this principle is an incontestable truism it can, itself alone, yield nothing. Whatever may be the laws of nature that it may please us to imagine, they can be made to conform to it.

" Einstein's attempt recalls that of Lagrange. Lagrange gave the equations of mechanics a form invariant under all spatial transformations that do not affect the time, but this form is merely another way of writing them. The followers of Einstein, for their part, *touch up* these equations in order to extend the invariance to the space-time transformations.

" Referred to axes $Oxyz$ of fixed directions relative to the stars, the motion of P will therefore be planar and, in polar coordinates r, and θ, its equations will be of the form (1) [p. 528] ; that is, will not depend explicitly on θ or t. In order to make progress classical mechanics assumes *Galileo's principle* (namely, that the acceleration of P at the time t only depends on the position of P) and thus arrives at the expression (A). Einstein's mechanics assumes that the equations of motion must be contained *in the very extensive, but nevertheless exceptional, class* of systems of equations of the second order which define the geodesics of a ds^2 in four dimensions. "

8a. Remark on the universes of Einstein and de Sitter.

By way of a complement to the summary of § 8, we note here that the ds^2 on Einstein's spatially closed universe can be put in the form

$$ds^2 = -R^2 [d\theta^2 + \sin^2 \theta (d\varphi^2 + \sin^2 \varphi d\omega^2)] + dt^2$$

(Einstein, *Sitzungsber.*, Berlin, 1917, p. 150). It is also possible to write it in the form

$$x_1^2 + x_2^2 + x_3^2 + x_4^2 = R^2$$

with

$$x_1 = R \cos \theta \qquad x_2 = R \sin \theta \cos \varphi$$
$$x_3 = R \sin \theta \sin \varphi \cos \omega \qquad x_4 = R \sin \theta \sin \varphi \sin \omega$$

$$\begin{pmatrix} 0 \leqq \theta \leqq \pi \\ 0 \leqq \varphi \leqq \pi \\ 0 \leqq \omega \leqq 2\pi \end{pmatrix}.$$

In particular, this representation allows the easy calculation of the volume of the spherical space. For

$$dV = R^3 \sin \theta \sin \varphi d\theta d\varphi d\omega$$

whence

$$V = 2\pi^2 R^3.$$

In Einstein's universe, if the space is *finite* the time is infinite and, moreover, the time is an *absolute time* (in the classical sense). This has resulted in the description of Einstein's universe as " *cylindrical.* "

The ds^2 of de Sitter's universe is given by

$$ds^2 = -R^2 [d\theta^2 + \sin^2 \theta (d\varphi^2 + \sin^2 \varphi d\omega^2)] + \cos^2 \theta dt^2$$

(de Sitter, *Proceedings Amsterdam*, Vol. 20, 1917, p. 230).

The space term is the same as the ds^2 pertaining to Einstein's universe, but the term in the time contains the factor $\cos^2 \theta$. It can also be written in the form

$$x_1^2 + x_2^2 + x_3^2 + x_4^2 + x_5^2 = -R^2$$

with

$$x_1 = iR \sin \theta \cos \varphi \qquad x_2 = iR \sin \theta \sin \varphi \cos \omega \qquad x_3 = iR \sin \theta \sin \varphi \sin \omega$$

$$x_4 = iR \cos \theta \operatorname{ch} \frac{t}{R} \qquad x_5 = R \cos \theta \operatorname{sh} \frac{t}{R}$$

$$\begin{pmatrix} 0 \leqq \theta \leqq \dfrac{\pi}{2} \\ 0 \leqq \varphi \leqq \pi \\ 0 \leqq \omega \leqq 2\pi \end{pmatrix}.$$

De Sitter's space is called *hyperbolic*, for it is situated on an imaginary hypersphere of radius iR, itself contained in an euclidean space of five dimensions and because it is possible to write, in real variables

$$X^2 + X'^2 + X''^2 - Y^2 - Z^2 = R^2.$$

Generalised relativity and Riemann spaces.

In the interpretative work that we have already cited (*Über die spezielle und die allgemeine Relativitätstheorie*), Einstein referred to Gauss' theory of surfaces in connection with his introduction of the mathematical tools of generalised relativity. Strictly speaking, in Gauss' work these surfaces were concrete entities in an euclidean space of three dimensions. A further step in the abstract direction was therefore necessary. This was taken by Riemann, and consists of the construction of spaces defined, *a priori*, by means of a ds^2 in n dimensions. Schlaefli's theorem ensures the possibility of immersing a space of n dimensions in an euclidean space of $\dfrac{n(n+1)}{2}$ dimensions. However, as this theorem is only concerned with a local realisation, it is preferable to consider a Riemann space as given *in itself*.

The notion of *tangent euclidean space at a point* in a given Riemann space plays, as we have seen, a fundamental part in the theory of generalised relativity. With E. Cartan it may be said " that the ds^2 of the Riemann space turns the neighbourhood of each point into a small portion of euclidean space, and that parallel displacement in the sense of Levi-Civita allows two infinitely close small portions to be linked together, so as to be integrated into one euclidean space. " E. Cartan, moreover, contemplated other laws of transport than that of Levi-Civita, restricting them by the only condition that they should preserve the length of a vector. In this way he defined spaces of *euclidean connection*. The theory of generalised relativity, by drawing the attention of geometers to the study of Riemann spaces, therefore stimulated profound studies with which we are unable to deal here.

E. Cartan, in agreement with Painlevé on this matter, deems that the principle of covariance " is insufficient in itself to allow the deduction of any physical law. In reality, the axioms which lie at the foundation of generalised relativity are the following—

" *a)* Space-time is a Riemann space of four dimensions and of *indefinite ds^2*.

" *b)* Space-time is not an arbitrary Riemann space ; it is included in a particular class of Riemann spaces.

" *c)* Together with a Riemann space, this class includes all others which are obtained from it by a simple transformation of coordinates.

" *It is in this last axiom that the principle of generalised relativity resides*. The analytical definition of the class of Riemann spaces which is sought constitutes the equations of gravitation in the vacuum. "
The choice $B_{ij} = 0$, made by Einstein, is one of an infinity of possible

solutions within the compass of the axioms a, b, c. In generalised relativity it is also possible to consider the space of 14 dimensions with the coordinates x_i and g_{ik}, and to define there a class of four-dimensional varieties which contains, together with one particular variety, all those that are deduced from it by the operations of the infinite group of point transformations in x_1, x_2, x_3, x_4.

CHAPTER THREE

THE DYNAMICS OF QUANTA IN BOHR'S SENSE

A. *PRESENTATION*

1. Bohr's first dynamical model.

In a quite different stream of thought, we return to the year 1913. At this time, in the *Philosophical Magazine*, there appeared a paper of Niels Bohr, called *On the constitution of Atoms and Molecules*,[1] which opened up a whole field of research into dynamics on the atomic scale.

Before the appearance of Bohr's paper, J. J. Thomson [2] and Rutherford [3] had already suggested models intended to account for the stability of atomic structures. In Rutherford's conception, a positively charged nucleus was surrounded by a system of electrons held by the attraction of the nucleus. In Thomson's model there was introduced a positively and uniformly electrified sphere, in the interior of which the electrons described circular orbits.

The inability of classical electrodynamics to account for the behaviour of systems on the atomic scale was universally recognised by physicists. It remained to introduce effectually into this study an element which was completely foreign to the classical conceptions—the *quantum of action* in the sense of Planck.

Here is a summary of Bohr's paper, provided by the author himself.

" In the present paper we shall attempt to develop a theory of the constitutions of atoms on the basis of the ideas introduced by Planck, with the intention of accounting for the black-body radiation, and the theory of the structure of atoms suggested by Rutherford to explain the dispersion of α-particles by matter. "

This statement is followed by Bohr's hypotheses and conclusions.

" 1) The energy of radiation is not emitted (or absorbed) continuously, as the classical electrodynamic theory assumes, but only during

---

[1] *Phil. Mag.*, Vol. 26, 1913, pp. 1 and 476.
[2] *Ibid.*, Vol. 7, 1904, p. 237.
[3] *Ibid.*, Vol. 21, 1911, p. 669.

the passage of an atomic system from one *stationary state* to another stationary state.

" 2) The dynamical equilibrium of a system in its stationary states is governed by the ordinary laws of mechanics, but these laws are not valid in the passage from one stationary state to another stationary state.

" 3) The radiation emitted during the transition of the system from one stationary state to another stationary state is monochromatic. The relation between its frequency ν and the total energy emitted is given by the relation

$$W = h\nu \qquad (h = \text{Planck's constant}).$$

" 4) The different stationary states of a system constituted of an electron rotating about a positively charged nucleus are determined by the following condition— the ratio of the total energy needed to realise a given configuration of the system to the mechanical frequency ω of the rotational motion of the electron is an integral multiple of $\frac{h}{2}$, or

$$\frac{W}{\omega} = n \frac{h}{2} \qquad (n, \text{integer}).$$

" If it is assumed that the orbit of the electron is circular, this condition is equivalent to the following— *the angular momentum of the electron is an integral multiple of* $\frac{h}{2\pi}$.

" 5) The ' *permanent* ' state of every atomic system, or the one which corresponds to the maximum energy emitted, is determined by the condition that the angular momentum of each electron about the centre of its orbit is equal to $\frac{h}{2\pi}$. "

We shall explain Bohr's calculation.

Let e be the charge of the electron, m its mass, a the radius of its circular orbit, ω the rotational frequency of the electron and E the nuclear charge. Ordinary dynamics, for any stationary state (hypothesis 2), yields the relation

$$\frac{eE}{a^2} = (2\pi\omega)^2 ma.$$

The energy W, the potential energy U and the kinetic energy T have the values

$$W = -\frac{1}{2}\frac{eE}{a} \qquad U = \frac{eE}{a} \qquad T = \frac{1}{2}\frac{eE}{a}$$

respectively.

Hypothesis 4 leads to the statement that the angular momentum of the electron,

$$p = 2\pi\omega m a^2$$

can only take the discrete values defined by the relation

$$p = \frac{nh}{2\pi} \qquad\qquad (n, \text{ integer}).$$

It follows that the energy of the system, the radius of the orbit and the mechanical frequency ω can only take discrete values, given by

$$- W = \frac{2\pi^2 m e^2 E^2}{n^2 h^2} \qquad a = \frac{n^2 h^2}{4\pi^2 m e E} \qquad \omega = \frac{4\pi^2 m e^2 E^2}{n^3 h^3}.$$

Among the possible electronic trajectories, the only ones that subsist are the discrete trajectories whose radii are proportional to the square of the *quantum numbers n*. When an electron *jumps* from one stationary trajectory to another, characterised by the numbers n_1 and n_2 respectively, the energy changes by

$$\delta W = \frac{2\pi^2 m e^2 E^2}{h^2} \left(\frac{1}{n_2^2} - \frac{1}{n_1^2} \right).$$

By hypothesis 3, the energy thus liberated (by contraction of the orbit) or absorbed (by expansion of the orbit) passes into space, or comes from it, in the form of a monochromatic radiation of frequency

$$\nu = \frac{\delta W}{h} = R \left(\frac{E}{e} \right)^2 \left(\frac{1}{n_2^2} - \frac{1}{n_1^2} \right).$$

The quantity R is a *universal constant* called Rydberg's constant.

In his first paper Bohr applied these results to the hydrogen atom $(E = e)$. He remarks on the perfect agreement of the frequency relation thus obtained with the series empirically discovered by Balmer as early as 1885, and which corresponds to $n_2 = 2$. The series corresponding to $n_2 = 1$, of which all the lines lie in the ultra-violet, had been observed by Lyman,[1] though only nine lines were known to Bohr. The series corresponding to $n_2 = 3$, of which only three lines had been indicated by Paschen,[2] was observed by Brackett,[3] together with the series corresponding to $n_2 = 4$. The other series lie in the far infra-red.

[1] *Nature*, Vol. 93, 1914, p. 314.
[2] *Ann. der Physik*, Vol. 27, 1908, p. 565.
[3] *Nature*, Vol. 109, 1922, p. 209.

The spectrum of hydrogen can be compared with that of ionised helium ($E = 2e$), which also contains no more than a single electron, and for which the frequency relation is

$$\nu = 4R\left(\frac{1}{n_2^2} - \frac{1}{n_1^2}\right).$$

The series $n_2 = 1$ lies in the far ultra-violet. Some lines of the series $n_2 = 2$ were observed by Lyman.[1] The series $n_2 = 3$ was observed by Fowler, Evans and Paschen between 1912 and 1916, and splits into two according to whether n_1 is even or odd. The series $n_2 = 4$ also splits up and gives, on the one hand, the Balmer series, and on the other, a series which Pickering [2] discovered in the spectrum of the star ζ of the Stern as early as 1896. For further details, the reader is referred to specialised works.

Motion of the nucleus. — The first model supposes the nucleus to be at rest or, alternatively, neglects the ratio $\dfrac{m}{M}$ of the mass of the electron to that of the nucleus. In order to take account of the motion of the nucleus it suffices to consider the motion of the system of the electron and the nucleus about its centre of gravity. As Bohr himself showed,[3] this leads simply to the multiplication of the Rydberg constant by the factor

$$\frac{M}{m + M}.$$

Thus it is possible to determine, as Fowler and Paschen did by comparing the spectra of helium and hydrogen ($M_{He} = 4M_H$), the ratio $\dfrac{m}{M}$ in the case of hydrogen. The result is about $1/1840$, in perfect agreement with Millikan's direct determination.

2. GENERALISATION OF THE FIRST MODEL— THE "QUANTUM CONDITIONS" FOR A SYSTEM OF SEVERAL DEGREES OF FREEDOM.

Bohr's first model does not immediately lend itself to an extension to more complicated mechanical systems. Nevertheless, in his first paper, Bohr had incidentally established the following theorem. " In every system composed of electrons and nuclei in which the nuclei are fixed and the electrons describe circular orbits (with velocities

---

[1] *Nature*, Vol. 104, 1919, p. 314.
[2] *Astrophysical Journal*, Vol. 4, 1896, p. 369.
[3] *Phil. Mag.*, Vol. 27, 1914, p. 509.

small compared with c), the kinetic energy is equal, apart from sign, to half the potential energy. "

This leads to the imposition, for each electron, of the condition

$$\frac{2T}{\omega} = nh.$$

Directly inspired by this result, W. Wilson established the *quantum conditions* for a conservative system of n degrees of freedom. The paper concerned is entitled *The quantum theory of radiation and line spectra*.[1]

Let q_i be the Lagrange coordinates and T, the kinetic energy, given by

$$2T = \sum_1^n p_k q_k'$$

Wilson assumes that, if necessary, by means of a suitable transformation of the q_i, this equation reduces to the form

$$2T = \sum_1^n A_k q_k'^2 = 2 \sum_1^n L_k.$$

Thus, for every degree of freedom,

$$2L_k = p_k q_k'.$$

It is then supposed that in each stationary state, for which ordinary dynamics remains valid, the system allows of a well determined period $\frac{1}{\omega_k}$ for each degree of freedom.

Given this, Wilson considers the integral

$$2 \int L_k dt = \int p_k dq_k$$

for each degree of freedom, where in integration is taken over the corresponding period $\frac{1}{\omega_k}$. He then supposes that this integral is necessarily equal to an integral multiple of Planck's constant; that is, he supposes n quantum conditions of the form [2]

$$\oint p_k dq_k = n_k h \qquad (n_k, \text{ integer}; \ k = 1, 2 \dots \text{ or } n).$$

---

[1] *Phil. Mag.*, Vol. 29, 1915, p. 795.
[2] The sign $\oint$, currently adopted after in BOHR's theory, expresses the fact that the integral is taken over the period $\frac{1}{\omega_k}$ corresponding to the variable q_k.

It was not long before Bohr's theory attracted the general attention of physicists, and its development rapidly became a collective task. Starting from the point of view of the *atomicity of energy* which he had held earlier, Planck himself studied the structure of *phase-space (Phasenraum)*, or the space of the variables q_k, p_k.[1] It is necessary to cite Epstein,[2] Schwartzschild [3] and Sommerfeld.[4] To the last named we owe, as well his personal contributions, the best known didactic work on the dynamics of quanta in Bohr's sense.[5]

We shall first consider the problem of the *linear oscillator*. Let m be the mass, ν the frequency and a the amplitude of the oscillator, so that

$$(1) \qquad \begin{cases} q = a \sin 2\pi\nu t \\ p = 2\pi\nu a m \cos 2\pi\nu t. \end{cases}$$

In the phase plane of coordinates (p, q) the representative point of the oscillator describes the ellipse

$$(2) \qquad \frac{q^2}{a^2} + \frac{p^2}{b^2} - 1 = 0 \qquad \text{with } b = 2\pi\nu a m.$$

Both the potential energy of the oscillator and q become zero at $t = 0$. Therefore the total energy may be written immediately as

$$(3) \qquad W = \frac{1}{2}\left[\frac{p^2}{m}\right]_{t=0} = 2\pi^2\nu^2 a^2 m = \pi a b \nu.$$

As W varies a family of ellipses is obtained, all these are similar, for $\dfrac{b}{a} = 2\pi\nu m$.

If it is decided to consider, of this family, only a *discrete series corresponding to Planck's condition of atomicity* the relation

$$(4) \qquad \qquad W_n = n\varepsilon = nh\nu \qquad \qquad (n, \text{ integer})$$

is obtained.

The area contained between two consecutive ellipses of this series is equal to $\dfrac{W_{n+1} - W_n}{\nu}$ or, by (3), to Planck's constant h.

The area of the nth trajectory is equal to nh, or

$$\iint dp\,dq = \int p\,dq = nh$$

---

[1] *Die physikalische Struktur des Phasenraumes*, Ann. der Physik, Vol. 50, 1916, p. 385.

[2] *Ann. der Physik*, Vol. 50, 1916, p. 489, and Vol. 51, 1916, p. 184.

[3] *Zur Quantenhypothese*, Sitzungsber., Berlin, 1916, p. 548. (This was SCHWARTZSCHILD's last publication. He died in 1916.)

[4] *Ann. der Physik*, Vol. 51, 1916, pp. 1 and 125.

[5] *Atombau und Spektrallinien*.

Thus Wilson's condition is retrieved— the phase integral (calculated over a period) is an integral multiple of Planck's constant. Sommerfeld writes, " *This necessary condition has the effect of distinguishing from the continuous ensemble of possible mechanical motions, a discrete infinity of motions that are possible in the quantum sense.* "

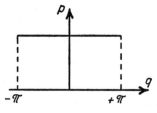

Fig. 116

We return to Bohr's initial problem—that of the *plane* oscillator. Take as the coordinate q the azimuth of the particle on the circular trajectory of radius a. The momentum

$$p = ma^2 q'$$

is constant with the angular velocity of rotation q'.

The trajectory of the representative point of the oscillator in the phase-plane (q, p) is therefore a straight line or, more accurately, a segment parallel to the axis of q and bounded at the points of the abcissa $q = \pm \pi$.

Then, from the continuous series of segments $p = $ const., a discrete series p_n is chosen such that the area contained between the segments p_n and p_{n+1} is equal to Planck's constant, or

$$2\pi (p_{n+1} - p_n) = h.$$

Whence

$$2\pi p_n = nh \qquad\qquad (n, \text{integer}).$$

This leads us back to the condition stated by Bohr (§ 1, number 4). The corresponding energy has the value

$$W_n = \frac{nh\omega}{2}$$

where ω denotes the frequency of the rotation of the plane oscillator, again in agreement with Bohr's hypothesis. It is apparent that Planck's energetic hypothesis is not directly applicable to the plane oscillator because of the presence of the factor $\frac{1}{2}$.

The investigation of systems of several degrees of freedom was carried out by Epstein and Schwartzschild [1] for every classical dynamical problem for which Jacobi's partial differential equation is amenable to solution by separation of the variables; that is, for which each partial derivative of the action can be expressed in the form

$$p_k = \frac{\partial S}{\partial q_k} = \sqrt{\varphi_k(q_k)}.$$

Then, for every degree of freedom the *phase integral*

$$J_k = \oint p_k dq_k,$$

can be formed. The symbol $\oint$ indicates that the variable q_k ranges over the whole domain of its variation. If q_k is a polar coordinate contained between the two limits $r_{\min}$ and $r_{\max}$, which are consecutive roots of the equation $\varphi_k(q_k) = 0$, the variation extends from $r_{\min}$ to $r_{\max}$ and returns to $r_{\min}$. If q_k is a cyclic coordinate (like the azimuth of the plane oscillator), it is allowed to vary from $-\pi$ to $+\pi$.

We owe to Schwartzschild, who introduced the methods of celestial mechanics into this subject, a particularly simple formulation in terms of " *angular coordinates.* "

We confine ourselves to a system whose hamiltonian, in classical dynamics, does not depend explicitly on the time. The phase integrals J_k are functions of the energy W and of the $n-1$ non-additive primary constants contained in a complete integral S of Jacobi's partial differential equation. Let this solution be

$$S = S(q_1, q_2, \ldots q_n, W, \alpha_1, \alpha_2, \ldots \alpha_{n-1}).$$

Conversely, such a complete integral can be written in the form

$$S(q_1, q_2, \ldots q_n, J_1, J_2, \ldots J_{n-1}).$$

Whence

$$\delta S = \frac{\partial S}{\partial q_k} \delta q_k + \frac{\partial S}{\partial J_k} \delta J_k = p_k \delta q_k + w_k \delta J_k \ [2]$$

if

$$w_k = \frac{\partial S}{\partial J_k}.$$

Therefore it is possible to pass from the variables (q, p) to the variables (w, J) by a contact transformation. The *angular variables* w_k and the

---

[1] *Loc. cit.*, p. 540.
[2] With summation over the dummy suffix k.

phase integrals J_k (which are the conjugate momenta in Hamilton's sense) therefore satisfy the canonical equations

$$\begin{cases} \dfrac{dw_k}{dt} = \dfrac{\partial H}{\partial J_k} = \text{const.} \\[2mm] \dfrac{dJ_k}{dt} = -\dfrac{\partial H}{\partial w_k} = 0. \end{cases}$$

Every angular coordinate is therefore a linear function of the time of the form

$$w_k = \nu_k t + \alpha_k.$$

Moreover, when the initial variable q_k alone ranges over the whole domain of its variation (in the sense explained above) the corresponding angular variable w_k increases by one unit. Indeed, under these conditions the action increases by the corresponding phase integral, so that

$$\Delta S = J_k.$$

Whence

$$\Delta w_k = \Delta \frac{\partial S}{\partial J_k} = 1.$$

Therefore the ν_k are identified as the mechanical frequencies of the system for the different degrees of freedom and the q_k are *periodic functions of unit period in the angular coordinates w_k.*

Inasmuch as the mechanical frequencies w_k are not commensurable among themselves the q_k are, in general, non-periodic functions of the time. The system is then called " *quasi-periodic.* "

Thus the q_k can be developed as Fourier series in n variables τ_1, $\tau_2, \ldots \tau_n$, in the form

$$q_k = \sum_{-\infty}^{+\infty} C^k_{\tau_1, \tau_2, \ldots \tau_n} e^{2\pi i (w_1 \tau_1 + w_2 \tau_2 + \ldots w_n \tau_n)}$$

or

$$q_k = \sum_{-\infty}^{+\infty} D^k_{\tau_1, \tau_2, \ldots \tau_n} e^{2\pi i (\nu_1 \tau_1 + \nu_2 \tau_2 + \ldots \nu_n \tau_n)}$$

The sign $\sum$ indicates that the quantities τ_i must take all values between $-\infty$ and $+\infty$.

The system is said, in Schwartzschild's sense, to be *degenerate* if there exist one or more linear and homogeneous relations, with integral coefficients, between the mechanical frequencies ν_k. In cases of degeneracy there are, simultaneously, several systems of variables q_k which allow the decomposition of Jacobi's equation into n particular equations in each variable. Such is the Kepler problem of a test body (a non-relativistic electron describing an elliptical trajectory about a fixed nucleus).

The polar and parabolic coordinates used by Epstein both allow the decomposition of Jacobi's equation for the keplerian problem. They do not give the same trajectories but lead to the same values of the quantised energy. The degeneracy disappears if the Coulomb field of the nucleus is supplemented by an external electric field (this corresponds to the Stark effect studied by Epstein) or if the variation of the mass with the velocity is taken into account (this is the problem of the relativistic electron studied by Sommerfeld).

For a quasi-periodic non-degenerate system the variables q_k which are necessary to achieve the separation of Jacobi's equation satisfy a simple geometrical criterion formulated by Epstein. The ensemble of possible trajectories can be represented in a space E_n of n dimensions by means of the variables $\nu_1, \nu_2, \ldots \nu_n$. Each trajectory passes as near as may be desired to all the points of a region of E_n defined by hyperspheres of $n-1$ dimensions. If the trajectory is varied by varying the constants of integration, it continues to be contained in a region defined by hyperspheres of $n-1$ dimensions. The hyperspheres defining the different trajectories form n families which can be represented, in a suitably chosen system of curvilinear coordinates, by $q_i =$ constant $(i = 1, 2, \ldots n)$. These are the variables q_i which allow the separation of Jacobi's equation.

From the purely mathematical point of view the property which characterises the phase integrals is that of *adiabatic invariance*— if the parameters which specify the strength of an external force-field applied to the system vary by an infinitesimal amount, the corresponding first order variation of a phase integral is zero. This invariance, which explains the part played by the phase integrals in the formulation of the quantum conditions, was investigated by Ehrenfest[1] and Burgers.[2]

3. EXAMPLE OF APPLICATION— THEORY OF FINE STRUCTURE.

By way of an example of the preceding generalities, we shall turn our attention to the relativistic model of the hydrogen atom, or more generally, of an atom of a single electron rotating about a fixed nucleus.

---

[1] *Ann. der Physik*, Vol. 51, 1916, p. 327, and *Phil. Mag.*, Vol. 33, 1917, p. 504.
[2] *Ibid.*, Vol. 52, 1917, p. 195.

As early as 1915 Bohr concerned himself with the relativistic treatment of the atomic model that he had put forward, and suggested that the doublets of the hydrogen spectrum be considered as an effect of the order of $\dfrac{v^2}{c^2}$.[1] But he confined himself to elliptical trajectories of very small eccentricity. The systematic development of this theory is due to Sommerfeld.[2]

In order not to delay ourselves with the formulation of this problem, we shall make use of the generalisation of Jacobi's classical equation which has been encountered above (p. 499) in the form

$$g^{ik}\frac{\partial S}{\partial x_i}\frac{\partial S}{\partial x_k} = 2J\,(U + W).$$

Let m_0 be the rest-mass of the electron. Since the nucleus is fixed at the origin and the potential energy is $\dfrac{eE}{r}$, Jacobi's equation, in the polar coordinates r, θ, becomes

$$\left(\frac{\partial S}{\partial r}\right)^2 + \frac{1}{r^2}\left(\frac{\partial S}{\partial \theta}\right)^2 = 2m_0\left(\frac{eE}{r} + W\right) + \frac{\left(\dfrac{eE}{r} + W\right)^2}{c^2}$$

and separates immediately.

The angular integral is written

$$p_\theta = \frac{\partial S}{\partial \theta} = \text{const.} = p$$

whence is obtained the azimuthal quantum condition

$$J_\theta = 2\pi p = nh \qquad\qquad (n,\ \text{integer}).$$

There remains

$$\left(\frac{\partial S}{\partial r}\right)^2 = A + 2\,\frac{B}{r} + \frac{C}{r^2}$$

with $\begin{cases} A = 2m_0 W + \dfrac{W^2}{c^2} \\[2mm] B = m_0\,eE + \dfrac{eEW}{c^2} \\[2mm] C = -\dfrac{n^2h^2}{4\pi^2} + \dfrac{e^2E^2}{c^2} = -\dfrac{n^2h^2}{4\pi^2}\left[1 - \dfrac{\alpha^2}{n^2}\left(\dfrac{E}{e}\right)^2\right] \end{cases}$ with $\alpha = \dfrac{2\pi e^2}{ch}$.

[1] Phil. Mag., Vol. 29, 1915, p. 332.
[2] Ann. der Physik, Vol. 51, 1916, pp. 1 and 125.

The phase integral

$$J_r = \oint \frac{\partial S}{\partial r}\, dr = n'h \qquad\qquad (n',\ \text{integer})$$

may be calculated, in the complex plane, by the method of residues. The quantised energy is obtained, given by

$$1 + \frac{W}{m_0 c^2} = \left[1 + \frac{\left(\alpha \dfrac{E}{e}\right)^2}{\left[n' + \sqrt{n^2 - \left(\alpha \dfrac{E}{e}\right)^2}\, \right]} \right]^{\frac{1}{2}}.$$

Sommerfeld develops the right hand side of this equation in powers of the constant α, called the fine-structure constant and, going over to frequencies by means of Planck's relation, arrives at

$$\nu = (n, n') - (k, k').$$

Whence

$$(n, n') = R \left(\frac{E}{e}\right)^2 \left[\frac{1}{(n + n')^2} + \frac{\alpha^2}{(n + n')^4} \left(\frac{E}{e}\right)^2 \left(\frac{1}{4} + \frac{n'}{n}\right) + \dots \right].$$

The constant R is Rydberg's constant; the unspecified terms, containing powers of α greater than two, are negligeable for light atoms (hydrogen, ionised helium).

The *relativistic correction* consists of two parts. The first,

$$\frac{1}{4} \frac{\alpha^2 R}{(n + n')^4} \left(\frac{E}{e}\right)^4$$

is common to circular and elliptical orbits. For the first term of the Balmer series of hydrogen ($E = e, n + n' = 2$) this correction represents a fraction of about $3 \cdot 10^{-6}$ of the corresponding term of the classical keplerian problem— namely $\dfrac{R}{(n + n')^2}.$

The second part of the relativistic correction,

$$\frac{n'}{n} \frac{\alpha^2 R}{(n + n')^4} \left(\frac{E}{e}\right)^4$$

depends on n and n' separately, and varies with the eccentricity of the orbit. *It is this second correction which determines the fine structure of lines.* The ratio of this corrective term to the term of the ordinary keplerian problem for the ellipse $n = 1$ and $n' = 1$ is, in the Balmer series of hydrogen, $1.3 \cdot 10^{-5}$.

Sommerfeld's book, which we have already cited, *Atombau und Spektrallinien*, develops the theory of fine structure (doublets, triplets, quartets, quintets in the spectra) and discusses the experimental corroboration at great length. We confine ourselves here to the reproduction of the table relevant to the *doublet separation constant* of the term $\dfrac{1}{2^2}$ of the Balmer series.

$\Delta\nu$ calculated		0.365	$(e = 4.77 \cdot 10^{-10},\ h = 6.55 \cdot 10^{-27})$
$\Delta\nu$ observed	H_α	0.32	(Michelson)
——	H_β	0.33	(Michelson)
——	H_α	0.306	(Fabry and Buisson)
——	H_α	0.288	(Meissner and Paschen)
——	H_α	0.272	(Gehrke and Lau)
——	H_β	0.283	(Gehrke and Lau)
——	H_γ	0.271	(Gehrke and Lau)

The experiments are rather difficult for the spectrum of hydrogen because of the lack of completeness of the lines. Paschen, by an indirect method applied to ionised helium, gave $\Delta\nu$ (hydrogen) as 0.3645, in perfect agreement with the theory.

4. BOHR'S CORRESPONDENCE PRINCIPLE.

The dynamics of Bohr's atomic models is a manifest departure from classical dynamics, where every acceleration produces radiation. This departure consists of the fact that Bohr's electron does not radiate as long as it describes a stationary trajectory, but only when it jumps from one stationary trajectory to another. As early as 1914, in a paper called *On the effect of electric and magnetic fields on spectral lines*, Bohr attempted to establish " a certain connection between this dynamics and ordinary electromagnetic theory. " He was able to achieve such an agreement, in the limit, for transitions from states characterised by large quantum numbers—that is, in the region of slow vibrations. In the later development *Bohr's correspondence principle* was to assume a quite special importance. Thus, as well as being the means by which *polarisation* and *intensity* were introduced into quantum theory, it was also to be made the foundation of Heisenberg's work in 1925—the work from which the mechanics called " quantum " mechanics has sprung.

We shall not delay ourselves here with the first form of the correspondence principle which Bohr gave,[1] or with his work, and that

[1] *Phil. Mag.*, Vol. 27, 1914, p. 506.

of Kramers, on the intensity and the polarisation of spectral lines. We shall follow a paper published by Bohr in 1923 under the title *Über die Anwendung der Quantentheorie auf dem Atombau*.[1]

Consider a quasi-periodic system of r degrees of freedom. For each degree of freedom the system has the *mechanical frequency*

$$\nu_k = \frac{\partial H}{\partial J_k}.$$

Consequently the system has the combined mechanical frequency

$$\tau_1 \nu_1 + \tau_2 \nu_2 + \ldots + \tau_r \nu_r$$

where $\tau_1, \tau_2, \ldots \tau_r$ are arbitrary integers that can be chosen in such a way that

$$\tau_k = n_k' - n_k'' \qquad (k = 1, 2, \ldots r).$$

In the classical electromagnetic theory the mechanical frequencies and the combined frequency are all *optical frequencies* of the system.

In Bohr's theory, on the contrary, the frequency of radiation emitted in the transition from the stationary state defined by the quantum numbers n_k' to the stationary state defined by the quantum numbers n_k'' has the value

$$\nu = \frac{1}{h}\{ H(n_1', n_2', \ldots n_r') - H(n_1'', n_2'', \ldots n_r'') \}$$

$$= \frac{1}{h} \int \sum_1^r \nu_k dJ_k.$$

If the quantum number n_k' and n_k'' are *large compared with their differences* $n_k' - n_k''$, it is possible to make the approximation, in the preceding integral, that the mechanical frequencies remain constant. Then it follows that

$$\nu \sim \sum_1^r \nu_k \frac{J_k' - J_k''}{h} = \sum_1^r \nu_k (n_k' - n_k'') = \sum_1^r \tau_k \nu_k.$$

That is, the optical frequency in Bohr's interpretation agrees, in the limit, with the combined frequency (mechanical or optical) in the sense of the classical theory. It is in this asymptotic agreement that the *principle of correspondence* resides.

[1] *Zeitschr. für Physik*, Vol. 13, 1923, p. 117.

B. *ANALYSIS AND INTERPRETATION*

1*a*. RETURN TO PLANCK'S FREQUENCY RELATION.

In the preceding presentation we have regarded Planck's frequency relation as a basic datum of the dynamics of quanta in Bohr's sense. It is however of some value to recall, briefly, the origin of this relation.

Kirchhoff showed that the structure of *black-body radiation*—the radiation emitted inside a sealed enclosure whose temperature is uniform—is independent of the nature of the bodies which constitute the enclosure, the nature of the bodies which are contained in it and the form and the dimensions of the enclosure. Therefore there must be a universal relation between temperature and wave-length.

Making use of the concept of linear oscillator due to Hertz, Planck arrived at a relation between the energy of a resonator of specified period and that of the corresponding radiation spectrum in the steady state.

Starting from the spectral distribution of energy which had been given by Wien in 1896, Planck transformed it into a relation between the entropy S and the energy W of a resonator— the reciprocal R of the second derivative $\dfrac{d^2S}{dW^2}$ is proportional to the energy. That is,

$$R = -bW.$$

But if this relation is satisfactorily confirmed in the region of short wave-lengths, the experiments of Lummer and Pringsheim and, later, those of Rubens and Kurlbaum (1900) showed that the quantity R was proportional to the square of the energy in the region of large wave-lengths.

Planck was then inspired to represent all the experimental results by the single formula

$$R = -bW - \frac{W^2}{c}.$$

Recalling that $\dfrac{1}{R} = \dfrac{d^2S}{dW^2}$ and that, by the definition of entropy, $dW = TdS$ (where T is the absolute temperature), an elementary integration leads to the relation

$$W = \frac{bc}{e^{\frac{b}{T}} - 1}$$

between temperature and energy.

In accordance with Boltzmann's ideas, Planck interpreted the entropy as a measure of the probability and undertook the calculation of the spectral distribution of the energy of a system composed of a large number of resonators.[1] This calculation led to an expression of the entropy identical with that required by the spectral distribution of radiation ; it was necessary, however, to assume—in obvious departure from the classical ideas—that there was an *atomicity of energy*. The energy of a resonator of frequency v is necessarily a multiple of the *quantum* $\epsilon = hv$, where h is a universal constant having the dimensions of action and the numerical value of

$$6.55 \cdot 10^{-27} \text{ erg. sec.}$$

Planck wrote, " this quantum represented something quite novel —unsuspected until that time—which seemed destined to overturn a physical philosophy founded on the very notion of continuity inherent in all causal relationships since the discovery of the infinitesimal calculus be Leibnitz and Newton. "

Outside the domain in which it was exploited by Bohr, this *quantum hypothesis* made it possible to explain the *photo-electric* effect. This is the phenomenon of the detachment of electrons from the surface of metals by radiation of high frequency. The velocity v of the electrons depends only on the frequency v of the radiation and, above a threshold v_0, is given by an equation due to Einstein,[2]

$$\frac{1}{2} mv^2 = h (v - v_0).$$

Experimented by Lenard (1902), Pohl and Pringsheim (1912) and by Millikan (1913), this effect enables the value of Planck's constant h to be found again. Einstein linked up this effect with the existence of *quanta of light* or *photons*, the only quantic form of radiant energy.

We shall recall the experiments made by Franz and Hertz on the losses of energy in the impact of electrons and atoms, in which the constant h comes into play on account of the determination of the critical velocity an electron must possess in order to provoke the emission of a quantum of light ; the photo-electric effect of X-rays and γ-rays studied by Maurice de Broglie and Ellis (1921), the measurements of the specific heats at low temperature made by Nernst in 1911 and interpreted, with the help of Planck's law, by Debye in 1912. All these phenomena lead to perfectly convergent determinations of the constant h.

---

[1] *Verh. der Deutsch. Physik. Gesellschaft*, December 1900, p. 237, also *Ann. der Physik*, Vol. 4, 1901, p. 556.

[2] *Ann. der Physik*, Vol. 17, 1905, p. 132 ; Vol. 20, 1906, p. 199 ; Vol. 22, 1907, p. 180.

1b. CRITICISM OF BOHR'S MODELS.

The clarity and precision of the postulates taken from the beginning by Bohr as the foundation of his new dynamics must not conceal from us the difficulties and even the paradoxes behind his mechanical models. Bohr himself has explained them with perfect frankness in his works of interpretation.

In the first place, from the very nature of the postulates accepted by Bohr, " certain features of the mechanical images, such as the periods of revolution and the forms of the orbits are inaccessible to direct observation. In particular, an atom in normal state does not radiate, although, from the mechanical images, its electrons are in motion ; it would be difficult to imagine a more marked contrast with the demands of the electro-magnetic theory, to underline the symbolic nature of these images ". [1] Moreover, in atoms possessing several electrons, the interactions of these latter are systematically neglected, without experiment contradicting this derogation from the laws of dynamics.

Besides, " the problem of the motion of the atomic particles is open to singular solutions which must be excluded from the multiplicity of the stationary states ". [2] This constitutes an artificial restriction of the rules of quantisation.

Lastly, " the concept of a stationary state implies the exclusion of any interaction with *individuals* not belonging to a system. " Bohr was led to endow these states with a " hyper-mechanical stability. " The atom is always in a well-determined stationary state before or after any external action.

When an atom is thus compared to an isolated system, it is in " making abstraction of the radiation which, even in the absence of external actions, puts an end to the average life of stationary states. If it is permissible to neglect this radiation in many applications, it is because the coupling between the atom and radiation field, calculated according to classical electro-dynamics, is generally weak compared with the coupling between the particles and the atom. " [3]

So it is the author himself of the theory—writing, it is true, in the light of the subsequent developments of quantum mechanics—who warns us against an unreserved acceptation of his models.

---

[1] BOHR, *La théorie atomique et la description des phénomènes*, Gauthier-Villars, 1932, p. 33.
[2] *Ibid.*, p. 41. The problem here is that of the hydrogen atom in crossed electrical and magnetic fields.
[3] *Ibid.*, p. 78.

2a. REMARKS ON BOHR'S THEOREM.

We have quoted Bohr's theorem which directly inspired Wilson in his enunciation of the conditions of the quanta for systems with several degrees of freedom.

This theorem applies to a system composed of fixed nuclei and electrons of constant mass describing circular orbits. The potential energy U is homogeneous and of degree -1 with respect to the coordinates, and is written

$$2 \sum T_i = U.$$

If the trajectory of each electron is elliptical and their masses remain constant, it is possible to write, *on the average*,

$$2 \sum \overline{T}_i = \overline{U}.$$

If the mass of each electron varies with its generalised kinetic energy according to the relativistic relationship $m_i = m_{0i} + \dfrac{\mathscr{C}_i}{c^2}$ and the trajectories remain circular, then

$$2\mathscr{C} = \sum \mathscr{C}_i \frac{1 + \dfrac{1}{2} \dfrac{\mathscr{C}_i}{m_{0i}c^2}}{1 + \dfrac{\mathscr{C}_i}{m_{0i}c^2}} = U.$$

If, finally, the system is composed of relativistic electrons actuated only by a quasi-periodic motion, it is possible to write *on the average*, for the duration of an approximate period,

$$2\overline{\mathscr{C}} = \overline{U}.$$

2b. ANTINOMY BETWEEN PLANCK'S FREQUENCY RELATION AND CLASSICAL DYNAMICS (POINCARÉ).

From the infinity of trajectories possible in the classical sense, Bohr's theory makes a selection which is based on the *atomicity of action*. The quantum conditions therefore appear to act by simple superposition, without causing any other modification of classical mechanics.

In truth, there is more than this— there is a *true antinomy* between Planck's frequency relation, which expresses the quantum conditions, and ordinary mechanics. Poincaré is credited with having called attention to this antinomy, in a paper to which we shall have occasion to return.[1]

Hamilton's canonical equations assume, in classical mechanics, that the last multiplier in Jacobi's sense is unity. According to Poincaré, this last multiplier may be interpreted as a *probability-density* in the phase-space (q_k, p_k).

Ordinary dynamics implies, because of the one fact that the last multiplier is unity, a complete homogeneity of the possibilities of the localisation of the representative point of the system in phase-space. This is the reason why every theory which supposes that the equations of dynamics are of the Hamilton-Jacobi form necessarily leads to the law of *equipartition* of the energy between the frequencies.

In accordance with the classical axiom of initial conditions, Poincaré retains the definition of the state of a mechanical system as a point of the phase space (q_k, p_k) or, tantamount to this, by a point of the *energy-phase* space (η_k, φ_k).

To retrieve Planck's relation Poincaré forms a last multiplier, W, which is uniform but *essentially discontinuous*— indeed, W is formed of factor like $w(\eta_k)$ which remain zero if the energy η_k differs from a multiple of the quantum ε, so that the integral

$$\int w(\eta_k) \, d\eta_k$$

is equal to the number of multiples of ε contained between its limits.

[1] *Journal de Physique*, Vol. II, January 1912, p. 1.

CHAPTER FOUR

WAVE MECHANICS IN THE SENSE
OF LOUIS DE BROGLIE AND SCHRÖDINGER

A. *PRESENTATION*

1. THE PHASE WAVE.

Wave mechanics, to which we now turn our attention, was created in 1923. The roots of this theory are contained in three notes published by Louis de Broglie in the *Comptes rendus*.[1] We shall follow the first systematic presentation of this mechanics ; namely, the thesis put forward by Louis de Broglie before the *Faculté des Sciences de Paris* in 1924.[2]

Basing his argument both on special relativity and Planck's relation, L. de Broglie associates with every "*fragment of energy*" of proper mass m_0 in the proper system of reference to which it is connected, a frequency ν_0 satisfying the *quantum relation*

$$h\nu_0 = m_0 c^2.$$

The proper system, connected to the particle, is actuated with velocity v with respect to a fixed system of reference. In the latter system the periodic phenomenon appears to be retarded, and its frequency becomes

$$\nu_1 = \frac{\nu_0}{\beta}. \qquad \left(\beta = \frac{1}{\sqrt{1 - \dfrac{v^2}{c^2}}}\right).$$

Similarly, in the fixed system the energy becomes $m_0 c^2 \beta$, and the quantum relation associates with it a frequency

$$\nu = \nu_0 \beta$$

---

[1] *Comptes rendus de l'Académie des Sciences* (Paris), Vol. 177, 1923, pp. 507, 548 and 630.

[2] *Recherches sur la théorie des quanta*, inserted in *Annales de Physique*, Series 10, Vol. III, 1925, p. 22.

which is essentially distinct from ν_1. " There is here, " declares L. de Broglie, " a difficulty which intrigued me for a considerable time. I have been able to remove it by demonstrating the following theorem, which I call the *theorem of the harmony of the phases.*

" *The periodic phenomenon associated with the particle, and whose frequency is equal to* $\nu_1 = \left(\dfrac{1}{h}\right) \cdot \dfrac{m_0 c^2}{\beta}$ *to a fixed observer, appears to the latter to be constantly in phase with a wave of frequency* $\left(\dfrac{1}{h}\right) \cdot m_0 c^2 \beta$ *which propagates in the same direction as the particle with velocity* $V = \dfrac{c^2}{v}.$ "

In fact, in the proper system the periodic phenomenon studied is represented by a sinusoidal function of $\nu_0 t_0$, where t_0 is the proper time of the particle. Now, because of the Lorentz transformation

$$t_0 = \beta \left(t - \frac{vx}{c^2}\right),$$

this phenomenon is represented in the fixed system, by the same sinusoidal function of

$$\nu_0 \beta \left(t - \frac{vx}{c^2}\right);$$

that is, by a wave of frequency $\nu_0 \beta$ propagating itself with velocity $\dfrac{c^2}{\nu}$ in the same direction as the particle.

The group velocity of the phase waves is equal to the velocity of the particle. For, according to Lord Rayleigh's principle, the group velocity of the phase waves considered is defined by

$$\frac{1}{U} = \frac{d}{d\nu}\left(\frac{\nu}{V}\right).$$

An elementary calculation, effected by expressing ν and V as functions of v, then gives

$$U = v.$$

2. RECAPITULATION OF THE PRINCIPLES OF HAMILTON AND MAUPERTUIS.

In Hamilton's sense, the principle of stationary action is written as

$$\delta \int_{t_1}^{t_2} (\mathcal{L} + U)\, dt = 0$$

where $\mathcal{L}$ is the Lagrange function and U the potential energy (force-function) which is supposed to be independent of the time. The energy integral is written as

$$T = U + W$$

where $2T$ is the kinetic energy and W, the total energy. The relation

$$T + \mathcal{L} = \frac{\partial \mathcal{L}}{\partial q_i'} q_i' = p_i q_i'$$

exists between the kinetic energy and the Lagrange function, where q_i are Lagrange's generalised coordinates and summation over the dummy suffix i for each degree of freedom of the system. If only varied trajectories corresponding to a fixed value of the energy are considered, Hamilton's principle reduces to that of Maupertuis ; namely, to

$$\delta \int_A^B p_i dq_i = 0$$

which refers to two given configurations A and B of the system.

3. APPLICATION OF THE PRINCIPLE OF LEAST ACTION TO THE DYNAMICS OF THE ELECTRON.

Louis de Broglie applied these two principles of Hamilton and Maupertuis to the dynamics of the electron, understanding by this term every particle of mass m_0 carrying an electric charge e. The electromagnetic field is defined by the vector potential $\vec{A}$ and the scalar potential φ. In the system of coordinates

$$x_1 = x \qquad x_2 = y \qquad x_3 = z \qquad x_4 = ct,$$

the *energy-momentum* vector *(Impulsvektor)* has the covariant components

$$\begin{cases} J_1 = -m_0 \beta v_x - e A_z = -p_x \\ J_2 = -m_0 \beta v_y - e A_y = -p_y \\ J_3 = -m_0 \beta v_z - e A_z = -p_z \\ J_4 = m_0 \beta c + \dfrac{e}{c} \varphi = \dfrac{W}{c}. \end{cases}$$

The principle of stationary action in Hamilton's sense is here written

$$\delta \int_P^Q J_i dx_i = 0$$

with summation from 1 to 4 in the dummy suffix i, and where P and Q

are two given points of the continuum (x, y, z, ct). Or again, explicitly,

$$\delta \int_{t_1}^{t_2} \left\{ -\frac{m_0 c^2}{\beta} + e\vec{A}\vec{v} - e\varphi \right\} dt = 0.$$

If the potentials do not depend on the time, the principle of least action can be written, because of the constancy of J_4, in the Maupertuis' form

$$\delta \int_A^B J_k dx_k = 0.$$

Here the summation is carried out, in the dummy suffix k, from 1 to 3, and A and B are the two points of the space (x, y, z) which correspond to the points P and Q of the continuum.

The components of the vector $\vec{p}$ differ from those of the momentum $m_0\beta\vec{v}$ if a magnetic field exists.

4. THE VECTOR "WORLD WAVE."

L. de Broglie defines, in the continuum (x, y, z, ct), a vector *world wave* by the relation

$$d\varphi = 2\pi O_i dx_i = 2\pi\nu \left(dt - \frac{dl}{V} \right).$$

Here l is the direction of propagation of the wave φ, whose frequency is ν and velocity of propagation, V. The O_i are given by

$$O_1 = -\frac{\nu}{V} \cos(x, l) \qquad O_2 = -\frac{\nu}{V} \cos(y, l) \qquad O_3 = -\frac{\nu}{V} \cos(z, l)$$

$$O_4 = \frac{\nu}{c}.$$

Thus O_1, O_2, O_3 are the components of a space vector $\vec{n}$ of length $\frac{\nu}{V} = \frac{1}{\lambda}$ (if λ is the wave-length) which is carried by the ray.

The principle of least action defining the world-ray is written in the hamiltonian form

$$\delta \int_P^Q O_i dx_i = 0$$

where P and Q are two given points of the continuum. If the component O_4—and therefore the frequency of the wave—is constant, the principle of least action reduces to Maupertuis' form

$$\delta \int_A^B O_k dx_k = 0$$

(where the summation is carried out over the index k, from 1 to 3) and then defines the ray of the wave. Taking account of the values of O_1, O_2, O_3, this form of Maupertuis' reduces to Fermat's principle ; namely,

$$\delta \int_A^B \frac{vdl}{V} = 0.$$

5. EXTENSION OF THE QUANTUM RELATION.

Given these principles, in order to determine the phase wave associated with a particle moving in an electromagnetic field, Louis de Broglie introduces an extension of the quantum relation

$$W = hv.$$

By means of the vectors $\vec{J}$ and $\vec{O}$, this relation is written as

$$J_4 = hO_4.$$

L. de Broglie generalises this by supposing that

$$\vec{J} = h\vec{O}.$$

Thus to Fermat's principle corresponds Maupertuis' principle, written in the form

$$\delta \int_A^B J_k dx_k = \delta \int_A^B p_k dq_k = 0.$$

In short, " *Fermat's principle applied to the wave is identical with Maupertuis' principle applied to the particle— the dynamically possible trajectories of the particle are identical with the possible rays of the wave.* "

Examples.

1) *Rectilinear and uniform motion of a free particle.*
Here

$$W = hv = m_0 c^2 \beta = mc^2$$
$$p_k dq_k = mv^2 dt = mvdl = \frac{hvdl}{V}.$$

Therefore

$$V = \frac{hv}{mv} = \frac{mc^2}{mv} = \frac{c^2}{v}$$

which is a result which has already been obtained.

2) *Motion of an electron in an electrostatic field* (the Bohr atom).

Here
$$W = h\nu = mc^2 + e\varphi$$
$$p_k dq_k = mvdl = \frac{h\nu dl}{V}.$$

Therefore
$$V = \frac{h\nu}{mv} = \frac{mc^2 + e\varphi}{mv} = \frac{c^2}{v}\left[1 + \frac{e\varphi}{c^2 m}\right].$$

3) *Motion of an electron in an electromagnetic field.*
Here
$$W = h\nu = mc^2 + e\varphi$$
$$p_k dq_k = mvdl + eA_l dl = \frac{h\nu dl}{V}.$$

Therefore
$$V = \frac{h\nu}{mv + eA_l} = \frac{mc^2 + e\varphi}{mv + eA_l} = \frac{c^2}{v}\left[\frac{1 + \dfrac{e\varphi}{c^2 m}}{1 + \dfrac{eA_l}{mv}}\right].$$

Thus the medium is no longer isotropic—the velocity V varies with the direction considered. The velocity $\vec{v}$ has a direction different from the normal to the wave, $\vec{n}\,(\vec{p} = h\vec{n})$. *The equality between the velocity of the particle and the group velocity of the waves remains.* In fact, if the axis of x is chosen to have the same direction as that of $\vec{v}$ at the point considered, then, in the first place,

$$h\frac{\nu}{V} = p_x$$

and then, by Hamilton's first canonical equation,

$$v = \frac{dx}{dt} = \frac{\partial W}{\partial p_x} = \frac{\partial(h\nu)}{\partial\left(h\dfrac{\nu}{V}\right)} = U.$$

6. RETURN TO THE BOHR-SOMMERFELD QUANTUM CONDITIONS.

Here Louis de Broglie starts from the form which, in 1917, Einstein gave to the Bohr-Sommerfeld quantum conditions.[1] This form is invariant under transformations of the coordinates and, for a closed trajectory, is written

$$\int p_i dq_i = nh$$

[1] *Verh. der Deutsch. Physik. Ges.*, Vol. 19, 1917, p. 77.

where the integral is taken over the closed trajectory, the index i is summed from 1 to 3 and n denotes an arbitrary integer. This integral is nothing else than Maupertuis' action integral.

It is evident that the variation of the phase, after a complete circuit of a closed trajectory, must be an integral multiple of 2π in order that the association of the waves and the particle may remain coherent.

For the path dl of the point on the trajectory, the phase varies by

$$d\varphi = - 2\pi\nu \frac{dl}{V}.$$

By integrating along a closed trajectory, it must therefore be that

$$\int \frac{\nu dl}{V} = n$$

or, by the relation between the vectors $\vec{O}$ and $\vec{J}$,

$$\int p_i dq_i = nh.$$

An argument which is a little more complicated, presented by L. de Broglie in his thesis (referred to above), makes it possible to derive from the preceding condition, the relations

$$\oint p_i dq_i = n_i h \qquad \text{(for each index } i\text{)}$$

which are applicable to quasi-periodic motions. These relations are the Bohr-Sommerfeld quantum conditions, related in this way to the resonance of the phase wave.

7. DIFFRACTION OF ELECTRONS BY MATTER.

It was natural to attempt to verify experimentally the existence of the wave associated with a material corpuscle by causing the diffraction of electrons by crystals.

For an electron, the wave-length of the associated wave has the value

$$\lambda = \frac{h}{m_0 v \beta}.$$

This quantity can be represented by $\dfrac{h}{m_0 v}$ if the velocity v is not too close to c. This wave-length is of the order of 10^{-8} to 10^{-9} for electrons of ordinary velocity. Therefore it is comparable with that of X-rays.

Von Laue's theory makes it possible to calculate the distance d between the reticular planes as a function of the angle of incidence and the wave-length λ of the incident rays. Conversely, the knowledge of d for a given grating makes it possible to determine λ. Davisson and Germer,[1] working at the Bell Telephone Laboratories (New York) in 1927, succeeded in obtaining a characteristic diffraction phenomenon by allowing a beam of monokinetic electrons to fall on a crystal of nickel, and collecting the diffracted beams in an ionisation chamber. Thus L. de Broglie's daring hypothesis was vividly confirmed. Quantitatively, the agreement between the observed and the calculated values of λ were confirmed to within 2 %.

The experiment was repeated by G. P. Thomson,[2] at the University of Aberdeen, using a method analogous to that of Debye and Scherrer—that is, by making a monokinetic beam of electrons traverse a crystalline powder. The result expresses itself as the creation of diffraction rings whose diameters are functions of the reticular distance of the crystals and of the wave-length of the incident beam. The experiment was continued by Rupp,[3] Kichuchi[4] Ponte[5] and by Dauvillier and Trillat.

In particular, Rupp demonstrated the existence of a factor $\dfrac{1}{\beta}$ in the wave-length of fast electrons.

8. SCHRÖDINGER'S WAVE EQUATION.

Erwin Schrödinger, directly inspired by the thesis of L. de Broglie, undertook the search for a general relationship between the dynamics of conservative systems in the classical sense and the phenomenon of wave propagation. Here we shall follow a paper called *Quantisation and Eigenvalues*.[6]

For the most general conservative system of classical dynamics, the Hamilton-Jacobi partial differential equation may be written as

$$\frac{\partial V}{\partial t} + T\left(q_k, \frac{\partial V}{\partial q_k}\right) - U(q_k) = 0.$$

Here V is the action, $2T$ the kinetic energy, U the potential energy and the q_k are the generalised Lagrange coordinates. Hamilton's equation may be replaced by the system of two equations—

[1] *Physical Review*, Vol. 30, 1927, p. 705.
[2] *Proc. Roy. Soc.*, Vol. 117, 1928, p. 600.
[3] *Ann. der Physik*, Vol. 85, 1928, p. 981.
[4] *Jap. Phys. Journal*, Vol. 5, 1928, p. 83.
[5] *Annales de Physique*, Vol. 13, 1929, p. 395.
[6] *Ann. der Physik*, Vol. 79, 1926, p. 489.

$$\begin{cases} \dfrac{\partial V}{\partial t} = -W \\[2mm] 2T\left(q_k, \dfrac{\partial V}{\partial q_k}\right) = 2(U + W). \end{cases}$$

Schrödinger distinguishes between the kinetic energy expressed in terms of the coordinates q_k and the momenta $p_k = \dfrac{\partial V}{\partial q_k}$ which he denotes by $2T$, and the kinetic energy expressed in terms of the q_k and their derivatives, q'_k, with respect to the time, which he denotes by $2\overline{T}$.

By means of the second quantity there may be defined, in the space of the q_k or *configuration space*, a non-euclidean metric of fundamental quadratic form

$$ds^2 = 2\overline{T}dt^2 = g_{ik}dq_idq_k.$$

It follows that

$$\begin{cases} 2\overline{T} = g_{ik}q'_iq'_k \\ 2T = g^{ik}p_ip_k \end{cases} \qquad (\text{by } p_i = g_{im}q'_m).$$

In the following argument the vectorial operations such as grad, div, div grad $= \varDelta_2$, are understood with respect to the non-euclidean metric defined above.

Thus the equation of the kinetic energy (the second of the two equations obtained by the separation of the Hamilton-Jacobi equation) becomes

$$g^{ik}\frac{\partial V}{\partial q_i}\frac{\partial V}{\partial q_k} = 2(U + W)$$

or

$$|\operatorname{grad} V| = \sqrt{2(U + W)}.$$

At the time t the surfaces $V =$ constant may be drawn. This can be accomplished by starting from one of them which, with the associated value of V, is chosen arbitrarily. To pass from the surface V_0 to the surface $V_0 + dV_0$ it suffices to move each point along the normal to V_0 through a distance

$$ds = \frac{dV_0}{\sqrt{2(U + W)}}.$$

The surface corresponding to the value V_0 at the time t will correspond to the surface $V_0 - Wdt$ at the time $t + dt$. It may be imagined that the surfaces V travel through the configuration space by transporting a determinate value of V. For this purpose it suffices to assign to each surface a normal velocity

$$u = \frac{ds}{dt} = \frac{W}{\sqrt{2(U+W)}}.$$

Indeed, under these conditions

$$0 = \frac{\partial V}{\partial t}\,dt + \frac{\partial V}{\partial s}\,ds = dV.$$

The ensemble of surfaces of equal action may thus be compared to an ensemble of wave surfaces whose velocity at each point is equal to u.

In this optical analogy the refractive index would have the value $\frac{1}{u}$; Hamilton's partial differential equation would express Huyghens' principle. By writing Fermat's principle in the form

$$0 = \delta \int_P^Q \frac{ds}{u} = \delta \int_P^Q \frac{ds\,\sqrt{2(U+W)}}{W} = \delta \int_P^Q \frac{2T}{W}\,dt = \delta \int_P^Q 2T\,dt\,,$$

Maupertuis' principle is retrieved (the variation being made at constant total energy).

The representative point of the mechanical system considered moves in configuration space with the velocity

$$v = \frac{ds}{dt} = \sqrt{2T} = \sqrt{2(U+W)}$$

which varies reciprocally with u. It cannot remain in contact with the same surface V.

Schrödinger insists on the fact that the analogy only exists between *geometrical optics* and mechanics. The analogy only retains the phase of the waves, and their amplitude, their wave-length and their frequency do not appear.

At this point Schrödinger introduces the hypothesis—whose arbitrary nature he emphasised—that the waves contemplated must be sinusoidal. The argument of the sine is a linear function of V. It is assumed that the coefficient of V is of the form $\frac{2\pi}{h}$, where h is a universal constant which is independent of W and of the nature of the mechanical system considered. Therefore an argument of the form

$$\sin\left(\frac{2\pi}{h}\,V + c^{te}\right) = \sin\frac{2\pi}{h}\,[-\,Wt + S(q_k) + c^{te}]$$

is chosen. Whence the frequency of the waves is

$$\nu = \frac{W}{h}.$$

In classical mechanics the energy W is only determined up to an arbitrary constant. On the other hand, the wave length λ is independent of this constant, for

$$\lambda = \frac{u}{\nu} = \frac{h}{\sqrt{2(U + W)}}.$$

The law of the dispersion of the waves may be written

$$u = \frac{h\nu}{\sqrt{2(h\nu + U)}}.$$

The group velocity of the waves,

$$\frac{d\nu}{d\left(\dfrac{\nu}{u}\right)}$$

here reduces to v and consequently coincides with the velocity of the representative point of the mechanical system considered. This result confirms that which L. de Broglie established by making appeal to the theory of relativity for the phase wave of an electron.

Schrödinger sets out to construct a *wave packet*, all of whose dimensions are very small, which may be capable of *replacing* the representative point. This is only possible if the wave-lengths are small compared with the dimensions of the trajectory and, in particular, with the curvature of the trajectory. Indeed, if it is desired that the wave packet should be approximately monochromatic—which is essential in order that it may move as a whole with a well defined group velocity, and that it may be able to correspond to a mechanical system having a well determined energy—it must extend over a domain containing a great number of wave-lengths.

Schrödinger arrives at the conclusion that *the motion of the point of concordance of phase for certain infinitesimal ensembles of waves of* n *parameters proceeds according to the same laws as the motion of the representative point of the corresponding mechanical system.*

Without representing Schrödinger's analysis, we shall confine ourselves to the following brief verification of his result.

Jacobi's classical method for the solution of Hamilton's equation is based on the knowledge of a complete integral of this equation. Such a complete integral depends on n primary constants a_i, of which

none is additive if n is the number of degrees of freedom of the system—
that is, the number of dimensions of configuration space. Once given
this complete integral, say

$$V(q_1, q_2, \ldots q_n, a_1, a_2 \ldots a_n, t)$$

Jacobi writes the known relations

(J) $$\begin{cases} \dfrac{\partial V}{\partial a_1} = \left(\dfrac{\partial V}{\partial a_1}\right)_0 \\ \cdots\cdots\cdots \\ \dfrac{\partial V}{\partial a_n} = \left(\dfrac{\partial V}{\partial a_n}\right)_0. \end{cases}$$

Here, since the system is conservative, it may be supposed that
$a_1 = W$, thus making the total energy play the part of a primary
constant. Then only the first of the equations (J) contains the time.

At each instant the equations (J) define the representative point
of the mechanical system considered. This point P, by the form
of the equations (J), is the point of concordance of phase of an infinitesimal ensemble with n parameters of the system of waves obtained
by making the a_i vary in a continuous way.

To develop the *wave conception* of mechanics, Schrödinger takes
as his starting point a wave equation in configuration space. The
only datum which can be utilised is the expression of the wave velocity
u in terms of the energy or the frequency.

It is also supposed that this equation is of the second order, so
that it must be written as

$$\text{div grad } \psi - \frac{1}{u^2} \frac{\partial^2 \psi}{\partial t^2} = 0.$$

This equation is only valid for phenomena which depend on the time
by means of a factor $e^{2\pi i \nu t}$; that is, for which

$$\frac{\partial^2 \psi}{\partial t^2} = -4\pi^2 \nu^2 \psi.$$

Whence, taking account of the value of u,

$$\boxed{\text{div grad } \psi + \frac{8\pi^2}{h^2} m(U + W)\psi = 0.}$$

This is Schrödinger's wave equation, which already embodies the
quantum conditions. " This equation automatically sorts out the
frequencies and energy levels and distinguishes those which can actually

appear in stationary phenomena, without other supplementary hypothesis than the following condition which every physical quantity
must almost naturally satisfy— *the function must be unique, finite and
continuous throughout the whole configuration space.* "

The quantum levels are all determined at the same time as the
eigenvalues of the wave equation which in itself contains its appropriate
conditions at the limits.

Example. — The linear oscillator.

We confine ourselves to the most simple of the examples treated
by Schrödinger in his paper— that of the linear oscillator.

The potential energy is here $U = -\dfrac{m\omega^2 q^2}{2}$ and Schrödinger's equation takes the form

$$\frac{d^2\psi}{dq^2} + \frac{8\pi^2 m}{h^2}\left(W - \frac{m\omega^2 q^2}{2}\right)\psi = 0.$$

We put

$$q = Q\sqrt{\frac{h}{2\pi m\omega}} \quad \text{and} \quad W = \frac{h\omega}{4\pi}w\,;$$

so that the equation becomes

$$\frac{d^2\psi}{dQ^2} + (w - Q^2)\psi = 0.$$

This equation is of the known form

$$y'' + (\lambda - x^2)y = 0.$$

It only has regular solutions which vanish at infinity if

$$\lambda = 2n + 1 \qquad\qquad (n,\ \text{integer}).$$

These solutions are written as

$$y = e^{-\frac{x^2}{2}}H_n(x)$$

where H_n is a Hermite polynomial.

Therefore it must be supposed that

$$w = 2n + 1$$

or

$$W = (2n + 1)\frac{h\omega}{4\pi} = \left(n + \frac{1}{2}\right)h\nu.$$

Schrödinger's quantisation here introduces " half-integer " numbers,
in contrast with Bohr's theory but in full accord with the result already
obtained by Heisenberg and to which we shall return in the next chapter.

Now the observation of the band spectra of diatomic molecules had revealed the necessity of considering such numbers.

Treating more complicated problems, such as the Stark effect and the normal Zeeman effect, Schrödinger also arrived at results in agreement with experiment.

B. *ANALYSIS AND INTERPRETATION*

1a. ON THE GUIDING IDEAS AND THE ORIGIN OF WAVE MECHANICS.

Louis de Broglie himself has explained the origin of wave mechanics.[1]

All the earlier work on black-body radiation had assumed its wave-like character. Louis de Broglie was inspired to adopt the corpuscular point of view by comparing this radiation to a " gas of photons." It was necessary to combine Boltzmann statistics with relativistic dynamics, for the velocity of the photons attained or approached that of light. This led, not to Planck's law, but to that of Wien. To obtain Planck's law it was necessary—as Bose was to verify later— to modify Boltzmann's analysis and to adopt the statistics that has become classical under the names of Bose and Einstein.

But it is fitting to let the author speak.

" The study of black-body radiation had strengthened my conviction that, to arrive at a more complete theory of light and the radiations, it was necessary to seek to unite the idea of corpuscles with the idea of waves. But, in reflecting on the matter, I suddenly had the intuition that such a union of waves and corpuscles was also necessary in the theory of matter. This is how I came to this conclusion. In the corpuscular conception of radiation the fundamental formula is Einstein's formula of ' light quanta, ' or $W = h\nu$. This relation connecting the energy of the corpuscle of radiation to the frequency of the corresponding wave establishes a kind of bridge between the conception of waves and that of corpuscles ; it creates a kind of correspondence between the two pictures. It follows that the constant h is in some way a feature of the union between the waves and the corpuscles.

" But if we turn to the theory of matter and, in particular, to that of atomic systems, what do we see ? We see the elementary particles of matter, in particular the electrons, describing quantised motions whose specification introduces the constant h and the integers. Now the intervention of the whole numbers suggests the phenomena of interference and resonance— that is, essentially wave-like phenomena.

[1] *Notice sur les travaux scientifiques de M. Louis de Broglie*, Paris (Hermann), 1931.

" May it not be thought, then, that there are grounds for uniting the waves with the corpuscles, the correspondence between the one and the other being defined by formulae in which Planck's constant will play an essential part ? And when this correspondence between waves and corpuscles is established for matter, perhaps it will reveal itself as identical with that which must be accepted between waves and corpuscles for light. Then a very beautiful conclusion will have been reached ; a general doctrine will have been formed which will establish the same correlation between waves and corpuscles both in the field of light and in that of matter.

" To bring this endeavour to a successful conclusion, I was guided by the formal analogy, which had been indicated for a considerable time, between the equations of analytical mechanics and those of geometrical optics—in particular, by the formal analogy between Maupertuis' principle of least action and Fermat's principle of the minimum time. I succeeded in establishing the correspondence that I had in mind by associating with every corpuscle of matter or of light, of energy W and momentum p, the propagation, according to the laws of geometrical optics, of a wave of frequency ν and wavelength λ, related to the quantities W and p by the relations

$$ \nu = \frac{W}{h} \qquad \lambda = \frac{h}{p}. $$

" Thanks to this correspondence between the mechanical and the wave-like properties, Maupertuis' principle for the corpuscle becomes equivalent to that of Fermat for the wave, and the possible trajectories of the corpuscles become identical with the rays in the optical sense. Applied to light, this correspondence requires the coexistence of photons and light-waves, the energy of a photon being determined by the frequency of the wave according to Einstein's relation.   Applied to matter, the same correspondence requires that every corpuscle of matter, every electron for example, is always associated with a wave which accompanies and controls its motion.   At that moment I had the idea that it must be possible to obtain interference or diffraction phenomena *with electrons*.   In this direction must be found the crucial experiment which would produce the direct demonstration of the existence of a wave associated with each electron. . . .

" Encouraged by the logical coherence of the result obtained, I continued my investigations by showing how the existence of a wave associated with the electron explains why the motion of an electron is quantised. The equation which expresses that the motion of an electron is quantised has, in effect, the following significance—

Maupertuis' integral $\int mvds$ taken along the closed trajectory is equal to an integral multiple of the constant h. But according to the correspondence that I had established between the momentum of the corpuscle and the wave-length that it is necessary to associate with it, the value, divided by h, of Maupertuis' integral along the trajectory is equal to the total variation of the phase of this wave along the trajectory, taking 2π as the unit.

" To write that the integral in question is a whole multiple of h amounts to the same as to write that the phase of the wave is a ' uniform ' function along the trajectory, that is to say that the wave is *stationary*. This latter word throws light on the deep meaning of the quantisation of electronic motions. An electronic motion is quantised, that is to say stable, when the corresponding wave is stationary. This fine explanation of quantisation finally convinced me that I was on the right road. "

Louis de Broglie states again, in another form—

" The theory of light was suffering from a stange illness, which took the form of an antagonistic dualism between the waves of Fresnel and Maxwell, on the one hand, and photons, on the other. Well ! in order to better the state of affairs a drastic remedy could be tried by seeking to communicate the same illness to the theory of matter, which up to then had been immune. In reality, there was a serious reason for so doing ; namely, that the theory of matter, too, had been showing disquieting symptoms for several years. The need for quantising the motions of particles of matter which had first manifested itself in the theory of black radiation and had won a triumphant success in that of Bohr's atom, indicated that the quantum of action did not allow, in the atomic field, the conceptions and equations of classical Mechanics to be maintained. The presence of whole numbers in the formulae of quanta gave these latter a certain resemblance with the formulae of interference of the wave theory, and the analogy of the principle of least action, the key-stone of classical Mechanics, with Fermat's principle, the key-stone of geometrical optics, suggested that classical mechanics might very well be an approximate form of a more general wave-Mechanics, playing in relation to the latter the part played by geometrical Optics in relation to wave-Optics. Thus appeared the idea of extending to matter the corpuscles-waves duality which was essential for light. "[1]

Such a revolutionary thesis as that of Louis de Broglie might have been greeted with some scepticism. But it came at the right time,

---

[1] *Voies anciennes et perspectives nouvelles en théorie de la lumière*, *Revue de Métaphysique et de Morale*, Vol. XLI, 1934, p. 448.

for the development of theoretical physics had then arrived at the necessity of a synthesis of the notions of wave and particle.

Moreover, the success was immediate and resounding. Einstein drew the attention of the scientific world to this thesis, and E. Schrödinger deduced from it, as early as 1926, a direct correlation between the dynamics of the most general conservative system and that of waves.

8a. ON SCHRÖDINGER'S EQUATION.

In the initial form which Louis de Broglie had given it, wave mechanics may have seemed to be linked with Einstein's special theory of relativity. Indeed, the Lorentz transformation appeared to determine, for example, the conditions assumed for the definition of the frequencies in a fixed system.

In Schrödinger's work, on the contrary, the question was that of a wave treatment of a problem of ordinary mechanics.

While Louis de Broglie had started from the principles of Fermat and Maupertuis, Schrödinger referred directly to Hamilton's optics which was, as we have seen, capable of a double interpretation in terms of emission and wave propagation.[1]

Thus thought in physics continually turned about Fermat's optics and the dynamics of Maupertuis, Euler and Lagrange ; about the dynamics and the optics of Hamilton, who formulated the first formal theory which admitted of both a corpuscular and a wave interpretation ; about the dynamics and optics of Hamilton in another aspect, for return to these created the essential mathematical tools of analytical mechanics. And we see these wanderings end in a true fusion of the two extremum principles of optics and dynamics.

---

[1] See Part IV, Chapter VI, p. 390. In connection with this return to HAMILTON, SCHRÖDINGER deplored the fact that, in most of the modern presentations, analytical mechanics and optics " found itself robbed of the magnificent intuitive dress that Hamilton had given it. "

CHAPTER FIVE

QUANTUM MECHANICS
IN THE SENSE OF HEISENBERG AND DIRAC

A. *PRESENTATION*

1. QUANTUM ANALOGUE OF CLASSICAL MECHANICS (HEISENBERG).

We return to the year 1925, to analyse W. Heisenberg's paper called *Über quantentheoretische Umdeutung kinematischer und mechanischer Beziehungen.*[1] In this paper the author laid the foundations of " quantum " mechanics. Heisenberg's guiding idea was that of constructing, by analogy with classical mechanics, a mechanics conforming with the theory of quanta in which only *observable quantities* (frequencies or energy levels) would occur, to the exclusion of the coordinates and velocities of electrons.

Heisenberg confined himself to mechanical problems of one degree of freedom. Adopting at first the kinematic point of view, he sought the quantum analogue of a classical quantity typified by $x(t)$.

In the classical field the mechanical frequency of a system of one degree of freedom is

$$\nu = \frac{dW}{dJ} = \frac{1}{h}\frac{dW}{dn}.$$

Here J is the phase integral, which may be supposed equal to nh. Therefore it is possible to write

$$(1) \qquad \nu(n,\alpha) = \frac{\alpha}{h}\frac{dW}{dn} = \alpha\nu(n).$$

The quantum analogue of formula (1) is written

$$(2) \qquad \nu(n,n-\alpha) = \frac{1}{h}\{W(n) - W(n-\alpha)\}$$

so that the frequency is associated with two energy levels.

---

[1] *Zeitschr. für Physik*, Vol. 33, 1925, p. 879.

In the classical field equation (1) implies the relation

$$\nu(n, \alpha) + \nu(n, \beta) = \nu(n, \alpha + \beta).$$

According to (2) the quantum analogue of this will be written as

$$\nu(n, n - \alpha) + \nu(n - \alpha, n - \beta) = \nu(n, n - \alpha - \beta)$$

or

$$\nu(n - \beta, n - \alpha - \beta) + \nu(n, n - \beta) = \nu(n, n - \alpha - \beta).$$

It is possible to develop a classical variable $x(n, t)$ in a Fourier series (which may be an integral) of the form

$$x(n, t) = \sum_{\substack{-\infty \\ (\alpha)}}^{+\infty} A_\alpha(n) e^{i\omega(n)\alpha t}.$$

To constitute the quantum analogue of $x(t)$, the terms of the preceding sum will be replaced by terms like

$$A(n, n - \alpha) e^{i\omega(n, n - \alpha)t}$$

associated with two energy levels.

To represent x^2 in the classical field use is made of the multiplication of Fourier series ; thus x^2 has the form of a series

$$x^2 = \sum_{\substack{-\infty \\ (\beta)}}^{+\infty} B_\beta(n) e^{i\omega(n)\beta t}$$

with

$$B_\beta e^{i\omega(n)\beta t} = \sum_{\substack{-\infty \\ (\alpha)}}^{+\infty} A_\alpha A_{\beta - \alpha} e^{i\omega(n)(\alpha + \beta - \alpha)t}.$$

The quantum analogue of x^2 will be constructed by means of a series of terms

$$B(n, n - \beta) e^{i\omega(n, n - \beta)t} = \sum_{-\infty}^{+\infty} A(n, n - \alpha) A(n - \alpha, n - \beta) e^{i\omega(n, n - \beta)t}.$$

Without the need for further explanation it can be understood that the procedure can be repeated so as to reach x^n and, consequently, any function $f(t)$ which can be expanded as a power series in x.

At this point Heisenberg remarks that if $x(t)$ is represented by factors like A and $y(t)$ by factors like B, the *quantum product* which is formed by analogy with the classical product given by

$$C_\beta = \sum_{\substack{-\infty \\ (\alpha)}}^{+\infty} A_\alpha(n) B_{\beta - \alpha}(n) \,,$$

will be
$$C(n, n-\beta) = \sum_{-\infty}^{+\infty} A(n, n-\alpha) B(n-\alpha, n-\beta)$$

In general this will not be commutative; that is, that $x(t)y(t)$ will in general differ from $y(t)x(t)$ in the quantum field.

Given these kinematic preliminaries, Heisenberg sets out to determine the quantum magnitudes A, ν, W in the quantum analogue of a problem of classical dynamics corresponding to

$$(3) \qquad x'' + f(x) = 0.$$

In the dynamics of quanta in Bohr's sense, the calculation proceeds by two stages; namely,

 a) the integration of the equation of motion (3)
 b) the selection of the solutions by the condition

$$(4) \qquad \oint p\,dq = \oint mx'\,dz = J = nh.$$

Heisenberg retains the equation of motion (3) of the classical problem *and substitutes the quantum analogue, in the sense defined above, for the Fourier solution of the classical problem.* He recognises that this procedure leads to a system of an infinite number of equations, with an infinite number of unknowns, which can only be reduced easily when a recurrence relationship can be distinguished.

The solution of the classical problem will be expressed in terms of the Fourier series (or integral)

$$x = \sum_{\substack{-\infty \\ (\alpha)}}^{+\infty} a_\alpha(n) e^{i\alpha\omega_n t}$$

whence

$$mx' = m \sum_{\substack{-\infty \\ (\alpha)}}^{+\infty} a_\alpha(n) i\alpha e^{i\alpha\omega_n t} \omega_n.$$

Then, by a classical calculation

$$\oint mx'\,dx = \oint mx'^2\,dt = 2\pi m \sum_{\substack{-\infty \\ (\alpha)}}^{+\infty} a_\alpha(n) a_{-\alpha}(n) \alpha^2 \omega_n.$$

The reality of x requires that

$$a_{-\alpha}(n) = a_\alpha^*(n)$$

where the quantity conjugate to a is denoted by a^*. Whence

$$(5) \qquad \oint mx'^2\,dt = 2\pi m \sum_{\substack{-\infty \\ (\alpha)}}^{+\infty} |a_\alpha(n)|^2 \alpha^2 \omega_n.$$

In the dynamics of quanta in Bohr's sense, this integral is equated to a multiple of h by a whole number, say nh. "*Not only does this condition arise in a quite artificial way in the classical dynamical problem, but it also appears arbitrary if the point of view of Bohr's correspondence principle is chosen. In effect, the phase integral J is only defined as an integral multiple of* h *apart from an arbitrary constant.*"

This suggests that instead of the equation

$$J = nh$$

(equation (4)) the equation

(4')
$$\frac{d}{dn}(nh) = \frac{d}{dn}\oint mx'^2 dt$$

should be used to express the correspondence principle. Therefore in the problem which concerns us we shall write, in the sense of the old dynamics of quanta and starting from equation (5), the condition

(5')
$$h = 2\pi m \sum_{\substack{-\infty \\ (\alpha)}}^{+\infty} \alpha \frac{d}{dn}(\alpha\omega_n |a_\alpha|^2).$$

Whence, by *quantum analogy*, Heisenberg infers the new *quantum condition*

(6) $$h = 4\pi m \sum_{\substack{0 \\ (\alpha)}}^{+\infty} \{|a(n, n+\alpha)|^2 \omega(n, n+\alpha) - |a(n, n-\alpha)|^2 \omega(n, n-\alpha)\}.$$

This condition had already been encountered by Kuhn[1] and Thomas.[2]

Heisenberg also introduces the hypothesis of a *normal state* without any radiation, which is characterised by the number n_0; that is, by

$$a(n_0, n_0 - \alpha) = 0 \qquad \text{if} \quad \alpha > 0.$$

Heisenberg applied these considerations to the anharmonic oscillator and to the harmonic oscillator. For the linear harmonic oscillator he demonstrated the existence of *semi-quanta* which it had been necessary to assume for the explanation of certain peculiarities of the structure of spectra (Katzer); thus he established the formula

$$W = \frac{\left(n + \frac{1}{2}\right)h\omega}{2\pi} = \left(n + \frac{1}{2}\right)h\nu$$

[1] *Zeitschr. für Physik*, Vol. 33, 1925, p. 408.
[2] *Naturw.*, Vol. 13, 1925.

which Schrödinger was to obtain, for his part, with the help of wave mechanics.[1]

Heisenberg then treated the case of the " rotator " (a particle of mass m describing a circle of radius a with constant angular velocity ω). Here the condition (4') is written in the " classical " field (this in the sense of Bohr's dynamics) in the form

$$h = 2\pi m \frac{d}{dn} (a^2 \omega)$$

and the quantum analogue is

$$h = 2\pi m \{ a^2 \omega (n + 1, n) - a^2 \omega (n, n - 1) \}.$$

Whence

$$\omega (n, n - 1) = \frac{h(n + \text{const})}{2\pi m a^2}$$

or, by the introduction of the normal state n_0,

$$\omega (n, n - 1) = \frac{hn}{2\pi m a^2}.$$

Here the energy has the value

$$W = \frac{mv^2}{2} = \frac{m}{2} a^2 \frac{\omega^2(n, n - 1) + \omega^2(n, n + 1)}{2} = \frac{h^2}{8\pi^2 m a^2} \left(n^2 + n + \frac{1}{2} \right)$$

and the frequency condition takes the form

$$\omega (n, n - 1) = \frac{2\pi}{h} [W(n) - W(n - 1)].$$

Again there appear the *semi-quanta*, whose existence it had been necessary to assume in order to account for certain peculiarities of the structure of spectra.

2. INTRODUCTION OF " MATRICES " (BORN, JORDAN).

We have retraced the inductive method by which Heisenberg, proceeding by analogy, attempted to construct a new mechanics of quanta based on a correspondence with the classical field. At this point there intervened a fruitful cooperation of the physicist with two mathematicians. Born and Jordan remarked that the mathematical tools which Heisenberg introduced were " matrices, " which have a non-commutative law of multiplication. Further, these matrices were " hermitian " matrices.[2]

[1] See above, p. 566.
[2] *Zeitschr. für Physik*, Vol. 34, 1925, p. 858.

A matrix is denoted by the symbol $a = a(n, m)$ and represents the infinite array

$$a(n, m) = \begin{pmatrix} a(0,0) & a(0,1) & a(0,2) \ldots \\ a(1,0) & a(1,1) & a(1,2) \ldots \\ a(2,0) & a(2,1) & a(2,2) \ldots \\ \cdots\cdots\cdots\cdots\cdots\cdots \end{pmatrix}.$$

Addition is defined by

$$a = b + c \qquad \text{where} \qquad a(n, m) = b(n, m) + c(n, m)$$

and multiplication by

$$a = bc \qquad \text{where} \qquad a(n, m) = \sum_{k=0}^{\infty} b(n, k)\, c(k, m).$$

The second rule is described by the epithet " *Zeilen mal Kolonnen.* "

Since this multiplication is associative, distributive with respect to addition, but not commutative, in general ab differs from ba. If $ab = ba$, the matrices a and b are said to be *commutative (vertauschbar)*.

The unit matrix is defined by $1 = \delta_{n, m}$ with $\delta_{n, m} = 0$ if $n \neq m$, and by $\delta_{n, n} = 1$.

This matrix commutes with all others, or $a1 = 1a = a$.

The matrices introduced in dynamics to represent the coordinates of position q and the Poisson-Hamilton momenta p are written

$$q = \left(p(n, m)\, e^{2\pi i \nu (n, m)t} \right)$$

$$p = \left(q(n, m)\, e^{2\pi i \nu (n, m)t} \right).$$

These matrices must be hermitian, so that

$$q(n, m)\, q(m, n) = [q(n, m)]^2$$

$$\nu(n, m) = - \nu(m, n).$$

In general a matrix g, a function of p and q, will have a derivative with respect to the time equal to

$$g' = 2\pi i \big(\nu(n, m)\, g(n, m) \big).$$

If $\nu(m, n) \neq 0$ for $n \neq m$, it is seen that the condition $g' = 0$ requires that $g(n, m) = \delta_{n, n} g(n, n)$, or that g must be *diagonal*.

We shall not delay ourselves further with the justification of the matrix calculus, but shall immediately present the framework which Heisenberg, with the collaboration of Born and Jordan, achieved. By comparison with the original presentation (above, § 1) the simplicity of this will be apparent.

3. FORMULATION OF MATRIX MECHANICS BY HEISENBERG, BORN AND JORDAN.[1]

Heisenberg starts from *Bohr's fundamental postulate*— The energy of a mechanical system can only take discrete values W_m, W_n, ... which are related to the frequency of the radiation emitted by the system by Planck's relation

(a) $$h\nu_{m,\,n} = W_m - W_n.$$

Each frequency is thus associated to two energy levels. Bohr's postulate is evidently in agreement with Ritz's purely experimental *combination principle*, expressed by the relation

$$\nu_{m,\,p} + \nu_{p,\,n} = \nu_{m,\,n}.$$

At this point Heisenberg recalls the development of the quasi-periodic systems of classical dynamics in Fourier series,[2] of the form

$$q_k = \sum D^k_{\tau_1,\,\tau_2,\,\ldots\,\tau_n} e^{2\pi i (\nu_1\tau_1 + \nu_2\tau_2 + \ldots + \nu_n\tau_n)t}$$

where

(b) $$\nu_k = \frac{\partial H}{\partial J_k} = \frac{dw_k}{dt}.$$

We recall that J_k is a phase integral and w_k the angular variable conjugate to J_k.

Heisenberg assumes that the equation (a) is the *quantum analogue* of the classical relation (b).[3]

Therefore if it is required to introduce the optical frequencies $\nu_{m,\,n}$ in the development of the quantum coordinates in the same way as the mechanical frequencies ν_k are introduced in the classical theory, one may write

$$q_k = \left| D^k_{m,\,n} e^{2\pi i \nu_{m,\,n} t} \right|.$$

Thus each coordinate is represented by a hermitian matrix. We remark that the rules of this matrix calculus are in accord with Ritz's combination principle.

Differentiating the matrix q_k with respect to the time and taking account of Bohr's postulate, it follows that

$$\frac{dq_k}{dt} = 2\pi i \nu_{m,\,n} \left| q^k_{m,\,n} \right| = \frac{2\pi i}{h} (W_m - W_n) \left| q^k_{m,\,n} \right|.$$

---

[1] *Cf. Zur Quantenmechanik*, II, by BORN, HEISENBERG and JORDAN, *Zeitschr. für Physik*, Vol. 35, 1926, p. 557.

[2] See above, Chapter III, p. 543.

[3] At least for those systems, which are the only ones considered here by HEISENBERG, for which the Hamiltonian H does not depend explicitly on the time.

But, by means of the discrete values of the energy the diagonal matrix

$$W_{m, n} = W_m \delta_{m, n}$$

can be constructed, so that the *quantum equations of motion* take the form

$$\frac{dq_k}{dt} = \frac{2\pi i}{h} (W q_k - q_k W)$$

on introducing the product of the matrices W and q_k.

4. Dirac's formulation.

We return to the year 1925. The reading of Heisenberg's original paper, which we have analysed in § 1, inspired Dirac [1] with the idea that the equations of classical analytical mechanics were in no way deficient, but that only the mathematical operations that were used in the attempt to deduce physical consequences from them required modifications. Thus inverting the problem, Dirac sought to what algorithm might correspond, in the classical field, a quantity like

$$xy - yx$$

in Heisenberg's sense.

We shall confine ourselves here to giving an outline of Dirac's paper, the reader can refer to the original for the details of the calculations.

In *quantum algebra*, Dirac adopted the same rules as Heisenberg, namely—

$$\{ x + y \}(nm) = x(nm) + y(nm)$$
$$xy(nm) = \sum_k x(nk)y(km)$$

which generally brings about the non-commutativity of the multiplication.

He granted that the *quantic derivation* satisfies the two rules—

(I) $$\frac{d}{dv}(x + y) = \frac{dx}{dv} + \frac{dy}{dv}$$

(II) $$\frac{d}{dv}(xy) = \frac{dx}{dv} \cdot y + x \cdot \frac{dy}{dv}$$

the order of x and y being safeguarded in rule (II).

The first rule (linearity) implies the possibility of the development

$$\frac{dx}{dv}(nm) = \sum_{n'm'} a(nm\,;\, n'm')x(n'm').$$

---

[1] *The Fundamental Equations of Quantum Mechanics*, Proc. Roy. Soc., A, 1925, Vol. 109, p. 642.

Dirac exhausts the algebraical consequences of the substitution of the above development in the t wo numbers of the equation (II) and arrives at the formula

$$\frac{dx}{dv} = xa - ax$$

in which the products are to be taken in the sense of Heisenberg.

Dirac then seeks the *classical analogy* of the quantum expression

$$xy - yx$$

by considering the case of the great quantum numbers, or more exactly, the case of quantum variables such as $x(n,\ n-a)$ in which n is very large compared with a, that is to say in the borderline case of the correspondence of Bohr. He arrives at the remarkable conclusion that the difference $xy - yx$ corresponds to the product by $\frac{ih}{2\pi}$ (where h is Planck's constant) of the Poisson brackets of the functions x and y as understood in classical mechanics, [1] and he thus writes

$$xy - yx = \frac{ih}{2\pi}\,[x, y].\ [2]$$

In the synthetic statement which he was to give later of quantum mechanics, Dirac omits the intermediaries ; he starts from the canonical equations of classical dynamics

$$\frac{dq_r}{dt} = \frac{\partial H}{\partial p_r} \qquad \frac{dp_r}{dt} = -\frac{\partial H}{\partial q_r}.$$

By means of the Poisson brackets he writes them in the form

$$\frac{dq_r}{dt} = [q_r, H] \qquad \frac{dp_r}{dt} = [p_r, H].$$

More generally, for any variable ξ which does not depend explicitly on the time, he obtains

$$\frac{d\xi}{dt} = \sum_r \left(\frac{\partial \xi}{\partial q_r} \cdot \frac{dq_r}{dt} + \frac{\partial \xi}{\partial p_r} \cdot \frac{dp_r}{dt} \right) = \sum_r \left(\frac{\partial \xi}{\partial q_r}\frac{\partial H}{\partial p_r} - \frac{\partial \xi}{\partial p_r}\frac{\partial H}{\partial q_r} \right) = [\xi, H].$$

The rules

$$[\xi_1 + \xi_2, \eta] = [\xi_1, \eta] + [\xi_2, \eta]$$

(a) $$[\xi_1 \xi_2, \eta] = \xi_1 [\xi_2, \eta] + [\xi_1, \eta]\,\xi_2$$

(b) $$[\xi, \eta_1 \eta_2] = [\xi, \eta_1]\,\eta_2 + \eta_1\,[\xi, \eta_2]$$

[1] See above, Book IV, p. 385.
[2] We have respected here the notation [ ] used by DIRAC, instead of POISSON's notation ( ).

are valid in the classical field and in this field the order in which the factors ξ_1, ξ_2 and η_1, η_2 appear is evidently irrelevant. But it can be agreed to write the formulae (a) and (b) *without ever inverting this order*. Thus, using either (a) + (b) or (b) + (a) to calculate the expression

$$[\xi_1\xi_2,\ \eta_1\eta_2]$$

and identifying the two results obtained, it is found that

$$[\xi_1,\ \eta_1]\,(\xi_2\eta_2 - \eta_2\xi_2) = (\xi_1\eta_1 - \eta_1\xi_1)\,[\xi_2,\ \eta_2].$$

Since the quantities ξ_1, ξ_2, η_1, η_2 are arbitrary ones it appears that in a general way, for any two variables, the ratio

$$\frac{\xi\eta - \eta\xi}{[\xi,\ \eta]}$$

has the properties of a universal constant. Dirac supposes this constant equal to $\dfrac{ih}{2\pi}$, so that

(1)
$$\xi\eta - \eta\xi = \frac{ih}{2\pi}\,[\xi,\ \eta].$$

Thus the symbol $[\xi,\ \eta]$ can be defined without the agency of a system of canonical variables.

In classical mechanics

$$[q_r,\ q_s] = 0 \qquad [p_r,\ p_s] = 0 \qquad [q_r,\ p_s] = \delta_{rs}.$$

Dirac assumes that these relations are still valid in quantum mechanics. With the definition of " quantum brackets " this leads to the quantum conditions

(2)
$$\begin{cases} q_rq_s - q_sq_r = 0 \\ p_rp_s - p_sp_r = 0 \\ q_rp_s - p_sq_r = \delta_{rs} \cdot \dfrac{ih}{2\pi}. \end{cases}$$

These conditions, with which it is appropriate to associate the *equations of motion* for each variable ξ which does not depend explicitly on the time, and which are written as

(3)
$$\frac{ih}{2\pi}\frac{d\xi}{dt} = \xi H - H\xi$$

thus take on a purely algebraic character.

The equations (2) and (3) were obtained by Dirac as early as 1925, in the original paper whose principal features we have analysed at the beginning of this paragraph. In this connection the author wrote, " The correspondence between the quantum and classical theories does not so must rest on their asymptotic agreement when $h \to 0$ as on the fact that the mathematical operations, in the two theories, usually obey the same rules. " Therefore Dirac's point of view was essentially formal.

We can only emphasise the remarkable achievement of such an abstract intuition and the part played in unifying an aptly chosen symbolism.

B. ANALYSIS AND INTERPRETATION

1a. ON THE ORIGIN OF QUANTUM MECHANICS IN HEISENBERG'S SENSE.

The evolution with which we are concerned was extraordinarily rapid. Heisenberg's first paper (summarised in § 1) is dated 1925. It is placed between Louis de Broglie's thesis (1924) and Schrödinger's first papers on wave mechanics (1926). Finally, Dirac's name appeared in the *Proceedings* of the Royal Society as early as 1925, before the paper of Heisenberg, Born and Jordan which has been referred to in § 3.

Louis de Broglie's wave mechanics evidently sprang from an independent body of thought. Heisenberg's quantum mechanics, wholly inspired by what the author himself called the " *Kopenhagener Geist der Quantentheorie* "—the teaching of Bohr—was at first completely independent of wave mechanics.

That it should be possible that these two streams of thought —concerned with corpuscles and with quanta—could be fused into a single doctrine was a fact that Schrödinger recognised in 1926. This fact, to which we shall have occasion to return, in no way prevented the wave and quantum disciplines from being, at their creation, completely distinct.

The reader will now appreciate why we have devoted a considerable time to Bohr's theory—which is now called " the old quantum theory. " Not only does this theory have an unquestionable historical value, but it was the source of the stream of quantum mechanics which Heisenberg inaugurated.

Bohr's correspondence principle was a connection of an asymptotic kind, for large quantum numbers, between his quantised theory and classical electromagnetic theory. Heisenberg's goal, and his achievement, was a true parallelism between the classical and quantum domains—a systematisation of Bohr's correspondence.

Moreover, as Louis de Broglie has remarked, Heisenberg adopted a " phenomenological " attitude in the philosophical sense of the term ; that is, he tried to dispense with elements that were not strictly *observable*, such as the position and the velocity or the trajectory of a Bohr electron, and to confine himself to directly accessible elements, such as frequencies and energy levels.

In this lies the profound originality of his thesis. It must be acknowledged, however, that the appearence of matrices representing the coordinates and momenta is, superficially, a departure from such a programme. And again, in this respect it would seem that as long as theoretical physics starts from experiment and returns to it, it should not be forbidden, in the interval, the use in the calculation of elements which are not themselves *observable*.

2a. COOPERATION OF PHYSICISTS AND MATHEMATICIANS.

Throughout the development of the modern physical theories of mechanics the fruitfulness of a cooperation of physicists with mathematicians is apparent. It is true that, in the formation of generalised relativity, Einstein had at his disposal, from the beginning, the absolute differential calculus and the geometry of Riemann spaces. But Schwartzschild profitably introduced the mathematical technique of celestial mechanics in the theory of generalised relativity, and also in Bohr's theory.

Heisenberg's work provides a still more striking example—at the beginning he performed matrix calculations without being aware of it. It is sufficient to compare the analyses of § 1 and § 3 of the present chapter to appreciate the extent to which cooperation may be valuable, and even necessary, in the formulation of the physicist's thought.

CHAPTER SIX

DEVELOPMENT OF THE PRINCIPLES
OF QUANTUM MECHANICS

A. *PRESENTATION*

1. MATHEMATICAL IDENTITY OF WAVE MECHANICS AND QUANTUM MECHANICS.

Wave mechanics, in the sense of Louis de Broglie and Schrödinger, and quantum mechanics, in the sense of Heisenberg, are mathematically equivalent. This essential fact was recognised by Schrödinger in a paper included in Volume 79 of the *Annalen der Physik* (1926).

Schrödinger observes that the rules of calculation of Heisenberg's quantum mechanics, applicable to functions of q_k and p_k, are identical with the rules of ordinary analysis applied to linear differential operators which only depend on the q_k.

To establish this correspondence, Schrödinger associated with every function F of the p_k and q_k an operator which he wrote as

$$[F,.].$$

This is obtained by replacing each p_k in F by the operator

$$K \frac{\partial}{\partial q_k}.$$

To avoid all ambiguity it is supposed that the function F is well ordered and that it can be developed as a power series in p_k.

Starting in this way from the function F, a matrix F^{ik} will be associated with it by considering a complete set of orthogonal and normalised functions in configuration space ; that is, a set

$$u_1(x) \quad u_2(x) \quad u_3(x) \ldots$$

such that

$$\int u_i(x) u_k(x) dx = \delta_{ik}$$

where x denotes the ensemble of the variables q_k.

This matrix will have the form

$$F^{ik} = \int u_i(x) \, [F, u_k(x)] \, dx.$$

Schrödinger verifies that the addition and multiplication of well ordered functions F or of their functions are expressed by the rules of the matrix calculus for the F^{ik}.

Moreover, the operation

$$\frac{\partial}{\partial q_r} q_r - q_r \frac{\partial}{\partial q_r}$$

applied to an arbitrary function merely reproduces this function. Therefore the operation

$$p_r q_r - q_r p_r$$

reduces to multiplication by a constant K and the corresponding matrix is

$$K\delta_{ik}.$$

To retrieve the *commutation rules* of quantum mechanics it is therefore sufficient to put

$$K = \frac{h}{2\pi i}.$$

Schrödinger next shows, by a method that we shall not describe, that the operators associated with the derivatives of every function F reduce to

$$\left[\frac{\partial F}{\partial q_r}, \cdot \right] = \frac{2\pi i}{h} \, [p_r F - F p_r, \cdot]$$

$$\left[\frac{\partial F}{\partial p_r}, \cdot \right] = \frac{2\pi i}{h} \, [F q_r - q_r F, \cdot].$$

These operations may be represented by matrix operations once the complete orthogonal and normalised set of functions $u_1, u_2, \ldots u_n,$ $\ldots$ is chosen.

Consider a specific mechanical system which is defined by its hamiltonian $H(p_k, q_k)$, supposed suitably " symmetrised " in such a way as to avoid all ambiguity in the order of the factors.   In quantum mechanics the equations of motion of this system are written in the matrix form

$$\left(\frac{dq_r}{dt} \right)^{ik} = \left(\frac{\partial H}{\partial p_r} \right)^{ik}$$

$$\left(\frac{dp_r}{dt} \right)^{ik} = - \left(\frac{\partial H}{\partial q_r} \right)^{ik}.$$

Here the differentiation with respect to the time has the significance given by

$$\left(\frac{dq_r}{dt}\right)^{ik} = 2\pi i\,(\nu_i - \nu_k)\,q_r^{ik}$$

$$\left(\frac{dp_r}{dt}\right)^{ik} = 2\pi i\,(\nu_i - \nu_k)\,p_r^{ik}.$$

To satisfy these equations by means of the operator calculus, Schrödinger *specifies the basic set* of functions $u_i(x)$.

For this purpose he chooses as the basic set $u_i(x)$ the eigenfunctions of the normal boundary value problem associated with the partial differential equation

(1) $$- [H, \psi] + W\psi = 0.$$

Here ψ is the unknown function and W, the parameter of the eigenvalues.

By the definition of the eigenfunctions and eigenvalues,

$$[H, u_r] = W_r u_r.$$

For the hamiltonian the matrix

$$H^{ik} = \tfrac{\kappa}{\zeta}\int u_i(x)\,[H, u_k(x)\,dx] = W_k\int u_i(x)\,u_k(x)\,dx = W_k \cdot \delta_{ik}$$

is deduced. Therefore H^{ik} is a diagonal matrix such that

$$H^{kk} = W_k.$$

By matrix multiplication

$$(Hq_r)^{ik} = \sum_m H^{im}q_r^{mk} = W_i q_r^{ik}$$

$$(q_r H)^{ik} = \sum_m q_r^{im} H^{mk} = W_k q_r^{ik}.$$

Therefore

$$\left(\frac{\partial H}{\partial p_r}\right)^{ik} = \frac{2\pi i}{h}\,[Hq_r - q_r H]^{ik} = \frac{2\pi i}{h}\,(W_i - W_k)\,q_r^{ik}.$$

To return to the first of the quantum equation of motion it is now sufficient to write

$$W_k = h\nu_{ik}$$

for all k.

The second of the quantum equations of motion would be retrieved in the same way.

As the partial differential equation (1) is nothing else than Schrödinger's wave equation, the identity of wave mechanics and quantum mechanics is established.

Briefly, Schrödinger writes—

" If, in a specific mechanical problem, the system of algebraic equations that relates the matrices of the coordinates and the momenta to the hamiltonian H is considered, then the solution of this system —that is, the equations of motion in the sense of quantum mechanics—is obtained by means of the choice of a *specific* orthogonal system ; namely, the system of eigenfunctions associated with the partial differential equation which is the basis of wave mechanics. The solution of the normal boundary value problem for this partial differential equation is absolutely equivalent to the solution of Heisenberg's algebraic problem. "

2. Probability interpretation of Schrödinger's wave function.

The *probability interpretation* of Schrödinger's wave function appeared in a paper of Max Born in 1926. This paper was devoted to the treatment of problems of impact in quantum mechanics.[1]

The author recalls that according to Heisenberg any accurate description of phenomena in space and time would appear to be impossible. According to Schrödinger, on the contrary, the waves of Louis de Broglie, which serve as a vehicle for atomic processes, would have some kind of physical meaning.

For his part, Born suggests a new interpretation which arises out of a remark of Einstein on the relation between photons and wave fields. Einstein said that waves only served to indicate the path to the particle and, in this connection, spoke of a virtual or " phantom " field *(Gespensterfeld)*. This field determines the probability that a photon, carrying energy and momentum, should take a certain path ; but it does not, itself, have energy or momentum. To Born the part played by the waves of quantum mechanics would, in an analogous way, be that of a pilot *(Führungsfeld)*.

This wave, represented by a scalar function ψ of the coordinates of the particles of the system and of the time, is propagated according to Schrödinger's partial differential equation. But this wave only determines the respective probabilities of the different possible trajectories for a single particle.

Born summarises this in the following statement, in which he himself appears to see some evidence of paradox. " *Die Bewegung der Partikeln folgt Wahrscheinlichkeitsgesetzen, die Wahrscheinlichkeit selbst aber breitet sich im Einklang mit dem Kausalgesetz aus* " or, in other

[1] *Quantenmechanik der Stossvorgänge, Zeitschr. für Physik*, Vol. 38, 1926, p. 803.

words, the motion of the particles is in accordance with the laws of probability, but the probability itself is propagated in accordance with the principle of causality. This is the sense in which Born uses the term *probability waves (Wahrscheinlichkeitswellen)*.

Without going in to the technical details of the paper from which we have taken the above interpretation, we shall give the principle of Born's analysis.

The eigenfunctions ψ_n, ψ_m, ... of the Schrödinger wave equation

$$(H - W)\,\psi = 0$$

are normalised by the relations

$$\int \psi_n(q)\,\psi_m(q)\,dq = \delta_{nm}.$$

It is possible to develop every wave function in a series of the eigenfunctions $\psi_n(q)$, and this will have the form

$$\psi(q) = \sum_n c_n\psi_n(q).$$

The integral relation

$$\int |\psi(q)|^2 dq = \sum_n |c_n|^2$$

leads to the consideration of the squares of the moduli of the coefficients c_n as the *respective weights* of each eigenstate ψ_n in the arbitrary state ψ. The evolution of the last state is governed by Schrödinger's equation.

3. HEISENBERG'S " UNCERTAINTY RELATIONSHIPS. "

We now come to a paper of Heisenberg which has played an extremely important part in the interpretation of quantum mechanics.[1]

Heisenberg believes that an intuitive understanding of a physical theory has been achieved when it is posssible, in all simple cases, to picture the consequences qualitatively and when it has been recognised that this theory contains no contradictions.

" The intuitive meaning of quantum theory is still full of internal contradictions ; notions borrowed from continuity and discontinuity —waves and corpuscles—are opposed to each other. It would seem that these contradictions cannot be avoided by means of the ordinary mechanical and kinematic concepts. "

It was for this reason that Heisenberg did his best to break away from these concepts, to consider only numerical relationships between measurable quantities. When this was accomplished, the mathe-

---

[1] *Über den anschaulichen Inhalt der quantentheoretischen Kinematik und Mechanik*, Zeitschr. für Physik, Vol. 43, 1927.

matical framework of quantum mechanics could be regarded as valid.

"But the necessity of a revision of the ordinary kinematical and dynamical concept appears as an immediate consequence of the fundamental equations of quantum mechanics. When we consider a specific mass m we immediately associate with it, in a classical way, a position and a velocity. But in quantum mechanics we have the relationship

$$pq - qp = \frac{h}{2\pi i}$$

between mass, position and velocity. Therefore there are good reasons to prevent us form using the terms 'position' and 'velocity' carelessly. . . . "

Heisenberg gives the following simple example. If, in place of a continuous trajectory possessing a tangent which defines the velocity, we have a discrete series of points defining the positions, it is meaningless to speak of velocity at a given place. For the velocity can only be defined in terms of two successive discrete positions ; or, conversely, to a given position there correspond two different velocities.

"Then arises the question of knowing whether, by a more thorough analysis of the kinematical and dynamical concepts, it is possible to resolve the contradictions indicated above and to arrive at an intuitive understanding of the relationships of quantum mechanics. . . .

"If it is sought to define clearly what must be understood by 'position of an object' (for example, an electron) it is necessary to picture an experiment which enables this quantity to be measured. . . .

"For example, the electron will be illuminated and observed with a microscope. The precision of this measurement is determined by the wave-length of the light used. If a γ-ray is used, it must be possible to obtain all the precision desired. But the picture is not clear— the Compton effect. Every observation of the light scattered by the electron involves a photoelectric effect. At the moment that its position is determined, or at the moment that the photon collides with the electron, the latter suffers a discontinuity of momentum. This discontinuity is greater as the wave-length of the light used is smaller ; that is, as the determination of the position is more accurate. Therefore the more precisely the position is known, the less accurately the momentum is known. The converse is also true. Here we see the direct interpretation of the relation

(1) $$pq - qp = \frac{h}{2\pi i}.$$

" If q_1 is the lack of the precision with which the position q is known and p_1 the lack of the precision with which the momentum p is known, the elementary theory of the Compton effect provides the relation

$$(2) \qquad p_1 q_1 \backsim h. \text{ "}$$

The determination of the position of the electron by a process of impact with particles—which are necessarily very fast—would lead to the same result.

We come to the concept of " trajectory " of an electron. " The commonly used expression [in the old quantum theory] of trajectory S_1 of the electron in the hydrogen atom is meaningless to us. To measure this trajectory it would be necessary to illuminate this trajectory with light of wave-length less than 10^{-8} cm. But a single quantum of light of this kind would suffice to throw the electron out of its trajectory. Therefore it is only possible to observe a single point of the trajectory.... However, this experiment can be repeated with a great number of atoms in the state S_1 since, thanks to the experiment of Stern and Gerlach, these atoms can be, in principle, isolated. Thus would be obtained a probability distribution of the position of the electron. According to Born this probability is $\psi \cdot \psi^*$ if ψ is the Schrödinger wave function in the state S_1. "

Thus quantum mechanics, unlike the classical theory, has a statistical appearance. In a determinate state of the atom the phases are essentially indeterminate. This is proved by the relation

$$Jw - wJ = \frac{h}{2\pi i}$$

in which J is an action variable and w an angular variable.

" When motions in the absence of forces are concerned, the word ' velocity ' can be defined by an experiment. For example, the object can be illuminated with red light and its velocity determined by the Doppler-Fizeau effect apparent in the light scattered by the object. The precision of the velocity will be greater as the wave-length of the light used is greater. And consequently the position of the object will be less precisely known, in agreement with the relation (2). "

Heisenberg also discusses the measurement of the energy. This is connected with the time by the relation

$$Et - tE = \frac{h}{2\pi i}.$$

The imprecision of the energy, say E_1, is correlated with the imprecision of the time by the relation

$$E_1 t_1 \backsim h.$$

Heisenberg summarises all these discussions in the following statement—

" All the concepts used in the classical theory to describe a mechanical
system may still be defined in similar fashion in the atomic field.   But
the experiments that use these definitions imply an uncertainty *(Un-
bestimmtheit)* when we wish to deduce from them the simultaneous values
of two quantities canonically conjugated.   The degree of this uncertainty (whatever the couple of conjugate quantities considered may be)
is given by the relation (2). "

Heisenberg thus interprets this result : all the experiments used to
define the words " position," " velocity," . . . in the quantum theory
necessarily include the uncertainty characterised by (2) even when they
allow an exact definition of the separate quantities $p, q, \dots$ If experiments existed which made possible a more accurate simultaneous determination of p and q than that which the relation (2) allows of, the
quantum theory would be invalidated.   The uncertainty established by
the relation (2) gives place to the verification of the relations of exchange (1), without the physical significance of the quantities p and q
being modified.   Basing himself on the Dirac-Jordan formulation,
Heisenberg demonstrates that there exists a logical connexion between
(1) and (2) [*loc. cit.*, p. 180].

And Heisenberg concludes :

" We have not admitted that the quantum theory, unlike the classical
theory, is a real statistical theory in the sense that, from exact data, it
can arrive only at statistical conclusions. . . .   Moreover, in all cases
where the quantities are simultaneously measurable, the classical
relations subsist wholly.   But in the ' strong ' formulation *(scharfe
Formulierung)* of the law of causality, ' The exact knowledge of the
present allows the future to be calculated ', it is not the conclusion but
the hypothesis that is false.   *We cannot* in principle know the present
in all its details.

That is why any experiment or any perception *(Wahrnehmen)* constitutes a choice from a number of possibilities and a limitation of what
remains possible in the future.   From the fact that the statistical character of quantum theory is intimately connected with the imprecision
of all perception, it is possible to ask whether there is still concealed,
behind the statistical universe of perception *(hinter der wahrgenommenen
statistischen Welt)*, a ' true ' universe *(wirkliche Welt)* in which the law
of causality would be valid.   But such speculation seems to us to be
valueless and meaningless, for physics must confine itself to the description of the relationships between perceptions.   Further, the true
problem is characterised in this way—since all experiments obey the

quantum laws and, consequently the relations (1), the incorrectness of the law of causality [in the sense explained above] is a permanently established consequence of quantum mechanics itself. "

4. EHRENFEST'S THEOREM.

Ehrenfest posed the question of how it might be possible, starting from quantum mechanics, to return to the newtonian law of motion.[1]

He replies to this question by starting from the Schrödinger's equation

$$\frac{h^2}{2m}\frac{\partial^2 \psi}{\partial x^2} + U(x)\psi = -ih\frac{\partial \psi}{\partial t}$$

for one degree of freedom, and the conjugate equation

$$\frac{h^2}{2m}\frac{\partial^2 \psi^*}{\partial x^2} + U(x)\psi^* = ih\frac{\partial \psi^*}{\partial t}.$$

If the mean value of the coordinate x is denoted by $Q(t)$, so that

$$Q(t) = \int_{-\infty}^{+\infty} x\psi^*\psi\, dx$$

and the derivative $\frac{dQ}{dt}$ is calculated by taking account of the wave equation and the boundary conditions, then the equation

$$\frac{\overline{dQ}}{dt} = \frac{ih}{m}\int_{-\infty}^{+\infty} \psi\frac{\partial \psi^*}{\partial x}\, dx$$

is obtained after an integration by parts.

In the same way, another derivation yields

$$m\frac{\overline{d^2Q}}{dt^2} = \int\left(\frac{\partial \bar{U}}{\partial x}\right)\psi^*\psi\, dx.$$

Ehrenfest expresses this result in the following way. " Whenever the dimension of the ' wave packet of probability ' $\psi^*\psi$ is sufficiently small compared with the macroscopic distances, the derivative of the momentum of the centre of gravity of the wave packet, in the sense of newtonian mechanics, is equal to the value of the force at the point at which this wave packet is localised. "

The discussion of the spreading-out of the wave packet of probability in the course of time is a difficult question which was studied by, among others, Kennard and Darwin.

[1] *Bemerkung über die angenäherte Gültigkeit der klassischen Mechanik innerhalb der Quantenmechanik*, Zeitschr. für Physik, Vol. 45, 1927, p. 455.

5. Dirac and the general theory of observation.

In this paragraph we shall follow the general theory of observation with which Dirac introduced his *Principles of Quantum Mechanics* (1929).

Dirac's fundamental concepts are those of *state, observation* and *observable*.

The definition of state, because of its extension, is a tricky matter. To say that a system is in a given state, after it has been suitable *prepared*, is to give all the data concerning its structure, its position in space-time and its internal motions. It is necessary to speak of space-time and not of space alone because a state is relative to the situation of a system during an unspecified interval of time— the evolution of a system which is *free of perturbation* is to be understood as one state. The classical analogue would be the motion of a system which is subject to given forces and starts from given initial conditions.

Therefore, if *perturbation is excluded*, the state continues to exist indefinitely. The notion of perturbation is itself *relative*, for the perturbing cause can be incorporated in the system. But it is assumed that the perturbation which consists of preparing a system in order to bring it into a given state has an *absolute* character or rather, an intrinsic significance, as has the perturbation necessarily caused, in general, by every observation made on a system in a given state.

Finally, it is assumed that it must be possible to consider any state as the result of the superposition of two or more different states. In this superposition both the " weight " and the " phases " of the component states appear.

We now come to the concept of *observation*. In general every measurement performed on a system, which is previously prepared in a suitable way, modifies the initial state of the system. The result of an observation is, in general, incompletely determinate and the repetition of an observation under *identical initial conditions* does not produce a unique result. It is only possible, by means of a large number of identical experiments, to infer the *probability of obtaining a given result*. This relative indeterminacy is related to the principle of superposition of states.

In general it is necessary to specify the interval of time which passes between the preparation of the system and the performance of the measurement (since one state is relative to space-time and does not exclude a determinate evolution which is able to modify the result of the measurement). Nevertheless, for certain states that are defined as *stationary*, this interval of time is irrelevant.

There is one case where the observation does not perturb the system; that is, for which there exists a *certainty*, not a probability, of obtaining a given result by means of this observation. Dirac assumes that this is also true of the *immediate repetition* (after the perturbation produced by a first measurement) of an observation which has given a first result— the second result is then identical with the first.

To appreciate the significance of the *postulate of repeatability*, it is necessary to bear in mind that the second observation is not situated on the same plane as the first— to produce the same observation under identical conditions it would be necessary to prepare the system anew so as to eliminate the perturbation caused by the measurement. Then, as a general rule, no more than a probability of obtaining a given result would remain.

Two observations are said to be *compatible* when the probability of obtaining a given result by means of the second observation is not modified by the perturbation due to the first (that is, it is equal to the probability of obtaining the given result *at the beginning* of the experiment, before the first result is known). This property is *reciprocal*.

The most important case is that of two or more compatible observations which are carried out simultaneously; n compatible observations can be considered as one and the same observation. If all of the greatest number of independent and compatible observations which is allowed by the system are performed simultaneously, the final state will be *defined* by this *maximal observation* independently of the initial state. By *repeating* this maximal observation immediately a complete and perfectly determined result is obtained with *certainty*, according to the postulate of repeatability.

The analogue of a classical dynamical variable is a quantum mechanical variable and it must be possible to consider it in an unspecified interval of time. The analogue of the *instantaneous value* of a classical dynamical variable is called an *observable*. Every observation, as in the classical theory, has the effect of providing number corresponding to the value of each observable.

Given these principles, Dirac develops what he calls " the *symbolic algebra of states and observables.* "

A state is represented by the symbol ψ accompanied by an index $(\psi_1, \psi_2, \dots \psi_n)$ characterising this state.

The postulate of superposition of states is written

$$\psi = c_1\psi_1 + c_2\psi_2 + \cdots$$

where the c_i are real or complex numbers.

The axioms of ordinary addition are valid for the symbolic addition of different states. But $\psi_1 + \psi_1$ must be regarded as equal to ψ_1 (the superposition of a state on itself reproduces the same state). More generally $c\psi_1$ must be regarded as identical with ψ_1 whatever the number c (different from zero) may be and whether or not it is real or complex.

With the symbols ψ are associated the symbols ψ^*, said to be the " imaginary conjugate " of the ψ, and given by

$$\psi^* = c_1{}^* \psi_1{}^* + c_2{}^* \psi_2{}^* + \ldots$$

($c_i{}^*$ is the complex conjugate of c_i).

A ψ and a ψ^* of the same index represent the *same state*. The addition of a ψ and a ψ^* has no meaning. Only a product like $\psi^*\psi$ has a meaning.

It is further supposed that $\psi_s{}^*\psi_r$ is the imaginary conjugate of $\psi_r{}^*\psi_s$ and that $\psi_r{}^*\psi_r$ is essentially positive. When $\psi_r{}^*\psi_r = 1$, it is said that the ψ are *normalised*.

The physical interpretation of the algebra of states is obtained in the following way. Apply the maximal observation of the state $\psi_r{}^*$ to the state ψ_s. It is no longer certain that the result of this maximal observation will be obtained on ψ_s. The *probability of agreement* of the two states, or the probability of obtaining, with ψ_s, the result of the maximum observation which defines $\psi_r{}^*$, is

$$\left| \psi_r{}^*\psi_s \right|^2$$

if $\psi_r{}^*$ and ψ_s are normalised.

An observable is represented by a symbol α. The product $\alpha\psi$ (or $\psi^*\alpha$) has the properties of a ψ (or a ψ^*). It therefore represents a state. Further

$$\alpha(\psi_1 + \psi_2) = \alpha\psi_1 + \alpha\psi_2$$
$$\alpha(c\psi) = c(\alpha\psi) \qquad (c, \text{number}).$$

The addition of two observables is defined by

$$(\alpha_1 + \alpha_2)\psi = \alpha_1\psi + \alpha_2\psi$$

and is commutative and associative.

Multiplication is defined by

$$(\alpha_1\alpha_2)\psi = \alpha_1(\alpha_2\psi)$$

with the property of being associative and distributive. In general this multiplication is not commutative.

The $\psi^*\alpha$ follow the same rules. Moreover

$$\psi^*(\alpha\psi) = (\psi^*\alpha)\psi = \psi^*\alpha\psi.$$

The quantities $\alpha_1 + \alpha_2$, $\alpha_1 \alpha_2$, and, in general, $\psi(\alpha)$, have the properties of observables.

The observable α can be defined by the knowledge of $\alpha\psi$ *for every* ψ. If there is available *a complete set of independent* ψ, such that every ψ may be a linear function of the ψ_r, the $\alpha\psi$, are sufficient to define α.

The connection between the algebra of states and that of observables is obtained by putting

$$\alpha\psi_r = a\psi_r.$$

This describes the fact that if the system is in a definite state ψ_r, the measurement of the observable α in that state *certainly* gives the (numerical) result a. This equation describes a *pure state*.

If α has the value a in the state ψ_r, it is assumed that $f(\alpha)$ has the value $f(a)$.

The number

$$\psi_r{}^*\alpha\psi_s$$

conjugates an observable with two distinct states.

The number

$$\psi_r{}^* \alpha\psi_r$$

only represents the mean *value* of α for a given state. In order to have the exact value of an observable it is in general necessary to conjugate it with two distinct states. It is assumed that $\psi_r{}^*\alpha\psi_r$ is on the average identical with the mean of the results obtained by making a large number of measurements of α in the state ψ_r where the system is of course *reprepared* on each occasion in order to get rid of the perturbation produced by the observation.

We now confine ourselves to *real observables*. The equation

$$\alpha\psi = a\psi$$

defines an *eigenstate* ψ of the observable α and an *eigenvalue* a of this observable.

The eigenvalues are the possible results of a measurement carried out on one observable. Because of the postulate of repeatability, the equation above is satisfied when, in place of ψ, the symbol of the state *immediately after* the observation is introduced, the observation having given the result a for this state.

Two states ψ_1 and ψ_2 are said to be *orthogonal* if $\psi_1{}^*\psi_2 = \psi_2{}^*\psi_1 = 0$ (zero probability of agreement). It is shown that the eigenstates corresponding to different eigenvalues of the same observable are orthogonal— a result which is required by physical considerations.

The symbols representing eigenstates belonging to different eigenvalues of the same observable are independent. It is assumed (postulate of series development) that every symbol ψ can be developed in a series of the eigensymbols of an arbitrary observable.

A state ψ can be a *simultaneous eigenstate* for two observables α and β ; that is

$$\alpha\psi = a\psi \quad \text{and} \quad \beta\psi = b\psi$$

whence

$$(\alpha\beta - \beta\alpha)\psi = (ab - ba)\psi = 0.$$

The existence of ψ when α and β themselves do not commute is exceptional. On the other hand, if $\alpha\beta - \beta\alpha = 0$, the simultaneous eigenstates of α and β may be used to develop an arbitrary ψ in a series. Conversely, if this happens, α and β commute.

With the help of these properties it is natural to consider an arbitrary function of commuting observables as a *unique observable*. This is expressed by

$$f(\alpha, \beta, \gamma, \ldots)\psi_{a, b, c} \ldots = f(a, b, c, \ldots)\psi_{a, b, c\ldots}$$

Then for each set of eigenvalues of the maximum number of independent and commutative observables there is only one simultaneous eigenstate. Each state is then determined by the set of eigenvalues of the complete set to which it belongs.

The physical significance of the symbolic algebra is also expressed by general theorems concerning probabilities.

To seek the probability of obtaining a given result by the measurement of an observable α in any state, the mean value

$$\psi^*f(\alpha)\psi \qquad (\psi^*\psi = 1)$$

of $f(\alpha)$ for that state is taken as the starting point.

By developing

$$\psi^* = \sum_a \psi_a^* \quad \text{and} \quad \psi = \sum_a \psi_a$$

in terms of the eigenstates of α corresponding to the eigenvalues a, the relation

$$\text{mean value of } f(\alpha) = \sum_a f(a)\psi_a^*\psi_a$$

is obtained.

According to the usual definition, if $P(a)$ is the probability of finding the result a, the mean value of $f(\alpha)$ will be

$$\sum_a P(a)f(a).$$

Therefore

$$P(a) = \psi_a^*\psi_a.$$

Clearly the relation

$$\sum_a P(a) = 1$$

may be verified.

If, on the other hand, ψ is an eigenstate corresponding to the value a, the probability that α should have the value a is equal to unity (pure state).

Dirac then turns to the representation of the preceding symbols in terms of ordinary numbers.

If ψ_p is the general member of a complete set of independent states, then every state ψ may be represented uniquely by

$$\psi = \sum_p a_p \psi_p.$$

The ψ_p are said to be the *fundamental states of the system*. Each state ψ is thus represented by a series of numbers a_p.

To represent an observable α, the symbol $\alpha\psi_q$ is developed as a series (ψ_q being one of the fundamental states) in the form

$$\alpha\psi_q = \sum_p \psi_p \alpha_{pq}$$

Thus each observable is represented by a matrix of two indices. The observable $\alpha + \beta$ is represented by $\alpha_{pq} + \alpha_{pq}$; the observable $c\alpha$ by $c\alpha_{pq}$ and finally, by definition,

$$(\alpha\beta)\,\psi_q = \sum_p \psi_p (\alpha\beta)_{pq}.$$

Now

$$(\alpha\beta)\,\psi_q = \alpha\,(\beta\psi_q) = \alpha \sum_r \psi_r \beta_{rq} = \sum_{pr} \psi_p \alpha_{pr} \beta_{rq}.$$

Thus Heisenberg's rule is retrieved.

An ordinary number c is represented by

$$c\psi_q = \sum_p \psi_p c_{pq}.$$

Whence $c_{pp} = c$ and $c_{pq} = 0$; say $c_{pq} = c\delta_{pq}$ (diagonal matrix).

The law of multiplication of the representations of an observable and a state is obtained by writing

$$\alpha\psi = \sum_q b_q \psi_q \qquad \text{(development of } \alpha\psi\text{)}$$

$$\alpha\psi = \alpha \sum_p a_p \psi_p \qquad \text{(development of } \psi\text{)}$$

$$= \sum_{pq} \psi_q \alpha_{qp} a_p.$$

Therefore

$$b_q = \sum_p \alpha_{qp} a_p.$$

Similarly the relation $\alpha\psi = a\psi$ is written as

$$\sum_q \alpha_{pq} a_q = a a_p.$$

Every relation between abstract elements of the symbolic algebra can thus be expressed by a relation between the representative numbers of these elements.

The ψ^* are treated in the same way by means of the relations

$$\psi^* = \sum_p a'_p \psi_p{}^* \quad \text{and} \quad \psi_p{}^*\alpha = \sum_q \alpha'_{pq}\psi_p{}^*.$$

If it is desired that the fundamental ψ and ψ^* (which are not in general imaginary conjugates) should give the same representation, it is necessary that $\alpha'_{pq} = \alpha_{qp}$. Then

$$\psi_p{}^*\alpha\psi_p = \sum_r \alpha_{pr}\psi_r{}^*\psi_q = \sum_r \psi_p{}^*\psi_r\alpha_{rq}.$$

In order that this should be true for all α it is necessary that

$$\psi_p{}^*\psi_q = 0 \quad \psi_p{}^*\psi_p = 1 \quad \text{say} \quad \psi_p{}^*\psi_q = \delta_{pq}.$$

A representation in which the ψ^* and the ψ provide the same representative numbers and are, if of the same index, imaginary conjugates, is said to be an *orthogonal representation*. If α is a *real observable*, then

$$(\psi_p{}^*\alpha\psi_r)^* = \psi_r{}^*\alpha^*\psi_p = (\alpha_{rp})^*$$

$$(\alpha_{rp})^* = \alpha_{pr}.$$

The matrices in an orthogonal representation are therefore *hermitian* (this is the case of Heisenberg's matrices).

The preceding analysis shows that the number of fundamental ψ is at most *infinitely denumerable*. But the total number of independent states may be non-denumerable. In such a case each fundamental ψ must be characterised by a continuous index p ; and for any state

$$\psi = \int a_p \psi_p dp \quad \left(\text{instead of } \sum_p a_p \psi_p\right).$$

To represent a fundamental state in these circumstances, Dirac makes use of the " δ function, " which he defines by

$$\begin{cases} \displaystyle\int_{-\infty}^{+\infty} \delta(x)\,dx = 1 \\ \delta(x) = 0 \text{ if } x \neq 0 \quad \text{or} \quad x\delta(x) = 0. \end{cases}$$

This function has the properties

$$\int_{-\infty}^{+\infty} f(x)\delta(x-a)\,dx = f(a)$$

$$\int_{-\infty}^{+\infty} \delta(a-x)\,dx\delta(x-b) = \delta(a-b)$$

$$\int_{-\infty}^{+\infty} f(x)\delta'(x-a)\,dx = -f'(a).$$

Thus the fundamental state ψ_q will be given by

$$\psi_q = \int_{-\infty}^{+\infty} \delta(p-q)\psi_p\,dp$$

and, more generally,

$$\alpha\psi_q = \int \psi_p\,dp\,\alpha_{pq} \qquad \text{(representation of an observable)}$$
$$(\alpha\beta)_{pq} = \int \alpha_{pr}\beta_{rq}\,dr \qquad \text{(product of matrices)}$$
$$c_{pq} = c\delta(p-q) \qquad \text{(representation of an ordinary number)}$$
$$\psi_q{}^*\psi_p = \delta(p-q) \qquad \text{(relation between fundamental states)}.$$

Dirac generalises these representations to include situations in which the continuous index p must be replaced by several indices $p_1, p_2, \ldots p_n$ which can take all possible values in a certain domain in a space of n dimensions.

He next considers representations where each fundamental ψ is a simultaneous eigenstate of a set of real commutative observables $\xi_1, \xi_2, \ldots \xi_n$. If this set of observables is supposed complete, the fundamental ψ are orthogonal. Each observable ξ_i is represented by a diagonal matrix (continuous if necessary) whose non-zero elements are the eigenvalues of this observable. Each fundamental ψ is itself determined apart from a phase factor of unit modulus.

If two representations are considered, one based on the eigenstates of ξ and the other on the eigenstates of the other variables η, it is natural to ask how it is possible to pass from one representation to the other for an arbitrary ψ.

This is accomplished by the consideration of the *transformation functions* $(\xi'\,|\,\eta')$ or $(\eta'\,|\,\xi')$.

The square of the modulus of these functions is a *probability amplitude* (Jordan) and the relation

$$|(\xi'\,|\,\eta')|^2 = |(\xi'\,|\,\eta')\,(\eta'\,|\,\xi')| = |(\eta'\,|\,\xi')|^2$$

expresses the fact that " the probability that the ξ should have the values ξ' when it is known that the η have the values η' is equal to the probability that the η should have the values η' when it is known that the ξ have the values ξ'. " This applies when the ξ and the η have discrete values. When the ξ or the η have continuous values, the interpretation requires more care. Thus $|(\xi'\,|\,\eta')|^2\,d\xi'$ is a measure of the probability that the ξ should be contained between ξ' and $\xi' + d\xi'$ when it is known that the η have the values η'. If the η themselves take continuous values, the preceding probability can only be relative.

6. WAVE FORMULATION OF THE GENERAL THEORY OF QUANTISATION (LOUIS DE BROGLIE).

In the last paragraph we have presented the essentials of Dirac's formulation without dealing with its applications ; this formalism can captivate or repel according to the reader's bent.

If the effort of abstraction which it demands is accepted, the logical vigour with which it unfolds cannot fail to be admired. If they are reflected upon, the premises on the subject of observation appear more natural—even though appreciably more complicated—than the classical ideas based on the independence between the observer and the system observed.

The mathematical tool itself has been specified by several authors (von Neumann, Weyl). In this connection it has been remarked that Dirac had worked with Hilbert spaces without knowing it, which may be considered as additional evidence of his originality.

It may however be useful, as Louis de Broglie has done, to approach this same general theory of quantisation by starting from the more intuitive point of view provided by wave mechanics, and by making use of the operational method that Schrödinger used, as we have seen, to establish the identity of wave mechanics and quantum mechanics.

Here we shall follow Louis de Broglie's *Théorie de la Quantification dans la nouvelle Mécanique*.[1] We consider complex functions $f(x, y, \ldots)$ in any number of variables $x, y, \ldots$ and suppose that they are defined in a certain domain D of these variables. The functions are supposed square summable in such a way that their norms, of the form

$$N(f) = \int_D f^*(x, y \ldots) f(x, y \ldots) d\tau$$

can be defined.

If $N(f) = 1$, f is said to be normalised (or standardised).

The scalar product of two functions f and g is written as

$$(f, g) = \int_D f^* g \, d\tau$$

and satisfies an inequality of the Schwarz type, of the form

$$|(f, g)|^2 \leqq N(f) N(g).$$

The functions f and g are said to be orthogonal if

$$(f, g) = 0.$$

[1] Paris (Hermann), 1932.

Every function f can be represented in the following way as a vector in Hilbert space.

Let $\varphi_1, \varphi_2, \ldots \varphi_n, \ldots$ be an infinity of orthogonal and normalised functions of the variables $x, y, \ldots$ defined in the domain D. Let ψ be a function of the same variables defined in the same domain. The quantities

$$c_i = (\varphi_i, \psi) = \int_D \varphi_i{}^* \psi \, d\tau$$

are called " the components of ψ in the system whose basis is formed by the φ_i."

If the system of the φ_i is complete, ψ can be developed as a series in the φ_i, to give

$$\psi = \sum_i c_i \varphi_i.$$

Whence

$$N(\psi) = \sum c_i^2$$

which supposes the convergence of the series on the right-hand side.

The matrices are introduced in this theory by means of *operators*. Let A be an operator (linear) which is expressed algebraically as a function of the coordinates and the derivatives with respect to these coordinates and let φ_i be a complete basic system of orthogonal and normalised functions. If the operator A is applied to one of the basic functions φ_i, the result, $A(\varphi_i)$, is a function which can be developed as a series in the φ_i, so that

$$A(\varphi_i) = \sum_k a_{ki} \varphi_k$$

with

$$a_{ki} = \int_D \varphi_k{}^* A(\varphi_i) \, d\tau.$$

The a_{ki} are the elements of a complex matrix corresponding to the operator A.

More generally, if ψ is a function of the Hilbert space, then $A(\psi)$ is another function of the Hilbert space. In the basic system φ_i the relations

$$\psi = \sum_i c_i \varphi_i \qquad A(\psi) = \sum_i c_i A(\varphi_i) = \sum_k \left(\sum_i a_{ki} c_i \right) \varphi_k$$

are written so that the components of $A(\psi)$ are

$$c_k' = \sum_i a_{ki} c_i.$$

The operation $A(\psi)$ has an intrinsic significance in the Hilbert space, but the matrix a_{ki} which represents this operation depends on the basic system chosen.

The operator A is said to be hermitian if the corresponding matrix is hermitian in all basic systems.   Wave mechanics is only concerned with hermitian operators.

The equation

$$A(\varphi) = a\varphi$$

in φ only in general admits finite, uniform and continuous solutions which vanish at the boundaries of D for certain real values of the real constant a. These solutions are said to be the eigenfunctions of the operator A. However, if D is infinite it may happen that there is a continuous sequence of eigenvalues a forming a " continuous spectrum. " The corresponding eigenfunction must strictly speaking be written as a " proper differential " in the interval Δa of the continuous spectrum (Weyl, Fues), of the form

$$\int_{\Delta a} \varphi(a)\, da.$$

Nevertheless it is possible to use the artifice of the Dirac δ-function for the study of continuous spectra.

Given these principles, wave mechanics can be reduced to the following two principles—

1) *Principle of quantisation.* — " To every mechanical quantity there corresponds an operator ; if the value of this mechanical quantity is accurately measured, this value can only be one of the eigenvalues of the corresponding operator. "

2) *Principle of spectral decomposition.* — " Let there be a mechanical quantity and a corresponding operator A.   If the eigenvalues and the eigenfunctions of this operator are denoted by a_i and φ_i and if the wave function (normalised) developed in terms of the φ_i is of the form

$$\psi = \sum c_i \varphi_i,$$

the probability that an observation should attribute a value a_i to the quantity considered is

$$c_i c_i^* = |c_i|^2.$$

" If the development of ψ is of the form

$$\psi = \int c(a)\,\varphi(a)\, da,$$

the probability that an observation should attribute a value contained in the interval a, $a + \Delta a$ to the quantity considered is

$$\int_{\Delta a} c(a)\, c^*(a)\, da. \text{ "}$$

The second statement completes the first.   The principle of quantisation indicates the possible values of a mechanical quantity—the principle of spectral decomposition gives the probabilities of the different possible values.

7. THE RELATIVISTIC AND QUANTISED ELECTRON IN DIRAC'S SENSE.

The relativistic treatment of quantum mechanical problems has proved to be very difficult. The only problem that will concern us here is that of an electromagnetic field. We shall follow Dirac's presentation.[1]

The author starts from a representation in which the cartesian coordinates x_t, y_t, z_t of the particle at the time t are diagonal. The representative of a state is then a function $(x_t, y_t, z_t|)$ of three variables depending on the time or, alternatively, is a function $(x, y, z, t|)$ of the four space-time variables. It is certainly possible to differentiate such a function so as to define operators like $\dfrac{\partial}{\partial x}$; but if this operation is applied to a wave function the result is a function which does not, in general, satisfy the wave equation. Consequently this operator *cannot be regarded as an observable*. Nevertheless Dirac introduces the operators

$$p_x = \frac{-ih}{2\pi}\frac{\partial}{\partial x} \quad p_y = \frac{-ih}{2\pi}\frac{\partial}{\partial y} \quad p_z = \frac{-ih}{2\pi}\frac{\partial}{\partial z} \quad W = \frac{ih}{2\pi}\frac{\partial}{\partial t}$$

and applies the algebraic rules of the calculus of observables to them. But quantities like $\psi^* p_x \psi$ and $\psi^* W \psi$ are no longer numbers.

In the absence of field the ordinary relativistic hamiltonian is written

$$H = W = c(m_0 c^2 + p_x^2 + p_y^2 + p_z^2)^{\frac{1}{2}}.$$

If the p and W are considered as operators, the preceding equation implies the wave equation

(a) $$\left(\frac{W}{c} - (m_0 c^2 + p_x^2 + p_y^2 + p_z^2)^{\frac{1}{2}}\right)\psi = 0.$$

On the other hand Schrödinger's equation, obtained by means of the generalised Jacobi equation, is written

(b) $$\left(\frac{W^2}{c^2} - m_0 c^2 - p_x^2 - p_y^2 - p_z^2\right)\psi = 0.$$

The last equation, quadratic in W, is invariant from the relativistic point of view. But it is not satisfactory from the quantum point of view, which requires linearity with respect to the operator $\dfrac{\partial}{\partial t}$. The equation (b) allows of all the solutions of (a), But only those solutions of (b) which correspond to positive values of the energy satisfy (a).

[1] *Principles of Quantum Mechanics*, Chapter XIII. For the original, see *Proc. Roy. Soc.*, Vol. 117, 1928, p. 610, and Vol. 118, 1928, p. 351.

Dirac looks for an equation similar to (a) and (b) which may be linear in the operator $\dfrac{\partial}{\partial t}$ like the first and invariant under a Lorentz transformation like the second. With this in view he writes the equation

(c) $$\left\{ \frac{W}{c} + \alpha_x p_x + \alpha_y p_y + \alpha_z p_z + \beta \right\} \psi = 0.$$

This equation is linear in W and the operators p and contains four coefficients $\alpha_x, \alpha_y, \alpha_z$ and β which commute with W and with the operators p and which obey the rules of the calculus of observables.

By multiplying *on the left* by

$$\frac{W}{c} - \alpha_x p_x - \alpha_y p_y - \alpha_z p_z - \beta$$

and taking account of the commutability conditions which have been supposed, it is found that

(d) $$\left\{ \frac{W^2}{c^2} - \sum_{xyz} (\alpha_x^2 p_x^2 + (\alpha_x \alpha_y + \alpha_y \alpha_x) p_x p_y + (\alpha_x \beta + \beta \alpha_x) p_x) - \beta^2 \right\} \psi = 0.$$

Putting $\beta = \alpha_0 m_0 c$, a sufficient condition that (d) should be identical with (b) is that

$$\alpha_i \alpha_j + \alpha_j \alpha_i = 2\delta_{ji} \quad \text{with } i, j = x, y, z \text{ or } 0.$$

Like equation (b), the equation (c) allows of negative energy solutions although W, which is here purely kinetic, is necessarily positive. We shall neglect them for the moment. Putting

$$\alpha_x = \varrho_1 \sigma_x \quad \alpha_y = \varrho_1 \sigma_y \quad \alpha_z = \varrho_1 \sigma_z \quad \alpha_0 = \varrho_2,$$

the equation becomes

(e) $$\left\{ \frac{W}{c} + \varrho_1 (\vec{\sigma}, \vec{p}) + \varrho_2 m_0 c \right\} \psi = 0,$$

where $\vec{\sigma}$ is the *spin* of the electron and $(\vec{\sigma}, \vec{p})$ a scalar product.

In the presence of an electromagnetic field, Dirac assumes that the wave equation can be obtained, in accordance with the accepted rule, by

replacing $\dfrac{W}{c}$ and $\vec{p}$ by $\dfrac{W}{c} + \dfrac{e}{c} A_0$ and $\vec{p} + \dfrac{e}{c} \vec{A}$ respectively. The quantities A_0 and $\vec{A}$ are the scalar and vector potentials of the field.

Thus in the presence of a field Dirac writes

(e$_1$) $$\left\{ \frac{W}{c} + \frac{e}{c} A_0 + \varrho_1 \left(\vec{\sigma}, \vec{p} + \frac{e}{c} \vec{A} \right) + \varrho_2 m_0 c \right\} \psi = 0.$$

The preceding equation is invariant under a Lorentz transformation. In the presence of a field the Schrödinger equation (b) becomes

(b$_1$) $$\left\{\left(\frac{W}{c}+\frac{e}{c}A_0\right)^2-m_0c^2-\left(p+\frac{e}{c}A\right)^2\right\}\psi=0.$$

It is natural to try to obtain equation (b$_1$) by multiplying (e$_1$) on the left by

$$\frac{W}{c}+\frac{e}{c}A_0-\varrho_1\left(\vec{\sigma},\vec{p}+\frac{e}{c}\vec{A}\right)-\varrho_2m_0c.$$

By means of the algebra of observables and Maxwell's equations, Dirac obtains the equation

(f) $$\left\{\left(\frac{W}{c}+\frac{e}{c}A_0\right)^2-\left(\vec{p}+\frac{e}{c}\vec{A}\right)^2-\left|m_0c^2-\frac{he}{2\pi c}(\vec{\sigma},\vec{H})-\frac{i\varrho_1he}{2\pi mc}(\vec{\sigma},\vec{E})\right\}\psi=0$$

where $\vec{H}$ and $\vec{E}$ denote the magnetic and electric fields.

The equations (f) and (b$_1$), unlike the corresponding equations for a free particle, are no longer equivalent. In the first approximation this result can be interpreted by saying that the electron has a magnetic moment $\dfrac{-he\sigma}{4\pi m_0c}$ and an electric moment (purely imaginary) $\dfrac{-ih\varrho_1e\sigma}{4\pi m_0c}$.

Returning to the *negative energy solutions*, Dirac observes that in ordinary relativity, without quantisation, the continuity of the values of the dynamical variables implies that if $W+eA_0$ is initially positive (then it is equal to m_0c^2 at least) it can never later become negative without a sudden change of at least $2m_0c^2$. In quantum mechanics on the contrary nothing excludes such a transition, so that the interpretation of the negative solutions is necessary.

In a suitably chosen representation of the α (α_x, α_y, α_z real and α_0 purely imaginary) every wave function which is the complex conjugate of a solution of the equation

(e$_1$) $$\left\{\frac{W}{c}+\frac{e}{c}A_0+\sum_{xyz}\alpha_x\left(p_x+\frac{e}{c}A_x\right)+\alpha_0m_0c\right\}\psi=0$$

satisfies the equation

(g) $$\left\{-\frac{W}{c}+\frac{e}{c}A_0+\sum_{xyz}\alpha_x\left(-p_x+\frac{e}{c}A_x\right)-\alpha_0m_0c\right\}\psi^*=0.$$

If a solution of (e$_1$) corresponds to a *negative* value of $\dfrac{W}{c}+\dfrac{e}{c}A_0$, the complex conjugate solution of (g) corresponds to a *positive* value of $\dfrac{W}{c}-\dfrac{e}{c}A_0$. But (g) is deduced from (e) by simply replacing e by $-e$.

Therefore the complex conjugate of a negative energy solution of (e) itself must be interpreted as representing the motion of a particle of charge $+ e$ in the same electromagnetic field.

" *Therefore the undesirable solutions of* (e$_1$) *are connected with the motion of an electron of charge* $+ e$. "

Thus Dirac was led to conclude that " the negative energy solutions of the wave equation (e$_1$) are related to the motion of protons or hydrogen nuclei, *although there still remains the difficulty of the large mass difference.*" He made the hypothesis that almost all the negative energy states were occupied, an electron in each state, in accordance with the *Pauli exclusion principle.*

Only a vacancy would be detectable in this homogeneous distribution $(- e, - W)$ and would appear with the properties $(+ e, + W)$. In the contemporary state of experimental knowledge, Dirac thought that this vacancy would be a proton. A perfect vacuum would then be a region where all positive energy states were *unoccupied* and all the negative energy states, occupied.

Dirac added, " the exclusion principle will come into play by preventing a positive energy electron from jumping into a negative energy state [occupied]. It will still be possible, however, for such an electron to fall into an *unoccupied* negative energy state. In this case an electron and a proton would disappear simultaneously, their energy being emitted in the form of radiation. Very probably such processes actually occur in nature. "

8. The Dirac electron and experiment.

We have seen that Dirac's theory introduces a magnetic moment associated with the electron. Here, in a new way, is confirmation of earlier hypotheses whose origins we must recall.

Experiments had revealed that the fine structure of optical and X-ray spectra was much more complicated that that calculated by Sommerfeld by means of the relativistic treatment of Bohr's theory. In order to account for the multiplicity of the lines observed it was necessary to make use of an " internal " quantum number of an empirical kind.

Bohr's theory also failed to account for the anomalous Zeeman effects and the gyromagnetic anomalies. This theory, considering the electron as a point charge, attributes to it a proper angular momentum $\dfrac{h}{2\pi}$ due to its orbital rotation. To this corresponds, apart from sign, a magnetic moment $\dfrac{he}{4\pi m_0 c}$ (called the Bohr magneton).

In 1925 Uhlenbeck and Goudsmit suggested attributing to the electron, likened to a small charged sphere, a rotational velocity of its own and consequently an intrinsic angular momentum *(spin)* having the value $\pm \dfrac{h}{4\pi}$. To this corresponded a magnetic moment whose absolute value was that of the Bohr magneton.

By means of this hypothesis Thomas and Frenkel succeeded in resolving, in terms of the old quantum theory, many of the difficulties thrown up by the experiments on fine structure and the anomalous Zeeman effects.

The application of Dirac's analysis also allowed the theory of fine structure to be put in order. But it is in the interpretation of negative energy states that this analysis most strikingly anticipated experiment.

As early as 1932 Anderson demonstrated, by means of a Wilson cloud chamber, the existence of high energy positively charged particles in cosmic rays. Blackett and Occhialini, using Geiger counters to control the expansion in the cloud chamber, succeeded in obtaining several photographs of the trajectories of charged particles in cosmic radiation.

In 1933 Anderson, comparing the density of ionised drops along the trajectories of positive and negative particles of the cosmic radiation, came to the conclusion that the charge of the positive particles was equal, to within 10%, to $+e$ and that, within 20%, their proper mass was that of the ordinary electron. At the same time (February, 1933) Blackett and Anderson decided on the existence of the positive electron, which is often called the " positron " or " positon. "

Shortly afterwards F. Joliot and I. Curie achieved the *creation of positive electrons*, which appeared in pairs with ordinary electrons during the bombardment of heavy atoms (lead for example) with very penetrating γ-rays of energy greater than $2m_0c^2$. The process is the following one—

$$\underbrace{\gamma\text{-ray on heavy atom}} \quad = \quad \underbrace{\text{electron} + \text{positron}} \quad + \quad \underbrace{2m_0c^2}$$

2.6×10^6 e. v.	1.6×10^6 e. v.	1×10^6 e. v.
(electron volts)	(kinetic energy)	(rest energy of electron pair).

The kinetic energy is unequally shared between the electron and the positron, the latter taking the greater part because of the repulsion of the nucleus of the heavy atom, which naturally increases with the atomic number. On the other hand, the energy of the incident radiation cannot be less than $2m_0c^2$, which is predicted by Dirac's theory.

If, after the positive electron has been created, it encounters matter, which contains a total of 3×10^{23} negative electrons per gram whatever the substance may be, there is every chance that it will *dematerialise* by encountering a negative electron. For this reason the mean life of the positive electron is 10^{-9} secs in water and 10^{-6} secs in air. Therefore the positive electron is only stable in a vacuum which explains why it was first discovered in cosmic rays.

If it does not have an appreciable kinetic energy the pair of electrons represents an energy of $2m_0c^2$. It creates two photons having equal and opposite momenta and energies of m_0c^2 according to the process of *dematerialisation*

$$\epsilon^+ + \epsilon^- = \text{rayons } \gamma.$$

F. Joliot and I. Curie also observed *positive electrons of transmutation*, which were not associated with negative electrons, in the interaction of α-particles (helium nuclei) from polonium and light atoms. Neutrons and protons were also produced. Here the process is

$$\underbrace{{}^{27}_{13}\text{Al}}_{\substack{\text{aluminium} \\ \text{nucleus}}} + \underbrace{{}^{4}_{2}\text{He}}_{\alpha\text{-particle}} = \underbrace{{}^{30}_{14}\text{Si}}_{\substack{\text{silicon} \\ \text{atom}}} + \underbrace{{}^{1}_{1}\text{H}}_{\text{proton}} = \underbrace{{}^{30}_{14}\text{Si}}_{} + \underbrace{{}^{1}_{0}\text{n}}_{\text{neutron}} + \underbrace{\epsilon^+}_{\text{positron}}.$$

The indices which appear above the atomic symbols are the mass numbers M and the indices below are the atomic numbers Z.

B. *ANALYSIS AND INTERPRETATION*

1a. ON THE IDENTITY OF WAVE MECHANICS AND QUANTUM MECHANICS.

It is of some interest to reproduce here Schrödinger's commentary on the demonstration that we have made the subject of § 1.

" If one reflects on the extraordinary diversity of the starting points and of the conceptions that characterise Heisenberg's theory on the one hand and the theory which we have called ' wave ' or ' physical ' mechanics on the other, one must confess that it is very strange to record that these two new quantum theories lead, at least for the particular cases so far known, to the same results, even when these results differ from those that the old quantum theory gives. In writing this, I think especially of the intervention of half-integer quantum numbers in the problems of the oscillator and the rotator. This is, indeed, very remarkable because everything—starting point, conception, method, mathematical apparatus used—appears radically different.

" It seems that these two theories depart from classical mechanics in diametrically opposed directions. In Heisenberg's mechanics the continuous classical variables are replaced by systems of discrete numbers depending on two integral indices (matrices) and which are determined by algebraic equation. The authors themselves call their thesis ' the true theory of the discontinuous.'

" On the contrary, wave mechanics shows progress with respect to classical mechanics in exactly the opposite direction— that is, towards a theory of the continuous.

" This theory replaces the mechanical phenomenon—which is described classically by a finite number of functions satisfying a finite number of ordinary differential equations—by a continuous field of phenomena in configuration space, governed by one partial differential equation which arises from a variation principle. This variation principle, or this partial differential equation, replace the ensemble of equations of motion and quantum conditions of the old ' theory of classical quanta.'

" ... I have indicated the intimate, extremely close, connections which exist between Heisenberg's quantum mechanics and my own. Formally, from a purely mathematical point of view, these two theories are identical."

3a. Reciprocal limitations of the validity of the corpuscular and wave-like representation (Heisenberg).

The fact of taking the wave-like representation to be entirely correct provides us with the limitations of the corpuscular representation.

The particle is then represented by a packet of plane waves of approximately equal wave-lengths. In order to construct this packet of plane waves use is made of the property according to which the plane waves may destroy each other by interference *outside* a field Δq as small as is desired if there exist any having $n + 1$ wave length in this field, n being appreciably equal to $\dfrac{\Delta q}{\lambda_m}$, in which λ_m is the average wave length of the packet.

The oscillation $\Delta \lambda$ of the wave length in Δq corresponds therefore to

$$\frac{\Delta q}{\lambda_m - \Delta \lambda} \geq n + 1.$$

We have then approximately—

$$\frac{\Delta q}{\lambda_m \left(1 - \frac{\Delta\lambda}{\lambda_m}\right)} \sim \frac{\Delta q}{\lambda_m}\left(1 + \frac{\Delta\lambda}{\lambda_m}\right) \geq \frac{\Delta q}{\lambda_m} + 1$$

or

$$\frac{\Delta q\,\Delta\lambda}{\lambda_m{}^2} \geq 1.$$

Now, from L. de Broglie's relation

$$p = mv = \frac{h}{\lambda_m}.$$

Therefore

$$\Delta p = \frac{h\Delta\lambda}{\lambda_m{}^2} \quad \text{and} \quad \Delta p\,\Delta q \geq h$$

Q.E.D.

Inversely, the fact of taking as correct the corpuscular representation provides us with the limits of undulatory representation.

Let us see, with Heisenberg,[1] what knowledge we can have, in these conditions of the amplitude of a wave, of the intensity of an electric or magnetic field, for example.

We consider in space a small element of volume $\delta v = (\delta l)^3$. In δv we have for the energy W and the quantity of motion $\vec{P}$ of the electromagnetic field the values—

$$W = \delta v\,\frac{1}{8\pi}\,(E^2 + H^2)$$

$$\vec{P} = \delta v\,\frac{1}{4\pi c}\,\vec{E} \wedge \vec{H}.$$

Since *a priori*, in the hypothesis we have taken, there is no lowest limit for δl, and in consequence for δv, it would seem that W and P may be made as small as possible.

But we know, from the corpuscular theory, that the energy and the quantity of motion are necessarily sums of discrete, finite quantities having the respective values, $h\nu$ and $\dfrac{h\nu}{c}$. On the other hand any wave length smaller than δl will escape observation ; therefore the frequence observable will have as a limit—

$$h\nu \leq \frac{hc}{\delta l} \quad \text{by} \quad \lambda \geq \delta l \quad \text{and} \quad \lambda = \frac{c}{\nu}.$$

[1] *Physikalische Prinzipien der Quantentheorie.*

From this it follows that the uncertainties of the values of the components of the electric and magnetic fields, namely ΔE_x, ΔE_y, ΔE_z and ΔH_x, ΔH_y, ΔH_z are given by

$$\begin{cases} \Delta E_x \, \Delta H_y \geqq \dfrac{hc}{\delta l \cdot \delta v} = \dfrac{hc}{(\delta l)^4} \\[2ex] \Delta E_y \, \Delta H_z \geqq \dfrac{hc}{(\delta l)^4} \\[2ex] \Delta E_z \, \Delta H_x \geqq \dfrac{hc}{(\delta l)^4}. \end{cases}$$

7a. ON THE DIRAC ELECTRON— RELATIVITY AND QUANTA.

In a paper called *Quelques remarques sur la théorie de l'électron magnétique de Dirac*,[1] Louis de Broglie illuminated the difficulties which are presented, from the relativistic point of view, by Dirac's mechanics.

In the first place the *mean value density* for a quantity A is given by

$$\sum_k \psi_k{}^* A \psi_k \qquad\qquad (k = 1, 2, 3, 4).$$

This same density cannot have a great deal of physical significance. In fact, it is merely the quantity which must be integrated over space to obtain the *mean values* in the strict sense, namely

(1) $$\overline{A} = \int\int\int_D \sum_k \psi_k{}^* A \psi_k dx dy dz.$$

Now in Dirac's theory only the mean value density has the tensorial properties of the quanties of classical relativity. " Thus we are in agreement with the theory of relativity, but only for quantities which have the significance of mean values and which, from the purely quantum point of view, do not seem to have physical meaning. . . . The point of view of pure quantum mechanics, which only has regard for eigenvalues, completely ignores the classical geometrical representation by means of vectors or tensors in space or space-time. Only when [the densities of mean values] are considered are the transformations of the tensorial type retrieved. Here we see the profound opposition that exists between quantum physics and classical physics and, at the same time, we see the possibilities of their statistical agreement. "

[1] *Archives des Sciences physiques et naturelles*, Geneva, Vol. 15, 1933, p. 465.

There is more than this. In the Dirac theory, just as much as in ·
simple wave mechanics, the time variable plays a part which is quite
different from that of the space variables. The determination of eigenvalues is set in the domain of space ; the mean values $\overline{A}$ themselves
are obtained by an integration in space. " Clearly such definitions
are not relativistic. It would be necessary to make use of space-time
domains for the definition of eigenvalues, and to make use of integrations
in space-time for the definition of mean values. " Mathematically
this is possible. Thus the mean value of A would be

$$\overline{A} = \int \int \int \int \psi^* A \psi dx dy dz dt.$$

But in this way there would obtained " an entirely static physics
in which all evolution in time would be forbidden. The quantum
theory needs an evolutionary parameter playing a part quite different
from that of the configuration variables to which the operators correspond. At present quantum theory takes the time to be the evolutionary
parameter and thus breaks up the relativistic symmetry between
space and time. "

Louis de Broglie adds, it is true, that even in the theory of relativity
in its classical form, the variables of space and time are far from being
equivalent. The time variable always varies in the same sense and
the world lines of all material units are inclined towards the direction
of the time. In other words, space-time has an essential " polarity. "

In short, if relativity and quanta are one day reconciled, it does
not seem that this will be accomplished in terms of an indissoluble
space-time in Minkowski's sense.

Elsewhere [1] Louis de Broglie calls our attention to another aspect
of the asymmetry between space and time in wave mechanics.

The symmetry between space and time requires that the uncertainty
relations

$$\Delta p_x \Delta q_x \geqq h \quad \text{etc...}$$

be supplemented by

$$\Delta W \cdot \Delta t \geqq h$$

for W is the time-component of the four vector " world momentum "
whose space components are p_x, p_y, p_z. But the time must be considered
as a parameter having a value which is specified without uncertainty.
Moreover, W corresponds to the hamiltonian operator and not to
$-\dfrac{ih}{2\pi} \dfrac{\partial}{\partial t}.$

[1] *L'électron magnétique de Dirac*, Paris (Hermann), 1934, p. 303.

To give a meaning to this fourth uncertainty relation, it is remarked that if the passage of a wave over a fixed point of space is considered during a time Δt, it is only possible to say that the wave has a frequency ν with the uncertainty

$$\nu \geqq \frac{1}{\Delta t}.$$

Therefore for a wave ψ, recalling that $W = h\nu$,

$$\Delta W \geqq \frac{h}{\Delta t}.$$

The relation $\Delta W \cdot \Delta t \geqq h$ therefore expresses the fact that an experiment or an observation made at a fixed point during an interval of time Δt cannot disclose the energy of a corpuscle with an uncertainty less than $\frac{h}{\Delta t}$. This fourth uncertainty relationship therefore has a meaning which is quite different from that of the first three.

CHAPTER SEVEN

DISCUSSION OF THE PRINCIPLES
OF QUANTUM MECHANICS

1. COMPLEMENTARITY IN THE SENSE OF BOHR.

For Bohr,[1] what characterizes the quantum theory is that it brings an *essential limitation* to the concepts of atomic phenomena.

Despite this limitation, we cannot dispense with these concepts, " these forms of intuition which in the last analysis constitute the framework of all our experiment and colours all our language. " [2]

The *quantum postulate* imparts to any atomic process a character of discontinuity, or rather individuality, completely foreign to the classical theories and which is characterized by Planck's quantum of action.

This postulate, Bohr declares, *obliges us to abandon a description at once causal and spatio-temporal of atomic phenomena.*[3]

Indeed, any observation of atomic phenomena implies a *finite inter-action* with the instrument of observation ; consequently it is impossible to attribute to the phenomena, or to the instrument of observation, an autonomous physical reality in the ordinary sense of the word.

In order to define the state of a physical system in the classical manner, abstraction must be made of all external actions. But then, in the atomic field, all possibility of observation is at once excluded. This means that the concepts of time and space lose their immediate meaning.

If, on the contrary, to make observation possible, we accept the possibility of interactions with the instruments of measure not belonging to a system, it is the univocal definition of the system that becomes impossible, " and there can no longer be any question of causality in the ordinary sense of the word. " [4]

---

[1] *La théorie atomique et la description des phénomènes*, french ed. Gauthier-Villars, 1932.
[2] *Ibid.*, p. 5.
[3] *Ibid.*, p. 50.
[4] *Ibid.*, p. 51.

" We must then, " Bohr writes, " contemplate a radical modification of the relation between the *spatio-temporal description* and the *principle of causality* . . . the union of which characterized the classical theories : by the very essence of the quantum theory we must indeed conceive them as *complementary but mutually exclusive aspects of our representation of experimental facts. . . .* In order to take the quantum postulate into account in the description of atomic phenomena, we must develop a *theory of complementarity*, whose non-contradiction can only be judged by confronting the possibilities of definition and the possibilities of observation. " [1]

As early as in the fundamental equations which constitute the common origin of Einstein's theory of photons and the wave theory of matter as understood by L. de Broglie—

$$(1) \qquad \frac{W}{\nu} = p\lambda = h$$

the two conceptions of light and matter are opposed. The couple (W, p) relates to a corpuscular image and the couple (ν, λ) to an undulatory image, or more exactly to a train of harmonic plane waves illimited in space and time.

By the interference of a group of elementary harmonic waves, a field of waves limited in time and in space can be obtained. The velocity of translation of the " individuals " associated with the waves is given by the velocity of the group of these latter. " The use of groups of waves implies an uncertainty in the definition of the period and length of the wave, and consequently in that of the energy and the impulsion which correspond to these relations (1). " [2]

Heisenberg's relations

$$(2) \qquad \Delta t \, \Delta W = \Delta x \, \Delta p_x = \Delta y \, \Delta p_y = \Delta z \, \Delta p_z = h$$

express the *minimum uncertainties* which affect the definition of the energy and the momentum of the individuals associated with the wave field.

" These relations can be considered as a very simple symbolic expression of the *complementary nature* of the spatio-temporal description and of the principle of causality. Their general form enables the use of the theorem of energy-momentum to be combined, to a certain extent, with the spatio-temporal representation of observation. Instead of the coincidence of two well defined events at a point of space-time, that of individuals defined with a certain limited accuracy in finite spatio-temporal domains may be considered. " [3]

[1] *La théorie atomique et la description des phénomènes*, p. 51.
[2] *Ibid.*, p. 56.
[3] *Ibid.*, p. 57.

Bohr's thesis can be summarised in the following way— As long as we retain the classical concepts (wave, particle) we shall find ourselves, just as much in the domain of light as in that of matter (since the experiment of Davisson and Germer), faced with a " dilemma which must nevertheless be considered as the accurate expression of the experimental data.   In reality we are not concerned with contradictions but, rather, with *complementary conceptions*, of which only the ensemble can form a natural generalisation of the classical method of description."[1]

The quantised universe is, in a sense, *too rich* to be amenable to a single method of intuitive description.   While rational mechanics only needs the corpuscular representation and physical optics, in a great many cases, is satisfied with the wave, quantum mechanics is obliged to have recourse to both of these two representations, neither of which rigourously suffices, and expresses their reciprocal limitation.

We must make clear that, to Bohr's mind, radiation in a vacuum, just as much as free material particles, are only abstractions or, to use his own expression, " *idealisations* " since, from the quantum postulate, their properties can only be defined or observed by their interaction with other systems.   However, " these abstractions are indispensable in order to reduce the expression of the experimental results to our ordinary forms of intuition." [2]

2. CLASSICAL " LEGALITY " AND THE " SEMI-LEGALITY " OF QUANTUM MECHANICS.

I beg the reader's pardon for raising a question of terminology here. We have seen [3] the meaning Painlevé attached to the " *principle of causality* " in classical mechanics.   We have also seen what might be understood by the " domain of causality " of ordinary mechanics or the mechanics of special relativity.

On the other hand, L. de Broglie gives us the following advice— " The physicist has not to contemplate determinism in its general and metaphysical aspect : he has to seek a precise definition of it within the framework of the facts he is studying." [4]

I shall not, then, speak here of " causality, " or of " determinism. " Adopting Meyerson's terminology, I shall use the word " legality, " but in a sense that I shall make clear.

To enounce certain previsions starting from an initial observation is

---

[1] *La théorie atomique et la description des phénomènes*, p. 53.
[2] *Ibid.*, p. 54.
[3] Book IV, p. 453.
[4] *Continu et discontinu en physique moderne*, p. 59.

the object of all mechanics. The nature itself of these previsions—certitude or probability—characterizes a given mechanics.

We shall call (after Dirac) a *maximal observation* one that it is necessary and sufficient to perform in order to specify the state of a mechanical system at a given time. This state will serve as a basis for the predictions made. It is the observation that occurs in the statement of the *axiom of initial conditions*.

In classical mechanics the simultaneous measurement of different quantities (or observables) connected with a system is unrestricted. That the state of a system is considered to be specified by the determination of the positions and the velocities at a given instant is a choice which is dictated by macroscopic experiment.

In quantum mechanics the maximal observation is the simultaneous measurement of the greatest possible number of observables connected with the system. Now two observables are only simultaneously measurable if they commute with each other. This condition cannot be satisfied by a coordinate x and the momentum p which is conjugate to it. *The maximal observation in quantum mechanics is therefore incomplete in the classical sense.*

For a system of one degree of freedom the maximal observation in quantum mechanics will be the measurement of x, or of p, or of a single function $F(x, p)$, the function F being otherwise arbitrary.

In all cases the uncertainties in x and p are connected by the relation

$$\Delta x\, \Delta p \geq h.$$

From the maximal observation specifying the state X_0 of the system at the time t_0 it must be possible—in all theories of mechanics—to make predictions about the system at the time t. These predictions will be denoted by $X(t)$.

According to M. J. L. Destouches,[1] the condition is expressed by the formula

(1) $$X(t) = U(t) X_0$$

where U is an operator applied to X_0. In addition, the following hypotheses are made—

a) homogeneity of the time in the absence of external actions depending on the time ;

b) property of the U of forming a group.

The condition (1) is to us of the nature of a postulate, which may be called the *postulate of predictability*. It expresses the knowledge

---

[1] *Bulletin de l'Académie Royale de Belgique*, Vol. 22, 1926, p. 525.

of a single observer. If different observers can agree on the same law, there will be a *principle of relativity* expressing the invariance of the operators U for the ensemble of corresponding reference frames.

In every theory of mechanics endowed with physical significance the small errors of measurement represented by a trivial uncertainty in X_0 must imply only small variations of $X(t)$.

The necessity of this condition or *postulate of stability* was advocated by Duhem.[1] It has been mathematically defined by M. Bouligand[2] and generalised to the case which concerns us by M. J. L. Destouches.[3]

The existence of neighbourhoods in the ensemble of the X_0 or of the X—in virtue of (1)—makes this ensemble an *abstract space* $(\mathscr{X})$. Then X must be stable in $(\mathscr{X})$ with respect to X_0.

M. J. L. Destouches, by supposing that $U(t)$ is differentiable with respect to t, arrives at the following equations—

$$(2) \qquad \frac{dX}{dt} = \mathscr{H}X \qquad \frac{dU}{dt} = \mathscr{H}U.$$

Here $\mathscr{H}$ is a new operator which is defined in a domain including the ensemble of the X.

By invoking only the two postulates above (predictability and stability) there are discovered, summarised in (2), Jacobi's equation of classical mechanics, Schrödinger's equation of wave mechanics and that which corresponds to them in the abstract point mechanics of the space $(\mathscr{X})$.

In the old mechanics the possibility of predicting with certainty, from X_0, the value of the different observables connected with the system at every time corresponds to what I shall call *legality*.

In quantum mechanics the observer has the *free choice* of measuring at the initial instant either x or p or an arbitrary quantity $F(x, p)$. One of these determinations suffices to define a motion of the system ; the different motions constructed in this way cannot simultaneously have experimental significance. This circumstance is connected with the incomplete character (in the classical sense) of the maximal observation in quantum mechanics.

In addition, apart from the case of first integrals, $X(t)$ will *almost never* be an eigenfunction of the observable α which is the subject of the initial measurement and whose eigenfunction is X_0.

The development of $X(t)$ as a series in the eigenfunctions of α will only make it possible to predict *at every time* the probabilities of the different possible values of α.

[1] *La Théorie physique*, Paris, 1906, p. 231.
[2] *Comptes rendus de l'Académie des Sciences*, Paris, Vol. 200, 1935, p. 1500.
[3] *Bulletin de l'Académie Royale de Belgique*, 1935, p. 780.

We say that from an initial measurement of the observable α, there is only *semi-legality* for this observable. This is in contrast with the legality which, because of (2), regulates the evolution of the system.

We now take up the case of first integrals that was excluded from the preceding discussion. According to Dirac, the necessary and sufficient condition that a quantity $A(x, p, t)$ should be a first integral of the system, conservative or not, that is characterised by the hamiltonian H, is written direcly as

$$(3) \qquad \frac{\partial A}{\partial t} + \frac{2\pi i}{h}[AH - HA] = 0.$$

This is obtained by analogy with the corresponding condition in ordinary mechanics by substituting the quantum square-bracket quantity for the Poisson bracket.

L. de Broglie [1] has stated a condition equivalent to (3) in the following form. Let

$$(4) \qquad \frac{h}{2\pi i}\frac{\partial \psi}{\partial t} = H\psi$$

be the wave equation for the system. Call $\psi_i(x, 0)$ the orthogonal and normalised eigenfunctions of the hamiltonian at the time $t = 0$. Consider the solution $\overline{\omega}_i(x, t)$ of (4) that reduces to $\overline{\omega}_i(x, 0)$ for $t = 0$. At the time t the functions $\overline{\omega}_i$ form a complete orthogonal normalised set which is, in general, different from the set of the eigenfunctions of H.

The quantity $A(x, p, t)$ *is a first integral when the elements of the matrix that represents it in the basis* $\overline{\omega}_i(x, t)$ *are independent of the time.*

By means of this new definition it is shown that (if A is a complete operator)

1) the proper values of A are independent of the time;

2) the respective probabilities $|c_i|^2$ of the different eigenvalues of A, which are fixed at the origin of time by means of the development $\psi_0 = \Sigma c_i \psi_i$ of the initial state in terms of the eigenfunctions of A, is not modified when the state $\psi = U\psi_0$ of the system evolves according to (4).

In particular—

If the initial state ψ_0 is an eigenstate of A corresponding to the eigenvalue a, the state $\psi = U\psi_0$ of the system at the time t is also an eigenstate of A corresponding to the same value a. *There is legality for the*

---

[1] *Théorie de la quantification dans la nouvelle mécanique*, Paris (Hermann), 1932, p. 227.

quantity A *(first integral) throughout the motion and this legality is expressed by an invariance.*

The semi-legality of observables, which is the general case, and the legality of first integrals are connected by a *theorem of Fermi* concerning the prediction, from a suitable measurement, of an *isolated certainty* for any observable.

Consider, as L. de Broglie [1] has done, the equation

$$(5) \qquad A_t{}^{\overline{\omega}_i(x,\,t)} = A_0{}^{\psi_i(x,\,0)}.$$

This expresses the fact that the matrix representing the first integral A at the time t in the basic set $\overline{\omega}_i(x, t)$ defined above has the same elements as the matrix representing A at the origin of time in the basic set $\overline{\omega}_i(x, 0)$ of the initial eigenfunctions of the hamiltonian.

Since $\overline{\omega}_i(x, t)$ is the solution of (4) which reduces to $\overline{\omega}_i(x, 0)$ for $t = 0$, then

$$(6) \qquad \overline{\omega}_i(x, t) = U\psi_i(x, 0).$$

Whence, by a change of basis,

$$(7) \qquad A_t{}^{\overline{\omega}_i(x,\,t)} = U^{-1}A_t{}^{\psi_i(x,\,0)}\,U$$

or, by (5),

$$A_t{}^{\psi_i(x,\,0)} = UA_0{}^{\psi_i(x,\,0)}\,U^{-1}.$$

The last equation makes it possible to form the first integral which reduces, at $t = 0$, to a quantity $A_0 = F(x, p)$ which may be chosen arbitrarily.

More generally, if $F(x, p)$ is an arbitrary function defined at the time t_1, there exists a first integral $A[F(x, p), t, t_1]$ reducing to $F(x, p)$ for $t = t_1$.

Suppose that at the time t_0, earlier than t_1, we measure the first integral and find the value a. Because of the legality of the first integral and its invariance, we are able to announce *with certainty* that the quantity $F(x, p)$ will have the value a at the time t_1. In other words—

If any physical quantity F(x, p) *is given, it is always possible to know the value it will have at any time* t > t_0 *by means of a suitable measurement carried out at the time* t_0.

Fermi has given a direct demonstration of this theorem.[2]

I must insist on the fact that the certainty that it is possible to acquire in this way, with respect to any quantity, is an *isolated* certainty.

---

[1] *Comptes rendus de l'Académie des Sciences*, Paris, Vol. 194, 1932, p. 693.
[2] *Nuovo Cimento*, 1930.

If this were not true, there would be a departure from the semi-legality which is the general rule.

We shall illustrate this by means of the example given by Fermi.

Let a free particle be constrained to move along the axis of x. The momentum p and the hamiltonian

$$H = \frac{p^2}{2m} = \frac{1}{2m} \left(\frac{h}{2\pi i} \right)^2 \frac{\partial^2}{\partial x^2}$$

are first integrals. In compensation x cannot be such (it does not commute with H).

As in ordinary mechanics, the operator

$$A_1 = x - \frac{p}{m}(t - t_1) = x + \frac{h(t - t_1)}{2\pi i m} \frac{\partial}{\partial x}$$

is a first integral and reduces to x for $t = t_1$. If therefore we measure A_1 at the time $t = 0$ and find the result a_1, we are able to predict with certainty that x will have the same value for $t = t_1$.

Similarly,

$$A_2 = x - \frac{p}{m}(t - t_2)$$

is a first integral. If we measure A_2 at the time $t = 0$ and find the result a_2, we are able to predict with certainty that x will have the same value for $t = t_2$.

But it is essential that the simultaneous measurement of A_1 and A_2 is impossible. Indeed, it is easily verified that

$$A_1 A_2 - A_2 A_1 = \frac{h}{2\pi i m}(t_2 - t_1)$$

or that A_1 and A_2 do not commute for any value of t_1. The observer will have the free choice of measuring *either* A_1 or A_2, but he will only acquire an *isolated certainty* of one of these operators and, consequently, of the value of x at a later *chosen* instant. It is clear that this ensures the semi-legality of x.

In short, quantum mechanics forces us to give up the classical axiom of initial conditions. We recall that, as Painlevé said, there is nothing in logic which imposes the " copernican " choice. As a particular instance, when we measure x with certainty we rediscover the scholastic axiom of initial conditions. The quantum axiom allows the initial state to be specified in an infinite number of ways corresponding to the *choice* of the arbitrary function $F(x, p)$.

The modifications of the classical legality of observables are situated on another plane. Certain interpretations see in them the " failure of determinism. " This is certainly an exaggeration. In the general analysis presented above, the appeal to a *postulate of predictability* is opposed to a circumstance in which the legality may be completely suspended. On the contrary, the conditions of stability, together with the first postulate, imply a complete legality in the evolution of the state of the system (apart from perturbation). If now the state itself is not considered, but rather the different observables connected with the system, the legality in the classical manner only subsists for first integrals alone. For any arbitrarily chosen observables quantum mechanics in general offers no more than a semi-legality (prediction at every instant of the distribution of the probabilities of the different possible values). Nevertheless, by means of a suitable measurement, it is possible to predict the measurement of any observable at an *isolated* later time which is arbitrarily chosen.

The preceding argument is true *in detail*. In addition, classical legality can reappear, through a statistical compensation, for an assembly of individuals x and p whose uncertainties are connected by Heisenberg's relation (Ehrenfest's theorem).

Moreover, the properties of *stability* and, should this be necessary, of *relativity*, are preserved in quantum mechanics. The former must be understood as the stability of a probability distribution in all cases of semi-legality.

3. The " probabilities of presence " in classical mechanics, in the old quantum theory, in wave mechanics and in Dirac's mechanics.

We return to the paper of Poincaré which was cited at the end of Chapter 3.[1] Let there be a system of first order differential equations

$$(1) \qquad \frac{dx_k}{dt} = X_k (x_1, x_2 \ldots, x_n) \qquad (k = 1, 2, \ldots n).$$

Let $Wd\tau$ be the *probability of presence* of the representative point of the state of the system in an elementary volume $d\tau$ of the space (x_k).

In a volume V of the same space the probability of presence of this point will be

$$\underbrace{\int \ldots \int}_{n} Wd\tau.$$

Poincaré writes— " By such a probability I understand the ratio

[1] *Journal de Physique*, Vol. II, January 1912, p. 1.

$\dfrac{t}{T}$, where T is a very long interval of time from the instant θ to the instant $\theta + T$ and t is the time for which, between these same two instants, the representative point is found in the volume V considered. Provided that t is very large, this probability has no meaning if $\dfrac{t}{T}$ cannot be considered as independent of θ and of T. If this condition is fulfilled and if W can be defined, it will necessarily satisfy the partial differential equation

$$(2) \qquad \sum_k \frac{\partial (WX_k)}{\partial x_k} = 0.$$

Therefore W will be a last multiplier of the equation (1). "

For the canonical equations of classical mechanics

$$(3) \qquad \frac{dq_k}{dt} = + \frac{\partial H}{\partial p_k} \qquad \frac{dp_k}{dt} = - \frac{\partial H}{\partial q_k}$$

the identity

$$\frac{\partial}{\partial q_k}\left(\frac{\partial H}{\partial p_k}\right) - \frac{\partial}{\partial p_k}\left(\frac{\partial H}{\partial q_k}\right) = 0$$

immediately provides

$$(4) \qquad W = 1.$$

The fact that unity is a last multiplier, in Jacobi's sense, of the canonical equations of classical mechanics represents the *complete homogeneity of the possibilities of localisation of the representative point of the state of the system in the phase space* (q_k, p_k).

The result (4) expresses at the same time the legality, the continuity and the homogeneity of the solutions which characterise classical dynamics in the space of (q_k, p_k).

Seeking to form a last multiplier or, rather, a probability of presence which leads to Planck's law and not, as the last multiplier $W = 1$ does, to the Rayleigh-Jeans law of equipartition, Poincaré obtains an *essentially discontinuous* function W containing factors $w(\eta_k)$ which are zero if the energy η_k differs from a multiple of the quantum ε. These discontinuities are inevitable if it is desired that the radiation should be finite.

" You know why the old theories lead us forcibly to the law of equipartition, which implies an infinite total of radiation and which is absolutely contradicted by experiment ; it is because they suppose that all the equations of mechanics are of Hamilton's form and, consequently, they assume unity as a last multiplier in Jacobi's sense.

It must then be supposed that the laws of impact between a free electron and a resonator are not of the same form, and that the equations which govern them admit a last multiplier other than unity. It is certainly necessary that they should have a *uniform* last multiplier, for otherwise the second principle of thermodynamics would not be true, but it is not necessary that this last multiplier should be unity." [1]

In another form, interpreting Planck's ideas, Poincaré writes—

" The probability of the continuous variable is obtained by considering independent elementary domains of equal probability. In classical dynamics, to find the elementary domains use is made of the theorem that two physical states which are such that one is the necessary effect of the other are equally probable. In a physical system, if one of the generalised coordinates is represented by q and the corresponding momentum by p, then according to Liouville's theorem the domain $\iint dpdq$ at any instant is an invariant with respect to the time as long as q and p vary in accordance with Hamilton's equations. On the other hand, q and p can, at a given instant, take all possible values independently of each other. Whence it follows that the domain of probability is infinitely small and of magnitude $dpdq$. . . .

" The hypothesis [of Planck] must serve the purpose of restricting the variability of p and q in such a way that these variables only vary in jumps or that they should be regarded as, in part, connected to each other. [2] Thus a reduction in the number of the elementary domains of probability is achieved, so that the extension of each of them is increased. The hypothesis of quanta of action consists of the supposition that these domains, all equal to each other, are no longer infinitely small but finite, and that for each of them

$$\iint dpdq = h$$

where h is a constant. " [3]

In quantum mechanics (simple wave mechanics) and because of the very definition of the state of a system by a function ψ which is a solution of the wave equation, it is only in an element $d\sigma$ of the *configuration space* (q_k) that the probability of presence of the representative point of the system is amenable to definition.

This probability is written $\psi^*\psi$ in accordance with Born's principle. The integral

$$\int \ldots \int \psi^*\psi d\sigma$$

[1] *Dernières Pensées*, Paris (Flammarion), p. 213.
[2] Note here that this could be called a sort of prediction of HEISENBERG's uncertainties.
[3] *Ibid.*, p. 183.

over all the configuration space is, because of the wave equation and the conditions imposed on the function ψ, invariant with respect to the time. The equation

$$(5) \qquad \frac{\partial(\psi^*\psi)}{\partial t} + \sum_k \frac{\partial}{\partial q_k}(\psi^*\psi \Xi_k) = 0$$

which expresses this invariance shows that $\psi^*\psi$ *is a last multiplier, in Jacobi's sense of the system*

$$(6) \qquad \frac{dq_k}{dt} = \Xi_k(q_1, \ldots q_k, t) = \frac{ih}{4\pi m_l}\frac{1}{\psi^*\psi}\left(\psi\frac{\partial\psi^*}{\partial q_k} - \frac{\partial\psi}{\partial q_k}\psi^*\right) \quad \begin{pmatrix} k = 1, 2 \ldots 3N \\ l = 1, 2 \ldots N \end{pmatrix}$$

relevant to a system of N particles of mass m_l.

The preceding analysis makes it possible to compare the *probability interpretations* of classical mechanics and simple wave mechanics.

We remark that the system (6) is true *in detail*. In quantum mechanics this system plays the part of the canonical equations in classical mechanics. It implies

$$(7) \qquad \begin{aligned} \text{mean value } \frac{dq_k}{dt} = \overline{\frac{dq_k}{dt}} &= \int \ldots_{3N} \int \psi^*\psi \frac{dq_k}{dt}\, d\sigma \\ &= \frac{ih}{4\pi m_l}\int \ldots_{3N} \int\left(\psi\frac{\partial\psi^*}{\partial q_k} - \frac{\partial\psi}{\partial q_k}\psi^*\right)d\sigma. \end{aligned}$$

A second differentiation, taking account of the wave equation, would give Ehrenfest's theorem expressing the statistical agreement of the classical equation in their usual from

$$(8) \qquad m_l\frac{\overline{d^2 q_k}}{dt^2} = \overline{\frac{\partial U}{\partial q_k}}.$$

In Dirac's mechanics the probability density is written

$$\psi_k^*\psi_k$$

(with summation of the dummy suffix k form 1 to 4).

The integral

$$\int\int\int \psi_k^*\psi_k\, d\sigma$$

over all the configuration space (q_1, q_2, q_3) is invariant with respect to the time.

The equation

$$(9) \qquad \frac{\partial}{c\partial t}(\psi_k^*\psi_k) - \frac{\partial}{\partial q_1}(\psi_k^*\alpha_1\psi_k) - \frac{\partial}{\partial q_2}(\psi_k^*\alpha_2\psi_k) - \frac{\partial}{\partial q_3}(\psi_k^*\alpha_3\psi_k) = 0$$

which expresses this invariance shows that $\psi^*\psi_k$ is a last multiplier, in Jacobi's sense, of the system

$$(10) \qquad \frac{c\,dt}{\psi_k{}^*\psi_k} = \frac{-\,dq_1}{\psi_k{}^*\alpha_1\psi_k} = \frac{-\,dq_2}{\psi_k{}^*\alpha_2\psi_k} = \frac{-\,dq_3}{\psi_k{}^*\alpha_3\psi_k}.$$

This system of equations implies the equation

$$(11) \qquad \iiint \psi_k{}^*\psi_k \frac{dq_r}{dt}\,d\sigma = \overline{\left(\frac{dq_r}{dt}\right)} = -\,c \iiint \psi_k{}^*\alpha_r\psi_k d\sigma$$

which, again, gives the motion of the centre of gravity of the probability. But, in contrast with simple wave mechanics, Ehrenfest's theorem breaks down here. In particular, in the absence of field the principle of inertia is not obtained as a statistical consequence. Instead of being actuated by a rectilinear and uniform motion, the centre of gravity oscillates about such a motion in virtue of the " *Schrödinger vibration,*"[1] a direct consequence of equation (11).

Thus for the probability density $\psi^*\psi$ of simple wave mechanics we establish a singularity (breakdown of the classical law $W = 1$) and a regularity (existence of a uniform last multiplier) in accordance with Poincaré's views but with the following essential distinctions—

a) the classical axiom of initial conditions is overthrown ;

b) the last multiplier $\psi^*\psi$, the corresponding system of differential equations and the state of a system are no longer studied in phase space but in configuration space.

In the quantisation, in Schrödinger's sense, of a system placed in a static field, the amplitude a of the wave is supposed continuous, uniform, finite and zero at the boundaries of the domain of configuration space considered. Because $a^*a = \psi^*\psi$, these regularities extend to the probability density which represents the last multiplier. *Correlative* discontinuities are introduced in the spectrum of the energy eigenvalues. It is the quantum mechanical definition of the state of a system which makes possible the regularisation of the last multiplier, while the classical definition of the state of a system forced Poincaré, in his attempt to preserve Planck's law, to adopt a probability density and, consequently, a last multiplier which are *essentially discontinuous*.

The statistical agreement with the old mechanics is obtained directly by means of the system of differential equations for which $\psi^*\psi$ is the last multiplier. In passing we have indicated how this statistical agreement subsists, apart from the " Schrödinger vibrations, " in Dirac's mechanics.

[1] *Annales de l'Institut Henri Poincaré*, Vol. II, p. 269.

4. ON THE " REALITY " OF QUANTUM MECHANICS.

Classical mechanics, in order to explain phenomena, appealed either to a unique picture (wave or particle) or to a model (mechanism, for example). In a general way the possibility of experimenting on a system without disturbing it was assumed. This set of circumstances let rather naturally to the conception of an objective reality independent of the observer.

In quantum mechanics, on the contrary, the dualism of the wave and the particle no longer permits of a concrete picture which might be supposed to represent an aspect of reality ; the simultaneous measurement of certain quantities becomes impossible. The result of this is to throw the problem of the relationship between the physical theory and " reality " into a new light.

This problem has been the concern of many people. In the first place, I shall analyse the thesis formulated by Einstein and two of his collaborators.[1]

A physical theory must be both *correct* and *complete*. Einstein does not concern himself with the correctness of quantum mechanics— that is, its agreement with experiment. By a " complete theory " he understands a theory in which each element of physical reality has its counterpart.

In the matter of reality, and without regarding this as anything but a sufficient condition, Einstein puts forward the following criterion.

" *If, without disturbing a system in any way, it is possible to predict with certainty (that is, with a probability equal to unity) the value of a physical quantity, then there exists an element of physical reality corresponding to this physical quantity.* "

We shall refer to this criterion by (R).

Einstein considers a free particle having only one degree of freedom, and which is in an eigenstate of the momentum (which can then be measured exactly). He establishes that the coordinate of this particle is, under these conditions, completely indeterminate. The converse is clearly true because of the symmetry of Heisenberg's uncertainty relationships.

From this, applying the criterion (R), Einstein concludes that when the momentum of a particle is known, its position has no physical reality. More generally, two quantities such that the corresponding operators do not commute cannot be simultaneously real in the sense of (R).

---

[1] *Can the quantum mechanical description of reality be considered complete ?* A. EINSTEIN, B. PODOLSKY and N. ROSEN, *Physical Review*, Vol. 47, 1935, p. 777.

Einstein indicates another conclusion— that which consists of the assumption that the description of reality offered by the wave equation is *incomplete* in the sense made clear above.

Indeed, Einstein considers two systems, I and II, originally in interaction for a finite time and whose interaction has ceased. He considers the total wave function of the system I + II after interaction. This function describes—perturbation apart—the later evolution of the total system.

To specify the wave function—or, alternatively, the state— of each of the systems after the interaction it is necessary to carry out further observations.

Suppose that each of the systems consists of a single particle of one degree of freedom.

Let p_1 and q_1 be the momentum and the coordinate of the particle which constitutes the system I and p_2, q_2 be the analogous quantities for the system II.

a) Measure p_1. After the measurement the state of the system I reduces to an eigenstate of the quantity p_1. The state of the system II is also specified.

b) Measure q_1 instead of measuring p_1. After this measurement the state of the system I reduces to an eigenstate of the quantity q_1. The state of the system II is also specified by this measurement.

Accordingly two different measurements carried out on the system I enable the assignment of two different states to the same reality (the system II) which, since all interaction has ceased, could not have been influenced by this measurement.

In the particular case studied by Einstein the two wave functions thus assigned to the same reality (system II) can be eigenfunctions of two magnitudes whose operations do not commute (namely p_2 and q_2).

There is a contradiction with the absence of simultaneous reality considered above for two magnitudes of this nature : quantity of motion and coordinate of a same particle.

To obviate this contradiction, Einstein aknowledges that the description of reality offered by quantum mechanics is *incomplete*.

Einstein justifies this conclusion as follows— " It could be objected to this conclusion that our criterion of reality is not sufficiently restrictive. It is true that this conclusion would not be arrived at if it were granted that two magnitudes must be considered as simultaneous elements of reality only on condition that they can be measured (or calculated) simultaneously. From this point of view, as soon as one or the other—but not the two simultaneously—of the magnitudes p_2 and q_2 can be measured (or calculated), these magnitudes are not simultaneously real. Reality would thus be made to depend on the magnitudes p_2 and q_2, belonging to the system II, of the observations carried out on the system I, and which could in no way disturb the system II. No reasonable definition of reality can admit such a circumstance."

Einstein's theory calls for a thorough discussion.

In the first place, the criterion (R) seems to link physical reality with the observable or magnitude endowed with physical meaning, instead of following the usual path of deducing the observable, which is already an abstract element, from reality.

The criterion (R) attaches significance only to the measurements effected *without perturbation*. It thus restricts the application of the quantum theory to *pure cases* alone.

Now these are exceptional : either they are encountered thanks to a suitable conjugation of the state of the system and of the observable to be measured (state suitable for observation), or they are obtained, on condition that we grant that any measurement is repeatable, by confirming immediately a first measurement without eliminating the perturbation that results in general from this latter and without leaving the observable the time to evolve.

In the example of interaction which Einstein develops, the wave function of the total system I+II appears as the integral of the product of the eigenfunctions of *either* q_1 and q_2 *or* of p_1 and p_2.

Thanks to this circumstance, the measurement of q_1 (or of p_1) makes possible the assignment of an exact value to q_2 (or to p_2). Only measurements without perturbation are encountered and certainties can be stated.

In the general case this is not true. Thus W. H. Furry[1] puts the following question—

" *If an arbitrary quantity* A *of the system* I *has been measured after the interaction of the systems* I *and* II *and the result* a *has been obtained, what can be said about an arbitrary quantity* B *belonging to the system* II *?* "

The following answer is obtained. In general it will only be possible to calculate the probability that the quantity B should have one of its possible values.

Accordingly, from a certainty about A it is only possible to state a probability for B. If the criterion (R) only takes results which are certain into consideration, the experiment carried out on the system I will not give information about the " reality " II.

---

[1] *Physical Review*, Vol. 49, 1936, p. 393.

In order that a certainty of A should provide a certainty of B, it is necessary that a certain connection should exist between the quantities A and B. This is realised in the particular example considered by Einstein but is not true in general.

Moreover, it seems that Einstein's argument provokes a more serious objection.

Given the symmetry of the total system I + II of two free particles which Einstein considers, the paradox in respect of the " reality " II must be reflected in a similar paradox in respect of the " reality " I.

Einstein supposes that experiments are made on the system I. Faced with the impossibility of measuring the conjugate quantities p_1 and q_1 simultaneously, the observer exercises a *free choice*— he measures p_1 or q_1.

The paradox which arises in respect of the system II follows exclusively from the particular properties attributed to the " reality " of this system.

All paradox in respect of I is ruled out by the impossibility of the simultaneous determination of the pair of observables p_1 and q_1. The paradox associated with II similarly disappears if the attribution of a special character to the " reality " of this system, apart from any measurement, is forgone.

Further, from the physical point of view it seems rather unnatural to attribute more " reality " to a system on which no observation is made than to one on which experiments are performed, and this merely because, in the absence of measurement, all alteration of the " reality " is excluded.

Finally we observe that in the example given by Einstein—and in analogous examples which could be constructed—the paradox appears when the interaction has ceased. But, because of the non-localisation of the particles, it is doubtful whether it may be said that the interaction has ceased.

We shall leave the question of interaction which has so far been our concern. More simply, we consider a system in any state ψ and an arbitrary quantity having the possible values $a, b, \ldots$ to which correspond the eigenfunctions $\psi_a, \psi_b. \ldots$

Quantum mechanics only discloses to us the *probabilities of agreement* of the states ψ and ψ_a, ψ and ψ_b, etc. . . . To the question " *If the system is in the state ψ, what will be the value of the quantity considered?* " it is not possible, in general, to reply with certainty. A large number of measurements made by returning the system to the state ψ after each measurement (by eliminating the perturbation due to this measurement) will only yield the weights of the different possible results $a, b. \ldots$

In other words, as a general rule if the system is in any state, an arbitrary quantity cannot have reality in the sense of (R). Here again the lack of generality of Einstein's criterion is apparent.

Opposed to Einstein's point of view—connecting the notion of "reality" with the occurrence of certainty—is the *pure probability* thesis developed, for example, by Henry Margenau.[1]

This author rejects the postulate according to which it is possible, *by means of a single measurement, to state a certainty.* He likens the state of a system in quantum mechanics to a set of numbers which are associated with a certain operator λ and which determine the probabilities of the different possible values of this operator (which itself represents a quantity endowed with physical meaning).

"A quantum state," writes Margenau, "is *synonymous* with a probability distribution. The latter cannot be determined by a single measurement, but requires a large number of observations."

Thus quantum mechanics would resemble a game of chance. "In throwing a die, the distribution of probabilities is the set of numbers $^1/_6, ^1/_6, \ldots ^1/_6$, for the possible result 1, 2, ... 6. If the die falls on the face 5, this provides a piece of evidence, but it does not fix the distribution of the probabilities of the different throws. A large number of throws is required for that : the knowledge of one isolated throw does not change the distribution of the whole of the initial probabilities."

There is a part of truth in this over-simple comparison. But in order to assimilate quantum mechanics to a game of dice, abstraction would have to be made of the *pure cases.* In such a case, the die is indeed guided in such a way as to fall on the same face always.

The pure cases, exactly satisfying the postulate that H. Margenau proposes to rule out, deserve on the contrary to be considered as privileged ones if only for the reason—which may have guided Einstein in the enunciation of the criterion (R)—that they constitute the only converging point of the classical and quantum doctrines in the consideration of physical measures.

As a matter of fact, it is necessary to distinguish between the certitude directly offered for a pure case and the certitude that may result, in a general way, from the immediate *confirmation* of any given measure.

The first is a certitude of the classical type. The second, on the contrary, is extracted from a superposition of states (generally infinite in number). From this infinity a state is chosen, but this choice does not influence the system and, without changing the course of the facts,

<hr />

[1] *Physical Review*, Vol. 49, 1936, p. 240.

modifies, according to Heisenberg, [1] only *our knowledge of these latter*.

The distinction may appear subtle ; however, it seems essential, although obviously a pure case may be considered mathematically as the limit of a mixture of states.

Thus then, *apart from pure cases*, the entity *state of a system* is distinct from the knowledge we can acquire of it, unless we make a great number of measurements which provide us with a distribution of the possibilities.   That is what makes W. H. Furry say : " *A mixture* [of states] *is essentially different from any pure state whatever.* "

Thus we arrive at a compromise between the pure probability conception and that of Einstein, which is confined to the consideration of pure states alone where the certainty obtains *directly*.   Quantum mechanical " reality " would be a complex participating in both these conceptions. We also note that a passage to the limit produces a statistical relation between the entity *state* and the mathematical fact *knowledge of a state*.

For his part, Bohr [2] rejects Einstein's criterion of reality.   He insists on the fact that after the interaction of the systems I and II we have the *free choice* of measuring one or the other of the two quantities p_1 and q_1, which does not interfere directly with the second particle constituting the system II.   This free choice corresponds to the discrimination between different experimental methods which allow the use, without ambiguity, of *complementary* classical concepts (coordinate or momentum).

And Bohr concludes, " Einstein's argument does not authorise us to conclude that the description of reality offered by quantum mechanics may be incomplete.   On the contrary, this description emerges as the rational utilisation of the possibilities of interpretation compatible with the finite and controllable interaction between objects and the instruments of measurement.   In fact it is only the mutual exclusion of two experimental processes (permitting the unambiguous definition of complementary physical quantities) which leaves room for the new laws whose simultaneous existence might, at first sight, seem irreconcilable with the very foundations of science. "

To conclude, we shall broach the epistemological interpretation of the above discussion.

For Einstein, there is a rupture between physical theory and reality :

" Any serious consideration of a physical theory must take into account the distinction between objective reality, which is independent

---

[1] *Physikalische Prinzipien der Quantentheorie.*
[2] *Physical Review*, Vol. 48, 1935, p. 696.

of any theory, and the concepts used by this theory. These concepts are created in order to correspond to objective reality and with their help we represent reality. "

However categorical it may be, this declaration is not, in the author's eyes, directed against the whole quantum domain. Indeed, Einstein grants " reality " only to the observations without perturbation. On the other hand, the wave—corpuscle duality seems to be opposed to a description other than symbolic of " reality. "

The quantum physicists are in disagreement over the existence of an objective reality, in the sense of the " exterior world " of classical physics.

For Bohr, the finite quantity of the quantum of action does not allow us to make, between the phenomenon and the instrument of observation, that clear distinction demanded by the ordinary concept of observation, and consequently by the classical ideas.

Heisenberg, it would seem, adopts this same point of view.

For Louis de Broglie, on the contrary, the fact that an experiment can involve a perturbation does not necessarily lead to the disappearance of the objective character of the system submitted to this experiment :

" Bohr believes that the physics of quanta makes uncertain the distinction between the subjective and the objective. Perhaps there is some misuse of words. In reality, the methods of observation, the instruments of observation and even the organs of our senses belong to the objective category and the fact that, in microscopic physics, it is longer possible to ignore their reactions on the parts of the external world that we wish to study can in no way abolish or even blur the traditional distinction between objective and subjective. "

Louis de Broglie concludes that quantum physics reveals the artificial character of the classical division between system observed and instrument of measurement, and that it proves " that a description of physical reality quite independent of the means with which we choose to observe it is strictly impossible. " [1]

Dirac, in his *Principles of Quantum Mechanics*, does not specify the properties that he attributes to physical reality but, basing himself on the wave-corpuscle duality, insists on the futility of efforts which would lead to a description of reality as something which contains both waves and particles interacting with each other and to an attempt to construct a mechanics expressing the relationships between these two concepts.

[1] *Les idées nouvelles introduites par la mécanique quantique*, l'Enseignement mathématique, 1933.

On this subject there remains a dispute which we are only able to leave to the philosophers.[1] Nevertheless we emphasise that it would be somewhat comforting to be able to continue, in spite of the quantum mechanical modification of classical ideas, to believe in the existence of an objective reality. To form a clear notion of it, we must accept a compromise between Einstein's thesis and the pure probability conception; we must also renounce to grasp simultaneously certain *complementary* aspects in Bohr's sense.

5. ON THE DIFFERENCE BETWEEN ABSTRACT REASONING AND INTUITIVE REASONING IN QUANTUM MECHANICS.

Louis de Broglie [2] reminds us that " since the time that, thanks to the progress of mathematical analysis, the theories of physics were able to take the form of coherent mathematical doctrines, two tendencies have been in conflict in the construction and the renovation of these physical theories. "

Scientists with intuitive tendencies *(les intuitifs)* have always sought to base themselves on concrete representations taken from the objects that surround us in every-day life. For them, the atom and even the electron were likened to a small ball endowed with shape, mass and durability ; the ether was a substratum necessary to the propagation of vibrations ; thermodynamics was clarified by statistical mechanics and the atomic hypothesis. They eschewed energetics. They hailed Bohr's atom as a true model in which point electrons rotate about a central nucleus, with the single addition of quantum jumps from one trajectory to the other.

On the contrary, to scientists with abstract tendencies *(les abstraits)* a physical theory reduces to a collection of mathematical relationships uniting the observable phenomena and making possible the statement of predictions on the basis of specified initial data. Their thermodynamics appeals to no intuitive representation. Their electromagnetism and their wave optics stem only from Maxwell's equations, laid down *a priori.* In their eyes, models are merely the imperfect and transitory features of a theory. This must be able to survive such representations " to emerge, finally freed from anthropomorphic blemishes, as being no more

---

[1] We recall that LEIBNITZ expressed these misgivings in the following terms— " When I was young I too was captivated by the Vacuum and by Atoms, but reason brought me back ; the imagination was agreeable ; its researches were confined there ; its meditation was fixed there as with a nail ; one thought that the primary elements, a *non plus ultra*, had been found. We could have wished that Nature should go no further ; that, like our mind, she should be limited. "

[2] *Continu et discontinu en physique moderne*, Paris (Albin Michel), 1941, p. 91 *et seq.*

than an abstract form. " Duhem, in his *Théorie physique*, made himself the champion of this tendency.

Louis de Broglie declares that, in principle, it is the " *abstraits* " who appear to be justified, at least in the atomic domain. Indeed, to interpret the properties of matter we must make use of a scale so small that the perceptions of our senses, corresponding to an infinitely greater scale, no longer have any value or any possibility of application.

But in practice, models have played an extremely valuable part in classical science.

Although it was created in an abstract form, thermodynamics has greatly benefited from the molecular hypothesis. Fresnel's ether preceded Maxwell's theory by forty years. In order that concrete representations—whose fallacious character is certain—should be able to render such service, it is very necessary that they contain some element of truth. In the progressive extrapolation that is necessary to pass from the macroscopic to the microscopic domain, the pictures formed of our perceptions are profitably used in the first stages, though there is a danger that they might completely mislead us on the atomic scale.

Bohr himself did not have for his own atomic model the " illusions " of the " *intuitifs.* " The discontinuous jumps that an electron makes from one trajectory to another are irreconcilable with the classical representations in the framework of space and time. Heisenberg's uncertainty relationships do not allow us to speak of trajectory in the ordinary sense. An electron makes its presence known throughout the extension of the atom and the distribution of its " potential presence " is dictated by a wave function whose character is completely symbolic.

Only the " *abstraits* " can declare themselves satisfied with a game in which only the relation of observable phenomena by algorithms is concerned. Also, in general, they must be content with probabilities where the classical science stated certainties.

Of the properties of the atom of Democritus and Lucretius, the elementary particles of atomic physics only retain the one characteristic of being " permanent units "—except when their " annihilation " appears possible. Then, of the traditional atom, there remains only " a simple arithmetical statement— the number of particles of a specified kind, which may be variable, is always integral. "

Nevertheless it seems that, without the atomic hypothesis and without Bohr's planetary model, it would never have been possible to formulate quantum mechanics. It is even possible to put the question whether science could still progress in a domain where all concrete representations had lost their validity. " We are only able to think with the help of pictures drawn from our sentient intuition. Without doubt abstract reasoning enables us, by schematisation and generalisation, to go outside this intuition, but does it allow us to free ourselves from it completely ? "

Louis de Broglie concludes this discussion with the opinion " that it is possible to justify both the ' *abstraits,* ' who have had strong grounds in the dispute, and the ' *intuitifs,* ' without whom progress would often have been difficult and perhaps impossible. "

6. CONCLUSION.

In the preceding pages we have been able to see the creation of the modern physical theories of mechanics. Quantum mechanics, in particular, has known and still knows, has experienced and still experiences, a proliferation of technical work. To mention only a few chapter headings, we have passed over in silence the quantum statistics, second quantisation, the relativistic quantum mechanics of systems, Louis de Broglie's theory of photons and the theories of the nucleus.

In penetrating into these fields, still in evolution, we would have run the risk of lacking the necessary perspective. More, to be frank, the technicalities would have found us wanting. We leave this task to the historian of tomorrow.

SOME REMARKS
BY WAY OF A GENERAL CONCLUSION

To write history, as to teach, is before everything, to choose.

The reader who has been willing to follow us to the end of this book will certainly be able to complain of some arbitrariness in the choice which has been made.

Above all, we have confined ourselves to the principles, in which it appears that the essential difficulty of mechanics lies ; we have not been concerned with the accumulation of facts.

As long as one remains in the paths opened up by the forerunners almost nothing is lost by this. For their efforts were directed almost exclusively to the isolation of principles that neither pure reason, nor their crude experiment, could inspire them to find.

On the other hand, none of the attempts of the early students has survived in its original form and the principles of many of them, later, have had to be forsaken. Therefore there is no risk that, in following them, the didactic field will have been duplicated.

This danger arises more frequently as the organisation of mechanics tends to become more developed— that is, from the XVIIIth Century onwards. We are aware of having encountered it more than once.

Until Huyghens and Newton the mathematical tools in mechanics were reduced to their simplest form and, in passing, the resources of the simple rule of three can be admired. Then the use of the differential calculus became common in mechanics. Indeed, it was indispensable for the expression of the effect of a force *in the first instant* and, in Leibniz's hands, for the connection of the living force with the static force. Total differentials appeared, with Clairaut, in hydrostatics and partial differential equations, with Euler and d'Alembert, in hydrodynamics. By the time of Lagrange the mathematical tools were highly perfected and became an essential feature of rational mechanics. In the modern physical theories of mechanics it has become necessary to use more elaborate procedures, like the absolute differential calculus and Riemann spaces in generalised relativity and abstract spaces in quantum mechanics.

This means that mechanics could not have evolved without having at its disposal, at each critical period, an adequate formulation and that, in this sense, it would appear linked to the progress of mathematics. It also means that, with the development of the formalism, there appears the danger of trusting in the tools of calculation and losing sight of the network of axioms. However *rational* it may be said to be, mechanics remains a branch of physics.

This branch only has a relative autonomy. Motion in a pure state does not exist. I have not treated, as being beside the strict purpose of this book, the relationship between mechanics and thermo-dynamics in the classical field.

In the modern physical theories of mechanics I have had to assume as given, without going back to their origins, the essential results of optics and electromagnetism. However, there exists *a point of view of the student of mechanics* which may be adopted without excessive arbitrariness. On the way, we have, to preserve historical accuracy, made some incursions into the domain of optics and electrodynamics. To recall only a single example, it was in optics, with Fermat, that the first minimum principle that was not trivial appeared. With Maupertuis, it was also an optical law (incorrect this time) that lay at the origin of the first form of the mechanical principle of least action. We note Hamilton's return to optics previous to his dynamical principle of stationary or varying action. With wave mechanics, with Louis de Broglie, appears a kind of fusion of the optical and mechanical principles of least action which, at least on the formal plane, recalls the dualistic aspect (emission and wave-propagation) of Hamilton's geometrical optics.

We have, in the course of this book, multiplied the quotations of original texts, only commenting on them for clarification when this, rightly or wrongly, appeared necessary to us. But we have restricted the length of these extracts to passages which seemed to us the most characteristic. The essential is that the reader, without being tired by repetitions and developments—which might occasionally make the original papers dull without adding anything really useful to their creative thought—should be taken back into the climate of the time and into the path, strewn with pitfalls, that the inventors *followed*. I emphasise this, for in the XVIIIth Century Clairaut, in his didactic works, was already speaking of the path that the inventors *should have followed*. This school of complaisance sees nothing in history. Shall I go as far as to say that I prefer the first classics —sometimes so difficult to read—for the very fact of the difficulty they offer in the process of making contact with a new idea? Genius is not as simple as the philosophers of the XVIIIth Century would have had us believe.

Also, I excuse myself from philosophising on the principles of mechanics that lie on the margin of history. Here is a subject of study that offers a real interest; but it emphasises the part of the critic at the expense of that of the players themselves. The personality of the historian is in danger of being encumbered, his true task being that of selection and not that of appreciation. I have not forbidden myself some incidental appreciation which some might have preferred that I should have omitted, but most often I have left the reader free to form his own opinion of the extracts. The discussions which I have retraced are, for the greater part, those of the actual creators. They have a constructive character to the extent that they proclaim, or even allow of, an extension of the principles. The periods in which science confines itself to the exploitation of determinate premises are periods of latent incomprehension. Through not continually questioning the premises, one ends by falling asleep in a deceptive security. This was the case at the time of the appearance of relativity. On the other side, the universal attraction was not passed of as a dogma in Newton's time.

I would detract from the lesson of history by attempting to comment of the evolution of mechanics in bold outlines; this would only be possible by schematising it. Now to schematise would, most often, distort the actual succession of things, which, in general, exhibits no regularity. Further, in this book I have not taken part—as I have done elsewhere—in the game of summarising, for example, the vicissitudes of the notion of force or those of the notion of kinetic energy. It is, indeed, a simple matter for the reader himself to indulge in this exercise by simply collating the material that we have put at his disposal. But he will quickly recognise that this game, however captivating it may seem, is often artificial. For the different keys to the problems of mechanics were not discovered independently, but are mutually interpenetrating.

I do not pretend to convince those who, on principle, feel that the history of science is an old-fashioned cult, and that each new generation, without looking back, must choose as quickly as possible the basic starting points of its progressive sciences. But from this to give all history of science the epithet of " old curiosity shop " would be too preposterous a step to take. Nothing is futile in scientific matters, not even the contemplation of the past. For this embodies the lesson of our vagaries, our scruples, our illusions and our errors. Science did not progress by that harmonious path, the illusion of which is easily created after the event. The direct knowledge of the old works, however they may be outstripped today, can only enrich the perspective of the future which opens up before us.

NOTES

ON THE MECHANICS OF THE MIDDLE AGES

When the French edition of this work was composed, we based ourselves mainly, as concerns the Middle Ages, on the researches carried
out by Duhem.

At a more recent date, fresh light has been thrown on medieval
scientific thought.

In an important book, *Robert Grosseteste and the Origins of Experimental Science*, [1] A. C. Crombie has shown how, as early as in the
beginning of the Thirteenth Century, the Oxford School had undertaken
to elaborate a method of physical science.

Robert Grosseteste, Chancellor of the University of Oxford (1214),
later Bishop of Lincoln (1235), progressed from the *Analytica posteriora*
of Aristotle by distinguishing the scientific knowledge of the reason for
a fact from the simple knowledge of this fact— *quia differt et propter
quid scire.*

Without entertaining any illusions as to the purely nominal character
of the definitions in use in the Schools, Grosseteste sought to establish
others which, while they expressed the empirical connexions observed
between the phenomena, would reveal the cause of the attributes ascer-
tained in the subjects. He set out to transform Aristotle's method,
which aimed at providing demonstrative proofs, into a process of
research to be submitted to verification or falsification of the experiment,
which at the same time sifts and judges the theories.

Grosseteste based his method of verification on the *principle of
uniformity* and the *principle of economy* in nature and this enabled him
to choose between definitions leading to the same facts. Thus, by
elimination, he arrived at the following definition of comets— "Subli-
mated fire assimilated to the nature of one of the seven planets. "

[1] Oxford, Clarendon Press, 1953.

If he dealt with physical theory in general terms, underlining the necessity of associating geometrical beings and the physical objects to define the physico-mathematical entities, Grosseteste seems to have been mainly preoccupied with optics, and had the merit of being the first to perceive a phenomenon of refraction in the rainbow. But he was the real leader of the school that directly inspired Roger Bacon, John Duns Scot, William of Ockham and Thomas Bradwardine.

On the other hand, the importance of the work of Thomas Bradwardine, who followed along the road opened up by Robert Grosseteste, and, in order the better to mathematise physics, sought to widen the field of functional relations by the development of algebra in his *Tractatus proportionum* (1328), has been emphasized by Anneliese Maier [1] in her remarkable researches.

For Thomas Bradwardine, it was a question of expressing the law of powers of Aristotle's dynamics : v being a velocity, p a power and r a resistance ; Bradwardine contemptuously dismissed the following formulations—

$$\frac{v_2}{v_1} = p_2 - r_2 - (p_1 - r_1)$$

$$\frac{v_2}{v_1} = \frac{p_2 - r_2}{p_1 - r_1}$$

$$v = \frac{p}{r}$$

and substituted for them a more scholarly functional relation, which Anneliese Maier translates into modern language thus—

$$v = \log \left(\frac{p}{r} \right).$$

A relation which is physically incorrect, as is the law of powers that he had set out to concretize, and which brings only a satisfaction of principle, but which presents the great interest of widening the field of algebraical relations, up to then limited to simple differences or proportions. A. Maier tells us that the age recognised in this algebra a real innovation ; hence its infatuation for the art of calculation (William Heytesbury, Richard Swineshead), which was to give birth, as we already know, in Oxford itself, to the abstract kinematics of motion, and in particular to the kinematics of accelerated motion. Thus becomes clear the filiation of the *veteres* of which Nicole d'Oresme considered himself the heir.

---

[1] *Die Vorläufer Galileis in 14. Jahrhundert*, Storia e Letteratura, Rome, 1949.

Anneliese Maier has, too, enriched and made clear the history of the doctrine of *impetus*, through the writings of Olivi, William of Ockham, Francesco de Marchia, Albert of Saxony and Marsilius of Inghen, at the same time giving us new information from the texts of Buridan and Nicole Oresme.[1]

In particular, A. Maier calls attention to the fact that, for Nicole Oresme, *impetus* is not something permanent, as Buridan understood it, but lasts only a certain time *(aliquandiu manet)*, like heat in water. Moreover, for Oresme, *impetus* increases and decreases like acceleration : the acceleration of falling bodies is not due to natural gravity, but to " accidental gravity, " synonymous here with *impetus*.

A. Maier strongly insists on the fact that the doctrine of *impetus* is opposed to the principle of inertia in this respect, that any violent motion continues, according to this doctrine—as in the physics of Aristotle—to require a *vis movendi*, which is simply transferred from the medium to the mobile.

It would appear, however, that Buridan rendered an immense service to mechanics in deducing from the most ordinary experience the first energy theory (after Joannes Philiponus), a theory that gave birth to a tradition which, often regressive in the sense that it was to insist—unlike the thought of the Paris master—on the perishable nature of *impetus*, was to continue up to the classical period.

NOTE II

Henri Poincaré and the principles of mechanics

The centenary of the birth of Henri Poincaré has provided us with an opportunity of defining in a more precise way than we did in this work the attitude of Henri Poincaré to the principles of mechanics, and in particular his contribution to the theory of relativity.

Already in the introduction to his book, *Électricité et Optique*, Poincaré was concerned with the following problem— what is the necessary and sufficient reason for a physical phenomenon to become the object of a mechanical explanation (in the classical sense) ?

Given a physical phenomenon, this latter depends upon the parameters q (n in number) which experiment achieves directly and allows us to measure.

---

[1] *Zwei Grundprobleme der scholastischen Naturphilosophie*, Storia e Letteratura, Rome, 1951.

The laws of variation of these parameters are known by experiment : these laws can generally be stated in the form of differential equations connecting between them the parameters q and the time t.

In order that this phenomenon should be amenable to explanation, it is necessary to find two functions $U(q)$ and $T(q, q')$ the second being a quadratic form in the q', the coefficients of which depend on the q, in such a way that by writing, with the help of these two functions, the principle of least action—

$$\delta \int_{t_1}^{t_2} (T - U)dt = 0$$

or, what amounts to the same, the corresponding equations of Lagrange, these latter become identified with the differential equations expressing the experimental laws of variation of the parameters q in terms of the time.

If such an identification is possible we can, in an infinity of ways, attribute the physical phenomenon in question to the motion of p isolated " molecules. "

Indeed, this amounts to determining p constants m_i and $3p$ unknown functions $\varphi_i(q)$, $\psi_i(q)$, $\theta_i(q)$ that we may consider as the masses and the coordinates of these molecules and which satisfy

$$U(q) = U(\varphi_i, \psi_i, \theta_i)$$

and

$$\frac{1}{2} \sum_1^p m_i \left[\varphi_i'^2 + \psi_i'^2 + \theta_i'^2 \right] = T(q, q')$$

in which

$$\varphi_i' = \frac{\partial \varphi_i}{\partial q_1} q_1' + \frac{\partial \varphi_i}{\partial q_2} q_2' + \ldots + \frac{\partial \varphi}{\partial q_n} q_n'$$

and the same for ψ_i' and θ'. Since the number p may be taken as large as we wish, we can always satisfy these conditions in an infinity of ways.

" *If therefore a phenomenon is amenable to a complete mechanical explanation, it will embody an infinity of others which will equally well account for all the particularities revealed by the experiment.* "

How then are we to choose from this infinity ? That is obviously a matter of personal appreciation. But, Poincaré adds, " there are solutions which everybody will reject on account of their strangeness and others which everybody will prefer on account of their simplicity." [1]

This brings us back to what the philosophers, not without a shade of scepticism, sometimes call Poincaré's " accomodatingness " *(commodisme)*.

---

[1] *La Science et l'Hypothèse*, p. 259.

About 1904, a wide public, the press and even fashionable circles became passionately interested in *La Science et l'Hypothèse*, convinced that they would find in it, to adopt the expression of the time, something like Latin without tears or Greek without groans. It was a source of incomprehension and misunderstanding. " I am beginning to be a bit annoyed, " wrote Poincaré, " with all the fuss a part of the Press is making over a few sentences taken from one of my books, and with the ridiculous opinions it attributes to me. " [1]   And again, " No experiment can disprove this principle : there is no absolute space, all the displacements we can observe are relative displacements.   I have sometimes had occasion to express these considerations, quite familiar to philosophers : they have even brought me a publicity with which I would willingly have dispensed ; according to all the reactionary French newspapers I have demonstrated that the Sun turns round the Earth ; in the famous trial between the Inquisition and Galileo, it is Galileo who would be completely in the wrong. " [2]

Let us hear what Léon Brunschwicg has to say— "By substituting the idea of *convenience (commodité)* for the classical notion of truth, Poincaré seemed to have destroyed the objectivity of geometry and physics, and in so doing to have returned to nominalist empiricism. " [3]   This was an error of interpretation, which Brunschwicg himself refutes in the following way— " For Poincaré, convenience is not only and solely logical simplicity ; it is also that which allows the intelligence to grasp the things themselves. " [4]

We shall make this clearer with the help of the analysis developed by Poincaré in *La Valeur de la Science.*

We start with a law of experimental origin expressing a relation between two objects A and B.   We introduce between these two objects two abstract intermediaries, fictitious if necessary, A' and B'.

Thus, for the initial law we substitute—

> a law between A and A'
> "   " A' and B'
> "   " B' and B

The law relating A' and B' can be raised to the dignity of a principle. To establish a law as a principle, as Poincaré understood it, is to adopt *conventions* such that this law is necessarily satisfied.   " The principle,

---

[1] *Bulletin de la Société française d'Astronomie*, May 1904, p. 216.
[2] *La Mécanique nouvelle* (Lecture delivered at Lille in 1909).
[3] *Revue de Métaphysique et de Morale*, Vol. 21, No. 5, 1913, p. 597.
[4] *Ibid.*, p. 601.

henceforth crystallized, so to speak, is no longer submitted to verification by experiment. It is not true or false, it is convenient. " [1]

Convenient here means satisfied by convention.

For such a principle to be useful, for it to be *convenient* in the widest sense that Poincaré generally uses this expression, it must be common to a large number of laws, and in consequence to a large number of experimental facts.

There is no doubt at all that, unlike the laws which, if they are not contingent, can at least be called in question again by a new experiment, a principle, in Poincaré's opinion, possesses a particular validity, namely the one we have conferred on it by convention.

But in order that such a principle should continue to deserve the qualification *convenient*, it must remain fertile from a practical scientific point of view.

We must not seek, although obviously it is always possible, " to mend the damaged principles by sand-papering them. " [2]   On the contrary, we must build anew— " If a principle ceases to be fertile, experiment, without directly contradicting it—(which is rightly excluded)— will however have condemned it. " [3]

It was the " crisis " in Physics, it seems, that convinced Poincaré, about 1904, of the necessity of a revision of certain principles.

As we have already seen, it was after the fundamental paper of Lorentz that Poincaré broached what he called the " new mechanics, " what, since Einstein, we call the theory of relativity.

Poincaré's work, *Sur la dynamique de l'électron*, published in *Rendi conti del Circolo matematico di Palermo* in 1906, is dated 23 July 1905. It is therefore posterior by less than a month to the paper in which Einstein formulated the principle of relativity (30 June 1905).

Everything tends to show that Poincaré's work is independent of Einstein's. Thus it is essential to note that Poincaré proved much more relativistic from the beginning than Lorentz ; for this latter, while he sought to explain the negative result of Michelson's experiment—which is why he had, as early as 1895, formulated the hypothesis of contraction, independently of Lodge and Fitzgerald—, still considered, in his paper of 1904, the contraction undergone by the electron in the direction of its motion *as the result itself of the application of a uniform translation.*

On the contrary, already in the introduction to his paper of 1905, Poincaré expressed himself quite clearly—

[1] *La Valeur de la Science*, p. 239.
[2] *Ibid.*, p. 207.
[3] *Ibid.*, p. 209.

" It seems that the impossibility of demonstrating experimentally the absolute motion of the Earth is a general law of Nature—that is to say a law of optical and electrical phenomena just as well as of mechanical phenomena ; we naturally tend to accept this law, which we shall call the *Postulate of Relativity*, and to accept it without restriction.

This postulate, which up to now agrees with experiment, must be confirmed or invalidated later by more accurate experiments ; it is interesting in any case to see what may be the consequences of it.

Moreover, if Lorentz was undeniably the first to establish most of the laws on which the Einsteinian doctrine was to be based, he was mistaken on the transformation of the density of an electric charge and on the composition of velocities.

Now these two errors were explicitly corrected by Poincaré in his paper (see § 1, eq. (4) and (4bis)) : he therefore shares with Einstein the merit of having established the new law of the composition of velocities.

Poincaré notes, too, that Lorentz's transformations form a continuous group, which he calls *Lorentz's group*, and from this he deduces that a certain parameter, introduced by Lorentz in the beginning of his analysis, is necessarily reduced to the unit.

For his part, Poincaré sought an explanation of the contraction of the electron and found it in a sort of constant external pressure applied to the deformable and compressible electron, the work of which would be proportional to the variation in volume of the electron. For an electron at rest, assimilated to a sphere superficially charged, this pressure, discovered by Poincaré, equilibrates the electrostatic repulsions. For an electron in motion, Poincaré's pressure, combined with the electrodynamic actions, inevitably produces the flattening required by the postulate of relativity.

Lorentz had already formulated the hypothesis that all sorts of forces are affected by a uniform translation, in the same way as electromagnetic forces.

Poincaré examines what modifications Lorentz's hypothesis necessarily brings about in the laws of gravitation. This leads him to suppose that gravitation is propagated with the velocity of light. Applying then to the forces of gravitation Lorentz's law of transformation for electromagnetic forces and combining it with Newton's law when the velocities of two attracting bodies are sufficiently weak to be negligible compared with the square of the velocity of light, Poincaré shows that the corrected attraction comprises two components, one parallel to the vector which joins the positions of the two bodies, the other parallel to the velocity of the attracting body.

Incidentally, in the course of this last research, while studying the lineary substitutions which conserve the quadratic form $x^2 + y^2 + z^2 - t^2$ (the velocity of light being taken as the unit), Poincaré was led to consider

$$x, y, z \quad \text{and} \quad t \sqrt{-1}$$

as the coordinates of a point in four-dimensional space, which makes him declare that " Lorentz's transformation is only a rotation of this space round the origin. " [1] This constitutes a perfectly clear anticipation of space-time as understood by Minkowski (1908).

In the posthumous work, to which was given the title *Dernières Pensées*, Poincaré returns to this same question and interprets as follows his calculation of 1905—

" Everything happens as if time were a fourth dimension of space ; and as if the four-dimensional space resulting from the combination of ordinary space with time could revolve not only round an axis of ordinary space, but round any axis. For the comparison to be mathematically correct, purely imaginary values would have to be attributed to this fourth coordinate of space. . . . But I shall not insist on this point ; the essential is to notice that in the new conception, space and time are no longer two entirely distinct entities that may be considered separately, but two parts of the same whole, two parts that are so closely interwoven that they cannot easily be separated. "

We pointed out above that everything tends to prove that Poincaré's work is independent of that of Einstein in which this latter formulated the principle of special relativity. Lorentz himself insisted, on this point, in paying a particularly clear tribute to Poincaré.[2]

With absolute sincerity, Lorentz acknowledges that in 1904—that is to say before the existence of the expression—he was not a relativist. " I had an idea that there is an essential difference between the systems x, y, z, t and x', y', z', t'. In the one, use is made—so I thought—of the axes of coordinates which have a fixed position in the ether and of what may be called *true time* ; in the other system, on the contrary, we have to do with simple auxiliary magnitudes, whose introduction is only a mathematical expedient. In particular, the variable t' could not be called *time* in the same sense as the variable t. "

Lorentz adds that it was only by the method of *trial and error* that he had arrived at his formulae of transformation, the exact invariance of which had escaped him— " My formulae remained encumbered with

---

[1] *Sur la dynamique de l'électron*, § 9.
[2] *Deux mémoires d'Henri Poincaré sur la physique mathématique, Acta mathematica*, Vol. 38, 1914, p. 293, and H. A. Lorentz, *Collected Papers*, Vol. VIII, p. 258.

certain terms that should have disappeared. These terms were too small to have a perceptible influence on the phenomena, and I was able therefore to explain the independence of the motion of the Earth that the observations had revealed ; but I did not establish the principle of relativity as being rigorously and universally true.

" On the contrary, Poincaré obtained a perfect invariance of the equations of electrodynamics and he formulated the *postulate of relativity*, an expression that he was the first to use. Indeed, adopting a point of view that had escaped me, he found the formulae [of the composition of velocities] and [of the transformation of the density of the electric charge]. We may add that when he thus corrected the imperfections of my work he never reproached me for them. "

Indeed, it must be emphasized that Einstein alone crossed the Rubicon by making a principle of what Poincaré qualified as a *postulate*. It is nonetheless true, as is testified by the great physicist himself whose work launched the relativistic movement, that Poincaré participated spontaneously in the 1905 revolution along parallel lines with Einstein.[1]

This has not prevented most of the authors opposed to relativity —and there still remain some after half a century—from invoking Poincaré's authority against Einstein, and from quoting in this connexion passages from a lecture delivered at Lille, 3 August 1909, entitled *La Mécanique nouvelle*.

When we re-read this text without preconceived ideas, we do not find in it any sign of a real opposition on the part of Poincaré to the new ideas that he himself had enriched by his own work. It is true that we find in it many proofs of his prudence and wisdom, for he knew better than anyone all that classical mechanics explain ; but, throwing himself into the game with complete liberty of thought and expression, and with all that humour which is one of the aspects of his genius so full of common-sense, Poincaré exposes in all their rigour the innovations brought to the classical ideas by the relativistic theories. " The principle of relativity, " he declares, "admits of no restriction in the new mechanics ; it has, if I may say, an absolute value. "

---

[1] In his *History of the Theories of Aether and Electricity*, Vol. II, Nelson, 1953, Sir Edmund Whittaker clearly attributes to Poincaré and Lorentz the paternity of the theory of relativity and he sees in Einstein's intervention only a systematization " which attracted much attention. "

On the other hand, it is said of Einstein's theory that it had its roots in tradition : " its heterodoxy lay in the strictness of its orthodoxy " (Herbert DINGLE, *Scientific and philosophical implications of the special theory of relativity*, in *Albert Einstein Philosopher-Scientist*, Tudor, New York, 1949, p. 537).

With the boldness of youth, Einstein was no doubt less aware than Poincaré of the sacrifices to which the abandon of the classical point of view led.

It is true that in his conclusion, Poincaré very clearly fixes the limits of the new science and considers that it would be premature, " in spite of the great value of the arguments and facts raised against classical science, " to consider this latter as being definitely condemned. He shows that classical mechanics will remain that " of our practical life and of our terrestrial technique, " and he emphasizes the necessity of a thorough knowledge of classical mechanics if we wish to understand the new mechanics.

But, in his *Dernières Pensées*,[1] Poincaré sums up clearly the position of the theory of relativity—

" On all the points on which it departs from Newtonian mechanics, the mechanics of Lorentz subsists. We continue to believe that a body in motion will never exceed the velocity of light, that the mass of a body is not constant, but depends on its velocity and the angle this velocity forms with the force acting on the body, that no experiment will ever be able to decide whether a body is at rest or in motion, either in relation to absolute space, or even in relation to the ether. "

In the last analysis, Poincaré thus gave his adhesion to the theory of relativity.

[1] Chap. VI, p. 165.

INDEX

The figures refer to page numbers in the text. Those in thick type **(148)** relate to passages where an author is considered at length; those in italics (*254*) indicate quotations from his work.

ABRAHAM (M.), 472.
 Ann. der Physik, vol. 10, 1903, p. 105.
ADRASTUS, 23, 24.
AILLY (P. d'), 68.
AIRY, 463.
ALBERTUS MAGNUS, 47, 52, 57, 82, 214.
ALBERT OF RICKMERSDORF, called Albert of Saxony, **51-58**, 61, 68, 72, 76, 81-87, 92, 103, 132, 213, 214, 222.
 Acutissimae Quaestiones: *52-56*.
ALEMBERT (J. le Rond d'), 28, 193, 233, 237, 243, **244-251**, 269, 274, **290-299**, 304, 307, 311, 313, 323, 324, 328, 332, 341, 344, 357, 368, 401, 423, 445, 522, 637.
 Traité de Dynamique, Paris, 1743; 1758 edition: *245-253, 269-270*.
 Traité de l'Équilibre et du Mouvement des Fluides, Paris, 1744 : *291-296*.
 Essai d'une nouvelle théorie de la résistance des Fluides, Paris, 1752 : *291*.
 Opuscules math., vol. V, ed. 1778, p. 132 :
 Paradoxe proposé aux Géomètres sur la résistance des Fluides : *296-299*.
ALEXANDER OF APHRODISIAS, 87.
AMONTONS (G.), 319, 321.
 Mémoires de l'Acad. des Sciences pour 1699.
AMPÈRE (A. M.), 374.
ANDERSON (C.), 607.
ANDRADE (J.), 440, 449.
 Leçons de mécanique physique, Soc. d'édit. scientifiques, Paris, 1898 : *441*.
ANONYMOUS AUTHOR OF THE XIIIth CENTURY (The), **41-46**, 75, 95, 103, 149.
 Liber Jordani de ratione ponderis : *42-43, 44-45*.
 De ponderibus.
APOLLONIUS OF PERGUM, 99.
ARAGO (F.), 382, 416.

ARCHIMEDES, **24-31**, 32-36, 38, 96, 99, 100, 123, 128, 146, 149-151, 166, 180, 188, 222, 225, 333.
 On the Equilibrium of Planes or on the Centres of Gravity of Planes, French translation by Peyrard, Paris, 1807 : *24-27*.
 On Floating Bodies, French translation by Peyrard, Paris, 1807 : *28-31*.
ARCHYTAS OF TARENTO, 99.
ARCY (d'), 341.
ARISTARCHUS OF SAMOS, 83.
ARISTOTELIAN, **19-24**, 29, 35, 36, 38, 47-49, 51, 53, 57, 58, 63, 65, 74, 82-91, 93, 97, 99, 103-105, 117, 129, 143, 145, 151, 157, 171, 208, 222, 224, 225, 333.
 Problems of Mechanics : *19*.
 Treatise on the Heavens : *20, 22, 23*.
 Physics : *20, 22*.
ARNOLD OF BRUSSELS, 69.
AVERROËS, 52, 57, 63, 87, 214.

BACHEM, 526.
BACON (Francis), 214.
BACON (Roger), 48, 52, 68, 214, 215.
BALDI (Bernardino), **106-107**, 149, 151, 163, 222.
 Exercitationes in mechanica Aristotelis problemata, 1582 : *106-107*.
BALMER, 537, 538, 546, 547.
BARTHOLIN (E.), 198.
BAZIN (H.), 318.
BEAUGRAND (J. de), 166, 167, 224.
 Géostatique, 1636.
BEECKMANN (I.), 158, 159.
 Journal : *159*.
BENEDETTI (J. B.), **103-105**, 149.
 Diversarum speculationum mathematicarum et physicarum, 1585 : *103-105*.

BERNOULLI (Daniel), **233-234**, 248, **287-290**, 291, 294, 304, 306, 308, 310, 333, 341.
Examen principiorum mechanicae et demonstrationes geometricae de compositione et resolutione virium, 1726.
Hydrodynamica, sive de viribus et motibus fluidorum commentarii, Argentorati (Strasbourg), 1738. *287-289*.
Mémoires de Petersbourg, 1727 and 1741.

BERNOULLI (Jacques), 193, **243-244**, 251.
Démonstration générale du Centre de balancement ou d'oscillation tiré de la nature du levier, 1703.

BERNOULLI (Jean), 40, **231-234**, 235, 290, 310, 334, 341, 361, 374.
Letter to Varignon (January 26th, 1717) : *231-233*.

BERTRAND (J.), 332, 372.
Journal de l'École polytechnique, vol. 32, 1848, p. 149.

BINET (J.), 382.

BITROGI (Al.), 47.

BLACKETT, 607.

BLASIUS OF PARMA, 69, 70.
Traité des Poids, 1476.
Quaestiones super tractatu de latitudinibus formarum, 1486.

BOHR (N.), **535-548**, 549-552, 558-560, 566, 568, 573-575, 577, 582, 606, 607, **614-616**, 632-635.
On the constitution of Atoms and Molecules, Phil. Mag., vol. 26, 1913, pp. 1 and 476 : *535-536*.
Phil. Mag., vol. 27, 1914, p. 509 and vol. 29, 1915, p. 332 : *538-539*.
Über die Anwendung der Quantentheorie auf dem Atombau, Zeits. für Physik, vol. 13, 1923, p. 117.
La théorie atomique et la description des phénomènes, Gauthier-Villars publ., Paris, 1932 : *551, 614-616, 633*.
Phys. Review, vol. 48, 1935, p. 696 : *632*.

BOLTZMANN (L.), 550, 567.

BORDA (Chevalier de), **305-308, 311-314**, 374.
Sur l'écoulement des fluides par les orifices des vases, Mémoires de l'Académie des Sciences pour 1766, p. 579 : *305-308*.
Mémoires de l'Académie des Sciences pour 1766, p. 147 : *308*.
Mémoires de l'Académie des Sciences pour 1763, p. 358.
Mémoires de l'Académie des Sciences pour 1767, p. 495 : *312, 313*.

BORN (M.), 575, 577, 581, **586-587**, 589, 624 (see Heisenberg, W.).
In collaboration with Jordan (P.) : Zeits. für Physik, vol. 34, 1925, p. 858.
Quantenmechanik der Stossvorgänge, Zeits. für Physik, vol. 38, 1926, p. 803 : *586, 587*.

BORELLI (J.), 216.

BORRO, 100.

BOSCOVITCH (R. G.), 456.

BOSE (S. N.), 567.
Zeits. für Physik, vol. 26, 1924, p. 178.

BOSSUT (The Abbé), **313-316**, 317.
Nouvelles expériences sur la résistance des fluides, Jombert publ., Paris, 1777 : *313-316*.

BOSWELL, 157.

BOUGUER (P.), 279.
Mémoires de l'Académie des Sciences pour 1734.

BOULIGAND (G.), 455, 618.
La causalité des théories mathématiques, Hermann publ., Paris, 1934.
Comptes Rendus de l'Acad. des Sciences, vol. 200, 1935, p. 1500.

BOULLIAU, 215.
Astronomia philolaïca, 1645.

BOUSSINESQ (J.), 456.

BRACKETT (F.), 537.
Nature, vol. 109, 1922, p. 209.

BRADLEY (J.), 463.

BRADWARDINE (Th.), 57, 66.
Tractatus proportionum, 1328.

BRAVAIS (A.), 382.

BROGLIE (L. de), 401, 501, **554-561**, 564, **567-570**, 581-583, 586, 600-602, 610-612, 615, 616, 619, 620, 633-636, 638.
Comptes Rendus de l'Acad. des Sciences, vol. 177, 1923, pp. 507, 548 and 630.
Recherches sur la théorie des quanta (Thèse), Annales de Physique, vol. III, 1925, p. 22 : *554, 555, 558*.
Notice sur les travaux scientifiques de M. Louis de Broglie, Hermann publ., Paris, 1931 : *567-569*.
Comptes Rendus de l'Acad. des Sciences, vol. 194, 1932, p. 693.
Théorie de la quantification dans la nouvelle Mécanique, Hermann, publ., Paris, 1932 : *602*.
Les idées nouvelles introduites par la mécanique quantique, l'Enseignement Math., 1933 : *633*.
Quelques remarques sur la théorie de l'électron magnétique de Dirac, Arch. de Genève, vol. 15, 1933, p. 465 : *611-612*.

L'électron magnétique de Dirac, Hermann publ., Paris, 1934.
Voies anciennes et perspectives nouvelles en théorie de la lumière, Revue de Métaphysique et de Morale, vol. 41, 1934, p. 448 : *568-569.*
Continu et discontinu en physique moderne, Albin Michel publ., Paris, 1941 : *634-636.*
BROGLIE (M. de), 550.
BRUNO (G.), 105, 109.
Cena de le Ceneri, 1584.
BUAT (P. du), **316-319.**
Principes d'hydraulique, ed. 1779 and 1786 : *316-319.*
BUISSON (H.), 547.
BULFINGER (G. B.), 235.
BURGERS, 544.
Ann. der Physik, vol. 52, 1917, p. 195.
BURIDAN (Jean I), **47-51,** 57, 59, 69, 72, 76, 83, 87-89, 93, 104, 117, 184.
Quaestiones octavi libri physicorum : *49-51.*
BURLEY (W.), 48, 54, 69.

CALCAGNINI, 215.
CAMPBELL, 527.
CAMUS (Ch.), 285.
CAMUS (de), 319.
CANONIO (DE), 37, 46.
CANTOR (M.), 59.
CARDAN (J.), **96-99,** 125, 129, 214.
De Subtilitate, 1551 : *96-98.*
Opus Novum, 1570 : *96-98.*
CARNOT (Lazare), 305, **323-331,** 356, 357, 436.
Essai sur les machines en général, Paris, 1783.
Principes généraux de l'équilibre et du mouvement, Paris, 1803 : *323-331.*
CARTAN (E.), 533.
CASSINI (J. D.), 286.
CASTELLI (Benedetto), 143, 166, 169.
Della misura dell'acque correnti, 1628.
CATELAN (The Abbé de), 193, 235, 236.
CAUCHY (A.), 304, 417, **418-422,** 423.
Sur les dilatations, les condensations et les rotations produites par un changement de forme dans un système de points matériels, Œuvres complètes, vol. XII, 2nd series, p. 343 : *422.*
CAVALIERI (B.), 166.
CAVENDISH (C.), 164.
CELAYA (J. de), 87, 90.
Expositio in libris Physicorum, 1517 : *89-90.*
CELSIUS (A.), 285.

CESALPIN (A.), 100.
CEVA (Father), 224.
CHAMBRE (C. de la), 254, 259.
CHARISTON (THE BOOK OF), 37, 38.
CHASLES (M.), 38.
CHATELET (The Marchioness of), 238.
CHAZY (J.), 496, 525, 527.
La théorie de la relativité et la mécanique céleste, 2 vol. Gauthier-Villars publ., Paris, 1928 and 1930.
CHRISTOFFEL (E.), 505, 508, 510, 519, 520.
Über die Transformation der homogenen Differentialausdrücke zweiten Grades, Journal de Crelle, vol. 70, 1869, p. 46.
CLAIRAUT (A.), 201, 238, 261, **279-285,** 301, 304, **370-372,** 373, 411, 637, 638.
Sur quelques principes donnant la solution d'un grand nombre de problèmes, Mémoires de l'Acad. des Sciences, 1742, p. 1 : *370-372.*
Théorie de la figure de la Terre tirée des principes de l'hydrodynamique, Durand publ., Paris, 1743 : *279-284.*
CLAPEYRON (B.), 436.
CLARKE, 235.
CLERSELIER (C.), 258-259.
Letter to Fermat, May 6th, 1662 : *257-259.*
COLBERT (J. B.), 181.
COLOMBE (L. delle), 143.
COMPTON (A. H.), 588, 589.
Phys. Review, vol. 25, 1925, p. 306.
CONDORCET (A. de), 313.
CONTI (The Abbé de), 220.
COPERNICUS (N.), 58, 62, 66, 72, **82-86,** 103-105, 108-111, 114, 118, 144, 145, 213, 215, 271, 444, 450, 451, 489, 493.
De revolutionibus orbium caelestium : *85.*
CORIOLIS (G.), 331, 370, 372, **374-379,** 380, 383.
Journal de l'École polytechnique, 21st cahier, 1832, p. 268 : *374-376.*
Ibid., 24th cahier, 1835, p. 142 : *377-379.*
CORONEL (L.), 87, 89.
Physicae perscrutationes, 1511 : *88.*
COTTINGHAM (E. T.), 527.
COULOMB (C. A. de), 318, **319-322,** 544.
Théorie des machines simples en ayant égard au frottement et à la raideur des cordages, Mémoires des Savants Étrangers, vol. X : *320, 321.*
CROMMELIN (A.), 527.
CUES (N. of), **71-72,** 76, 77, 88, 105, 110, 117.
De docta ignorantia.

De ludo globi.
Dialogus trilocutorius de Possest : *71.*
CURIE (I.), 607, 608.
CURTIUS TROJANUS, 41, 95.
CURTZE, 38.

DARBOUX (G.), 408, 523.
DARCY (H.), 318.
DARMOIS (G.), 526.
DARWIN (C.), 591.
DAUNOU (P.), 38.
DAUVILLIER (A.), 561.
DAVIDSON (C.), 527.
DAVISSON (C.), 561, 616.
Phys. Review, vol. 30, 1927, p. 705.
DAVY, 436.
DEBYE (P.), 550, 561.
DEMOCRITUS, 635.
DESAGUILLERS (J.), 319.
DESCARTES (R.), 40, 59, 72, 100, 107, 150,
154-169, 171, 172, 174, 176, 178, 180,
193, 205, 218-222, 224, 225, 231, 254,
255, 257-262, 268, 269, 274, 295, 333,
334, 340, 359, 456.
Letter to Constantin Huyghens,
October 5th, 1637 : *154-156.*
Letters to Mersenne (Cf. Œuvres com-
plètes) : *157-160, 163, 164, 166, 167.*
Dioptrique, 1637 : *160.*
Principia Philosophiae, Amsterdam,
1644 : *161-162.*
Letter to Boswell, 1646 : *157.*
Œuvres complètes, publiées par C. Adam
et P. Tannery, Cerf publ., Paris, 1897:
163, 164, 165, 167, 168.
DESTOUCHES (J. L.), 617, 618.
Bulletin de l'Acad. Royale de Belgique,
1935, p. 780 et 1936, p. 525.
DINOSTRATUS, 225.
DIOCLES, 225.
DIRAC (P. A. M.), **578-581**, 582, 590,
592-600, 602, **603-606**, 607, 611, 612,
617, 619, 622, 625, 626, 633.
The Fundamental Equations of Quan-
tum Mechanics, Proceed. Royal Society,
A, 1925, vol. 109, p. 642 : *581.*
Proceed. Royal Society, A, vol. 117,
1928, p. 610 et A, vol. 118, 1928,
p. 351.
Principles of Quantum Mechanics,
Cambridge, 1930. French translation
by Proca and Ullmo, Presses Universi-
taires de France, Paris, 1931 : *593, 606.*
DOPPLER (Ch.), 463, 468, 589.
DROSTE, 526.
DUGAS (R.).

Sur la pensée dynamique d'Hamilton :
origines optiques et prolongements mo-
dernes, Revue Scientifique, vol. 79,
1941, p. 15.
Sur l'origine du théorème de Coriolis,
Revue Scientifique, vol. 79, 1941, p. 267.
Le principe de la moindre action dans
l'œuvre de Maupertuis, Revue Scientifique, vol. 80, 1942, p. 51.
L'énergie cinétique à travers l'histoire
de la mécanique, Revue Scientifique,
vol. 84, 1946, p. 67.
Vicissitudes de la notion de force, Revue
Scientifique, vol. 84, 1946, p. 451.
Axiome des conditions initiales et léga-
lité en mécanique quantique, Bull. de
l'Acad. Royale de Belgique, vol. 22,
1936, p. 1318.
Sur l'évolution de la mécanique des
quanta, Journal de l'Ecole polytechnique, 1937, pp. 64, 144 and 216.
Etude comparée de méthode quantique,
Revue philosophique, vol. 63, 1938,
p. 317.
DUHEM (P.), 14, 15, 20, 22, 36-38, 41, 43,
46-48, 55, 56, 58, 59, 62, 67, 74, 93, 96,
104, 107, 143, 158, 160, 169, 208, 213,
214, 225, 226, **456-457**, 618, 635.
L'Évolution de la Mécanique, Joanin
publ., Paris, 1903 : *457.*
Les Origines de la Statique, 2 vol.
Hermann publ., Paris, 1905 : *22, 40,
41, 96, 101, 158, 226.*
Études sur Léonard de Vinci, 3 vol.
Hermann publ., Paris, 1906, 1909 and
1913 : *58, 71, 107.*
La Théorie physique, Chevalier et Ri-
vière publ., Paris, 1903.
DULLAERT DE GAND (J.), 87-90.
DUMAS, 382.

EDDINGTON (Sir A. E.), 495, 496, 501,
506, 527.
Space, time, gravitation, Cambridge,
1920 ; French translation, Hermann
publ., Paris, 1921 : *495, 496.*
EHRENFEST (P.), 544, 591, 622, 625, 626.
Annal. der Physik, vol. 51, 1916, p. 327
and Phil. Mag., vol. 33, 1917, p. 504.
Bemerkung über die angenäherte Gültigkeit der klassischen Mechanik innerhalb der Quantenmechanik, Zeits. für
Physik, vol. 45, 1927, p. 455 : *591.*
EICHENWALD (A.), 463.

EINSTEIN (A.), **472-482**, 485, 488-497, 501, **502-517**, 521, 522, 524, 528-534, 550, 559, 567-570, 581, 586, 615, 627-630, 632-634.
Annal. der Physik, vol. 17, 1905, p. 132.
Zur Elektrodynamik bewegter Körper, Annal. der Physik, vol. 17, 1905, p. 891 : *473, 474, 482*.
Annal. der Physik, vol. 20, 1906, p. 199 and vol. 22, 1907, p. 180.
Jahrbuch für Radioaktivität und Elektronik, vol. 4, 1907.
Annal. der Physik, vol. 35, 1911.
Die Grundlage der allgemeinen Relativitätstheorie, Annal. der Physik, vol. 49, 1916, p. 769 : *503-504, 507, 508-509*.
Über die spezielle und die allgemeine Relativitätstheorie (gemeinverständlich), Vieweg publ., Brunschweig, 1916 : *491-493*.
Kosmologische Betrachtungen zur allgemeinen Relativitätstheorie. Sitzungsber. der Preuss. Akad. der Wissenschaften, 1917, p. 150 : *515*.
Verh. der Deutsch. Physik. Gesellschaft, vol. 19, 1917, p. 77.
Sitzungsber. der Preuss. Akad. der Wissenschaften, 1924, p. 261.
Can quantum mechanical description of physical reality be considered complete? (in collaboration with B. Podolsky and N. Rosen), Phys. Review, vol. 47, 1935, p. 777 : *627, 629, 632-633*.
ELLIS, 550.
ENGELHARDT, 271.
EPSTEIN (P.), 540, 542, 544.
Annal. der Physik, vol. 50, 1916, p. 489 and vol. 51, 1916, p. 184.
ERASME (D.), 90.
Colloquia, 1522.
ERATOSTHENES, 52.
EUCLID, 99.
EUCLID (FRAGMENTS ATTRIBUTED TO), 36, 38, 41.
Livre d'Euclide sur la balance, transl. Wœpke, Journal Asiatique, 1851.
Liber Euclidis de gravi et levi.
Liber Euclidis de ponderibus secundum terminorum circonferentiam.
EULER (L.), 182, 233, **239-242**, 248, 251, 252, 271, 272, **273-274**, 275, **276-278**, 295, **300-304**, 324, 326, 332, 339-342, 346, 347, 374, 391, 393, 407, 418, 570, 637.
Mechanica, sive motus scientia analytice exposita, St-Petersbourg, 1736 :

239-241. Methodus inveniendi lineas curvas maximi minimive proprietate gaudentes, Bousquet publ., Lausanne, 1744 : *273, 274*.
Dissertation sur le principe de la moindre action, Berlin, 1753 : *271-273*.
Theoria motus corporum solidorum seu rigidorum, 1760 : *276-278*.
Principes généraux de l'état d'équilibre des fluides, Mémoires de l'Acad. de Berlin, 1755, p. 217 : *299-301*.
Principes généraux du mouvement des fluides, Mémoires de l'Acad. de Berlin, 1755, p. 274 : *301-302, 303, 304*.
Continuation des recherches sur la théorie du mouvement des fluides, Mémoires de l'Acad. de Berlin, 1755, p. 316 : *305*.
EVANS, 538.
EVERSHED, 526.

FABRI (Father), 222.
Tractatus physicus de motu locali, 1646.
FABRY (Ch.), 547.
FERMAT (P. de), 166, 167, 169, 224, **254-260**, 261, 262, 264, 268, 274, 275, 391, 394, 452, 558, 563, 568, 570, 638.
Œuvres complètes, éditées par P. Tannery et Ch. Henry, Gauthier-Villars publ., Paris, 1891-1894, 1896-1912 and 1922 : *167*.
Letter to C. de la Chambre, January 1st, 1662 : *254-255, 257*.
Synthesis ad refractiones, 1662 : *256*.
Letter to Clerselier, May 21st, 1662 : *258-259*.
Letter to an unknown person, 1664 : *259*.
FERMI (M.), 620, 621.
Nuovo Cimento, 1930.
FERNEL (J.), 86.
Cosmotheoria, Paris, 1528.
FITZGERALD (G.), 466.
FIZEAU (H.), 463, 468, 497, 589.
FORLI (J. de), 70.
De intentione et remissione formarum, Venise, 1496.
FOUCAULT (L.), **380-383**, 450, 454.
Démonstration physique du mouvement de la Terre au moyen du pendule, Comptes Rendus de l'Acad. des Sciences, vol. 32, 1851, p. 135 : *380-383*.
Comptes Rendus de l'Acad. des Sciences, vol. 35, 1852, p. 421 : *382-383*.
FOURIER (J. B.), 28, **361-366**, 543, 572, 573, 577.

Mémoire sur la Statique, contenant la démonstration du principe des vitesses virtuelles et la théorie des moments, Journal de l'École polytechnique, 5th cahier, an VI : *361, 362, 363, 365, 366.*
FRANK (J.), 550.
Verh. der Deutsch. Physikal. Gesellschaft, vol. 15, 1913, p. 613.
FRANK (P.), 491.
Die Stellung des Relativitätsprinzips im System der Mechanik und Elektrodynamik, Wiener Sitzungsber., vol. 118, 1909, p. 337.
FRANKLIN (B.), 314.
FRASCATOR (G.), 214.
FRENKEL, 607.
FRESNEL (A.), 463, 490, 492, 493, 497, 569, 635.
FUES, 602.
FULLENIUS, 194.
FURRY (W. H.), 615, 618.
Phys. Review, vol. 49, 1936, p. 393 : *629, 632.*

GALILEO (G.), 24, 35, 46, 52, 100, **129-145**, 146-149, 156-160, 166, 181-184, 195, 200, 205, 215, 222, 231, 240, 332, 333, 339, 444, 453-455, 467, 493, 531.
Discorso intorno alle cose che stanno in su l'acqua o che in quella si muovono, Florence, 1612 : *143.*
Les Mécaniques de Galilée (French translation by P. Mersenne), Paris, 1634 : *129, 130, 131.*
Discorsi e dimostrazioni matematiche intorno a due nove scienze attenanti alla meccanica ed i movimenti locali ; 1638 and 1655 : *135-142.*
Dialogo sopra i due massimi sistemi del Mondo, Tolemaico, e Copernicano, Florence, 1632 : *144-145.*
Della scienza meccanica, Ravenne, 1649.
Opere, national italian edition, Florence, 1908 : *132-133.*
GASSENDI (P.) 147.
Tres epistolae de motu impresso a motore translato, 1640.
GAUSS (C. F.), **367-369**, 447, 505, 533.
Über ein neues Grundgesetz der Mechanik, Journal de Crelle, vol. V, 1829 : *367-369.*
GEHRKE, 547.
GEIGER (H.), 607.
GERLACH, 589.
GERMER (L.), 561, 616.

GILBERT (W.), 214.
De Magnete, London, 1660 : *214.*
GIRARD, 414.
GLYMI ESHEDI (R. de), 67.
De medio uniformiter difformi : *68.*
GOUDSMITH (S.), 607.
GRAVESANDE (J. s'), 235, 271.
GRAZIA (V. di), 143.
GREBE, 526.
GRISOGONE, 215.
GUIDO UBALDO, 35, 41, **100-101**, 106, 149, 151, 155, 156, 169, 333, 334.
Mechanicorum liber, 1577 : *100-101.*
GUYE (C. E.), 472.
Arch. de Genève, vol. 41, 1916, p. 286.

HALLEY (E.), 200, 216, 217.
HAMILTON (Sir W. Rowan), 275, **390-401**, 402, 403, 406, 407, 452, 461, 499, 543, 555, 556, 559, 561-564, 570, 576, 623, 624, 638.
On a general method of expressing the paths of light and of the planets, by the coefficients of a charasteristic function, Dublin University Review, 1833 : *391-392.*
Third Supplement to an Essay on the Theory of Systems of Rays, 1832.
First Essay on a general method in Dynamics, 1834 : *395.*
On the application to Dynamics of a general mathematical method previously applied to Optics, 1834 : *399, 400.*
Second Essay on a general method in Dynamics, 1835.
The mathematical papers of Sir William Rowan Hamilton. Vol. I, Geometrical Optics, Cambridge University Press, 1931. — Vol. II, Dynamics, Cambridge University Press, 1940.
HARRESS (F.), 463.
HEISENBERG (W.), 547, 566, **571-577**, 578, 579, 581-583, 586, **587-590**, 599, **609-611**, 615, 617, 622, 624, 628, 632, 633, 635.
Über quantentheoretische Umdeutung kinematischer und mechanischer Beziehungen, Zeits. für Physik, vol. 33, 1925, p. 879 : *574.*
(In collaboration with Born and Jordan) Zur Quantenmechanik II, Zeits. für Physik, vol. 35, 1926, p. 557.
Über den anschaulichen Inhalt der quantentheoretischen Kinematik und Mechanik, Zeits. für Physik, vol. 43, 1927, p. 172 : *587-591.*

Les principes physiques de la théorie des quanta, French translation, Gauthier-Villars publ., Paris, 1932 : *632.*

HELMHOLTZ (H. von), **434-436,** 451.

Mémoire sur la Conservation de la force, French translation Pérard, Masson publ., Paris, 1869 : *435, 436.*

HENRY, 436.

HÉRIGONE (P.), 149, 158.

HERMAN (J.), 238, 252, 270.

Phoronomia, sive De Viribus et Motibus corporum solidorum et fluidorum, Amsterdam, 1716.

HERMITE (Ch.), 566.

HERO OF ALEXANDRIA, **32-33,** 34, 38, 100.

Les Mécaniques, French translation Carra de Vaux, Journal Asiatique, 1894 : *19.*

HERTZ (G.), 550 (see Frank, J.).

HERTZ (H.), **444-447,** 457, 544.

Die Prinzipien der Mechanik in neuem Zusammenhang dargestellt, Leipzig, 1894 : *444-447.*

HEVELIUS (J.), 150.

HEYTESBURY (W.), **66-67,** 70, 81, 87, 138.

Regulae solvendi sophismata.

De tribus praedicamentis : *66, 67.*

HILBERT (D.), **526,** 582, 600, 601.

HOOKE (R.), 216.

An attempt to prove the annual motion of Earth, London, 1674 : *216.*

HUGONIOT (H.), **423-433.**

Mémoire sur la propagation du mouvement dans les corps et spécialement dans les gaz parfaits, Journal de l'École polytechnique, 57th cahier, 1887, p. 3, and 58th cahier, 1889, p. 1 : *424, 428-430.*

HUYGHENS (Constantin), 154.

HUYGHENS (Christian), 166, 172, **176-199,** 200, 201, 205, 207, 210, 216, 219, 221, 222, 231, 239, 262, 268, 279, 280, 286, 287, 310, 332, 333, 339, 340, 345, 370, 390, 438, 461, 563, 637.

Horologium oscillatorium, sive de motu pendulorum ad horologia aptato demonstrationes geometricae, Paris, 1673 : *181-193, 197-198.*

Traité de la Lumière, Paris, 1679.

De motu corporum ex percussione, 1700 : *176-180.*

De vi centrifuga, 1703 : *195-196.*

Œuvres complètes, publiées par la Société hollandaise des Sciences, La Haye, 1934 : *198.*

INGHEN (M. d'), 68, 92.

JACOBI (C. G.), **401-408,** 499, 500, 523, 542, 544, 545, 553, 561, 564, 565, 603, 618, 623-625.

Journal de Crelle, vol. 17, p. 97 : *401.*

Vorlesungen über Dynamik, Reimer publ., Berlin, 1866 : *402-404, 407.*

JANDUN (J. de), 48, 57, 213.

JEANS (J. H.), 623.

JOHN OF ALEXANDRIA (Philopon), 47.

Erudissima commentaria in primis quatuor Aristotelis de naturali auscultatione libris, Venise, 1532 : *47.*

JOLIOT (F.), 607, 608.

JORDAN (P.), (see Born M. and HEISENBERG W.), 576, 581, 590.

JORDANUS OF NEMORE, **38-42,** 43, 45, 46, 56, 67, 103, 149, 158, 222, 231.

Elementa Jordani super demonstrationem ponderis : *39, 40.*

JOUGUET (E.), 14, 15, 33, 54, 135, 140, 163, 165, 173, 180, 205, 241, 366, 437, 441, 447.

Lectures de Mécanique, 2 vol., Gauthier-Villars publ., Paris, 1922 : *163, 205, 366, 437.*

JOULE (J.), 436.

KAUFMANN (W.), 472.

Physical. Zeitschrift, vol. 4, 1902, p. 55.

KENNARD, 591.

KEPLER (J.), 72, 82, 88, 108, **110-119,** 209, 213-216, 271, 453, 492, 493, 529.

Mysterium Cosmographicum, Tubingen, 1596.

Astronomia nova αἰτιολογητός, seu Physica tradita commentariis de motibus stellae Martis ex observationibus G. V. Tychonis Brahe, Prague, 1609 : *110, 111, 112, 113, 114, 115, 116, 118, 119.*

Harmonices Mundi, Linz, 1619 : *116.*

Tabulae Rudolphinae, 1627.

Opera omnia, Heyden and Zimmer publ., Francfort and Erlangen, 1858 : *117, 118.*

KICHUCHI, 561.

Jap. Phys. Journal, vol. 5, 1928, p. 83.

KIRCHHOFF (G.), **441-443,** 446, 449, 457, 549.

Mechanik, Teubner, Leipzig, 1876 : *441-442.*

KOENIG (S.), 238, 270-273.

Acta eruditorum, Leipzig, 1751 : *270.*

KRAMERS, 548.

KUHN, 574.

Zeits. für Physik, vol. 33, 1925, p. 408.

KURLBAUM, 549.

LAGRANGE (L.), 100, 194, 227, 231, 247, 252, 253, 275, 290, 295, 303, 323, **332-349**, 353, 367, 372, 374, 384, 387, 389, 391, 393, 395, 396, 398-400, 407, 418, 423, 447, 452, 456, 458, 499, 523, 531, 539, 555, 561, 570, 637.
Mécanique analytique, Paris, 1788 ; *posthumous* ed. 1853-1855 : *290, 296, 304, 334-349.*
LALANDE (J. de), 86.
LAMÉ (G.), 416, 417, 456.
Leçons sur la théorie mathématique de l'Élasticité des corps solides, Bachelier publ., Paris, 1852 : *416, 417.*
LAMY (Father), **222-224**, 226, 334.
Traitez de Méchanique, 1679.
Letter to Dieulamant, 1687 : *222-223.*
LAPLACE (P. S.), 331, **354-360**, 389, 396, 423, 428, 498.
Mécanique céleste, Duprat et Bachelier publ., Paris, an VII : *354-358.*
Exposition du Système du Monde, 4th edition, Vve Courcier publ., Paris, 1813 : *359-360.*
LAU, 547.
LAUE (M. von), 494, 561.
LAVANCHY, 472.
LEGENDRE (A. M.), 313.
LEIBNIZ (G. W. von), 67, **219-221**, 260, 262, 268, 270, 271, 287, 341, 456, 550, 634, 637.
Acta eruditorum, Leipzig, 1682.
Ibid., Leipzig, 1686 : *220.*
Specimen dynamicum, 1695 : *221.*
LEJEUNE DIRICHLET (P.), 347.
LENARD (P.), 487-490.
Über Äther und Uräther, Hirzel, Leipzig, 1921 : *474.*
LEVI-CIVITA (T.), 408, 505, 518, 524, 534.
Math. Annalen, vol. 54, 1899 : *505-506.*
LIOUVILLE (J.), 408, 523, 624.
LIPSCHITZ, 408, 523.
LOCKE (J.), 323.
LOCKERT (G.), 87.
LODGE (O.), 466.
Philosophical Transactions, vol. 184, 1893, p. 727.
LONGOMONTANUS, 114.
LORENTZ (H. A.), 455, 463, **464-472**, 477, 481-485, 488-490, 493-495, 503, 555, 570, 604, 605.
Versuch einer Theorie der elektrischen und optischen Erscheinungen in bewegten Körpern, Leyde, 1895 : *465.*
Electromagnetic phenomena in a system moving with any velocity smaller than that of light, Proceed. Acad. Sc. Amsterdam, vol. 6, 1904, p. 809 : *469.*

LUCRETIUS, 635.
LUMMER, 549.
LYMAN, 537.

MACH (E.), 14, 15, 28, 144, 201, 205, 274, **443-444**, 447, 493.
Die Mechanik in ihrer Entwicklung, historisch kritisch dargestellt, Brokhaus, Leipzig, 1883.
French translation, Hermann publ., nouvelle édition, Paris, 1925 : *201, 274, 443-444.*
MAC LAURIN (C.), 235, 285, 290, 339.
Treatise of Fluxions, Edimbourg, 1742.
MAGELLAN (F. de), 85.
MAIRAN (J. de), **236-238**, 262.
Dissertation sur l'estimation et la mesure des forces motrices des corps, 1726 : *236-237.*
MAJORIS (J.), 84, 87, 88, 90.
De infinito.
Disputationes theologiae.
MALEBRANCHE (Father of), 271.
MARGENAU (H.), 631.
Phys. Review, vol. 49, 1936, p. 240 : *631.*
MARICOURT (P. de), 231.
MARIOTTE (The Abbé), 173, **199**, 207, 243, 286, 290, 310.
Traité de la percussion et du choc des corps, Paris, 1677 : *211.*
Traité du mouvement des eaux et des autres corps fluides, Paris, 1684.
MAUPERTUIS (P. L. Moreau de), 238, **260-272**, 273-275, 285, 329, 330, 341, 346, 367, 391, 452, 555, 558, 563, 568-570, 638.
Mémoires de l'Académie des Sciences pour 1740 : *260, 261.*
Mémoires de l'Académie des Sciences pour 1744 : *260-265.*
Mémoires de l'Académie de Berlin, 1747 : *264-267.*
Œuvres complètes, Walter publ., Dresde, 1752 and Bruyset publ., Lyon, 1756 : *267-269, 270-272.*
MAURO OF FLORENCE, 86.
MAXWELL (J. Clerk), 463, 464, 472, 473, 479, 480, 490, 519, 521, 569, 605, 634, 635.
MAZIÈRE (Father), 235.
MEISSNER, 547.
MELANCHTON (P.), 84.
MERSENNE (Father), 41, 103, 107, 129, **149-150**, 151, 156-160, 163, 166, 167, 169, 222.

Synopsis mathematica (Mechanicorum libri), Paris, 1626.
Harmonicorum libri, Paris, 1636.
MEUSNIER (Ch.), 316.
MEYERSON (E.), 616.
MICHAUD (J.), 38.
MICHELSON (A. A.), 461, 464-466, 471, 484, 487, 488, 547.
American Journal of Science, vol. 22, 1881, p. 120.
American Journal of Science, vol. 34, 1887, p. 333.
MILLIKAN (R. A.), 538, 550.
MINKOWSKI (H.), 482-486, 490, 494, 500, 501, 504, 612.
Raum und Zeit, Phys. Zeitschrift, vol. 10, 1909, p. 104 and in Lorentz-Einstein-Minkowski : Das Relativitätsprinzip, Teubner 4th ed., Leipzig-Berlin, 1922 : *484-485, 500, 501.*
MONGE (G.), 313, 423.
MONNIER (Le), 285.
MONTUCLA (J. F.), 38.
MORLEY (E. W.), 465.
MOTTE (A.), 201.
MUSCHENBROEK (P. van), 319.

NAVIER (L.), 374, **409-417**, 456.
Mémoires de l'Acad. royale des Sciences de l'Institut de France, 1823, p. 389 : *409-412, 414.*
Sur les lois de l'équilibre et du mouvement des corps solides élastiques, 1821 : *414-415.*
Annales de chimie et de physique, vol. 38, 1828, p. 304 : *415, 416.*
Annales de chimie et de physique, vol. 40, 1829, p. 103 : *416.*
NERNST (W.), 550.
NEUMANN (J. von), 600.
Mathematische Grundlage der Quantenmechanik, Springer publ., Berlin, 1932.
NEWCOMB (S.), 527.
NEWTON (Sir I.), 82, 110, 148, 182, **200-218**, 222-224, 226, 231, 239, 261, 274, 279, 285, 286, 309, 310, 332, 334, 339, 341, 359, 360, 391, 442-445, 450-454, 461, 502, 510, 513, 514, 550, 637, 639.
Philosophiae naturalis principia mathematica, London, 1687 : *200-213.*
NICETE OF SYRACUSE, 83.
NICODEMUS, 225.
NICOMEDES, 333.
NIFO (A.), 86.

NOBLE (H. R.), 472.
Proceed. Royal Society, vol. 72, 1903, p. 132.
NOËL (Father), 171.

OCCHIALINI, 607.
OCKAM (G. d'), 48, 58, 83, 84, 99, 213.
ORESME (N.), **58-66**, 68, 71, 81, 83, 88-90, 94, 106, 109, 138, 159.
Traité du Ciel et du Monde, 1377 : *58, 62-66.*
Tractatus de figuratione potentiarum et mensurarum difformitatum : *60-62.*
OUTHIER (The Abbé), 285.

PAINLEVÉ (P.), 15, **453-455**, 462, 487, 493, 497, 498, 521, **528-531**, 533, 616, 621.
Les Axiomes de la Mécanique et le principe de causalité, Bull. de la Soc. française de Philosophie, vol. V, 1905.
Les Axiomes de la Mécanique, Gauthier-Villars publ., Paris, 1922 : *453-455, 493.*
La Mécanique et la théorie de la relativité, Comptes Rendus de l'Acad. des Sciences, vol. 173, 1921, p. 677.
La gravitation dans la Mécanique de Newton et dans celle d'Einstein, *ibid.*, p. 873 : *530-531.*
PAPPUS, **33-35**, 38, 100, 101.
Collections, Livre VIII.
PARDIES (Father), 310.
PASCAL (B.), 72, 82, 143, 144, **169-171**, 186.
Expériences nouvelles touchant le vide, 1647.
Traités de l'équilibre des liqueurs et de la pesanteur de la masse de l'air, 1663 : *169-170.*
PASCAL (E.), 167.
PASCHEN (F.), 537, 538, 547.
Annal. der Physik, vol. 27, 1908, p. 565.
PAULI (W.), 606.
PEREIRA (B.), 99.
De communibus omnium rerum naturalium principiis, Rome, 1652.
PÉROT (A.), 525.
PETIT (P.), 169.
PHILLIPS (E.), 423.
PHILOLAUS OF CRETE, 83.
PICARD (J.), 217, 286.
PICKERING, 538.
Astrophysical Journal, vol. 4, 1896, p. 369.
PLANA, 382.

PLANCK (M.), 482, 535, 536, 539-541, **549-550**, 552-554, 567, 568, 579, 614, 623, 624, 626.
Verh. der Deutsch. Physik. Gesellschaft, 1900, p. 237.
Annal. der Physik, vol. 4, 1901, p. 566.
Die physikalische Struktur des Phasenraumes, Annal. der Physik, vol. 50, 1916, p. 385.
PLATO, 87.
PLINY THE ELDER, 51.
PODOLSKY (B.), 627.
POINCARÉ (H.), 434, **447-452**, 488, 489, 552, 553, 622-624, 626.
Science et Hypothèse, Flammarion publ., Paris, 1902 : *448-452*.
Rendiconti di Palermo, vol. 21, 1906, p. 129 : *489*.
Journal de Physique, vol. II, January, 1912, p. 1 : *623*.
Dernières Pensées, Flammarion publ., Paris, 1913 : *623, 624*.
POINSOT (L.), 382.
POISEUILLE (J. L.), 414.
POISSON (S.), 233, 382, **384-389**, 396, 415, 416, 423, 456, 514, 576, 579, 619.
Journal de l'École polytechnique, 15th cahier, p. 266 : *384, 387, 389*.
POMPONAZZI (P.), 87.
De intentione et remissione formarum, 1514.
De reactione, 1514.
PONCELET (J.), 382, 423, 428.
PONTE, 561.
Annales de Physique, vol. 13, 1929, p. 395.
PRINGSHEIM, 549, 550.
PRONY (G. de), 317.
PTOLEMY, 51, 52, 63, 108-111, 114, 144, 215, 444, 451, 489.
PYTHAGORAS, 83.

QUET, 382.

RANKINE (J.), 435.
RAYLEIGH (Lord), 394, 555, 623.
REECH (F.), **438-442**, 449.
Cours de Mécanique d'après la nature généralement flexible et élastique des corps, Paris, 1852 : *438-440*.
REICH (F.), 380.
REISCH (G.), 85, 86.
Margarita philosophica, 1496.
RICCI (G.), 505, 524.
Math. Annalen, vol. 54, 1899 : *505*.
RICCIOLI (Father), 84.
RIEMANN (B.), 423, 505, 508, 510, 520, 533, 582, 638.

RITZ (W.), 577.
ROBERVAL (G. Personne de), 107, 149, **150-153**, 157, 158, 165-167, 186, 193, 207, 216, 333.
Traité des mouvements composés, ed. 1693 : *150*.
Traité de mécanique (Ms) : *150-151*.
Traité annexé aux Harmonicorum libri du P. Mersenne : *152, 153*.
Letter to Cavendish, May, 1646 : *164, 165*.
ROENTGEN (W. C.), 463.
ROSEN (N.), 627.
ROUX (J. Le), 488, 495.
Relativité restreinte et systèmes ondulatoires, Journal de Math., vol. I, fasc. 2, 1922 : *488*.
ROWLAND, 463.
RUBENS, 549.
RUPP, 561.
Annal. der Physik, vol. 85, 1928, p. 981.
RUTHERFORD (E.), 535.
Phil. Mag., vol. 21, 1911, p. 669.
RYDBERG, 537, 546.

SACCHERI (Father), 169, **224**.
Neo Stattica, 1703.
SACRO BOSCO (J. de), 52.
De Sphaera.
SAGNAC (G.), 463.
SAINT JOHN, 527, 528.
SAINT THOMAS AQUINAS, 47, 52, 57, 83, 87, 88.
Commentaria in libros Aristotelis de Caelo et Mundo, vol. II, lecture VII : *47-48*.
SAINT VENANT (A. Barré de), 331, 423, **436-437**, 456.
Principes de Mécanique fondés sur la Cinématique, Paris, 1851 : *436, 437*.
SOLOMON OF CAUX, 128.
Les raisons des forces mouvantes avec diverses machines tant utiles que plaisantes auxquelles sont adjoints plusieurs desseigns de grotes et fontaines, Francfort, 1615.
SARPI, 132.
SCALIGER (J. C.), **99-100**, 129, 214.
Exotericarum exercitationum libri XV, Paris, 1557 : *99*.
SCHAEFLI, 533.
SCHERRER, 561.
SCHRÖDINGER (E.), 399, 401, **561-567**, 570, 581, **583-586**, 587, 589, 600, 603, 605, **608-609**, 618, 626.
Quantisierung als Eigenwertproblem, II, Annal. der Physik, vol. 79, 1926, p. 489 : *566*.

Über das Verhältnis der Heisenberg-Born-Jordanschen Quantenmechanik zu der meinen, Annal. der Physik, vol. 79, 1926, p. 734 : *586, 608, 609.*
Annales de l'Institut Henri Poincaré, vol. II, p. 269.
SCHWARTZ, 600.
SCHWARTZSCHILD (K.), 514, 524-527, 530, 540, 542, 543, 582.
Sitzungsber. der Preuss. Akad. der Wissenschaften, 1916, p. 189 and p. 548.
SCOT (M.), 83.
SCOTT (J. Duns), 54, 99.
Doctor subtilis.
SIMPLICIUS, 51, 87.
SITTER (W. de), 515, 526, 532.
Monthly Notices, vol. 77, 1916, p. 712.
Proceed. Amsterdam, vol. 20, 1917, p. 230.
SNELLIUS, 262.
SOMMERFELD (A.), 540, 541, 544, **545-547,** 559, 560, 606.
Annal. der Physik, vol. 51, 1916, pp. 1 and 125.
Atombau und Spektrallinien, 2 vol., French translation, Blanchard publ., Paris, 1923 : *526.*
SOTO (Dominic de), **91-94,** 138.
Quaestiones in libros Physicorum.
Theologi ordinis praedicatorum super octo libri Physicorum Aristotelis quaestiones, Salamanca, 1572 : *91, 92, 93-94.*
STARK (J.), 544, 567.
STERN, 589.
STEVIN (S.), 46, **123-128,** 149, 157, 171, 207, 222, 333.
De Beghinselen der Weegconst, Leyden, 1586.
Hypomnemata mathematica, Leyden, 1608 ; French translation, 1634 : *123, 124-125, 127.*
STIRLING, 235.
STOKES (Sir G.), 317, 414.
STURM (C.), 382.
SWINESHEAD (The " Calculator "), 67, 70 90, 99.

TAIT (P. G.), 408, 523.
TANNERY (P.), 159.
TARTAGLIA (N.), 41, **95-96,** 99, 101, 149.
Nova Scientia, 1537 : *95.*
Quesiti et inventioni diversi, 1546.
Jordani opusculum de ponderosite, Nicolai Tartaleae studio correctum, 1565.
TEMPIER (E.), 83, 84.

THEMON, 68.
THEON OF SMYRNA, 24.
THOMAS, 574, 607.
Naturw., vol. 13, 1925.
THOMSON (G. P.), 561.
Proceed. Royal Society, vol. 117, 1928, p. 600.
THOMSON (J. J.), 535.
Phil. Mag., vol. 7, 1904, p. 237.
THOMSON (W.), 408, 457, 523.
TIÈNE (G. of), 70.
Recollaectae, Venice, 1494.
TORNI (B.), 70, 88.
TORRICELLI (E.), **145-148,** 169, 170, 182, 187, 226, 227, 286, 333, 334.
De motu gravium naturaliter descendentium et projectorum, Florence, 1644 : *145, 146.*
TRILLAT (J. J.), 561.
TROUTON (F.), 472.
Proceed. Royal Society, vol. 72, 1903, p. 132.
TRUMPLER, 527.
TURGOT (A. de), 313.
TYCHO-BRAHÉ (G. V.), **108-109,** 110, 111, 114.
Astronomiae instauratae progymnasmata, 1582 : *108-109.*

UHLENBECK (G.), 607.

VALLA (G.), 87.
VARIGNON (P.), 148, 223, **224-227,** 231, 233, 286, 334.
Projet d'une Nouvelle Mécanique, Paris, 1687.
Nouvelle Mécanique ou Statique, Paris, 1725 : *224, 225.*
VERRIER (U. Le), 509, 514, 527.
VILLALPAND (J. B.), **101-102.**
Apparatus Urbis ac Templi Hierosolymitani, Rome, 1603 : *101, 102.*
VINCI (L. da), 41, 46, **72-82,** 97-99, 103, 106, 125, 143, 222.
Les Manuscrits de Léonard de Vinci, publiés par Ch. Ravaisson-Mollien, Paris, 1890 : *72, 73-74, 75, 76-78, 79, 80-81, 82.*
Traité de la Peinture : *81.*
Del moto e misura dell'acqua.
VIVES (J. L.), 70, 90, 91.
De prima philosophia, 1531.
De philosophiae naturae corruptione, 1531 : *90, 91.*
De medicina : *91.*
In pseudo dialecticos : *91.*
VIVIANI (V.), 147.

VOIGT (W.), 468, 484, 494.
Göttinger Nachrichten, 1887, p. 41.
VOLDER (de), 194.
VOLTAIRE, 218, 238, 271.
Élemens de la philosophie de Neuton mis à la portée de tout le monde, Amsterdam, 1738 : *218*.
Diatribe du Dr Akakia, médecin du Pape : *271*.

WALLIS (J.), 67, **172-175**, 180, 186, 199, 207, 224, 231, 333, 340.
Mechanica, sive de Motu, London, 1669-1671 : *172-175*.
WEYL (H.), **517-521**, 525, 600, 601.
Sitzungsber. der Preuss. Akad. der Wissenschaften, 1918.
Raum, Zeit, Materie, Springer publ., Berlin, 1918 ; French translation, Blanchard, publ., Paris, 1922.
WIEN (W.), 463, 549, 567.
WILSON (C. T. R.), 607.

WILSON (H. A.), 463.
WILSON (W.), 539, 541, 552.
The quantum theory of radiation and line spectra, Phil. Mag., vol. 29, 1915, p. 795.
WOLF (A.)
A History of Science, Technology and Philosophy :
XVIth and XVIIth Centuries, George Allen and Unwin publ., London, 1935.
XVIIIth Century, George Allen and Unwin publ., London, 1938.
WOLF (J. C.), 235, 271.
WREN (Ch.), 172, **175**, 178, 186, 199, 207, 268.
Philosophical Transactions, 1669 : *175*, *176*.

ZACCHI (Father), 222.
Nova de machinis philosophia, Rome, 1649.
ZEEMAN (P.), 567, 606, 607.

A CATALOG OF SELECTED
DOVER BOOKS
IN SCIENCE AND MATHEMATICS

A CATALOG OF SELECTED
DOVER BOOKS
IN SCIENCE AND MATHEMATICS

QUALITATIVE THEORY OF DIFFERENTIAL EQUATIONS, V.V. Nemytskii and V.V. Stepanov. Classic graduate-level text by two prominent Soviet mathematicians covers classical differential equations as well as topological dynamics and ergodic theory. Bibliographies. 523pp. 5⅜ × 8½. 65954-2 Pa. $10.95

MATRICES AND LINEAR ALGEBRA, Hans Schneider and George Phillip Barker. Basic textbook covers theory of matrices and its applications to systems of linear equations and related topics such as determinants, eigenvalues and differential equations. Numerous exercises. 432pp. 5⅜ × 8½. 66014-1 Pa. $9.95

QUANTUM THEORY, David Bohm. This advanced undergraduate-level text presents the quantum theory in terms of qualitative and imaginative concepts, followed by specific applications worked out in mathematical detail. Preface. Index. 655pp. 5⅜ × 8½. 65969-0 Pa. $13.95

ATOMIC PHYSICS (8th edition), Max Born. Nobel laureate's lucid treatment of kinetic theory of gases, elementary particles, nuclear atom, wave-corpuscles, atomic structure and spectral lines, much more. Over 40 appendices, bibliography. 495pp. 5⅜ × 8½. 65984-4 Pa. $12.95

ELECTRONIC STRUCTURE AND THE PROPERTIES OF SOLIDS: The Physics of the Chemical Bond, Walter A. Harrison. Innovative text offers basic understanding of the electronic structure of covalent and ionic solids, simple metals, transition metals and their compounds. Problems. 1980 edition. 582pp. 6⅛ × 9¼. 66021-4 Pa. $15.95

BOUNDARY VALUE PROBLEMS OF HEAT CONDUCTION, M. Necati Özisik. Systematic, comprehensive treatment of modern mathematical methods of solving problems in heat conduction and diffusion. Numerous examples and problems. Selected references. Appendices. 505pp. 5⅜ × 8½. 65990-9 Pa. $11.95

A SHORT HISTORY OF CHEMISTRY (3rd edition), J.R. Partington. Classic exposition explores origins of chemistry, alchemy, early medical chemistry, nature of atmosphere, theory of valency, laws and structure of atomic theory, much more. 428pp. 5⅜ × 8½. (Available in U.S. only) 65977-1 Pa. $10.95

A HISTORY OF ASTRONOMY, A. Pannekoek. Well-balanced, carefully reasoned study covers such topics as Ptolemaic theory, work of Copernicus, Kepler, Newton, Eddington's work on stars, much more. Illustrated. References. 521pp. 5⅜ × 8½. 65994-1 Pa. $12.95

PRINCIPLES OF METEOROLOGICAL ANALYSIS, Walter J. Saucier. Highly respected, abundantly illustrated classic reviews atmospheric variables, hydrostatics, static stability, various analyses (scalar, cross-section, isobaric, isentropic, more). For intermediate meteorology students. 454pp. 6⅛ × 9¼. 65979-8 Pa. $14.95

RELATIVITY, THERMODYNAMICS AND COSMOLOGY, Richard C. Tolman. Landmark study extends thermodynamics to special, general relativity; also applications of relativistic mechanics, thermodynamics to cosmological models. 501pp. 5⅜ × 8½. 65383-8 Pa. $12.95

APPLIED ANALYSIS, Cornelius Lanczos. Classic work on analysis and design of finite processes for approximating solution of analytical problems. Algebraic equations, matrices, harmonic analysis, quadrature methods, much more. 559pp. 5⅜ × 8½. 65656-X Pa. $12.95

SPECIAL RELATIVITY FOR PHYSICISTS, G. Stephenson and C.W. Kilmister. Concise elegant account for nonspecialists. Lorentz transformation, optical and dynamical applications, more. Bibliography. 108pp. 5⅜ × 8½. 65519-9 Pa. $4.95

INTRODUCTION TO ANALYSIS, Maxwell Rosenlicht. Unusually clear, accessible coverage of set theory, real number system, metric spaces, continuous functions, Riemann integration, multiple integrals, more. Wide range of problems. Undergraduate level. Bibliography. 254pp. 5⅜ × 8½. 65038-3 Pa. $7.95

INTRODUCTION TO QUANTUM MECHANICS With Applications to Chemistry, Linus Pauling & E. Bright Wilson, Jr. Classic undergraduate text by Nobel Prize winner applies quantum mechanics to chemical and physical problems. Numerous tables and figures enhance the text. Chapter bibliographies. Appendices. Index. 468pp. 5⅜ × 8½. 64871-0 Pa. $11.95

ASYMPTOTIC EXPANSIONS OF INTEGRALS, Norman Bleistein & Richard A. Handelsman. Best introduction to important field with applications in a variety of scientific disciplines. New preface. Problems. Diagrams. Tables. Bibliography. Index. 448pp. 5⅜ × 8½. 65082-0 Pa. $12.95

MATHEMATICS APPLIED TO CONTINUUM MECHANICS, Lee A. Segel. Analyzes models of fluid flow and solid deformation. For upper-level math, science and engineering students. 608pp. 5⅜ × 8½. 65369-2 Pa. $13.95

ELEMENTS OF REAL ANALYSIS, David A. Sprecher. Classic text covers fundamental concepts, real number system, point sets, functions of a real variable, Fourier series, much more. Over 500 exercises. 352pp. 5⅜ × 8½. 65385-4 Pa. $10.95

PHYSICAL PRINCIPLES OF THE QUANTUM THEORY, Werner Heisenberg. Nobel Laureate discusses quantum theory, uncertainty, wave mechanics, work of Dirac, Schroedinger, Compton, Wilson, Einstein, etc. 184pp. 5⅜ × 8½. 60113-7 Pa. $5.95

INTRODUCTORY REAL ANALYSIS, A.N. Kolmogorov, S.V. Fomin. Translated by Richard A. Silverman. Self-contained, evenly paced introduction to real and functional analysis. Some 350 problems. 403pp. 5⅜ × 8½. 61226-0 Pa. $9.95

PROBLEMS AND SOLUTIONS IN QUANTUM CHEMISTRY AND PHYSICS, Charles S. Johnson, Jr. and Lee G. Pedersen. Unusually varied problems, detailed solutions in coverage of quantum mechanics, wave mechanics, angular momentum, molecular spectroscopy, scattering theory, more. 280 problems plus 139 supplementary exercises. 430pp. 6½ × 9¼. 65236-X Pa. $12.95

ASYMPTOTIC METHODS IN ANALYSIS, N.G. de Bruijn. An inexpensive, comprehensive guide to asymptotic methods—the pioneering work that teaches by explaining worked examples in detail. Index. 224pp. 5⅜ × 8½.   64221-6 Pa. $6.95

OPTICAL RESONANCE AND TWO-LEVEL ATOMS, L. Allen and J.H. Eberly. Clear, comprehensive introduction to basic principles behind all quantum optical resonance phenomena. 53 illustrations. Preface. Index. 256pp. 5⅜ × 8½.
65533-4 Pa. $7.95

COMPLEX VARIABLES, Francis J. Flanigan. Unusual approach, delaying complex algebra till harmonic functions have been analyzed from real variable viewpoint. Includes problems with answers. 364pp. 5⅜ × 8½.   61388-7 Pa. $8.95

ATOMIC SPECTRA AND ATOMIC STRUCTURE, Gerhard Herzberg. One of best introductions; especially for specialist in other fields. Treatment is physical rather than mathematical. 80 illustrations. 257pp. 5⅜ × 8½.   60115-3 Pa. $5.95

APPLIED COMPLEX VARIABLES, John W. Dettman. Step-by-step coverage of fundamentals of analytic function theory—plus lucid exposition of five important applications: Potential Theory; Ordinary Differential Equations; Fourier Transforms; Laplace Transforms; Asymptotic Expansions. 66 figures. Exercises at chapter ends. 512pp. 5⅜ × 8½.   64670-X Pa. $11.95

ULTRASONIC ABSORPTION: An Introduction to the Theory of Sound Absorption and Dispersion in Gases, Liquids and Solids, A.B. Bhatia. Standard reference in the field provides a clear, systematically organized introductory review of fundamental concepts for advanced graduate students, research workers. Numerous diagrams. Bibliography. 440pp. 5⅜ × 8½.   64917-2 Pa. $11.95

UNBOUNDED LINEAR OPERATORS: Theory and Applications, Seymour Goldberg. Classic presents systematic treatment of the theory of unbounded linear operators in normed linear spaces with applications to differential equations. Bibliography. 199pp. 5⅜ × 8½.   64830-3 Pa. $7.95

LIGHT SCATTERING BY SMALL PARTICLES, H.C. van de Hulst. Comprehensive treatment including full range of useful approximation methods for researchers in chemistry, meteorology and astronomy. 44 illustrations. 470pp. 5⅜ × 8½.   64228-3 Pa. $10.95

CONFORMAL MAPPING ON RIEMANN SURFACES, Harvey Cohn. Lucid, insightful book presents ideal coverage of subject. 334 exercises make book perfect for self-study. 55 figures. 352pp. 5⅜ × 8¼.   64025-6 Pa. $9.95

OPTICKS, Sir Isaac Newton. Newton's own experiments with spectroscopy, colors, lenses, reflection, refraction, etc., in language the layman can follow. Foreword by Albert Einstein. 532pp. 5⅜ × 8½.   60205-2 Pa. $9.95

GENERALIZED INTEGRAL TRANSFORMATIONS, A.H. Zemanian. Graduate-level study of recent generalizations of the Laplace, Mellin, Hankel, K. Weierstrass, convolution and other simple transformations. Bibliography. 320pp. 5⅜ × 8½.   65375-7 Pa. $8.95

CATALOG OF DOVER BOOKS

THE ELECTROMAGNETIC FIELD, Albert Shadowitz. Comprehensive undergraduate text covers basics of electric and magnetic fields, builds up to electromagnetic theory. Also related topics, including relativity. Over 900 problems. 768pp. 5⅜ × 8¼. 65660-8 Pa. $18.95

FOURIER SERIES, Georgi P. Tolstov. Translated by Richard A. Silverman. A valuable addition to the literature on the subject, moving clearly from subject to subject and theorem to theorem. 107 problems, answers. 336pp. 5⅜ × 8½. 63317-9 Pa. $8.95

THEORY OF ELECTROMAGNETIC WAVE PROPAGATION, Charles Herach Papas. Graduate-level study discusses the Maxwell field equations, radiation from wire antennas, the Doppler effect and more. xiii + 244pp. 5⅜ × 8½. 65678-0 Pa. $6.95

DISTRIBUTION THEORY AND TRANSFORM ANALYSIS: An Introduction to Generalized Functions, with Applications, A.H. Zemanian. Provides basics of distribution theory, describes generalized Fourier and Laplace transformations. Numerous problems. 384pp. 5⅜ × 8½. 65479-6 Pa. $9.95

THE PHYSICS OF WAVES, William C. Elmore and Mark A. Heald. Unique overview of classical wave theory. Acoustics, optics, electromagnetic radiation, more. Ideal as classroom text or for self-study. Problems. 477pp. 5⅜ × 8½. 64926-1 Pa. $12.95

CALCULUS OF VARIATIONS WITH APPLICATIONS, George M. Ewing. Applications-oriented introduction to variational theory develops insight and promotes understanding of specialized books, research papers. Suitable for advanced undergraduate/graduate students as primary, supplementary text. 352pp. 5⅜ × 8½. 64856-7 Pa. $8.95

A TREATISE ON ELECTRICITY AND MAGNETISM, James Clerk Maxwell. Important foundation work of modern physics. Brings to final form Maxwell's theory of electromagnetism and rigorously derives his general equations of field theory. 1,084pp. 5⅜ × 8½. 60636-8, 60637-6 Pa., Two-vol. set $19.90

AN INTRODUCTION TO THE CALCULUS OF VARIATIONS, Charles Fox. Graduate-level text covers variations of an integral, isoperimetrical problems, least action, special relativity, approximations, more. References. 279pp. 5⅜ × 8½. 65499-0 Pa. $7.95

HYDRODYNAMIC AND HYDROMAGNETIC STABILITY, S. Chandrasekhar. Lucid examination of the Rayleigh-Benard problem; clear coverage of the theory of instabilities causing convection. 704pp. 5⅜ × 8¼. 64071-X Pa. $14.95

CALCULUS OF VARIATIONS, Robert Weinstock. Basic introduction covering isoperimetric problems, theory of elasticity, quantum mechanics, electrostatics, etc. Exercises throughout. 326pp. 5⅜ × 8½. 63069-2 Pa. $7.95

DYNAMICS OF FLUIDS IN POROUS MEDIA, Jacob Bear. For advanced students of ground water hydrology, soil mechanics and physics, drainage and irrigation engineering and more. 335 illustrations. Exercises, with answers. 784pp. 6⅛ × 9¼. 65675-6 Pa. $19.95

NUMERICAL METHODS FOR SCIENTISTS AND ENGINEERS, Richard Hamming. Classic text stresses frequency approach in coverage of algorithms, polynomial approximation, Fourier approximation, exponential approximation, other topics. Revised and enlarged 2nd edition. 721pp. 5⅜ × 8½.
65241-6 Pa. $14.95

THEORETICAL SOLID STATE PHYSICS, Vol. I: Perfect Lattices in Equilibrium; Vol. II: Non-Equilibrium and Disorder, William Jones and Norman H. March. Monumental reference work covers fundamental theory of equilibrium properties of perfect crystalline solids, non-equilibrium properties, defects and disordered systems. Appendices. Problems. Preface. Diagrams. Index. Bibliography. Total of 1,301pp. 5⅜ × 8½. Two volumes. Vol. I 65015-4 Pa. $14.95
Vol. II 65016-2 Pa. $14.95

OPTIMIZATION THEORY WITH APPLICATIONS, Donald A. Pierre. Broad-spectrum approach to important topic. Classical theory of minima and maxima, calculus of variations, simplex technique and linear programming, more. Many problems, examples. 640pp. 5⅜ × 8½.
65205-X Pa. $14.95

THE MODERN THEORY OF SOLIDS, Frederick Seitz. First inexpensive edition of classic work on theory of ionic crystals, free-electron theory of metals and semiconductors, molecular binding, much more. 736pp. 5⅜ × 8½.
65482-6 Pa. $15.95

ESSAYS ON THE THEORY OF NUMBERS, Richard Dedekind. Two classic essays by great German mathematician: on the theory of irrational numbers; and on transfinite numbers and properties of natural numbers. 115pp. 5⅜ × 8½.
21010-3 Pa. $4.95

THE FUNCTIONS OF MATHEMATICAL PHYSICS, Harry Hochstadt. Comprehensive treatment of orthogonal polynomials, hypergeometric functions, Hill's equation, much more. Bibliography. Index. 322pp. 5⅜ × 8½.
65214-9 Pa. $9.95

NUMBER THEORY AND ITS HISTORY, Oystein Ore. Unusually clear, accessible introduction covers counting, properties of numbers, prime numbers, much more. Bibliography. 380pp. 5⅜ × 8½.
65620-9 Pa. $9.95

THE VARIATIONAL PRINCIPLES OF MECHANICS, Cornelius Lanczos. Graduate level coverage of calculus of variations, equations of motion, relativistic mechanics, more. First inexpensive paperbound edition of classic treatise. Index. Bibliography. 418pp. 5⅜ × 8½.
65067-7 Pa. $11.95

MATHEMATICAL TABLES AND FORMULAS, Robert D. Carmichael and Edwin R. Smith. Logarithms, sines, tangents, trig functions, powers, roots, reciprocals, exponential and hyperbolic functions, formulas and theorems. 269pp. 5⅜ × 8½.
60111-0 Pa. $6.95

THEORETICAL PHYSICS, Georg Joos, with Ira M. Freeman. Classic overview covers essential math, mechanics, electromagnetic theory, thermodynamics, quantum mechanics, nuclear physics, other topics. First paperback edition. xxiii + 885pp. 5⅜ × 8½.
65227-0 Pa. $19.95

HANDBOOK OF MATHEMATICAL FUNCTIONS WITH FORMULAS, GRAPHS, AND MATHEMATICAL TABLES, edited by Milton Abramowitz and Irene A. Stegun. Vast compendium: 29 sets of tables, some to as high as 20 places. 1,046pp. 8 × 10½.   61272-4 Pa. $24.95

MATHEMATICAL METHODS IN PHYSICS AND ENGINEERING, John W. Dettman. Algebraically based approach to vectors, mapping, diffraction, other topics in applied math. Also generalized functions, analytic function theory, more. Exercises. 448pp. 5⅜ × 8¼.   65649-7 Pa. $9.95

A SURVEY OF NUMERICAL MATHEMATICS, David M. Young and Robert Todd Gregory. Broad self-contained coverage of computer-oriented numerical algorithms for solving various types of mathematical problems in linear algebra, ordinary and partial, differential equations, much more. Exercises. Total of 1,248pp. 5⅜ × 8½. Two volumes.   Vol. I 65691-8 Pa. $14.95
                                                                 Vol. II 65692-6 Pa. $14.95

TENSOR ANALYSIS FOR PHYSICISTS, J.A. Schouten. Concise exposition of the mathematical basis of tensor analysis, integrated with well-chosen physical examples of the theory. Exercises. Index. Bibliography. 289pp. 5⅜ × 8½.
                                                                          65582-2 Pa. $8.95

INTRODUCTION TO NUMERICAL ANALYSIS (2nd Edition), F.B. Hildebrand. Classic, fundamental treatment covers computation, approximation, interpolation, numerical differentiation and integration, other topics. 150 new problems. 669pp. 5⅜ × 8½.   65363-3 Pa. $14.95

INVESTIGATIONS ON THE THEORY OF THE BROWNIAN MOVEMENT, Albert Einstein. Five papers (1905–8) investigating dynamics of Brownian motion and evolving elementary theory. Notes by R. Fürth. 122pp. 5⅜ × 8½.
                                                                          60304-0 Pa. $4.95

CATASTROPHE THEORY FOR SCIENTISTS AND ENGINEERS, Robert Gilmore. Advanced-level treatment describes mathematics of theory grounded in the work of Poincaré, R. Thom, other mathematicians. Also important applications to problems in mathematics, physics, chemistry and engineering. 1981 edition. References. 28 tables. 397 black-and-white illustrations. xvii + 666pp. 6⅛ × 9¼.
                                                                          67539-4 Pa. $16.95

AN INTRODUCTION TO STATISTICAL THERMODYNAMICS, Terrell L. Hill. Excellent basic text offers wide-ranging coverage of quantum statistical mechanics, systems of interacting molecules, quantum statistics, more. 523pp. 5⅜ × 8½.   65242-4 Pa. $12.95

ELEMENTARY DIFFERENTIAL EQUATIONS, William Ted Martin and Eric Reissner. Exceptionally clear, comprehensive introduction at undergraduate level. Nature and origin of differential equations, differential equations of first, second and higher orders. Picard's Theorem, much more. Problems with solutions. 331pp. 5⅜ × 8½.   65024-3 Pa. $8.95

STATISTICAL PHYSICS, Gregory H. Wannier. Classic text combines thermodynamics, statistical mechanics and kinetic theory in one unified presentation of thermal physics. Problems with solutions. Bibliography. 532pp. 5⅜ × 8½.
                                                                          65401-X Pa. $11.95

CATALOG OF DOVER BOOKS

ORDINARY DIFFERENTIAL EQUATIONS, Morris Tenenbaum and Harry Pollard. Exhaustive survey of ordinary differential equations for undergraduates in mathematics, engineering, science. Thorough analysis of theorems. Diagrams. Bibliography. Index. 818pp. 5⅜ × 8½. 64940-7 Pa. $16.95

STATISTICAL MECHANICS: Principles and Applications, Terrell L. Hill. Standard text covers fundamentals of statistical mechanics, applications to fluctuation theory, imperfect gases, distribution functions, more. 448pp. 5⅜ × 8½. 65390-0 Pa. $9.95

ORDINARY DIFFERENTIAL EQUATIONS AND STABILITY THEORY: An Introduction, David A. Sánchez. Brief, modern treatment. Linear equation, stability theory for autonomous and nonautonomous systems, etc. 164pp. 5⅜ × 8¼. 63828-6 Pa. $5.95

THIRTY YEARS THAT SHOOK PHYSICS: The Story of Quantum Theory, George Gamow. Lucid, accessible introduction to influential theory of energy and matter. Careful explanations of Dirac's anti-particles, Bohr's model of the atom, much more. 12 plates. Numerous drawings. 240pp. 5⅜ × 8½. 24895-X Pa. $6.95

THEORY OF MATRICES, Sam Perlis. Outstanding text covering rank, non-singularity and inverses in connection with the development of canonical matrices under the relation of equivalence, and without the intervention of determinants. Includes exercises. 237pp. 5⅜ × 8½. 66810-X Pa. $7.95

GREAT EXPERIMENTS IN PHYSICS: Firsthand Accounts from Galileo to Einstein, edited by Morris H. Shamos. 25 crucial discoveries: Newton's laws of motion, Chadwick's study of the neutron, Hertz on electromagnetic waves, more. Original accounts clearly annotated. 370pp. 5⅜ × 8½. 25346-5 Pa. $10.95

INTRODUCTION TO PARTIAL DIFFERENTIAL EQUATIONS WITH APPLICATIONS, E.C. Zachmanoglou and Dale W. Thoe. Essentials of partial differential equations applied to common problems in engineering and the physical sciences. Problems and answers. 416pp. 5⅜ × 8½. 65251-3 Pa. $10.95

BURNHAM'S CELESTIAL HANDBOOK, Robert Burnham, Jr. Thorough guide to the stars beyond our solar system. Exhaustive treatment. Alphabetical by constellation: Andromeda to Cetus in Vol. 1; Chamaeleon to Orion in Vol. 2; and Pavo to Vulpecula in Vol. 3. Hundreds of illustrations. Index in Vol. 3. 2,000pp. 6⅛ × 9¼. 23567-X, 23568-8, 23673-0 Pa., Three-vol. set $41.85

CHEMICAL MAGIC, Leonard A. Ford. Second Edition, Revised by E. Winston Grundmeier. Over 100 unusual stunts demonstrating cold fire, dust explosions, much more. Text explains scientific principles and stresses safety precautions. 128pp. 5⅜ × 8½. 67628-5 Pa. $5.95

AMATEUR ASTRONOMER'S HANDBOOK, J.B. Sidgwick. Timeless, comprehensive coverage of telescopes, mirrors, lenses, mountings, telescope drives, micrometers, spectroscopes, more. 189 illustrations. 576pp. 5⅜ × 8¼. (Available in U.S. only) 24034-7 Pa. $9.95

CATALOG OF DOVER BOOKS

SPECIAL FUNCTIONS, N.N. Lebedev. Translated by Richard Silverman. Famous Russian work treating more important special functions, with applications to specific problems of physics and engineering. 38 figures. 308pp. 5⅜ × 8½.
60624-4 Pa. $8.95

OBSERVATIONAL ASTRONOMY FOR AMATEURS, J.B. Sidgwick. Mine of useful data for observation of sun, moon, planets, asteroids, aurorae, meteors, comets, variables, binaries, etc. 39 illustrations. 384pp. 5⅜ × 8¼. (Available in U.S. only)
24033-9 Pa. $8.95

INTEGRAL EQUATIONS, F.G. Tricomi. Authoritative, well-written treatment of extremely useful mathematical tool with wide applications. Volterra Equations, Fredholm Equations, much more. Advanced undergraduate to graduate level. Exercises. Bibliography. 238pp. 5⅜ × 8½.
64828-1 Pa. $7.95

POPULAR LECTURES ON MATHEMATICAL LOGIC, Hao Wang. Noted logician's lucid treatment of historical developments, set theory, model theory, recursion theory and constructivism, proof theory, more. 3 appendixes. Bibliography. 1981 edition. ix + 283pp. 5⅜ × 8½.
67632-3 Pa. $8.95

MODERN NONLINEAR EQUATIONS, Thomas L. Saaty. Emphasizes practical solution of problems; covers seven types of equations. ". . . a welcome contribution to the existing literature. . . ."—*Math Reviews.* 490pp. 5⅜ × 8½. 64232-1 Pa. $11.95

FUNDAMENTALS OF ASTRODYNAMICS, Roger Bate et al. Modern approach developed by U.S. Air Force Academy. Designed as a first course. Problems, exercises. Numerous illustrations. 455pp. 5⅜ × 8½. 60061-0 Pa. $9.95

INTRODUCTION TO LINEAR ALGEBRA AND DIFFERENTIAL EQUATIONS, John W. Dettman. Excellent text covers complex numbers, determinants, orthonormal bases, Laplace transforms, much more. Exercises with solutions. Undergraduate level. 416pp. 5⅜ × 8½. 65191-6 Pa. $9.95

INCOMPRESSIBLE AERODYNAMICS, edited by Bryan Thwaites. Covers theoretical and experimental treatment of the uniform flow of air and viscous fluids past two-dimensional aerofoils and three-dimensional wings; many other topics. 654pp. 5⅜ × 8½. 65465-6 Pa. $16.95

INTRODUCTION TO DIFFERENCE EQUATIONS, Samuel Goldberg. Exceptionally clear exposition of important discipline with applications to sociology, psychology, economics. Many illustrative examples; over 250 problems. 260pp. 5⅜ × 8½. 65084-7 Pa. $7.95

LAMINAR BOUNDARY LAYERS, edited by L. Rosenhead. Engineering classic covers steady boundary layers in two- and three-dimensional flow, unsteady boundary layers, stability, observational techniques, much more. 708pp. 5⅜ × 8½.
65646-2 Pa. $18.95

LECTURES ON CLASSICAL DIFFERENTIAL GEOMETRY, Second Edition, Dirk J. Struik. Excellent brief introduction covers curves, theory of surfaces, fundamental equations, geometry on a surface, conformal mapping, other topics. Problems. 240pp. 5⅜ × 8½. 65609-8 Pa. $7.95

ROTARY-WING AERODYNAMICS, W.Z. Stepniewski. Clear, concise text covers aerodynamic phenomena of the rotor and offers guidelines for helicopter performance evaluation. Originally prepared for NASA. 537 figures. 640pp. 6⅛ × 9¼.
64647-5 Pa. $15.95

DIFFERENTIAL GEOMETRY, Heinrich W. Guggenheimer. Local differential geometry as an application of advanced calculus and linear algebra. Curvature, transformation groups, surfaces, more. Exercises. 62 figures. 378pp. 5⅜ × 8½.
63433-7 Pa. $8.95

INTRODUCTION TO SPACE DYNAMICS, William Tyrrell Thomson. Comprehensive, classic introduction to space-flight engineering for advanced undergraduate and graduate students. Includes vector algebra, kinematics, transformation of coordinates. Bibliography. Index. 352pp. 5⅜ × 8½. 65113-4 Pa. $8.95

A SURVEY OF MINIMAL SURFACES, Robert Osserman. Up-to-date, in-depth discussion of the field for advanced students. Corrected and enlarged edition covers new developments. Includes numerous problems. 192pp. 5⅜ × 8½.
64998-9 Pa. $8.95

ANALYTICAL MECHANICS OF GEARS, Earle Buckingham. Indispensable reference for modern gear manufacture covers conjugate gear-tooth action, gear-tooth profiles of various gears, many other topics. 263 figures. 102 tables. 546pp. 5⅜ × 8½. 65712-4 Pa. $14.95

SET THEORY AND LOGIC, Robert R. Stoll. Lucid introduction to unified theory of mathematical concepts. Set theory and logic seen as tools for conceptual understanding of real number system. 496pp. 5⅜ × 8¼. 63829-4 Pa. $10.95

A HISTORY OF MECHANICS, René Dugas. Monumental study of mechanical principles from antiquity to quantum mechanics. Contributions of ancient Greeks, Galileo, Leonardo, Kepler, Lagrange, many others. 671pp. 5⅜ × 8½.
65632-2 Pa. $14.95

FAMOUS PROBLEMS OF GEOMETRY AND HOW TO SOLVE THEM, Benjamin Bold. Squaring the circle, trisecting the angle, duplicating the cube: learn their history, why they are impossible to solve, then solve them yourself. 128pp. 5⅜ × 8½. 24297-8 Pa. $4.95

MECHANICAL VIBRATIONS, J.P. Den Hartog. Classic textbook offers lucid explanations and illustrative models, applying theories of vibrations to a variety of practical industrial engineering problems. Numerous figures. 233 problems, solutions. Appendix. Index. Preface. 436pp. 5⅜ × 8½. 64785-4 Pa. $10.95

CURVATURE AND HOMOLOGY, Samuel I. Goldberg. Thorough treatment of specialized branch of differential geometry. Covers Riemannian manifolds, topology of differentiable manifolds, compact Lie groups, other topics. Exercises. 315pp. 5⅜ × 8½. 64314-X Pa. $8.95

HISTORY OF STRENGTH OF MATERIALS, Stephen P. Timoshenko. Excellent historical survey of the strength of materials with many references to the theories of elasticity and structure. 245 figures. 452pp. 5⅜ × 8½. 61187-6 Pa. $11.95

GEOMETRY OF COMPLEX NUMBERS, Hans Schwerdtfeger. Illuminating, widely praised book on analytic geometry of circles, the Moebius transformation, and two-dimensional non-Euclidean geometries. 200pp. 5⅜ × 8¼.
63830-8 Pa. $8.95

MECHANICS, J.P. Den Hartog. A classic introductory text or refresher. Hundreds of applications and design problems illuminate fundamentals of trusses, loaded beams and cables, etc. 334 answered problems. 462pp. 5⅜ × 8½. 60754-2 Pa. $9.95

TOPOLOGY, John G. Hocking and Gail S. Young. Superb one-year course in classical topology. Topological spaces and functions, point-set topology, much more. Examples and problems. Bibliography. Index. 384pp. 5⅜ × 8¼.
65676-4 Pa. $9.95

STRENGTH OF MATERIALS, J.P. Den Hartog. Full, clear treatment of basic material (tension, torsion, bending, etc.) plus advanced material on engineering methods, applications. 350 answered problems. 323pp. 5⅜ × 8½. 60755-0 Pa. $8.95

ELEMENTARY CONCEPTS OF TOPOLOGY, Paul Alexandroff. Elegant, intuitive approach to topology from set-theoretic topology to Betti groups; how concepts of topology are useful in math and physics. 25 figures. 57pp. 5⅜ × 8½.
60747-X Pa. $3.50

ADVANCED STRENGTH OF MATERIALS, J.P. Den Hartog. Superbly written advanced text covers torsion, rotating disks, membrane stresses in shells, much more. Many problems and answers. 388pp. 5⅜ × 8½. 65407-9 Pa. $9.95

COMPUTABILITY AND UNSOLVABILITY, Martin Davis. Classic graduate-level introduction to theory of computability, usually referred to as theory of recurrent functions. New preface and appendix. 288pp. 5⅜ × 8½. 61471-9 Pa. $7.95

GENERAL CHEMISTRY, Linus Pauling. Revised 3rd edition of classic first-year text by Nobel laureate. Atomic and molecular structure, quantum mechanics, statistical mechanics, thermodynamics correlated with descriptive chemistry. Problems. 992pp. 5⅜ × 8½. 65622-5 Pa. $19.95

AN INTRODUCTION TO MATRICES, SETS AND GROUPS FOR SCIENCE STUDENTS, G. Stephenson. Concise, readable text introduces sets, groups, and most importantly, matrices to undergraduate students of physics, chemistry, and engineering. Problems. 164pp. 5⅜ × 8½. 65077-4 Pa. $6.95

THE HISTORICAL BACKGROUND OF CHEMISTRY, Henry M. Leicester. Evolution of ideas, not individual biography. Concentrates on formulation of a coherent set of chemical laws. 260pp. 5⅜ × 8½. 61053-5 Pa. $6.95

THE PHILOSOPHY OF MATHEMATICS: An Introductory Essay, Stephan Körner. Surveys the views of Plato, Aristotle, Leibniz & Kant concerning proposi- tions and theories of applied and pure mathematics. Introduction. Two appen- dices. Index. 198pp. 5⅜ × 8½. 25048-2 Pa. $7.95

THE DEVELOPMENT OF MODERN CHEMISTRY, Aaron J. Ihde. Authorita- tive history of chemistry from ancient Greek theory to 20th-century innovation. Covers major chemists and their discoveries. 209 illustrations. 14 tables. Bibliog- raphies. Indices. Appendices. 851pp. 5⅜ × 8½. 64235-6 Pa. $18.95

DE RE METALLICA, Georgius Agricola. The famous Hoover translation of greatest treatise on technological chemistry, engineering, geology, mining of early modern times (1556). All 289 original woodcuts. 638pp. 6¾ × 11.
60006-8 Pa. $18.95

SOME THEORY OF SAMPLING, William Edwards Deming. Analysis of the problems, theory and design of sampling techniques for social scientists, industrial managers and others who find statistics increasingly important in their work. 61 tables. 90 figures. xvii + 602pp. 5⅜ × 8½.
64684-X Pa. $15.95

THE VARIOUS AND INGENIOUS MACHINES OF AGOSTINO RAMELLI: A Classic Sixteenth-Century Illustrated Treatise on Technology, Agostino Ramelli. One of the most widely known and copied works on machinery in the 16th century. 194 detailed plates of water pumps, grain mills, cranes, more. 608pp. 9 × 12.
25497-6 Clothbd. $34.95

LINEAR PROGRAMMING AND ECONOMIC ANALYSIS, Robert Dorfman, Paul A. Samuelson and Robert M. Solow. First comprehensive treatment of linear programming in standard economic analysis. Game theory, modern welfare economics, Leontief input-output, more. 525pp. 5⅜ × 8½.
65491-5 Pa. $14.95

ELEMENTARY DECISION THEORY, Herman Chernoff and Lincoln E. Moses. Clear introduction to statistics and statistical theory covers data processing, probability and random variables, testing hypotheses, much more. Exercises. 364pp. 5⅜ × 8½.
65218-1 Pa. $9.95

THE COMPLEAT STRATEGYST: Being a Primer on the Theory of Games of Strategy, J.D. Williams. Highly entertaining classic describes, with many illustrated examples, how to select best strategies in conflict situations. Prefaces. Appendices. 268pp. 5⅜ × 8½.
25101-2 Pa. $7.95

MATHEMATICAL METHODS OF OPERATIONS RESEARCH, Thomas L. Saaty. Classic graduate-level text covers historical background, classical methods of forming models, optimization, game theory, probability, queueing theory, much more. Exercises. Bibliography. 448pp. 5⅜ × 8¼.
65703-5 Pa. $12.95

CONSTRUCTIONS AND COMBINATORIAL PROBLEMS IN DESIGN OF EXPERIMENTS, Damaraju Raghavarao. In-depth reference work examines orthogonal Latin squares, incomplete block designs, tactical configuration, partial geometry, much more. Abundant explanations, examples. 416pp. 5⅜ × 8¼.
65685-3 Pa. $10.95

THE ABSOLUTE DIFFERENTIAL CALCULUS (CALCULUS OF TENSORS), Tullio Levi-Civita. Great 20th-century mathematician's classic work on material necessary for mathematical grasp of theory of relativity. 452pp. 5⅜ × 8½.
63401-9 Pa. $9.95

VECTOR AND TENSOR ANALYSIS WITH APPLICATIONS, A.I. Borisenko and I.E. Tarapov. Concise introduction. Worked-out problems, solutions, exercises. 257pp. 5⅜ × 8¼.
63833-2 Pa. $7.95

THE FOUR-COLOR PROBLEM: Assaults and Conquest, Thomas L. Saaty and Paul G. Kainen. Engrossing, comprehensive account of the century-old combinatorial topological problem, its history and solution. Bibliographies. Index. 110 figures. 228pp. 5⅜ × 8½. 65092-8 Pa. $6.95

CATALYSIS IN CHEMISTRY AND ENZYMOLOGY, William P. Jencks. Exceptionally clear coverage of mechanisms for catalysis, forces in aqueous solution, carbonyl- and acyl-group reactions, practical kinetics, more. 864pp. 5⅜ × 8½. 65460-5 Pa. $19.95

PROBABILITY: An Introduction, Samuel Goldberg. Excellent basic text covers set theory, probability theory for finite sample spaces, binomial theorem, much more. 360 problems. Bibliographies. 322pp. 5⅜ × 8½. 65252-1 Pa. $8.95

LIGHTNING, Martin A. Uman. Revised, updated edition of classic work on the physics of lightning. Phenomena, terminology, measurement, photography, spectroscopy, thunder, more. Reviews recent research. Bibliography. Indices. 320pp. 5⅜ × 8¼. 64575-4 Pa. $8.95

PROBABILITY THEORY: A Concise Course, Y.A. Rozanov. Highly readable, self-contained introduction covers combination of events, dependent events, Bernoulli trials, etc. Translation by Richard Silverman. 148pp. 5⅜ × 8¼. 63544-9 Pa. $5.95

AN INTRODUCTION TO HAMILTONIAN OPTICS, H. A. Buchdahl. Detailed account of the Hamiltonian treatment of aberration theory in geometrical optics. Many classes of optical systems defined in terms of the symmetries they possess. Problems with detailed solutions. 1970 edition. xv + 360pp. 5⅜ × 8½. 67597-1 Pa. $10.95

STATISTICS MANUAL, Edwin L. Crow, et al. Comprehensive, practical collection of classical and modern methods prepared by U.S. Naval Ordnance Test Station. Stress on use. Basics of statistics assumed. 288pp. 5⅜ × 8½. 60599-X Pa. $6.95

DICTIONARY/OUTLINE OF BASIC STATISTICS, John E. Freund and Frank J. Williams. A clear concise dictionary of over 1,000 statistical terms and an outline of statistical formulas covering probability, nonparametric tests, much more. 208pp. 5⅜ × 8½. 66796-0 Pa. $6.95

STATISTICAL METHOD FROM THE VIEWPOINT OF QUALITY CONTROL, Walter A. Shewhart. Important text explains regulation of variables, uses of statistical control to achieve quality control in industry, agriculture, other areas. 192pp. 5⅜ × 8½. 65232-7 Pa. $7.95

THE INTERPRETATION OF GEOLOGICAL PHASE DIAGRAMS, Ernest G. Ehlers. Clear, concise text emphasizes diagrams of systems under fluid or containing pressure; also coverage of complex binary systems, hydrothermal melting, more. 288pp. 6½ × 9¼. 65389-7 Pa. $10.95

STATISTICAL ADJUSTMENT OF DATA, W. Edwards Deming. Introduction to basic concepts of statistics, curve fitting, least squares solution, conditions without parameter, conditions containing parameters. 26 exercises worked out. 271pp. 5⅜ × 8½. 64685-8 Pa. $8.95

TENSOR CALCULUS, J.L. Synge and A. Schild. Widely used introductory text covers spaces and tensors, basic operations in Riemannian space, non-Riemannian spaces, etc. 324pp. 5⅜ × 8¼. 63612-7 Pa. $8.95

A CONCISE HISTORY OF MATHEMATICS, Dirk J. Struik. The best brief history of mathematics. Stresses origins and covers every major figure from ancient Near East to 19th century. 41 illustrations. 195pp. 5⅜ × 8½. 60255-9 Pa. $7.95

A SHORT ACCOUNT OF THE HISTORY OF MATHEMATICS, W.W. Rouse Ball. One of clearest, most authoritative surveys from the Egyptians and Phoenicians through 19th-century figures such as Grassman, Galois, Riemann. Fourth edition. 522pp. 5⅜ × 8½. 20630-0 Pa. $10.95

HISTORY OF MATHEMATICS, David E. Smith. Nontechnical survey from ancient Greece and Orient to late 19th century; evolution of arithmetic, geometry, trigonometry, calculating devices, algebra, the calculus. 362 illustrations. 1,355pp. 5⅜ × 8½. 20429-4, 20430-8 Pa., Two-vol. set $23.90

THE GEOMETRY OF RENÉ DESCARTES, René Descartes. The great work founded analytical geometry. Original French text, Descartes' own diagrams, together with definitive Smith-Latham translation. 244pp. 5⅜ × 8½. 60068-8 Pa. $6.95

THE ORIGINS OF THE INFINITESIMAL CALCULUS, Margaret E. Baron. Only fully detailed and documented account of crucial discipline: origins; development by Galileo, Kepler, Cavalieri; contributions of Newton, Leibniz, more. 304pp. 5⅜ × 8½. (Available in U.S. and Canada only) 65371-4 Pa. $9.95

THE HISTORY OF THE CALCULUS AND ITS CONCEPTUAL DEVELOPMENT, Carl B. Boyer. Origins in antiquity, medieval contributions, work of Newton, Leibniz, rigorous formulation. Treatment is verbal. 346pp. 5⅜ × 8½. 60509-4 Pa. $8.95

THE THIRTEEN BOOKS OF EUCLID'S ELEMENTS, translated with introduction and commentary by Sir Thomas L. Heath. Definitive edition. Textual and linguistic notes, mathematical analysis. 2,500 years of critical commentary. Not abridged. 1,414pp. 5⅜ × 8½. 60088-2, 60089-0, 60090-4 Pa., Three-vol. set $29.85

GAMES AND DECISIONS: Introduction and Critical Survey, R. Duncan Luce and Howard Raiffa. Superb nontechnical introduction to game theory, primarily applied to social sciences. Utility theory, zero-sum games, n-person games, decision-making, much more. Bibliography. 509pp. 5⅜ × 8½. 65943-7 Pa. $12.95

THE HISTORICAL ROOTS OF ELEMENTARY MATHEMATICS, Lucas N.H. Bunt, Phillip S. Jones, and Jack D. Bedient. Fundamental underpinnings of modern arithmetic, algebra, geometry and number systems derived from ancient civilizations. 320pp. 5⅜ × 8½. 25563-8 Pa. $8.95

CALCULUS REFRESHER FOR TECHNICAL PEOPLE, A. Albert Klaf. Covers important aspects of integral and differential calculus via 756 questions. 566 problems, most answered. 431pp. 5⅜ × 8½. 20370-0 Pa. $8.95

CATALOG OF DOVER BOOKS

CHALLENGING MATHEMATICAL PROBLEMS WITH ELEMENTARY SOLUTIONS, A.M. Yaglom and I.M. Yaglom. Over 170 challenging problems on probability theory, combinatorial analysis, points and lines, topology, convex polygons, many other topics. Solutions. Total of 445pp. 5⅜ × 8½. Two-vol. set.
Vol. I 65536-9 Pa. $7.95
Vol. II 65537-7 Pa. $6.95

FIFTY CHALLENGING PROBLEMS IN PROBABILITY WITH SOLUTIONS, Frederick Mosteller. Remarkable puzzlers, graded in difficulty, illustrate elementary and advanced aspects of probability. Detailed solutions. 88pp. 5⅜ × 8½.
65355-2 Pa. $4.95

EXPERIMENTS IN TOPOLOGY, Stephen Barr. Classic, lively explanation of one of the byways of mathematics. Klein bottles, Moebius strips, projective planes, map coloring, problem of the Koenigsberg bridges, much more, described with clarity and wit. 43 figures. 210pp. 5⅜ × 8½. 25933-1 Pa. $5.95

RELATIVITY IN ILLUSTRATIONS, Jacob T. Schwartz. Clear nontechnical treatment makes relativity more accessible than ever before. Over 60 drawings illustrate concepts more clearly than text alone. Only high school geometry needed. Bibliography. 128pp. 6⅛ × 9¼. 25965-X Pa. $6.95

AN INTRODUCTION TO ORDINARY DIFFERENTIAL EQUATIONS, Earl A. Coddington. A thorough and systematic first course in elementary differential equations for undergraduates in mathematics and science, with many exercises and problems (with answers). Index. 304pp. 5⅜ × 8½. 65942-9 Pa. $8.95

FOURIER SERIES AND ORTHOGONAL FUNCTIONS, Harry F. Davis. An incisive text combining theory and practical example to introduce Fourier series, orthogonal functions and applications of the Fourier method to boundary-value problems. 570 exercises. Answers and notes. 416pp. 5⅜ × 8½. 65973-9 Pa. $9.95

THE THEORY OF BRANCHING PROCESSES, Theodore E. Harris. First systematic, comprehensive treatment of branching (i.e. multiplicative) processes and their applications. Galton-Watson model, Markov branching processes, electron-photon cascade, many other topics. Rigorous proofs. Bibliography. 240pp. 5⅜ × 8½. 65952-6 Pa. $6.95

AN INTRODUCTION TO ALGEBRAIC STRUCTURES, Joseph Landin. Superb self-contained text covers "abstract algebra": sets and numbers, theory of groups, theory of rings, much more. Numerous well-chosen examples, exercises. 247pp. 5⅜ × 8½. 65940-2 Pa. $7.95

Prices subject to change without notice.
Available at your book dealer or write for free Mathematics and Science Catalog to Dept. GI, Dover Publications, Inc., 31 East 2nd St., Mineola, N.Y. 11501. Dover publishes more than 175 books each year on science, elementary and advanced mathematics, biology, music, art, literature, history, social sciences and other areas.